PERGAMON INTERNATIONAL LIBRARY
of Science, Technology, Engineering and Social Studies

The 1000-volume original paperback library in aid of education,
industrial training and the enjoyment of leisure

Publisher: Robert Maxwell, M.C.

AN INTRODUCTION TO CHEMICAL EQUILIBRIUM AND KINETICS

Pergamon Series in Analytical Chemistry

Volume 1

General Editors:

R. Belcher (*Chairman*), D. Betteridge and L. Meites

Some Other Pergamon Titles of Interest

Books*

BECKEY:
Principles of Field Ionization and Field Desorption
Mass Spectrometry

ELWELL & GIDLEY:
Atomic Absorption Spectrophotometry, 2nd edition

JACKMAN & STERNHELL:
Applications of Nuclear Magnetic Resonance
Spectroscopy in Organic Chemistry, 2nd edition

LAIDLER:
Reaction Kinetics—
Volume 1: Homogeneous Gas Reactions
Volume 2: Reactions in Solution

Journals†

Progress in Analytical Atomic Spectroscopy

Ion-selective Electrode Reviews (Applications,
Theory and Development)

*Available under the Pergamon textbook inspection copy service
†Free specimen copy available on request.

AN INTRODUCTION TO CHEMICAL EQUILIBRIUM AND KINETICS

BY

LOUIS MEITES

Professor of Chemistry
Chairman, Department of Chemistry
Clarkson College of Technology

PERGAMON PRESS

OXFORD · NEW YORK · TORONTO · SYDNEY · PARIS · FRANKFURT

U.K.	Pergamon Press Ltd., Headington Hill Hall, Oxford OX3 0BW, England
U.S.A.	Pergamon Press Inc., Maxwell House, Fairview Park, Elmsford, New York 10523, U.S.A.
CANADA	Pergamon of Canada, Suite 104, 150 Consumers Road, Willowdale, Ontario M2J IP9, Canada
AUSTRALIA	Pergamon Press (Aust.) Pty. Ltd., P.O. Box 544, Potts Point, N.S.W. 2011, Australia
FRANCE	Pergamon Press SARL, 24 rue des Ecoles, 75240 Paris, Cedex 05, France
FEDERAL REPUBLIC OF GERMANY	Pergamon Press GmbH, 6242 Kronberg-Taunus, Hammerweg 6, Federal Republic of Germany

First edition 1981

British Library Cataloguing in Publication Data
Meites, Louis
An introduction to chemical equilibrium and
kinetics. — (Pergamon international library).
— (Pergamon series in analytical chemistry; vol. 1)
1. Chemical equilibrium 2. Chemical reaction,
Rate of
I. Title
541′.39   QD503   79-40743

ISBN 0-08-023802-5 (Hardcover)
ISBN 0-08-023803-3 (Flexicover)

Printed in Hungary by Franklin Printing House

Preface

DURING the last 25 years there has been an explosive growth of chemical knowledge and understanding, and it has had profound effects on the teaching of chemistry at the college and university level. It has been, and will continue to be, necessary to add new material, and to make room for it by discarding what seems to be outdated or only marginally important. Courses on qualitative inorganic and organic analysis have disappeared from many curricula, and have been replaced by others that deal with instrumental analysis and physical organic chemistry or biochemistry. Other courses common in 1950 have survived, but for the most part they have survived in name only and their contents have undergone radical changes. There has been much experimentation with the order in which different courses are given, and much concern with the development of integrated laboratory courses that not only provide practice in chemical experimentation but also illustrate the interaction of ideas taught in separate lecture courses.

All these changes have been attempts to give students better preparation for the world into which they will graduate than their instructors got from the curriculum and courses of an earlier day. Almost anything might be risked in such a cause—except the danger that the world will be irretrievably changed for the worse by a distortion in the students' view of it. The quintessence of chemistry, or any other science, is its reliance on induction—on the use of observed facts and numbers in synthesizing models and the use of models in developing theories, as well as the use of theories in guiding experimentation that will lead to increased confidence or broader understanding. The process is difficult, and so is the teaching of it. It would be so much easier to describe the theories now accepted, and to show how they account for the observations already made! Doing so produces graduate students who think that silver chloride is a pale-green gas, and instructors who believe that prediction is both more important and more reliable than measurement. At different levels of sophistication both express the notions that experiment and observation are no longer very important in chemistry and that what is important is working out the consequences of well-established theory in the deductive fashion of Euclidean geometry or Aristotelian logic. Under the domination of one or two generations thus deluded, chemistry might suffer the fate of Carthage at the hands of Rome.

When my colleagues and I revised the chemistry curriculum at Clarkson College of Technology several years ago, our thinking was much influenced by such considerations. They led us to design a program having some unusual features. Two successive one-semester lecture courses, of which the first deals with structure and bonding while the second deals with equilibrium and kinetics, form the backbone of the freshman year. They are accompanied by a year-long integrated laboratory course and a course on computer programming. Those who know of the proposals made by Professor George S. Hammond some years ago will easily detect our indebtedness to him. The course on structure and bonding deals with both inorganic and organic compounds; the course on equilibrium and kinetics stresses physical and analytical chemistry. The laboratory course is designed to illustrate many of the most important chemical operations and includes analysis, synthesis, and physical measurement; the course on computer programming draws examples from a wide spectrum of chemical problems. We feel that this program provides the student with a sound and well-rounded basis for the more advanced and specialized courses that follow.

This textbook was written for the course on equilibrium and kinetics and is thus intended for use at the freshman or sophomore level. Its structure and aims are discussed at some length in Chapter 1 and need not be reviewed here. Of necessity these reflect my own judgments of what is important, as do the choice of material included in this text and the ways in which it is presented. There has not been enough experimentation with courses of this sort to produce general agreement on these matters. I hope that this book will help to stimulate more.

Most of the book's characteristics are for the reader to discover and evaluate, but there are two that I think specially important. One is that it is not intended to be an "easy" text. It is aimed at the students who will be the leaders, not the followers, of the future. It attempts to provide them with challenge and food for thought, and also with a sense of growing understanding. The other is that I have attempted to convey some sense of the intensely personal and human nature of science. Anyone thoroughly familiar with a scientific field can often tell that a piece of work imagined, carried out, interpreted, and described in a certain way could only have been done by one particular colleague in all the world, and instructors are well aware of the importance of imagination and intuition and the pure pleasure of feeling that an observation or interpretation is right. Beginning students, who do not know these things, tend to overemphasize the scientist's concern with inanimate substances and objects. In using pen and paper to write a poem or a symphony, the poet or composer also deals with inanimate objects, and personal involvement and artistic creation are no more essential to good poetry or music than they are to good science. These convictions are responsible for several features of the book: for its tone and construction, for its digressions into scientific philosophy, and for its revelations of some of my own enthusiasms and prejudices.

The book owes its existence to a number of people and institutions. Foremost among these are the colleagues and students who have helped to shape my ideas about science and its teaching. They are far too numerous to name individually, but I must acknowledge the special debts I owe to Professor James J. Lingane, who guided my introduction to the frustrations and joys of scientific investigation, and to Professor

I. M. Kolthoff, my scientific grandfather. I am grateful to Clarkson College of Technology for the opportunity to take a sabbatical leave that made the writing possible, to my colleagues there for enabling me to take advantage of the opportunity by assuming my duties during my absence, and to Professor Paolo Papoff and the Italian Consiglio Nazionale delle Ricerche for their generous hospitality and support while it was being written. Professors Raymond R. Andrews, George L. Jones, Jr., Egon Matijević, Donald Rosenthal, Henry C. Thomas, and Petr Zuman read the manuscript and commented extensively on it, and it is a pleasure to thank them for their help and advice. I am much indebted to the successive generations of freshmen who witnessed the nucleation and crystallization of many of the ideas and formulations recorded here. I am deeply grateful to Mrs. Joan Mackey, partly for having typed the manuscript and, in doing so, having found and corrected many of its errors and oversights, but chiefly for her generous gifts of cheerfulness, warmth, and friendship.

In the hope that it will help them to begin happy and productive careers in the service of science and their fellow human beings, this book is dedicated to the students who will use it.

Potsdam, New York   LOUIS MEITES

Contents

CHAPTER 1

What This Book is About

SCIENCE is mankind's attempt to understand the natural world and its behavior, and the purpose of science is to enable human beings to predict natural phenomena and to influence them in ways that seem desirable. Chemistry is the branch of science that deals with the properties and behaviors of the innumerable different kinds of matter in the universe. Understanding the properties of these enables us to detect their presence, either in raw materials where they are desirable or in wastes where they are not. It enables us to select the ones that serve beneficial purposes, such as the cure of disease or the improvement of crop yields. It enables us to guard against any harmful effects they may have. Understanding how they react with each other enables us to find ways of disposing of them when they are no longer wanted, to find ways of conserving the world's resources while making them in the quantities we need, and to create new ones that will serve new purposes.

We cannot control the universe unless we understand it, and we cannot achieve our goals unless we know what effects our attempts will have. The understanding that chemistry seeks, and the opportunities that will arise from that understanding, will play leading roles in mankind's struggle toward a healthier, safer, more comfortable, and happier life.

Understanding and knowledge are not the same. Knowledge deals with facts and is obtained in the laboratory; understanding arises in the human brain and is obtained by asking questions, finding the answers to them, and combining those answers with theory, experience, and intuition. The process is complex, and much of your further education in chemistry will be devoted to showing you how to do it. Here a simple example will have to suffice. Imagine a chemist who has obtained a sample of a colorless gas and wishes to identify it. The first question might be "What is its density?" On measuring the volume, temperature, pressure, and mass of the gas it might be found that the answer was "0.671 mg cm^{-3}† at a temperature of 15° C and a pressure of 752 torr". This is knowledge but not yet understanding. On applying theory as represented by the ideal gas law, it may be calculated that the density would

---

†That is, 0.671 mg/cm³ or 0.671 milligram per cubic centimeter. The exponential notation is especially useful when the dimensions are more complicated than they are here. For example, a rate of change of concentration expressed in moles per cubic decimeter of solution per second would be written as mol dm^{-3} s^{-1}, because either "mol/dm³/s" or "mol/dm³ s" might be misunderstood.

be 0.715 mg cm^{-3} under standard conditions, which corresponds to a molecular weight of 16.03. Knowing that this is almost exactly equal to the molecular weight of methane might lead the chemist to suspect that the gas was methane. However, experience or intuition should qualify that suspicion by introducing the realization that the same density might be obtained with a mixture of two or more gases, of which one had a density lower than that of methane while another had a density higher than that of methane. The sample might be a mixture of hydrogen and ethane, or of helium and nitrogen. It might also contain some methane, and there are many other possibilities. What question would you ask next?

As the preceding paragraph suggested, the first general question that the chemist asks when confronted with a substance is "What is it?" It may be a material so nearly pure that any contaminants in it are difficult, or even almost impossible, to detect; or it may be a mixture of two or more components. In any case its properties and behavior will depend on the elements or compounds that it contains and on the proportions in which they are present. The question has two kinds of answers. If the sample were completely pure, an appropriate answer might be "sodium chloride". The chemist would arrive at this conclusion by finding that the material contained sodium and chloride but nothing else that could be detected, and that some convincingly large number of its properties were indistinguishable from those recorded in the literature for pure sodium chloride. However, if the sample is a mixture the answer must be more complicated, and more work must usually be done to obtain it. A qualitative answer like "a mixture of water and ethanol" would not be useful, for we would not know whether we could burn the sample, drink it, or pour it into the river. Quantitative information about the relative proportions of the mixture would be needed as well. In addition to showing that both water and ethanol were present and that nothing else could be detected, the chemist would have to perform some measurement or operation that would reveal their percentages.

Another possibility is that the material is pure but has never been made before. Synthetic chemists have to face this situation very often. They deal with it by comparing the composition and behavior of the unknown substance with those of known classes of compounds. If an unknown compound contains carbon, hydrogen, and oxygen, but no other element, it might be an alcohol (ROH, where R represents some group of carbon and hydrogen atoms) or an aldehyde (RCHO) or an ether (ROR', where R and R' represent two groups that may or may not be identical), or any of a number of other things. However, it could not be a hydrocarbon or an amine (RNH$_2$, RR'NH, or RR'R''N) or a thiol (RSH). Compounds containing the C=O group can be distinguished from most other kinds of compounds by virtue of the fact that they absorb infrared radiation having a wavelength of about 5.8 μm. If the unknown compound does not absorb at this wavelength it might be an alcohol or an ether (or a peroxy compound, ROOR', or an epoxy compound,

$$R-\underset{\diagdown\ \ \diagup}{CH}-\underset{O}{CH}-R'\Bigg),$$

but it could not be an aldehyde or a ketone or a carboxylic acid. On adding other bits of information obtained in other ways it eventually becomes possible to identify the various groups of atoms that are present and to tell how they must be arranged. If the molecule is large and complex it may have to be decomposed into smaller molecules that can be identified more easily because their individual properties are already known. Then the chemist has to imagine how the original molecule must have been arranged to cause it to decompose into those particular fragments.

The preceding question has led naturally into the next one, "What is its structure?" This is most often asked with pure materials, for which it may be answered in terms of either molecular structure or crystal structure or both, but it is also applicable to mixtures. Liquid water has structural features that are affected by the addition of solute, and these features have important influences on the properties and behaviors of aqueous solutions. Much research is now being carried out to discover whether the atoms of metals dissolved in mercury are uncombined or are bound to each other or to atoms of mercury.

Once the nature, composition, and structure of the material have been found, the next general question is "What are its properties?" There are a great many properties that may be of interest: a very few of them are the density, the index of refraction, the heat capacity, the melting and boiling points, the heats of fusion and vaporization, and the degree of opacity to radiation of various wavelengths. Even for a pure substance each of these depends on the temperature and pressure at which it is measured. For mixtures and solutions they depend on composition as well. The electrical conductivity of a crystal of sodium chloride is not the same as that of an aqueous solution of sodium chloride, different aqueous solutions of sodium chloride have different conductivities, and solutions of sodium chloride in water and in methanol have different conductivities even if the concentrations of sodium chloride in them are the same.

Even for a material as simple as pure sodium chloride there are so many different properties, each of which might be measured under so many different conditions, that all of the world's chemists could not hope to measure them all if they spent their whole lives in the attempt. Theory is a powerful aid in dealing with the subsidiary questions, "How do these properties depend on the experimental conditions?" and "How do they depend on the structure?" There are different levels, of which the lowest is purely empirical. For example, we might measure the temperatures at which sodium chloride melts under several different pressures, observe that the melting point T_f changes continuously and smoothly with the pressure P, and plot T_f against P. We would then be able to find the melting points at other pressures by interpolation on this curve instead of having to measure them.

This would be convenient, but we would not have obtained much understanding at this point. Generalization and insight are needed to obtain it. Such plots show that changing the pressure affects T_f differently for different substances: the variation is much larger for some substances than for others. Its magnitude must depend on other properties of the substance. Le Chatelier's principle might occur to us. It leads us to expect that the melting point will decrease if the pressure rises, which is true for most substances, if a given amount of material occupies less volume in the liquid state than it does when solid. It is only a short step from this to the guess that the effect of pressure on the melting point depends on the difference between the density of the liquid and that of the solid. An equation describing the dependence was obtained long ago and may be found in any textbook on physical chemistry. If it had not already been obtained we would have to try to infer it, either from data on the properties of many different compounds or by more abstract reasoning. Once we have it, and have confirmed it by checking it against the best data available, we can save much experimental labor, for we would not need to make separate measurements of properties that we knew to be related.

There are higher levels of theory. We might suspect that the densities of solid and liquid sodium chloride are related to other properties of the sodium and chloride ions, such as their radii and charges, and that the other factors appearing in the equation might also be deduced from more fundamental ones. Then it might be possible to deduce these in turn from still more fundamental ones, and so on.

Everything that has been described so far is aimed at providing a background for what follows. We are about to begin considering chemical reactions and the questions that chemists ask about them. To obtain useful experimental answers to these new questions and useful understanding of those answers, we must have asked and answered the questions that have already been raised. There would be no point to studying a reaction between two materials of unknown natures, compositions, structures, and properties, and there would be even less point to trying to make any practical use of such a reaction.[†] The number of possible combinations of such materials is indefinitely large, and it would not even be possible to store and retrieve the information that might be obtained. Only by correlating information about the reactants with information about the reaction can we hope to bring order out of chaos, to understand what we observe, and to make useful predictions about combinations of reactants and conditions that have not yet been tried.

This book is chiefly concerned with the questions that follow. Some of its chapters consider how answers to them can be obtained in the laboratory and how understanding can be obtained from the answers. Others are devoted to ways in which those questions can be asked and answered in dealing with the questions that have already been raised.

[†] This is so obvious that it is really surprising how often it is ignored. One of the classic examples of its importance was provided by the German chemist Fritz Haber. Shortly after World War I he persuaded the German government, which was struggling with runaway inflation and other economic problems, to equip a ship to extract gold from sea water. This was expected to be a cheap and almost endless source of wealth, but the venture failed at great expense because the percentage of gold in sea water is several orders of magnitude smaller than Haber thought. Both knowledge and understanding are indispensable in any serious attempt to mold the world to our desires.

Certainly the first question that must be asked about a reaction is "What is the product?" We are back at our starting point! Some reactions yield only a single product. It may be obtained in a fairly pure state, or it may be contaminated with unreacted starting material or solvent. Such contamination is easy to suspect and check. More difficulty arises when different products are formed. There may be one chief product contaminated with small amounts of others, and because these were not present in the original material they may be much less easy to imagine and identify. Some reactions involve intermediates that can react with each other or with the starting materials in a number of different ways, and a dozen or more different products may result. The problem of discovering what is present, and in what proportions, has to be faced all over again.

Fortunately it helps to know what the reactants and conditions were. This enables the chemist to make some reasonable guesses that provide a starting point. If the reaction being studied is the decomposition of pure sodium hydrogen carbonate when it is heated, the solid product might conceivably contain any or all of half a dozen sodium compounds, but the possibility that it contains any potassium compounds is too far-fetched to consider. A little knowledge of inorganic chemistry leads us to think that sodium carbonate, oxide, and hydroxide are much more likely possibilities than, say, sodium carbide, and from this as a starting point several reasonable lines of investigation can be more or less easily imagined.

There are several subsidiary questions. One is "Does the nature or composition of the product depend on the experimental conditions?" This may be answered by experiment, and the further question "Why?" (or "Why not?") will be considered in a later paragraph. Another, "What reactants might be combined to obtain a particular product?", is the theme song of the synthetic chemist—as is still another, "How can the substance that is wanted be isolated from the product?", which arises whenever the product is a mixture.

Next we ask "How far does the reaction proceed?" It may be so nearly complete that no remaining trace of starting material can be detected, or so incomplete that only just barely detectable traces of product are formed. Its extent of completion may depend on many experimental conditions, including the temperature, the pressure, the identity and properties of the solvent, the nature and concentration of any electrolyte that is present, the acidity, and many others. If we have enough understanding of these effects we may be able to make a desired reaction more complete, so that a useful product can be obtained more easily and cheaply, or to make an undesired one less complete, so that the yield of useless or toxic by-products is decreased and the world becomes a little more livable for us and our descendants.

We must also ask "What is its rate?" and the closely related question "What is its mechanism?" Some reactions are almost instantaneous, while others drag on for days or centuries. Very fast reactions are those in which almost every collision of molecules of the reactants leads to reaction. Often, however, colliding molecules cannot react unless their energies are exceptionally large. Then only a tiny fraction

2*

of the collisions will result in reaction, and formation of the product will be very slow. It may be possible to find a catalyst, which accelerates the reaction by lowering the energy that is required. Smog contains hydrocarbons together with oxygen and oxides of nitrogen, which are capable of oxidizing the hydrocarbons. In the absence of a catalyst these reactions are so slow that their rates cannot be measured at ordinary temperatures, but in the catalytic converter of an automobile they become much faster and a much cleaner exhaust gas results.

Like almost everything else we have mentioned, the rates of reactions are affected by the experimental conditions. The reactions of hydrocarbons with oxygen, though extremely slow at ordinary temperatures, become so fast at higher temperatures that they can be used to heat houses or propel automobiles. There may be changes that are even more drastic. Under one set of conditions a desired product may be obtained in a nearly pure state because the reactions by which it is formed are much faster than other competing reactions that lead to the formation of an unwanted contaminant. Changing the conditions may increase the rates of the side reactions much more than it increases the rates of the desired reactions, and then the unwanted contaminant may become the principal product.

The equation that represents a chemical reaction describes only the starting materials and the final products; the mechanism of the reaction is a detailed map of the path that is followed. Some reactions are simple and uncomplicated, like those that involve only the transfer of a proton or an electron from one of the reactants to the other. Others involve the breaking of a great many bonds and the formation of many others, and these take place by a great many successive steps, of which some produce new substances or intermediates that are consumed in later steps. Polymerization reactions are extreme examples of this sort. The rate and mechanism of a reaction are inextricably interwoven. Measuring the rate of a reaction enables us to infer the natures of the steps by which it occurs, the identities of the intermediates that are formed in those steps, and the various energy relationships that are involved. Conversely, knowing the mechanism of a reaction enables us to predict the effects that changes of the experimental conditions will exert on its rate or on the proportions in which different products are formed.

Other questions can now be asked. How do the rate, mechanism, and extent of a reaction depend on the structures and properties of the reactants and products? Why do they depend on them in one way rather than another?

Limitless vistas open up, but we must end this introduction because we have already raised many more questions than we can begin to answer here. They will reappear, in various forms, in many courses that follow the one for which this book is intended. The happiness, the security, and even the lives of our fellow human beings depend largely on the abilities of chemists, now and in the future, to answer them.

In the chapters that follow, it will be assumed that you have a fair working knowledge of algebra, are reasonably familiar with the basic ideas of chemistry, and know or are learning something about physics, elementary calculus, and computer programming. All of these are indispensable to the chemist, and it would be a disservice to you to pretend that they are not. There are several short appendixes that enable you to review some things taken for granted in the text or that provide numerical values of some important constants.

Different chapters of the book serve different purposes. Some, like the ones on thermodynamics and kinetics, are relatively brief introductions to topics about which there is so much to be said that they must reappear in later courses. In these chapters the emphasis is on the fundamental concepts, and if you look back at them in a few years you will see that they are very much simplified. Few real reactions, for example, are as simple as the ones discussed in Chapter 3. However, enough is said there to introduce you to the basic ideas and their meanings, and you will develop the ability to handle more complex situations as you come across them. Other chapters deal with material so fundamental that your later texts will assume that you are familiar with it, just as this one assumes that you are familiar with algebra. In these chapters the use of the concepts in solving chemical problems receives as much stress as the concepts themselves.

Many American high schools overemphasize the importance of knowledge and memory in mathematics and science and underemphasize that of understanding and insight. By doing this they create the impression that the study of chemistry consists chiefly of memorizing facts and equations and using them in mechanical ways. It is true that you will not be well regarded as a chemist if you cannot remember the symbol for sodium or if you think that calcium chloride can be represented by the formula CaCl. However, it is far more important to understand how an equation is derived, and what assumptions are made in deriving it, than it is to memorize the equation. Similarly, it is far more important to understand the thought processes that are used in solving a problem than to try to remember the details of a particular solution. It is fatally easy to confuse one equation with another that is almost, but not quite, identical with it, and no amount of memory will equip you for coping with a situation that you have not seen before. Success in chemistry reflects, not one's ability to remember facts, figures, and equations, but one's ability to combine seemingly different ideas. Almost every chemical problem can be solved, and almost every chemical phenomenon can be studied, in several different ways. The chemist who is not imaginative enough to think of alternatives would rarely be lucky enough to light on the best or easiest. Most chemical observations can be interpreted in several different ways, and inspired guesswork is often needed to find the one that is most probable and most satisfying. There are two other questions that you should keep in mind as you go through this book, and during all the remainder of your chemical education. They are "Why ...?" and "What would happen if ...?"

CHAPTER 2

An Introduction to Chemical Thermodynamics and Equilibrium

2.1. Why do chemical reactions occur?

> PREVIEW: This section asks why different reactions occur to different extents and introduces the idea of a spontaneous reaction.

The reaction that occurs in a lead storage battery can be represented by the equation

$$Pb(s) + PbO_2(s) + 2 H^+ + 2 HSO_4^- = 2 PbSO_4(s) + 2 H_2O (l)^\dagger$$

Why does it occur? And why does it always occur in the same direction unless energy is supplied to the battery from an external source?

Some reactions, like this one, proceed to completion, leaving no detectable trace of one or more of the reactants. Others, like the one

$$AgCl (s) = Ag^+ + Cl^-$$

that occurs when silver chloride dissolves in water, hardly occur at all. What does a reaction of either of these two kinds have in common with other reactions of the same kind? What is the difference between these two kinds of reactions?

Questions like these deal with chemical equilibria and lie within the province of chemical thermodynamics, which includes a study of the changes of energy that accompany chemical reactions. There are many different kinds of energy—thermal energy, kinetic energy, potential energy, electrical energy, and many others—but some of them are much more interesting to chemists than others. Thermal energy, or heat energy, is important because chemical reactions evolve or absorb heat and because their rates and extents vary as the temperature changes. Kinetic energy is less interesting because changing the kinetic energy of a reaction mixture, perhaps by dropping

---

† The symbol "(s)" denotes a solid and "(l)" denotes a liquid. We write "2 H⁺ + 2 HSO₄⁻" because hydrogen ion and hydrogen sulfate ion are the species that predominate in aqueous solutions of sulfuric acid at the concentrations used in lead storage batteries. It would be misleading to write "2 H₂SO₄" because there is very little undissociated sulfuric acid in such solutions, or to write "4 H⁺ + SO₄²⁻" because very little of the hydrogen sulfate ion undergoes further dissociation to yield sulfate ion.

8

the beaker containing it onto the floor, will ruin the beaker but will not affect the rate or extent of the reaction. However, we shall see that changing the kinetic energies of individual molecules—or, more accurately, the relative kinetic energies of two or more molecules that may react—does have interesting effects on the rates of most reactions.

Physicists deal with questions somewhat similar to those asked here. Why do objects always fall down? Why does water always run downhill? It is of course the idea of spontaneity that we are discussing. Objects fall down spontaneously, but they can also be propelled upward. Water runs downhill spontaneously, but it can also be pumped uphill. Note that if a process is spontaneous in one direction it must be non-spontaneous in the other direction. Some kind of energy must be supplied from an external source to bring about a non-spontaneous process or to reverse a spontaneous one. By supplying electrical energy to the terminals of a battery we can reverse the spontaneous reaction that occurs as the battery is discharged. Let us rephrase our questions. Why are some chemical reactions spontaneous? Things fall down because their potential energies decrease when they do. What is the quantity analogous to potential energy for a chemical reaction?

> SUMMARY: Thermodynamics deals with the changes of energy that occur in chemical reactions and will lead us to a quantity that describes the spontaneity of a reaction.

2.2. Thermochemistry and calorimetry

> PREVIEW: This section deals with the changes of enthalpy—the quantities of heat that are evolved or absorbed—that accompany chemical reactions and explains how those changes are measured and combined.

The most striking fact about nearly every highly spontaneous reaction is that it evolves an enormous amount of heat. A few such reactions are the combustions of hydrocarbons, hydrogen, magnesium and many other metals, and the hydrazine derivative

$$H_2NN\begin{cases} CH_3 \\ H \end{cases}$$

that is used as a rocket fuel. We may suspect that there is a close connection between the spontaneity of a reaction and the amount of heat that it evolves or absorbs, and indeed there is. Other factors are also involved, but the amounts of heat involved in chemical reactions are interesting to chemists for so many different reasons that we shall discuss them at some length.

The experimental measurement of these amounts of heat is called *calorimetry*. A *calorimeter* is a device used for making such measurements. One common kind of calorimeter consists essentially of a vessel in which the reaction takes place and which is surrounded by a shield designed to prevent heat from leaking into or out of the vessel. A Thermos bottle would be a serviceable, though crude, calorimeter for studying reactions that occur in solutions. At the start of an experiment the vessel contains a solution of one of the reactants, and a syringe or ampule containing a solution of the other reactant is immersed in it. Some provision is made for detecting and measuring changes of the temperature of the solution in the vessel. Thermometers were once used, but are no longer considered to be sensitive enough, and electrical devices such as thermocouples and thermistors are now used instead. They permit the measurement of temperature changes as small as a few microdegrees. The temperature may change for some time after the apparatus is assembled, because the vessel, the solution in it, and the second solution in the syringe or ampule may all have been at different temperatures initially. When these have been equalized and the temperature has become constant, the two solutions are mixed by depressing the plunger of the syringe or by breaking the ampule, and the change of temperature caused by the occurrence of the reaction is measured. To correlate the change of temperature with the amount of thermal energy that has been liberated or absorbed, a known amount of electrical energy is then added to the mixture and the resulting further change of temperature is measured. The electrical energy may be added by passing a known electrical current I for a measured length of time t through a platinum wire of known resistance R immersed in the reaction mixture. The added electrical energy is given by I^2Rt, and will have the units of joules if I is expressed in amperes, R in ohms, and t in seconds. Since each of the two measured changes of temperature must be proportional to the amount of energy that caused it, the number of joules involved in the chemical reaction is easy to deduce. Because the number of joules is proportional to the amount of reaction that took place, it is uninteresting by itself. Saying that 4 joules were liberated in a certain reaction would be meaningless if we did not know how much of some substance had been consumed or produced.

We therefore speak of the *change of enthalpy that accompanies a reaction.* It is represented by the symbol ΔH, and it is defined as *the number of joules absorbed during the consumption of one mole of a reactant or the formation of one mole of a product.* It has the units J mol^{-1}.[†]

Reactions that absorb heat are called *endothermic* and have positive values of ΔH. Reactions that evolve heat are called *exothermic* and have negative values of ΔH.

---

† Values of ΔH have long been expressed in calories (or kilocalories) per mole, but the International Union of Pure and Applied Chemistry (IUPAC) and the International Union of Pure and Applied Physics (IUPAP) have recently recommended that the calorie be dropped as a unit of energy and that the joule be used instead. 1 J (joule) = 0.239 cal = 1 W s (watt second) = 1 V A s (volt ampere second), and of course 1 kJ (kilojoule) = 10^3 J. Note that the abbreviation "s" (second) is not the same as the abbreviation "(s)" (solid).

Example 2.1. An ester is a compound having the formula $RC\!\!\diagdown\!\!{}^{O}_{OR'}$. The symbols R and R' represent groups of atoms that may, but need not, be the same. Some typical esters are methyl acetate, $CH_3C\!\!\diagdown\!\!{}^{O}_{OCH_3}$, and ethyl benzoate, $C_6H_5C\!\!\diagdown\!\!{}^{O}_{OC_2H_5}$.

An ester reacts with hydroxide ion to give the alcohol R'OH and the anion $RC\!\!\diagdown\!\!{}^{O}_{O^-}$ of the acid RCOOH:

$$RC\!\!\diagdown\!\!{}^{O}_{OR'} + OH^- = RC\!\!\diagdown\!\!{}^{O}_{O^-} + R'OH$$

A mixture containing 2.00×10^{-4} mole of a certain ester and an excess of hydroxide ion reacts in a calorimeter, and there is an increase of 0.0225° in the temperature. After the reaction is complete a current of 0.0975 A is passed for 12.75 s through a 101.0-Ω resistor immersed in the reaction mixture. This causes an increase of 0.0261° in the temperature. What is the value of ΔH for the reaction?

Answer. The electrical energy that was added was equal to $I^2Rt = (0.0975)^2 \times 101.0 \times 12.75 = 12.24$ J. Since it produced an increase of 0.0261° in the temperature, 12.24/0.0261 J would have been needed to raise the temperature 1°. The rise of 0.0225° that occurred during the reaction therefore corresponds to the evolution of $0.0225 \times 12.24/0.0261$ J. The number of joules per mole (of ester or any other substance taking part in the reaction, since all of the coefficients in the balanced equation for the reaction are equal to 1) is $0.0225 \times 12.24/(0.0261 \times 2.00 \times 10^{-4})$. Because the reaction evolved heat, ΔH is negative, and so we write

$$\Delta H = -0.0225 \times 12.24/(0.0261 \times 2.00 \times 10^{-4}) = -5.2_8 \times 10^4 \text{ J mol}^{-1} = -52._8 \text{ kJ mol}^{-1}.$$

We are not entitled to three significant figures in the final result (Appendix I).

The value of ΔH for any reaction depends on the temperature and pressure, and if the reaction takes place in solution it also varies slightly with the concentrations. However, it does not depend on the path that is followed in obtaining the products from the reactants; this generalization is called Hess' law. If dilute solutions containing silver and iodide ions are mixed, the exothermic reaction $Ag^+ + I^- = AgI(s)$ takes place directly, and the change of enthalpy that accompanies it is found to be -112.34 kJ mol^{-1} (of silver iodide or of either ion). This means that 112.34 kJ is evolved for each mole of silver iodide that is formed, or for each mole of either ion that reacts. The same information can be conveyed in less space by writing

(1) $Ag^+ + I^- = AgI(s); \quad \Delta H = -112.34$ kJ mol^{-1}.

We can obtain the same final product by treating silver ion with chloride ion to precipitate silver chloride, then adding iodide ion. The silver chloride will be transformed into silver iodide, and the chloride ion that was consumed in the first step will be set free again. The changes of enthalpy for these two steps are represented by

(2) $Ag^+ + Cl^- = AgCl(s); \quad \quad \Delta H = -65.47$ kJ mol^{-1}.
(3) $AgCl(s) + I^- = AgI(s) + Cl^-; \quad \Delta H = -46.87$ kJ mol^{-1}.

If we add the chemical equations for these two steps, the silver chloride and the chloride ion disappear, and the sum is identical with the equation for step (1) above. The sum of the changes of enthalpy for steps (2) and (3), $-65.47+(-46.87) = -112.34$ kJ mol^{-1}, is also identical with the value for step (1), in which the same overall process occurs directly instead of proceeding through silver chloride as an intermediate. Hess' law means that, if one process C is the sum of two others A and B, the change of enthalpy accompanying C is equal to the sum of the changes of enthalpy accompanying A and B.

Values of ΔH for different reactions are not listed in tables. There are too many reactions, and the tables would have to be too long. Tables of chemical thermodynamic properties list values of the standard enthalpies of formation of individual compounds and ions instead. The standard enthalpy of formation of a compound, represented by the symbol H_f^0, is the change of enthalpy that accompanies the formation of one mole of the compound from its elements at a specified temperature[†] when each reactant and each product is in its standard state. We shall define the standard state in Section 2.3.

The value of H_f^0 is equal to zero for any element in its standard state. Calorimetric measurements, made at 298 K with chlorine gas at a pressure of 760 torr and with pure crystals of silver and silver chloride (which are the standard states of all these substances) give

$$Ag(s) + \tfrac{1}{2}Cl_2\,(g,\,760\text{ torr}) = AgCl(s); \quad \Delta H = -127.03 \text{ kJ mol}^{-1}$$

and therefore the standard enthalpy of formation of silver chloride is -127.03 kJ mol^{-1} at 298 K.

Values of H_f^0 can be used to calculate the changes of enthalpy for chemical reactions in the following way. Suppose that we wish to calculate the change of enthalpy for the reaction $CH_4(g)+Cl_2(g) = CH_3Cl(g)+HCl(g)$ at 298 K. The standard enthalpies of formation of the three compounds involved in this reaction may be found in tables: they are -74.85 kJ mol^{-1} for methane, -82.01 kJ mol^{-1} for methyl chloride, and -92.30 kJ mol^{-1} for hydrogen chloride. The first of these values means that, at 298 K and with each substance in its standard state, the change of enthalpy that occurs during the reaction $C(s)+2 H_2(g) = CH_4(g)$ is -74.85 kJ mol^{-1} (of methane). Let us apply Hess' law to this reaction as process A and the same reaction, but proceeding in the opposite direction, as process B. We shall have

Process A: $C(s)+2 H_2(g) = CH_4(g); \quad \Delta H_A = -74.85$ kJ mol$^{-1} \quad (= H_{f,\,CH_4(g)}^0)$.

Process B: $CH_4(g) = C(s)+2 H_2(g); \quad \Delta H_B = ?$

Process C: \qquad\qquad nothing happens; $\quad \Delta H_C = 0$.

---

† Throughout this book the temperature should be understood to be 298 K (= 25 °C) unless another temperature is specified.

Since $\Delta H_C = \Delta H_A + \Delta H_B$, it is not difficult to conclude that $\Delta H_B = -\Delta H_A$. Reversing the direction of any process changes the sign, but does not alter the magnitude, of the enthalpy change. We are now in a position to imagine a sequence of reactions, each involving the formation of a compound from its elements or the decomposition of a compound into its elements, and having the desired reaction as the overall sum:

Process A: $CH_4(g) = C(s) + 2 H_2(g)$; $\qquad\qquad \Delta H_A = -H^0_{f,\, CH_4(g)}.$

Process B: $C(s) + \frac{3}{2} H_2(g) + \frac{1}{2} Cl_2(g) = CH_3Cl(g)$; $\quad \Delta H_B = H^0_{f,\, CH_3Cl(g)}.$

Process C: $\frac{1}{2} H_2(g) + \frac{1}{2} Cl_2(g) = HCl(g)$; $\qquad\qquad \Delta H_C = H^0_{f,\, HCl(g)}.$

Process D (the sum of A, B, and C):

$$CH_4(g) + Cl_2(g) = CH_3Cl(g) + HCl(g); \quad \Delta H_D = \Delta H_A + \Delta H_B + \Delta H_C$$

$$\Delta H_D = -H^0_{f,\, CH_4(g)} + H^0_{f,\, CH_3Cl(g)} + H^0_{f,\, HCl(g)} \quad (= -99.46 \text{ kJ mol}^{-1}).$$

More generally, for a reaction that involves n_i moles of the ith reactant and n_j moles of the jth product,

$$\Delta H^0 = \sum_j n_j (H^0_f)_{j,\, \text{products}} - \sum_i n_i (H^0_f)_{i,\, \text{reactants}}. \qquad (2.1)$$

The superscript zeros, of which the one on the left-hand side has been added for consistency, remind us that we are dealing with substances in their standard states (Section 2.3).

Values of the standard enthalpies of formation of individual ions are also given in tables and are used in the same way. To find the change of enthalpy that accompanies the reaction of silver ion with chloride ion, we would write

$$Ag^+ + Cl^- = AgCl(s); \quad \Delta H^0 = H^0_{f,\, AgCl(s)} - (H^0_{f,\, Ag^+} + H^0_{f,\, Cl^-}).$$

To find the change of enthalpy that accompanies the precipitation of magnesium hydroxide, we would write

$$Mg^{2+} + 2OH^- = Mg(OH)_2(s); \quad \Delta H^0 = H^0_{f,\, Mg(OH)_2} - (H^0_{f,\, Mg^{2+}} + 2H^0_{f,\, OH^-}).$$

Here the factor of 2 in the last term is necessary because 2 moles of hydroxide ion are needed to obtain 1 mole of magnesium hydroxide, and the standard change of enthalpy accompanying the formation of 2 moles of hydroxide ion is twice as large as the standard change of enthalpy for the formation of (1 mole of) hydroxide ion.

SUMMARY: The change of enthalpy, ΔH, that accompanies a reaction is equal to the number of joules that are absorbed for each mole of a reactant that is consumed, or each mole of a product that is formed. The value of ΔH is positive for an endothermic reaction, in which heat is absorbed, and negative for an exothermic one, in which heat is evolved. Values of ΔH may be calculated from enthalpies of formation, and may be combined by using Hess' law.

✗ 2.3. Standard states

> PREVIEW: Since the reactivity of a substance depends on factors such as pressure and composition, we speak of a standard state, in which its reactivity is fixed, reproducible, and dependent only on its chemical identity. This section defines the standard states of solids, liquids, gases, and dissolved substances.

The reactivity of a substance—in the context of this chapter, the changes of energy that accompany its chemical reactions—depends on its form and also on the conditions under which it is studied. A solid is more reactive if it is amorphous than if it is crystalline, and if it can exist in different crystalline forms these will differ in reactivity. A pure liquid or solid becomes less reactive if it is diluted with another liquid or solid that is miscible with it, and a gas becomes less reactive as its partial pressure decreases, while the reactivity of a solute decreases as its concentration decreases.

To characterize the behavior of any particular substance it is convenient to focus our attention on one fixed set of conditions, which we call the standard state. Different choices of standard states may be made to serve different purposes, but the ones we shall use may be defined and illustrated as follows:

1. If the substance is a solid at the temperature of interest, its standard state is the pure crystalline solid under a pressure of 760 torr. If there are two or more crystalline forms, the standard state is the one that has the lowest energy and is therefore the most stable. At 25° the standard state of silver chloride is pure crystalline silver chloride, while that of carbon is pure crystalline graphite, which is more stable than diamond.

2. If the substance is a liquid at the temperature of interest, its standard state is the pure liquid under a pressure of 760 torr. At 298 K the standard state of water is pure liquid water. At 248 K it would be crystalline ice I, which is more stable at this temperature and pressure than any of the half-dozen other crystalline forms of ice.

3. If the substance is a gas at the temperature of interest, its standard state is the gasous form at a partial pressure of 760 torr.

4. If the substance is a solute, its standard state is a solution in which its activity is equal to 1 mole per cubic decimeter.[†] More will be said about the activity in Section 2.6, and also in Chapter 13. For the time being we shall pretend that the activity of a solute is equal to its concentration. This would make a 1 *M* (molar) solution the standard state of any solute.

> SUMMARY: The standard state of a substance is an arbitrary condition in which its reactivity is fixed. Definitions are given of the standard states of solid, liquid, gaseous, and dissolved substances.

[†] The liter was defined many decades ago in such a way that it corresponded to 1000.027 cm³. The resulting difference between 1 ml and 1 cm³ was just large enough to be irritating in very accurate work. Consequently an international Conférence Générale des Poids et Mesures in 1964 redefined the liter as 1000 cm³ exactly. That gave us two names (1 ml and 1 cm³) for the same thing, exactly one more than we need. The latest IUPAC recommendation is that the liter be abandoned and that the cubic decimeter be used instead. 1 dm (decimeter) = 10 cm; 1 dm³ = 1000 cm³.

2.4. Entropy

PREVIEW: The entropy of a system is a measure of its disorder. The change of entropy that accompanies a chemical process is a measure of the increase of disorderliness that results from the process.

Although a reaction is more likely to be spontaneous if it evolves heat than if it absorbs heat, the value of ΔH alone does not enable us to decide whether a reaction is spontaneous or not. There are many endothermic processes that are spontaneous, and many exothermic ones that are not. The effect of temperature is especially important: many reactions evolve very large amounts of heat and are spontaneous at low temperatures, but become non-spontaneous at very high temperatures.

The freezing of water is exothermic: for the process $H_2O(l) = H_2O(s)$ the value of ΔH^0 is equal to -6.01 kJ mol^{-1}, and it is almost, though not completely, independent of temperature. The superscript zero in the symbol ΔH^0 means that we are discussing the freezing of pure liquid water to give pure crystalline ice. Although ΔH^0 is negative and water does freeze spontaneously at 268 K ($-5°$ C), it is the melting of ice, for which ΔH^0 is positive, that is spontaneous at 278 K ($+5°$ C).

Something else has to be taken into account, and this is the change of entropy that accompanies the reaction. The change of entropy is represented by the symbol ΔS, and has the units J K^{-1} mol^{-1}. Like the change of enthalpy ΔH, it depends on the reactants and products but not on the way in which the process occurs. It is convenient to consider only the change of entropy that occurs when all the products and reactants are in their standard states; this particular value of ΔS for any reaction is called the standard change of entropy and given the symbol ΔS^0. As with ΔH^0, and for the same reason, values of ΔS^0 are not tabulated for complete reactions; values of the standard entropies of the individual products and reactants, S^0, must be combined to find ΔS^0. Values of S^0 and H_f^0 for a number of common substances are given in Appendix II.

The value of ΔS^0 for any process can be calculated by combining the tabulated values of S^0 with the equation

$$\Delta S^0 = \sum_j n_j S^0_{j,\,\text{products}} - \sum_i n_i S^0_{i,\,\text{reactants}} \tag{2.2}$$

which is exactly like eq. (2.1). However, S^0 is not in general equal to zero for the elements, and therefore free elements must be included in calculations based on eq. (2.2) although they can be ignored in those based on eq. (2.1). This is because H_f^0 is defined as the change of enthalpy for a reaction in which the substance of interest, in its standard state, is formed by the combination of its constituent elements in their standard states, whereas S^0 is the total change of entropy accompanying the formation of that substance, in its standard state, from an imaginary perfect crystal of the same substance at 0 K.

Example 2.2. What change of entropy accompanies the reaction $Mg(s) + \frac{1}{2} O_2$ (g, 760 torr) = $= MgO(s)$ at 298 K?

Answer. Tables of standard entropies give $S^0 = 26.78$ J K^{-1} mol^{-1} for magnesium oxide, 32.51 J K^{-1} mol^{-1} for elemental magnesium, and 205.02 J K^{-1} mol^{-1} for oxygen at 298 K. The desired value of ΔS^0 is given by $S^0 = S^0_{MgO(s)} - (S^0_{Mg(s)} + \frac{1}{2} S^0_{O_2(g)}) = 26.87 - (32.51 + 102.51) = = -108.24$ J K^{-1} mol^{-1} (of magnesium or magnesium oxide). The entropy decreases as the reaction proceeds. Almost all of the decrease results from the disappearance of the disorder that exists in gaseous oxygen.

The entropy of a system is a measure of its randomness or disorder. In a typical crystal the disorder is low, because the atoms or molecules are neatly ranked in rows, columns, and files. If we know how they are arranged in one portion of the crystal we have quite a good chance of predicting accurately how they will be found to be arranged in another portion a little distance away. However, imperfections in the crystal will make our predictions wrong, and thermal vibrations of the atoms or molecules around their average positions will make them imprecise. Because there is some disorder in their arrangement, the value of S^0 is positive, even for a crystalline element, at any temperature above 0 K. An increase of temperature increases the disorder, and therefore the value of S^0 for any substance increases as the temperature increase. Melting a crystal, or evaporating a liquid, also increases the disorder, and therefore the value of S^0 increases sharply at the melting or boiling point. In a large molecule, such as that of a protein or a synthetic polymer, there is usually a great deal of regularity. One carbon atom in a chain is likely to be followed by another at an average distance that does not vary from one part of the chain to another. Certain functional groups, such as carbonyl $\left(\rangle C{=}O \right)$ or amino $(-NH_2)$ groups, or certain sequences of amino acid residues, are likely to recur at more or less fixed distances. On breaking the chemical bonds to form simpler molecules, many structural features will vanish, and disorder may be greatly increased.

The entropy of the universe tends to increase. As every housekeeper knows, randomness comes naturally, while order has to be striven for. Processes in which disorder increases tend to be spontaneous, while processes in which order is created out of chaos tend to be non-spontaneous. The value of ΔS is positive if disorder increases during a process, negative if disorder decreases.

A positive value of ΔS is therefore conducive to spontaneity, as is a negative one of ΔH. However, for a reaction to be spontaneous it does not suffice for ΔS to be positive, just as it does not suffice for ΔH to be negative. For the process $H_2O(l) = H_2O(s)$, which was mentioned above, ΔS is negative because the arrangement of the molecules is more orderly in ice than in liquid water. Its numerical value is -22.00 J K^{-1} mol^{-1} at 273 K ($0°$ C). Because it is negative the process tends to be non-spontaneous, and yet the tendency becomes fact only if the temperature is above 273 K.

SUMMARY: The standard change of entropy, ΔS^0, that accompanies a reaction can be calculated from the standard entropies of the reactants and products. A positive value of ΔS^0 means that the universe becomes more disorderly as the process occurs, and is conducive to spontaneity.

2.5. Free energy

PREVIEW: A negative value of ΔH, or a positive one of ΔS, suggests that a reaction is likely to be spontaneous, but neither of these pieces of information alone is a guarantee of spontaneity. The change of free energy ΔG is defined by the equation $\Delta G = \Delta H - T\Delta S$. It provides the fundamental criterion of spontaneity.

We began this discussion by envisioning some sort of energy, analogous to the potential energy for mechanical processes, which would decrease during spontaneous reactions. We have so far identified two quantities, ΔH and ΔS, that are related to the spontaneity of a reaction. One of them, ΔH, represents a change of energy even though we have chosen to give it the units of energy per mole so that it will be independent of the physical size of a system to which we apply it. The other, ΔS, is a change of energy per degree per mole. Let us create a quantity having the same dimensions as ΔH by multiplying ΔS by the temperature T (in Kelvins, not degrees Centigrade or Celsius). Now let us compare the values of ΔH with those of $T\Delta S$ for the process $H_2O(l) = H_2O(s)$ at three different temperatures, say 268, 273, and 278 K ($-5°$ O, and $+5°$ C):

T $=$	268	273	278	K
The process is	spontaneous	at equilibrium	non-spontaneous	
The quantity we seek is	negative	zero	positive	
ΔH $=$	-6.008	-6.008	-6.008	kJ mol^{-1}
$T\Delta S$ $=$	$(268\ \Delta S =)-5.898$	$(273\ \Delta S =)-6.008$	$(278\ \Delta S =)-6.118$	kJ mol^{-1}
ΔH is	smaller than	equal to	larger than	$T\Delta S$
$\Delta H - T\Delta S$ is	negative	zero	positive	

All the values of ΔH are negative, and so are all those of $T\Delta S$, but those of ΔH are all the same while those of $T\Delta S$ become more negative as the temperature rises, At 273 K, where the two are equal, the process is at equilibrium: we can freeze the water in a mixture of water and ice at this temperature by withdrawing heat from it, or melt the ice in it by adding heat, but nothing will happen if we leave it strictly alone. Here the quantity we seek must be zero, just as the difference between the potential energies of an object at two different points must be equal to zero if the object does not fall spontaneously from either point to the other. The fact that ΔH and $T\Delta S$ are equal when the process is at equilibrium suggests a possible way of evaluating ΔS. If we could find some conditions under which a process is at equilibrium (or, so that we can have something to measure, very near it), we would need only to measure ΔH under those conditions and divide it by the temperature T.

At any other temperature the values of ΔH and $T\Delta S$ for the process $H_2O(l) = H_2O(s)$ must be unequal, for although ΔH depends only slightly on the temperature, $T\Delta S$ is approximately proportional to the temperature (and would be exactly so if ΔS^0 were completely independent of temperature). Since the freezing of water is spontaneous below 273 K but non-spontaneous above it, and since we seek a quantity that will be negative for a spontaneous process but positive for a non-spontaneous one, we cannot escape the combination $\Delta H - T\Delta S$.

We call this combination the *change of free energy that accompanies the process* and we assign it the symbol ΔG:

$$\Delta G = \Delta H - T\Delta S. \tag{2.3}$$

The symbol honors J. Willard Gibbs, who was the first to tread this path. We can summarize the entire discussion up to this point by saying that a process is spontaneous if the value of ΔG for it is negative, is at equilibrium if $\Delta G = 0$, and is non-spontaneous if ΔG is positive.

From eq. (2.3) we can easily see why many common processes behave as they do. Although ΔS is negative for the combustion of magnesium (Example 2.2), ΔH is so enormously negative (-601.8 kJ mol^{-1}) that ΔG must also be negative at ordinary temperatures. However, because ΔS is negative ΔG must become less negative as the temperature rises, and even this highly exothermic reaction becomes non-spontaneous at extremely high temperatures. For the hydrogenation of benzene in the gas phase to give cyclohexane at 298 K:

$$\bigcirc (g) + 3\,H_2(g) = \bigcirc (g)$$

the standard change of entropy is -362.8 J K^{-1} mol^{-1}. This is a large negative value, but nevertheless the process is spontaneous because under these conditions ΔH ($= -206.1$ kJ mol^{-1}) is also large and negative, and together these two values give

$$\Delta G = -206.1 - (298)(-0.3628) = -206.1 + 108.1 = -98.0 \text{ kJ mol}^{-1}.$$

The value of ΔG tells us only whether a process will occur *spontaneously;* it does not tell us whether the process *will* occur. Similarly, the fact that the potential energy of a car decreases when it falls from a bridge into a river below the bridge means that the car will fall *spontaneously* but not that it *will* fall: if the car, driver, and guard rail are all in good condition it need not be suicidal to drive across a bridge. For the reaction $2\,Al(s) + \frac{3}{2}O_2$ (g, 0.2 atm) $= Al_2O_3(s)$ the value of ΔG at 298 K is close to -1.5 MJ[†] mol^{-1}, which is negative and staggeringly large, but it is quite practical to expose an aluminum ladder or storm window to the open air. Whether or not a reaction does occur at a finite rate is the province of chemical kinetics (Chapter 3); thermodynamics tells us only what will happen if anything happens. With benzene,

[†] 1 MJ (megajoule) $= 10^3$ kJ $= 10^6$ J.

hydrogen, and cyclohexane gases in their standard states, the hydrogenation of benzene is extremely slow at 298 K. Because ΔG is negative under these conditions, chemists knew that it would be possible to prepare cyclohexane by this reaction if a catalyst (Section 3.9) could be found. It was discovered that the reaction is quite rapid at a moderately elevated temperature in the presence of finely divided metallic nickel, and cyclohexane is made in this way in very large amounts. If ΔG were positive instead of negative, a search for a catalyst would have been a waste of time.

> SUMMARY: A process is spontaneous if it involves a decrease of free energy, so that ΔG is negative. It is at equilibrium if ΔG is equal to zero, and is non-spontaneous if ΔG is positive. A spontaneous process will occur in the forward direction rather than in the reverse direction, but may occur too slowly to be observed: the spontaneity of a process is unrelated to its rate.

2.6. How free energies behave

> PREVIEW: This section introduces the standard change of free energy, ΔG^0, and shows how it can be calculated from the standard free energies of formation of the reactants and products. If any of these is not in its standard state, its free energy differs from its standard free energy and ΔG for the process differs from ΔG^0. The differences are in accord with le Chatelier's principle.

In Section 2.2 we saw that reversing the direction of a reaction changes the sign, but does not affect the magnitude, of the change of enthalpy associated with the reaction. Exactly the same thing is true of the changes of entropy and free energy:

$$\Delta H_{\text{backward}} = -\Delta H_{\text{forward}}; \quad \Delta S_{\text{backward}} = -\Delta S_{\text{forward}};$$
$$\Delta G_{\text{backward}} = -\Delta G_{\text{forward}}. \tag{2.4}$$

This means that if a reaction is spontaneous in one direction it is non-spontaneous in the other.

The most important special form of eq. (2.3) is the one that describes the change of free energy when all of the reactants and products are in their standard states:

$$\Delta G^0 = \Delta H^0 - T \Delta S^0. \tag{2.5}$$

Like values of ΔH^0 and ΔS^0, values of ΔG^0 are not tabulated for complete reactions. For every substance there is a certain change of free energy that accompanies its formation, in its standard state, from its constituent elements in their standard states. This change of free energy is called the standard free energy of formation and is represented by the symbol G_f^0. Like H_f^0, G_f^0 is equal to zero for any element in its standard state. Values of G_f^0 can be combined to obtain values of ΔG^0 for chemical reactions by using an equation that is completely analogous to eqs. (2.1) and (2.2):

$$\Delta G^0 = \sum_j n_j (G_f^0)_{j, \text{ products}} - \sum n_i (G_f^0)_{i, \text{ reactants}}. \tag{2.6}$$

As this implies, the value of ΔG^0 is independent of the path that the reaction follows. Since this is true for both ΔH^0 and ΔS^0, it must also be true for ΔG^0.

Example 2.3. The standard free energies of formation of silver ion, chloride ion, and pure crystalline silver chloride at 298 K are $+77.11$, -131.17, and -109.70 kJ mol^{-1}, respectively. What is the standard change of free energy for the reaction $Ag^+ + Cl^- = AgCl(s)$?

Answer. $\Delta G^0 = G^0_{f,\,AgCl(s)} - (G^0_{f,\,Ag^+} + G^0_{f,\,Cl^-}) = -109.70 - (77.11 + (-131.17)) = -55.64$ kJ mol^{-1}. Because this value is negative, the reaction will proceed spontaneously in a solution that contains 1 M silver ion and 1 M chloride ion and that is in equilibrium with pure crystalline silver chloride, and the solid silver chloride will not dissolve spontaneously in such a solution.

What if a reactant or product is not in its standard state? Its reactivity is affected, as was argued at length in Section 2.3, and the value of ΔG for the reaction changes accordingly. The reaction $Ag^+ + Cl^- = AgCl(s)$ is spontaneous if the concentrations of silver ion and chloride ion are both 1 M and if the silver chloride is crystalline and pure; under these conditions ΔG is equal to ΔG^0 and its value is -55.64 kJ mol^{-1} (Example 2.3). If the concentration of silver ion were increased, there would be an increase of the tendency for the reaction to occur in the forward direction. This increase is in accordance with le Chatelier's principle,[†] and corresponds to a change of the value of ΔG from -55.64 kJ mol^{-1} to some more negative value. The free energy of the product is not affected, because that is equal to the standard free energy of formation of silver chloride regardless of how much silver chloride there is. (Remember that the free energy and the standard free energy are really energies per mole.) The reason why raising the concentration of silver ion affects the value of ΔG for the reaction must be that it affects the free energy of silver ion. For any individual substance—say, the ith one—in a reaction mixture, the free energy is described by the equation

$$G_i = G_i^0 + RT \ln a_i \qquad (2.7)$$

where G_i represents the free energy of formation of the substance, at the activity a_i, from its elements in their standard states; G_i^0 is the standard free energy of formation. The usual subscript "f" has been dropped from the symbols for the free energy and standard free energy of formation to simplify them. The product RT is needed to convert the natural logarithm ($\log_e$ or $\ln$) of the activity, which is a dimensionless number, to units of J mol^{-1} or kJ mol^{-1} so that the second term on the right-hand side can be combined with the first. R is the gas constant, so called because it also

---

† Le Chatelier's principle states that a system will respond to a change or stimulus in a way that tends to counteract that change or to decrease its magnitude. If heat is added to a mixture or ice and water, ice will melt: the process $H_2O(s) = H_2O(l)$ absorbs some or (if there is enough ice) all of the added heat, and the resulting rise of temperature is smaller than it would be if this process did not occur. If hydrochloric acid is added to a solution containing acetate ion, the reaction $H_3O^+ + OAc^- = HOAc(aq) + H_2O(l)$ will occur in the forward direction, and the increase of the concentration of hydronium ion is smaller than it would be if this reaction did not occur.

appears in the ideal-gas law $PV = nRT$, and its value in the units we want here is 8.3143 J K^{-1} mol^{-1}; T is the temperature (K). At 298 K the value of RT is 2479.0 J mol^{-1} or, more conveniently because we usually wish to express free energies and standard free energies in kJ mol^{-1}, 2.4790 kJ mol^{-1}.

We must now say what we mean by the activity a_i. The substances that will be most important to us are solutes, and for these we have already said (in Section 2.3) that we shall take activities and concentrations to be the same until much later in the book. For gases we shall say that·activities are equal to partial pressures (in atmospheres). For a liquid we shall say that the activity is equal to the mole fraction, which is the ratio of the number of moles of the particular liquid in which we are interested to the total number of moles of all the different materials present in the liquid phase. In pure water we would accordingly say that the activity of water is equal to 1, but in a mixture that contained one mole of ethanol for each mole of water we would say that the activity of water is equal to 0.5. When we have occasion to speak about solid solutions we shall define the activities of the solids dissolved in them in the same way that we do for the liquids that are contained in a liquid mixture: the activity of each component of a solid solution is equal to its mole fraction.

Equation (2.6) described the change of free energy that accompanies a reaction when all of the reactants and products are in their standard states. If any of them is not, we shall simply have

$$\Delta G = \sum_j n_j G_{j,\,\text{products}} - \sum_i n_i G_{i,\,\text{reactants}} \qquad (2.8)$$

in which each of the free energies appearing on the right-hand side will be given by eq. (2.7).

Example 2.4. What is the change of free energy at 298 K for the reaction $Ag^+(2\,M) + Cl^-(1\,M) = AgCl(s)$?

Answer. The chloride ion and silver chloride are in their standard states, and their free energies are therefore equal to their standard free energies of formation, -131.17 and -109.70 kJ mol^{-1}, respectively (Example 2.3). Because silver ion is not in its standard state, its free energy must first be obtained from eq. (2.7). Since $G^0_{f,\,Ag^+} = +77.11$ kJ mol^{-1},

$$G_{Ag^+} = 77.11 + 2.4790 \ln(2) = 77.11 + 2.4790(0.693) = 78.83 \text{ kJ mol}^{-1}$$

Now

$$\Delta G = G_{AgCl} - (G_{Ag^+} + G_{Cl^-}) = 109.70 - (78.83 + (-131.17)) = -57.36 \text{ kJ mol}^{-1}$$

This is more negative than the result obtained in Example 2.3 with a silver-ion concentration of 1 M.

SUMMARY: The value of ΔG for a process is numerically equal to, but has the opposite sign from, the value of ΔG for the reverse process. Values of ΔG^0, the standard change of free energy for a process in which all the reactants and products are in their standard states, can be calculated by combining tabulated values of the standard free energies of formation of the reactants and products, or from values of ΔH^0 and ΔS^0. The free energy of a substance, G, is related to its activity a by the equation $G = G^0 + RT \ln a$, from which we can calculate the effects of changes in the activities of reactants and products on the change of free energy that accompanies a process.

2.7. Free energies and equilibrium constants

PREVIEW: Equations appearing in earlier sections are combined to obtain a relationship between the standard change of free energy that accompanies a process and the equilibrium constant of the process.

Examples 2.3 and 2.4 showed that, in accordance with le Chatelier's principle, increasing the concentration of a reactant increases the spontaneity of a reaction. Increasing the concentration of silver ion from 1 M to 2 M makes the reaction $Ag^+ + Cl^-(1\ M) = AgCl(s)$ more spontaneous. Decreasing the concentration of silver ion would make it less so. The value of ΔG, which is -55.64 kJ mol^{-1} with 1 M silver ion (Example 2.3), would become less and less negative as the concentration of silver ion decreased. There would be some particular concentration of silver ion at which ΔG became equal to zero. With this concentration of silver ion the reaction would be at equilibrium; if $\Delta G = 0$ the reaction is not spontaneous in either direction. We can write

$$\Delta G = 0 \quad \text{at equilibrium.} \tag{2.9}$$

This is indeed the fundamental definition of chemical equilibrium. A mixture whose composition does not change over a long period of time might be at equilibrium or it might not. Perhaps two or more of its constituents react spontaneously, but the reaction is so slow that we cannot observe it. In situations like this the application of eq. (2.9) enables us to recognize that equilibrium has not been reached.

Equations (2.7), (2.8), and (2.9) enable us to describe the composition of an equilibrium mixture in a very general way. Let us imagine a reaction that is represented by the equation

$$aA + bB = yY + zZ.$$

The change of free energy that accompanies it is given by eq. (2.8), which for these particular reactants and products can be written

$$\Delta G = yG_Y + zG_Z - [aG_A + bG_B]. \tag{2.10}$$

Each of the individual free energies can be described by eq. (2.7):

$$\Delta G = y(G_Y^0 + RT \ln a_Y) + z(G_Z^0 + RT \ln a_Z) - [a(G_A^0 + RT \ln a_A) + b(G_B^0 + RT \ln a_B)].$$

The right-hand side consists of four standard free energies of formation and four terms that involve activities. Separating these, and noting that each logarithm is multiplied by RT:

$$\Delta G = yG_Y^0 + zG_Z^0 - aG_A^0 - bG_B^0 + RT(y \ln a_Y + z \ln a_Z - a \ln a_A - b \ln a_B)$$

$$= \Delta G^0 + RT \ln \frac{a_Y^y a_Z^z}{a_A^a a_B^b}.$$

The second equality recalls eq. (2.6) and a fundamental property of logarithms. If the mixture is at equilibrium, then $\Delta G = 0$ in accordance with eq. (2.9), and when this is true we must have

$$\Delta G^0 = -RT \ln \frac{a_Y^y a_Z^z}{a_A^a a_B^b}. \tag{2.11}$$

The argument of the logarithmic term $(a_Y^y a_Z^z / a_A^a a_B^b)$ is the equilibrium constant of the reaction, which is always represented by the symbol K. In short,

$$\Delta G^0 = -RT \ln K. \tag{2.12}$$

Example 2.5. What is the value of the equilibrium constant for the reaction $Ag^+ + Cl^- = AgCl(s)$ at 298 K?

Answer. $\Delta G^0 = -55.64$ kJ mol^{-1} (Example 2.3). Using eq. (2.12), with $RT = 2.4790$ kJ mol^{-1}, $\ln K = -55.64/(-2.4790) = 22.44$. If you have a pocket calculator with an "e^x" key, you can now obtain a value of K from the relation $e^{\ln x} = x$, or $e^{22.44} = 5.59 \times 10^9 = K$. If you do not, you need to know that $\log_{10} x = 0.4343 \ln x$. This gives $\log_{10} K = (0.4343)(22.44) = 9.748$, so that $K = 10^{0.748} \times 10^9$. Since $10^{\log_{10} x} = x$ and since 0.748 is the decadic logarithm of 5.59, $K = 5.59 \times 10^9$ mol^{-2} dm^6. Section 2.8 will indicate why the units are important.

If a reaction is spontaneous when all its products and reactants are in their standard states, it has a negative value of ΔG^0. Because of the minus sign in eq. (2.12), $\ln K$ will be positive, and K will exceed 1. For a reaction that is non-spontaneous with all products and reactants in their standard states, ΔG^0 is positive, $\ln K$ is negative, and K is smaller than 1. Knowing that silver and chloride ions react in 1 M solutions to yield a precipitate of silver chloride tells us that K is larger than 1 for the reaction $Ag^+ + Cl^- = AgCl(s)$.

Every chemical process has two directions. In Section 2.2 we saw that the change of enthalpy for any reaction is equal in magnitude but opposite in sign to that for the same reaction proceeding in the opposite direction. The same thing is true of the changes of entropy and free energy. It would be true of the change of any quantity, such as mass or cost, that was additive and depended only on the initial and final compositions of a reaction mixture. Since $\Delta G^0_{backward} = -\Delta G^0_{forward}$, eq. (2.12) shows that $\ln K_{backward} = -\ln K_{forward}$, which is equivalent to $K_{backward} = 1/K_{forward}$. This result could also be proved by writing expressions for $K_{backward}$ and $K_{forward}$ and observing that each was the reciprocal of the other.

SUMMARY: Since ΔG is equal to zero at equilibrium, the values of ΔG^0 and the equilibrium constant K are related by the equation $\Delta G^0 = -RT \ln K$.

2.8. Expressions for equilibrium constants

PREVIEW: This section tells how algebraic expressions for equilibrium constants are written.

You should obtain some practice in writing expressions for equilibrium constants, and there are some problems at the end of this chapter that will help. You should learn to do some things automatically in writing them, and the foregoing discussion should have enabled you to understand the reasons for these. Here are the important points, with an illustration of each:

1. You must begin with a balanced chemical equation for the reaction. You should check it to make sure that its two sides contain equal numbers of each of the different kinds of atoms that appear in it, and also that the sums of the electrical charges are the same for the two sides. If your "equation" is something like $PbBr_2(s) = Pb^{2+} + Br^-$ you should repair it (by changing its right-hand side to "$Pb^{2+} + 2\,Br^-$") before going on.

2. An equilibrium constant is a ratio. Activities of products belong in the *numerator;* activities of reactants belong in the *denominator*. For the process $H_2O(l) = H_2O(s)$ you must write $K = a_{H_2O(s)}/a_{H_2O(l)}$, not $K = a_{H_2O(l)}/a_{H_2O(s)}$.

3. If either the numerator or the denominator contains the activity of more than one substance, it must be a *product* of activities. For the reaction $AgCl(s) = Ag^+ + Cl^-$ you must write $K = a_{Ag^+}a_{Cl^-}/a_{AgCl(s)}$, not $K = (a_{Ag^+} + a_{Cl^-})/a_{AgCl(s)}$.

4. If n ions or molecules of any one kind appear in the balanced equation for the reaction, the activity of that ion or molecule must be raised to the nth power. For the reaction $2\,Cu^+ = Cu^{2+} + Cu(s)$, you must write $K = a_{Cu^{2+}}a_{Cu(s)}/a^2_{Cu^+}$, not $K = a_{Cu^2}a_{Cu(s)}/a_{Cu^+}$.

The things that were said in Section 2.6 about the activities of different kinds of substances often make it possible to simplify the resulting expression. You will be more comfortable and accurate at first if you write the expression according to the above rules and then check the points below. After you have gained some experience and self-confidence you will begin combining these steps subconsciously. The things to look for are:

5. If a reactant or a product is a solid, you should assume that it is the pure crystalline solid (unless there is some reason to believe that it is not) and replace its activity by 1. The three illustrative expressions above would become $K = 1/a_{H_2O(l)}$, $K = a_{Ag^+}a_{Cl^-}$, and $K = a_{Cu^{2+}}/a^2_{Cu^+}$.

6. If a reactant or a product is a liquid, you should similarly assume that it is pure (again, unless there is some reason to believe that it is not) and replace its activity by 1. This gives $K = 1$ for the process $H_2O(l) = H_2O(s)$; if you find this surprising you should read Sections 2.3 and 2.6 again.

Even the presence of quite substantial amounts of dissolved materials should not deter you from making the last of these assumptions. One dm³ of pure water at 298 K contains 55.35 moles of water ($= 997.1$ g dm⁻³/18.02 g mol⁻¹). If we dissolve one mole of potassium chloride in this amount of water, the solution will contain approximately[†] $1\ M\ (= 1$ mol dm⁻³) potassium ion and $1\ M$ chloride ion. The mole fraction or activity of water will be equal to the ratio of the number of moles of water (55.35) to the total number of moles of material present ($55.35+2 = 57.35$), and $55.35/57.35 = 0.97$. The addition of the potassium chloride has decreased the activity of water only 3%, which is smaller than the uncertainties in the great majority of equilibrium constants. Of course the effect is larger than this in more concentrated solutions, but even in a saturated solution of potassium chloride the activity of water is still about 0.86. As long as the concentrations of solutes do not exceed several mol dm⁻³ there is very little risk of being misled by taking the activity of water as equal to 1 in any aqueous solution.

Large proportions of other liquids with which water is miscible do make the assumption risky. You can get some idea of this from the following numbers. Those in the first line are the percentages by weight of ethanol in some mixtures with water; below these are the mole fractions of water in these mixtures:

% C_2H_5OH	50	70	80	90	95	98
N_{H_2O}	0.72	0.52	0.39	0.22	0.12	0.05

In very concentrated solutions you should represent the activity of a liquid by its mole fraction N, which is always accompanied by a subscript that identifies the liquid, as in the example just given.

7. If a gas appears in the equation for the reaction, and if it makes sense to assume that the reaction mixture will be in equilibrium with the gas phase, you should replace the activity of the gas by its partial pressure. This rule and the preceding one lead to the expression $K = p_{O_2}^{1/2}/a_{H_2O_2}$ for the equilibrium constant of the reaction $H_2O_2(aq) = H_2O(l)+\frac{1}{2}O_2(g)$. The partial pressure of a pure gas is equal to its total pressure. The partial pressure of any one gas in a mixture may be taken as equal to the product of the total pressure of the mixture and the mole fraction of that particular gas. Partial pressures are always expressed in atmospheres.

It is also possible to treat a dissolved gas in the same way as any other solute, using rule 8 below, and this must be done if equilibrium with the atmosphere will not be maintained. In carrying out the reaction $NH_4^+ + OH^- = NH_3 + H_2O(l)$ we might not wish to wait indefinitely for the ammonia to volatilize out of the reaction mixture and become distributed throughout the atmosphere, and therefore we would write $K = [NH_3]/[NH_4^+][OH^-]$ rather than $K = p_{NH_3}/[NH_4^+][OH^-]$. To make our in-

---

† These concentrations are approximate because the volume of the solution will not be quite the same as the original volume of water, but the difference is much too small to alter the argument.

tention perfectly clear we would do well to write $NH_3(aq)$ in the equation for the reaction. The symbol "(aq)" is used to denote a substance dissolved in an aqueous solution. If the same process occurred in liquid ammonia as the solvent, we should write its equation as $NH_4^+ + OH^- = NH_3(l) + H_2O$ and represent its equilibrium constant by $K = [H_2O]/[NH_4^+][OH^-]$. Examples like this show that the expression that is written for an equilibrium constant may contain a good deal of information about the conditions under which the reaction is carried out, and this is why such expressions should be written with care.

8. The activity of a solute should be replaced by its concentration in mol dm^{-3}. If the solute is X its concentration may be written as either [X] or c_X. Both symbols are used in this book; we write $K = [Ag^+][Cl^-]$ for the equilibrium constant of the reaction $AgCl(s) = Ag^+ + Cl^-$, but use dc_{Ag^+}/dt to represent the rate of change of the concentration of silver ion. Combining this rule with the preceding one gives $K = p_{O_2}^{1/2}/[H_2O_2]$ for the reaction $H_2O_2(aq) = H_2O(l) + \frac{1}{2}O_2(g)$.

In using these rules the chief difficulty arises in deciding whether a substance is a solute or something else. Ions are certainly solutes, but molecular formulas are less easy to interpret. Of course, the best guides are the symbols "(s)", "(l)", "(g)", and "(aq)" that chemists ought to use more frequently than they do. If you could be sure of finding these where they belong, you could easily tell that the equation $AgCl(s) = Ag^+ + Cl^-$ describes the dissolution of silver chloride while the one $AgCl(aq) = Ag^+ + Cl^-$ describes the behavior of dissolved but undissociated molecules of silver chloride in an aqueous solution. You would write $K = [Ag^+][Cl^-]$ for the first, but $K = [Ag^+][Cl^-]/[AgCl]$—or even $K = [Ag^+][Cl^-]/[AgCl(aq)]$—for the second. If you saw the equation $AgCl = Ag^+ + Cl^-$ you might not be sure which of the two expressions was intended. Sometimes you can decide such questions from the context. The chemist writing about the decomposition of pure liquid hydrogen peroxide in contact with the air might write simply $H_2O_2 = H_2O + \frac{1}{2}O_2$ and trust you to figure it out that $K = [H_2O]p_{O_2}^{1/2}$, but if you saw the same equation all by itself your safest course would be to write $K = N_{H_2O}p_{O_2}^{1/2}/N_{H_2O_2}$.

Chemists omit the units of equilibrium constants more often than they should. For many reactions the omission is excusable because concentrations of dissolved substances are so generally expressed in mol dm^{-3} that the units of K can be taken for granted. For others the omission may lead to ambiguity that could be avoided. In the example just cited, you could decide which of the two expressions for K was the intended one if you knew whether the units were mol dm^{-3} $atm^{1/2}$ or $atm^{1/2}$. In this book each numerical value of K will be accompanied by its units.

You can deduce the equation for a chemical reaction from the expression for its equilibrium constant by working backward through the whole process. If you were given the expression $K = [Cu^{2+}]/[Cu^+]^2$ you would begin by imagining that the equation was $2\,Cu^+ = Cu^{2+}$. As this is not balanced, an atom of copper must have been omitted from the expression for the equilibrium constant, and because it was omitted its activity must have been equal to 1. You would conclude that the equation was $2\,Cu^+ = Cu^{2+} + Cu(s)$. Similarly, you could tell that the expression $K = [NH_4^+][NH_2^-]$ belonged to the reaction $2\,NH_3(l) = NH_4^+ + NH_2^-$ in liquid ammonia as the solvent.

SUMMARY: Rules are given for writing, simplifying, and interpreting algebraic expressions for equilibrium constants.

2.9. Different kinds of equilibrium constants

PREVIEW: This section describes a number of special kinds of equilibrium constants.

There are many different common kinds of reactions. The expressions for their equilibrium constants have different forms, which the chemist should be able to recognize immediately. Different symbols or subscripts are generally used for convenience in writing and understanding equations that contain two or more equilibrium constants. The commonest ones are:

1. The dissolution of a non-electrolyte. The expression for the equilibrium constant depends on whether the non-electrolyte is a solid or liquid or a gas. For the dissolution of solid iodine in water $K = [I_2(aq)]$, while for that of liquid benzene in water $K = [C_6H_6(aq)]$. These expressions have the same form because the activity of either solid iodine or liquid benzene is equal to 1. They mean that at equilibrium (when the solution is saturated) the concentration of dissolved iodine or benzene is constant at whatever temperature we have specified. An equilibrium constant of this sort is often called a *solubility* and given the symbol S. The dissolution of a gas, such as oxygen, in water is described by an expression like $K = [O_2(aq)]/p_{O_2}$, which means that its solubility is proportional to its partial pressure.

2. The distribution of a solute between two immiscible liquids. If an aqueous solution of iodine is shaken with carbon tetrachloride, which is immiscible with water, some of the iodine leaves the aqueous phase and dissolves in the carbon tetrachloride. The process can be represented by the equation $I_2(aq) = I_2(CCl_4)$, and its equilibrium constant is given by $K = [I_2(CCl_4)]/[I_2(aq)]$. Such an equilibrium constant is often called a *distribution coefficient* and given the symbol D.

3. The dissolution of an electrolyte. When calcium fluoride dissolves in water the reaction that takes place is $CaF_2(s) = Ca^{2+} + 2\,F^-$, and its equilibrium constant is given by $K = [Ca^{2+}][F^-]^2$. The equilibrium constant for the dissolution of an electrolyte is always the product of two or more concentrations, and is therefore called a *solubility product*. Solubility products are given symbols like K_{CaF_2}, with the formula of the electrolyte as a subscript to permit distinguishing among different solubility products appearing in the same equation.

4. The "dissociation" of an acid. The behavior of acetic acid (CH_3COOH, abbreviated as $HOAc^\dagger$) in aqueous solutions may be represented by the equation $HOAc(aq) + H_2O(l) = H_3O^+ + OAc^-$. Before the role of the solvent in such reactions was recognized, the equation was written as $HOAc(aq) = H^+ + OAc^-$, and many chemists still call the process dissociation. The equilibrium constant for the reaction is therefore called a *dissociation constant* or, since there are different kinds of dissocation constants, an *acidic dissociation constant*, and it is given the symbol K_a. For acetic acid it is given by the expression $K_a = [H_3O^+][OAc^-]/[HOAc(aq)]$, which is often abbreviated to $K_a = [H^+][OAc^-]/[HOAc(aq)]$.

In a molecule of acetic acid there is only one proton that can be transferred to water or another base. Acids of which this is true are called *monobasic* or *monofunctional*. Phosphoric acid is a tribasic or trifunctional acid, and the reactions that can take place in its solutions, together with expressions for their equilibrium constants are

$$H_3PO_4(aq) + H_2O(l) = H_3O^+ + H_2PO_4^- ; \quad K_1 = [H_3O^+][H_2PO_4^-]/[H_3PO_4(aq)];$$
$$H_2PO_4^- + H_2O(l) = H_3O^+ + HPO_4^{2-} ; \quad K_2 = [H_3O^+][HPO_4^{2-}]/[H_2PO_4^-];$$
$$HPO_4^{2-} + H_2O(l) = H_3O^+ + PO_4^{3-} ; \quad K_3 = [H_3O^+][PO_4^{3-}]/[HPO_4^{2-}].$$

Each equilibrium constant corresponds to the dissociation of an acid, but K_a could not be used for all of them, and consequently subscript numbers are used as shown. K_1 always corresponds to the loss of the first proton.

5. The "dissociation" of a base. The equation $NH_3(aq) + H_2O(l) = NH_4^+ + OH^-$ represents the "dissociation" of the base ammonia in exactly the same sense that a similar equation in the preceding subsection represents the "dissociation" of acetic acid. For a long time the equilibrium constant for such a reaction was called a *basic dissociation constant*, given the symbol K_b, and described by an expression like $K_b = [NH_4^+][OH^-]/[NH_3]$. Values of K_b are disappearing from the chemical literature, for reasons that are given in Section 6.1, and are not used in this book.

6. The formation of a complex. Metal ions can react with certain anions or uncharged molecules which are called *ligands,* to form complexes. Almost the last step in developing a photographic film or print is its treatment with a solution containing thiosulfate ion to dissolve any unreacted silver compounds by forming the complex ion $Ag(S_2O_3)_2^{3-}$. The stability of a complex can be described either by its dissociation constant, which is the equilibrium constant for the reaction $Ag(S_2O_3)_2^{3-} = Ag^+ + 2 S_2O_3^{2-}$ in which the complex dissociates, or by its *formation constant* or *stability constant,* which is the equilibrium constant for the reaction $Ag^+ + 2 S_2O_3^{2-} =$

$\dagger$ Some chemists abbreviate acetic acid as HAc (and acetate ion as Ac^-), but organic chemists use the symbol Ac to denote the acetyl group $CH_3C\underset{}{\overset{O}{\diagup}}$ and are likely to think that HAc means not acetic acid but acetaldehyde, $CH_3C\overset{O}{\underset{H}{\diagdown}}$.

$Ag(S_2O_3)_2^{3-}$ in which the complex is formed from its constituents. Although the formation constant is simply the reciprocal of the dissociation constant and contains exactly the same information, dissociation constants are slowly losing ground to formation constants in the chemist's vocabulary.

This is happening because formation constants are easier and more convenient to use when, as is usually true, the coordination of a metal ion with a complexing agent occurs in successive steps. That of copper(II) ion with ammonia may be represented by the equations

$$Cu^{2+} + NH_3(aq) = Cu(NH_3)^{2+}; \quad K_1 = [Cu(NH_3)^{2+}]/[Cu^{2+}][NH_3(aq)];$$
$$Cu(NH_3)^{2+} + NH_3(aq) = Cu(NH_3)_2^{2+}; \quad K_2 = [Cu(NH_3)_2^{2+}]/[Cu(NH_3)^{2+}][NH_3(aq)];$$
$$Cu(NH_3)_2^{2+} + NH_3(aq) = Cu(NH_3)_3^{2+}; \quad K_3 = [Cu(NH_3)_3^{2+}]/[Cu(NH_3)_2^{2+}][NH_3(aq)]$$

and so on until six molecules of ammonia have been added. Contrary to what is done for acids and bases, the numbering of these *stepwise formation constants* starts with the *addition* of the first ion or molecule of the ligand. For such systems one often wants to correlate the concentration of a particular complex with that of the free metal ion. They are related by equations like

$$Cu^{2+} + 3NH_3 = Cu(NH_3)_3^{2+}; \quad \beta_3 = [Cu(NH_3)_3^{2+}]/[Cu^{2+}][NH_3(aq)]^3 K_1 K_2 K_3.$$

The symbol β is used to denote an *overall formation constant* like this one, and it is given a subscript that is equal to the total number of ions or molecules of the ligand added to a single metal ion in the overall reaction. The value of β_5 would describe the formation of $Cu(NH_3)_5^{2+}$ from Cu^{2+} and ammonia; the formation of $Ag(S_2O_3)_2^{3-}$ from Ag^+ and thiosulfate ion would be described by the overall formation constant β_2.

7. *The autoprotolysis of a solvent.* An *amphoteric* substance is one that is both acidic and basic. As an acid it can donate a proton to a base; as a base it can accept a proton from an acid. One molecule or ion of an amphoteric substance, behaving as an acid, can donate a proton to another molecule or ion, behaving as a base. Such a reaction is called *autoprotolysis.* The equations $2H_2O(l) = H_3O^+ + OH^-$ and $2CH_3COOH(l) = CH_3COOH_2^+ + CH_3COO^-$ represent two common autoprotolytic reactions. With just one exception, the equilibrium constant for the autoprotolysis of a solvent is given the symbol K_s and is called the *autoprotolysis constant* of the solvent. The autoprotolysis constant of acetic acid is given by $K_s = [CH_3COOH_2^+][CH_3COO^-]$. The exception is of course the autoprotolysis constant of water, which is always given the symbol $K_w (= [H_3O^+][OH^-])$ and is usually called the ion-product constant of water.

8. *Other processes.* The symbol K_t is widely used for the equilibrium constant of a reaction that occurs during a titration, and will appear in later chapters. Equilibrium constants for other kinds of processes are often needed, but are generally defined and given subscripts as the need arises. The chemist speaking about the dimerization of nitrogen dioxide, $2NO_2(g) = N_2O_4(g)$, might call its equilibrium constant a dimerization constant and give it the symbol K_d.

We shall describe the behaviors of some of these different kinds of reactions in detail in later chapters. As we do so, we shall show how their equilibrium constants can be evaluated experimentally and used to find the equilibrium compositions of reaction mixtures.

You should not leave this chapter with the impression that equilibrium constants are generally calculated from values of ΔG^0. We made such calculations because they helped to show why and how changes of the enthalpy, entropy, and free energy are related to chemical equilibria. The calculations are easy and convenient, but they would not be possible if one or more of the necessary values of G_f^0 could not be found in a table. We would then have to evaluate K experimentally. The result might be used to calculate ΔG^0 from eq. (2.12). If the values of G_f^0 were known for all but one of the reactants and products, we could go on to calculate G_f^0 for the remaining one and add it to the table so that we, and other chemists, could later use it to calculate the equilibrium constants of other reactions involving that substance. If H_f^0 was not known for that substance but was known for all the others appearing in the reaction we could go back to the laboratory, evaluate ΔH^0 for our reaction, and obtain a new value of H_f^0 that we could add to the table. Knowing the values of ΔG^0 and ΔH^0, we could calculate the change of entropy involved in the reaction, and so on .

SUMMARY: This section gives the special names and symbols used to denote the equilibrium constants of reactions belonging to a number of different important classes.

Problems

Answers to some of these problems are given on page 520.

2.1. A solution containing 4.55×10^{-4} mole of hydronium ion is added to one containing an excess of hydroxide ion. The reaction $H_3O^+ + OH^- = 2 H_2O(l)$ occurs and consumes almost all of the hydronium ion. As it does so, there is an increase of $0.1028°$ in the temperature of the calorimeter containing the mixture. A current of 0.1000 A is then passed through a 252.7-ohm resistor immersed in the calorimeter for 10.75 s, and there is a further increase of $0.1087°$ in the temperature. What is the value of ΔH for the reaction between hydronium and hydroxide ions?

2.2. The value of ΔH for the reaction $Ag^+ + Cl^- = AgCl(s)$ is -65.48 kJ mol^{-1}. A solution containing an unknown amount of chloride ion is mixed with a solution containing an excess of silver ion in a calorimeter, and the rise of temperature resulting from the reaction is found to correspond to 2.933 J. What weight of chloride ion did the unknown solution contain?

2.3. The standard enthalpy of formation, H_f^0, is -415.9 kJ mol^{-1} for Na$_2$O(s) and -504.6 kJ mol^{-1} for Na$_2$O$_2$(s). What is the value of ΔH^0 for the reaction $2 Na_2O_2(s) = 2 Na_2O(s) + O_2(g)$?

2.4. Lead(II) oxide, PbO, can exist in two different crystalline forms, of which one is red and the other yellow. The standard enthalpies of formation are -219.24 kJ mol^{-1} for the red form and -217.86 kJ mol^{-1} for the yellow form. Is the transformation of yellow PbO into red PbO favored or hindered by raising the temperature?

2.5. Pure crystalline hydrogen peroxide melts at $-1.7°$ C, and its heat of fusion at that temperature is given by

$$H_2O_2(s) = H_2O_2(l); \qquad \Delta H^0 = 10.53 \text{ kJ mol}^{-1}.$$

What is the value of ΔS^0 for this process?

2.6. The standard entropy, S^0, is 186.2 J K^{-1} mol^{-1} for methane (CH$_4$), 219.8 J K^{-1} mol^{-1} for ethylene (C$_2$H$_4$), and 269.9 J K^{-1} mol^{-1} for propane (CH$_3$CH$_2$CH$_3$), all in the gaseous state at 298 K. What is the value of ΔS^0 for the reaction CH$_4$(g) + C$_2$H$_4$(g) = CH$_3$CH$_2$CH$_3$(g)?

2.7. The values of H_f^0 and S^0 for liquid and gaseous bromine at 298 K are given by

Substance	H_f^0, kJ mol^{-1}	S^0, J K^{-1} mol^{-1}
Br$_2$(l)	0.00	152.3
Br$_2$(g)	30.7	245.3

(a) Is the process Br$_2$(l) = Br$_2$(g, 1 atm) spontaneous at 298 K?

(b) Assuming that these values are independent of temperature, what is the boiling point of pure liquid bromine under a pressure of 1 atm?

2.8. The values of H_f^0 and S^0 for gaseous chlorine and bromine monochloride (BrCl) at 298 K are given by

Substance	H_f^0, kJ mol^{-1}	S^0, J K^{-1} mol^{-1}
Cl$_2$(g)	0.00	223,0
BrCl(g)	14.69	239.9

(a) Calculate the values of ΔH^0 and ΔG^0 for the reaction Br$_2$(l) + Cl$_2$(g) = 2 BrCl(g) at 298 K. Use the values of H_f^0 and S^0 for Br$_2$(l) given in the preceding problem.

(b) Is the process Br$_2$(l) + Cl$_2$(g, 1 atm) = 2 BrCl(g, 1 atm) endothermic or exothermic at 298 K?

(c) Is it spontaneous or non-spontaneous?

(d) At what temperature would the value of K for this reaction be equal to 1?

2.9. Using the values of G_f^0 given in Appendix II (page 526), calculate the value of ΔG^0 for the reaction Ag$^+$ + Br$^-$ = AgBr(s) and compare it with the one obtained in Example 2.3 (page 20), for the reaction Ag$^+$ + Cl$^-$ = AgCl(s).

2.10. Write an expression for the equilibrium constant of each of the following reactions, assuming that it takes place in a dilute aqueous solution.

(a) Fe^{3+} + Cu$^+$ = Fe^{2+} + Cu^{2+}.

(b) Cu^{2+} + Cl$^-$ = CuCl$^+$.

(c) Cu$^+$ + Cl$^-$ = CuCl(s).

(d) C$_2$O$_4^{2-}$ + H$_3$O$^+$ = HC$_2$O$_4^-$ + H$_2$O(l).

(e) Ag$^+$ + 2 NH$_3$(aq) = Ag(NH$_3$)$_2^+$.

(f) 2 HPO$_4^{2-}$ = H$_2$PO$_4^-$ + PO$_4^{3-}$.

(g) 2 HgI$_2$(s) = HgI$^+$ + HgI$_3^-$.

(h) 2 Hg(l) + Br$_2$(aq) = Hg$_2$Br$_2$(s).

(i) H$_2$(g) + 2 AgCl(s) = 2 Ag(s) + 2 H$^+$ + 2 Cl$^-$.

(j) Pb(s) + I$_2$(s) = PbI$_2$(s).

2.11. Write an expression for the equilibrium constant of each of the following reactions under the conditions indicated:

(a) C$_2$H$_5$OH + OH$^-$ = C$_2$H$_5$O$^-$ + H$_2$O   in nearly pure water.

(b) C$_2$H$_5$OH + OH$^-$ = C$_2$H$_5$O$^-$ + H$_2$O   in nearly pure ethanol.

(c) 2 NH$_3$ = NH$_4^+$ + NH$_2^-$ in pure liquid ammonia.

(d) CO$_2$(g) + OH$^-$ = HCO$_3^-$ in a dilute aqueous solution.

(e) CO$_2$(g) + OH$^-$ = HCO$_3^-$ in fused sodium hydroxide.

2.12. Write a balanced chemical equation for the reaction that corresponds to each of the following equilibrium-constant expressions. Assume that each reaction takes place in a dilute aqueous solution.

(a) $K = $ [Mg^{2+}] [CO$_3^{2-}$].

(b) $K = $ [Mg^{2+}] [OH$^-$] [HCO$_3^-$].

(c) $K = $ [H$_2$O$_2$(aq)]^2/p_{O_2}.

(d) $K = $ [Fe^{2+}]^3/[Fe^{3+}]^2.

(e) $K = $ 1/[BrO$_3^-$] [Br$^-$]^5[H$^+$]^6.

2.13. The reaction $Cu(s) + AgBr(s) = CuBr(s) + Ag(s)$ takes place when metallic copper is brought into contact with a saturated aqueous solution of silver bromide.

(a) Write an expression for its equilibrium constant.
(b) At 298 K the standard free energy of formation of $AgBr(s)$ is -93.68 kJ mol^{-1} and that of $CuBr(s)$ is -99.62 kJ mol^{-1}. What is the value of ΔG^0 for the above reaction at 298 K?
(c) The answers to parts (a) and (b) do not agree with eq. (2.12). What does this mean?

CHAPTER 3

An Introduction
to Chemical Kinetics

3.1. Introduction

Thermodynamics asks why a reaction occurs and what total changes of energy are
involved in it, and answers these and other questions by considering only the starting
materials and the final products. This chapter poses other questions that thermodynamics cannot answer. How rapidly does a reaction occur, and why? What are the
factors that affect its rate, and why do they affect it in the ways they do? How does a
reaction occur: what is its path? What are the structures and properties of the intermediate substances that are formed and consumed, and what energies are involved
in their formations and reactions?

Chemical kinetics asks these questions and provides ways of answering them.
If road maps contained information resembling what thermodynamics and chemical
kinetics tell us about reactions, a thermodynamic map might show only New York
and Los Angeles. It would give a good deal of information about these, including
their heights above sea level, but it would not indicate whether the highway between
them went through Canada, Missouri, or Mexico. A kinetic map would show something about that highway and the mountain ranges that it crossed, and would indicate
how long the trip takes. Both thermodynamic and kinetic information are needed to
obtain a full understanding of a reaction.

3.2. The rates of chemical reactions

PREVIEW: This section tells how the concentrations of reactants and products change during a
chemical reaction, and defines the rate of a reaction and the rate law or rate equation.

Data on the rates of chemical reactions are the raw material of chemical kinetics.
It is by interpreting those rates and the ways in which they depend on the experimental conditions that the above questions are answered. However, the rates of
reactions are not easy to measure. There are some fairly new techniques that make

33

such measurements possible, but what is almost always done is to follow the changing concentration of some reactant or product while the reaction is taking place.

Figure 3.1 shows how the concentrations of the reactants and products varied during a reaction between ethyl acetate and excess hydroxide ion in an aqueous solution at 298 K. The reaction was

$$CH_3C\diagup^O_{OC_2H_5} + OH^- = CH_3C\diagup^O_{O^-} + C_2H_5OH(aq) \qquad (3.1)$$

$$\text{ethyl acetate} \qquad\qquad\qquad \text{acetate ion} \quad \text{ethanol}$$

At the start of the reaction the mixture contained 0.006 26 M ethyl acetate and 0.2744 M hydroxide ion, but no acetate ion or ethanol. Curve a shows how the concentration of ethyl acetate decreased as the reaction proceeded. Curve b shows how the concentrations of acetate ion and ethanol increased. These two concentrations were always the same because the reaction produced one acetate ion along with each molecule of ethanol, and because neither of these products was present initially. If the right-hand side of eq. (3.1) were, say, $CH_3COO^- + 2\,C_2H_5OH(aq)$, or if the reaction were carried out in a mixture of ethanol and water

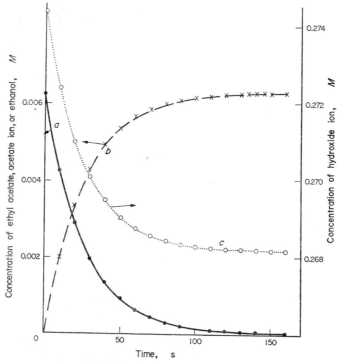

FIG. 3.1. Effects of time on the concentrations of the reactants and products in the reaction of 0.006 26 M ethyl acetate with 0.2744 M hydroxide ion. (a) Ethyl acetate (left-hand ordinate scale), (b) acetate ion or ethanol (left-hand ordinate scale), (c) hydroxide ion (right-hand ordinate scale).

as the solvent, the concentrations of ethanol and acetate ion would not be the same. Curve c shows how the concentration of hydroxide ion changed. It is drawn to the same scale as curves a and b, but its ordinate axis is different because the initial concentration of hydroxide ion was much larger than that of ethyl acetate.

Figure 3.1 contains several different kinds of information:

1. The concentration of ethyl acetate approaches zero as the reaction proceeds. The reaction is said to go to completion. In reality there is an equilibrium, but it lies so far to the right that ethyl acetate cannot be detected in the reaction mixture when the reaction is over.

2. The decrease of the concentration of hydroxide ion from the beginning of the reaction to its end is equal to the decrease of the concentration of ethyl acetate, and both of these changes are equal to the increases of the concentrations of acetate ion and ethanol. Of course, this is what eq. (3.1) says, but if we did not know very much about the reaction we might wonder whether eq. (3.1) told the whole truth: perhaps there is another reaction that takes place at the same time and leads to the formation of other products. These things are true for the whole reaction, and they are also true for any interval of time $\Delta t^{\dagger}$. If during some interval the change of the concentration of acetate ion is Δc, then that of the concentration of ethanol is also Δc, and the changes of the concentrations of ethyl acetate and hydroxide ion are both $-\Delta c$. These changes are positive for acetate ion and ethanol, but negative for ethyl acetate and hydroxide ion, because the first two are produced by the reaction whereas the last two are consumed. Therefore $\Delta c_{OAc^-}/\Delta t = \Delta c_{EtOH}/\Delta t = -\Delta c_{EtOAc} - \Delta c_{OH^-}/\Delta t$, in which we have abbreviated acetate ion as OAc^-, ethanol as $EtOH$, and ethyl acetate as $EtOAc$ for compactness. If Δt were vanishingly small Δc would also be vanishingly small, and we would have

$$dc_{OAc^-}/dt = dc_{EtOH}/dt = -dc_{EtOAc}/dt = -dc_{OH^-}/dt. \qquad (3.2)$$

3. At the beginning of the reaction ethyl acetate and hydroxide ion disappear fairly rapidly and the slopes of curves a and c are large and negative, while acetate ion and ethanol are produced fairly rapidly and the slope of curve b is large and positive. As time goes on two things happen. One is that the concentrations of the products and reactants change, and approach constant values as the reaction nears completion. The other is that the rates of change of these concentrations change as well, and approach zero as the reaction nears completion. Chemical kinetics is largely concerned with the relation between the concentrations and the rates at which they change.

The *rate of a reaction* is defined by the equation

$$\text{rate} = \mp \, dc_i/dt \qquad (3.3)$$

---

† They would be true for the whole reaction even if some intermediate substance were formed at a high concentration during the first part of the reaction and then disappeared as the reaction approached completion, but would not then be true in any particular interval of time.

where c_i is the concentration of the ith reactant or product. The negative sign is used if the ith substance is a reactant, and the positive sign is used if it is a product. We say that the rate of a reaction is positive when it proceeds in the forward direction, so that dc/dt is negative for each of the reactants and positive for each of the products. Equations (3.2) and (3.3) would have to be a little more complicated for a reaction such as $Sn^{2+} + 2\,Fe^{3+} = Sn^{4+} + 2\,Fe^{2+}$, in which the stoichiometric coefficients are not all the same. At any instant during this reaction the concentration of Fe^{2+} would be increasing twice as fast as that of Sn^{4+}, and we would say that the rate was equal to either $dc_{Sn^{4+}}/dt$ or $\frac{1}{2}\,dc_{Fe^{2+}}/dt$.

The *rate law* or *rate equation* for a reaction is an equation that relates its rate at any instant to the concentrations, at the same instant, of all the substances that affect its rate. For the reaction described by eq. (3.1) the rate law is

$$\text{rate} \ (= -dc_{EtOAc-}/dt = -dc_{OH-}/dt, \ \text{etc.}) = kc_{EtOAc}c_{OH-} \tag{3.4}$$

The constant of proportionality k is called a *rate constant*. A rate law may have only a single term on its right-hand side, or it may have two or more.

SUMMARY: The rate of a reaction is a positive quantity given at any instant by $\mp dc/dt$, where c is the concentration of some reactant or product. The rate law is a differential equation showing how the rate of a reaction depends on the concentrations of the substances that affect it, and includes a constant of proportionality called the rate constant.

3.3. Elementary and rate-determining steps

PREVIEW: This section defines three important terms: elementary step, molecularity, and rate-determining step.

A reaction may take place in a single step, which is called an *elementary step* or *elementary reaction*. Every elementary step can be described by a chemical equation, and a single-headed arrow is used to represent an elementary step whose rate in the reverse direction is being neglected. For the elementary step $B \to C$ the rate is given by

$$\text{rate} \ (= -dc_B/dt = dc_C/dt) = kc_B \tag{3.5}$$

The *molecularity* of an elementary step is equal to the number of particles that react in it. An elementary step that involved only a single ion or molecule of the reactant, as this one appears to do, would be called *unimolecular*. One that involves two reacting ions or molecules is called *bimolecular*. The elementary step $B^1 + B^2 \to C$ is bimolecular and its rate is given by

$$\text{rate} \ (= -dc_{B^1}/dt = dc_C/dt, \ \text{etc.}) = kc_{B^1}c_{B^2} \tag{3.6a}$$

Another bimolecular elementary step is $2\,B \to C$, for which the rate is given by

$$\text{rate} \ (= dc_C/dt) = kc_B^2 \tag{3.6b}$$

An elementary step in which three ions or molecules react is called *termolecular*. The elementary step $B^1 + B^2 + B^3 \rightarrow C$ is termolecular and its rate is given by

$$\text{rate} \left(= -dc_{B^1}/dt = dc_C/dt, \text{ etc.} \right) = kc_{B^1}c_{B^2}c_{B^3} \tag{3.7a}$$

If B^1, B^2, and B^3 were all the same this would be identical with $3\,B \rightarrow C$, for which

$$\text{rate} \left(= dc_C/dt \right) = kc_B^3 \tag{3.7b}$$

Bimolecular elementary steps are very common because collisions between two particles are very frequent. Termolecular ones are much less common because the simultaneous collision of three particles is comparatively rare. The collision of four or more particles is so wildly improbable that quadrimolecular and higher elementary steps are unknown. Hence a reaction such as the one between iron(II) and permanganate ions, for which the equation is $MnO_4^- + 5\,Fe^{2+} + 8\,H^+ = Mn^{2+} + 5\,Fe^{3+} + 4\,H_2O$ cannot possibly occur in a single step. Instead it must comprise a fairly large number of elementary steps. One conceivable sequence of elementary steps begins with the one $MnO_4^- + H^+ = HMnO_4$. The permanganic acid might then accept an electron from an iron(II) ion ($HMnO_4 + Fe^{2+} = HMnO_4^- + Fe^{3+}$), yielding a hydrogen manganate(VI) ion that could then accept another proton ($HMnO_4^- + H^+ = H_2MnO_4$). These three steps together account for the consumption of only two hydrogen ions and one iron(II) ion, and for the reduction of the permanganate only to manganese(VI), but show how it is possible to account for a very complicated overall reaction by a succession of elementary steps, each of low molecularity.

In any sequence of steps there are some that are faster than others, and the differences among their rates may be very large. At one extreme, proton-transfer steps, such as $H_3O^+ + OAc^- \rightarrow H_2O + HOAc$ or $H_3O^+ + OH^- \rightarrow 2\,H_2O$, may be extremely fast. Many were actually thought to be instantaneous until very sophisticated techniques were developed for following the rates of very rapid reactions. Even for these, however, the rates depend on the strengths of the bonds that have to be broken: so much energy is required to break a hydrogen–oxygen bond in a molecule of hydrogen peroxide that the elementary step $H_2O_2(aq) + H_2O(l) = H_3O^+ + HO_2^-$ is not very fast. Electron-transfer steps may also be fast if they involve nothing more than the transfer of electrons, but if the reacting species have charges of the same sign electrostatic repulsion will make their collisions less frequent, and in addition many bonds to water molecules or other ligands may have to be broken or rearranged.

It often happens that one elementary step is much slower than all the others involved in a reaction, and an elementary step of which this is true is called the *rate-determining step* of the reaction. An analogy may help you to understand the significance of a rate-determining step. Imagine that a large number of people are standing on the sidewalk outside an empty theater waiting to enter it, but that the door of the theater is very narrow. It is difficult to squeeze through the door, and only one person can do so at a time. The rate at which people enter the theater will be limited by the rate at which they can get through the door; it will not increase if more people stand

4*

outside the theater waiting to get in, or if everyone inside the theater runs to a seat instead of walking. The step

person outside theater → person inside theater

is the rate-determining step in the overall process

person at home = person watching movie.

We could find the rate at which people got through the door by observing the rate of increase of the number of people watching the movie, and we could do this even if the door were in total darkness so that we could not see whether anyone is getting through it or not.

If these processes could occur in only one direction, as is implied by the single-headed arrows in the equations

A → B	fast	person at home → person outside theater
B → C	rate-determining step	person outside theater → person inside theater
C → P	fast	person inside theater → person watching movie

a great deal of B would be formed, and the line outside the theater would become very long, before the rate-determining step had gone very far. There are indeed many chemical reactions in which an intermediate like B accumulates rapidly at first and then disappears slowly as it is consumed in the rate-determining step. Most enzyme-catalyzed reactions behave in this way under certain conditions (Section 3.9). Often, however, there is an equilibrium between A and B: it may be very unfavorable to B, and at any instant the concentration of B may be far too small to detect. But as B is consumed in the rate-determining step more A reacts to form B so that the equilibrium is always maintained. A scheme that is applicable to many reactions is

$$\left. \begin{array}{ll} A = B & \text{fast prior equilibrium,} \\ B \rightarrow C & \text{rate-determining step,} \\ C \rightarrow P & \text{fast subsequent step(s).} \end{array} \right\} \qquad (3.8)$$

There are many possible variations. The equilibrium constant of the first step may have almost any value, and the fraction of the A that is transformed into B will vary accordingly. The transformation of C into P may occur in a number of fast steps rather than in a single one, or there may be a rapidly established equilibrium between C and P. In the latter case some C will be left in the solution at the end of the reaction unless the equilibrium constant is very large. The rate at which A is consumed depends on the value of the rate constant for the rate-determining step, and it also depends on the position of the prior equilibrium because that governs the concentration of the reactant in the rate-determining step [eqs. (3.6)], but it is not affected by any fast step that follows the rate-determining one. From the rate of a chemical reaction we can obtain information about the rate-determining step and the steps that precede it, but not about anything that happens afterwards.

There are other possibilities as well. The transformation of C into P may involve another slow step, whose rate is comparable to that of the step B → C. There may be two or more different ways in which B can react, giving rise to different final products. We shall not discuss these but shall confine our attention to the common scheme depicted by eqs. (3.8).

SUMMARY: A reaction may occur in a single step, or in a succession of steps called elementary steps. The molecularity of an elementary step is equal to the number of ions or molecules that react in that step. If one elementary step is much slower than all the others involved in a reaction, its rate governs the rate of the overall reaction.

3.4. Rate laws and reaction orders

PREVIEW: This section defines and discusses the order of a reaction and the order with respect to a particular substance.

Equation (3.4) on p. 36 gave the rate law for the reaction between ethyl acetate and hydroxide ion. The *order of a reaction* is the sum of the exponents of all the concentrations that appear on the right-hand side of its rate equation. The reaction between ethyl acetate and hydroxide ion is a second-order reaction: the sum of the exponents of c_{EtOAc} and c_{OH^-} on the right-hand side of eq. (3.4) is equal to 2. Here are the overall equations, the rate laws, the conditions under which the rate laws are applicable, and the orders of some other reactions:

Overall equation	Rate law	Conditions	Order of the reaction
$2 N_2O_5 = 2 N_2O_4 + O_2$	$dc_{N_2O_5}/dt = -kc_{N_2O_5}$	solutions in CCl_4	first order
$OCl^- + I^- = Cl^- + OI^-$	$dc_{OCl^-}/dt = -kc_{H^+}c_{OCl^-}c_{I^-}$	alkaline aqueous solutions	third order
$V^{2+} + VO^{2+} + 2 H^+ = 2 V^{3+} + H_2O(l)$	$dc_{V^{2+}}/dt = -kc_{V^{2+}}c_{VO^{2+}}c_{H^+}$	strongly acidic aqueous solutions	third order

Sixth- and even higher-order reactions are known. The rate constant of an nth-order reaction is called an nth-order rate constant.

From these examples it should be clear that the overall equation for a reaction does not enable us to predict its order or the form of its rate law. We could not have expected the rate of the vanadium(II)–vanadium(IV) reaction to depend on the first power of the concentration of hydrogen ion, or the rate of the hypochlorite–iodide reaction to depend in any way on the concentration of hydrogen ion. Rate laws and reaction orders must be obtained experimentally. This section and the two following ones explain how they are obtained, and Section 3.7 explains how they are interpreted.

The experiments generally consist of mixing two solutions, of which one contains one reactant while the other contains all the rest, and observing how the concentration of some reactant or product changes with time. This is called the *batch technique*. Other techniques are available, and have been especially valuable in studying very fast reactions, which approach completion within the time that is needed to mix the solutions.

It is not possible to measure a concentration directly, but it is possible to measure many quantities that are proportional to concentration. Which one is chosen for any particular reaction depends on the properties of the reactants and products. Often it is much easier to follow the concentration of one of these than it would be to follow any of the others. One may absorb ultraviolet or visible radiation of some wavelength while the others do not, and if so it is convenient to measure the ability of the reaction mixture to absorb at that wavelength. One may be ionic and therefore able to conduct electricity through the reaction mixture while the others are not, and if so it is convenient to measure the conductance of the mixture. One may fluoresce at a certain wavelength when the solution is irradiated with ultraviolet or visible light of a different wavelength, and if so its concentration can be followed by measuring the intensity of the fluorescence. Whatever the measured quantity may be, its value can be read and written down in a notebook at each of a number of instants after the solutions have been mixed. A recording instrument, which draws a plot of the measured quantity against time as the reaction proceeds, is more expensive but also more convenient, especially if the reaction is fairly fast. Devices that automatically transfer the numerical value of the measured quantity to the core memory of a computer at preselected intervals are still more expensive but still faster and more convenient because they greatly facilitate the analysis of the data.

There are three especially common ways in which the concentration of a substance can affect the rate of a reaction. If the substance is a reactant, so that its concentration decreases as the reaction proceeds, these are

$$dc_A/dt = -k(c_B\ldots) \qquad \text{zeroth order in A,} \qquad (3.9a)$$
$$dc_A/dt = -kc_A(c_B\ldots) \qquad \text{first order in A,} \qquad (3.9b)$$
$$dc_A/dt = -kc^2_A(c_B\ldots) \qquad \text{second order in A.} \qquad (3.9c)$$

The right-hand side of a rate equation does not often contain only a single concentration, but it is so convenient to focus attention on one concentration at a time that all the concentrations except that of A, which is the one being followed, are lumped together into the quantity $(c_B \ldots)$. The experimental significance of this will appear in Section 3.5. The *order in A* (or the *order with respect to A*) is the exponent of the concentration of A on the right-hand side of the rate equation. The overall reaction $ClO_3^- + 6\,Fe^{2+} + 6\,H^+ = Cl^- + 6\,Fe^{3+} + 3\,H_2O(l)$ is said to be third order with respect to hydrogen ion: eqs. (3.9) do not exhaust all the possibilities, but understanding how to handle them will enable you to cope with any others you may meet.

Figure 3.2 shows how the concentration of A would change with time if each of these three rate laws were obeyed. The curve for a reaction that is third order in A is also included to help you to see the point. The initial concentration of A is assumed to be 0.001 M ($= 1$ mM) for each curve, and the values of $k(c_B \ldots)$ are assumed to be constant throughout each reaction but differ for the different curves in ways that make the initial rates equal.

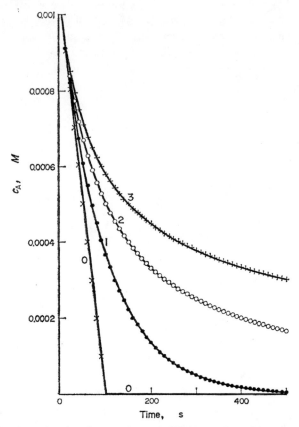

FIG. 3.2. Variations of c_A with time for reactions of different orders with respect to A. (0) Zeroth order, $k(c_B \ldots) = 1 \times 10^{-5}$ mol dm^{-3} s^{-1}; (1) first order, $k(c_B \ldots) = 1 \times 10^{-2}$ s^{-1}; (2) second order, $k(c_B \ldots) = 10$ dm^3 mol^{-1} s^{-1}; (3) third order, $k(c_B \ldots) = 1 \times 10^4$ dm^6 mol^{-2} s^{-1}.

The plot of c_A against time is a straight line for a reaction that is zeroth order in A. This is because its slope, which is of course equal to dc_A/dt, does not change as the reaction proceeds and the concentration of A decreases. Consequently this concentration decreases at a uniform rate until it becomes equal to zero, and then it stops decreasing because it cannot become negative.

All of the other plots in Fig. 3.2 are non-linear. They are all concave upward because their slopes approach zero as the concentrations of A decrease and the reactions approach completion. Their shapes are not exactly the same, but the differences among them are very difficult to recognize. The difficulty is illustrated by Fig. 3.3. Curve 1 in Fig. 3.3 is for a reaction that is first order in A; curve 2 is for a reaction that is second order in A. For each curve the initial concentration of A is 0.001 M (= 1 mM), but the values of $k(c_B \ldots)$ differ. The two curves do not have exactly the same shape (curve 2 is steeper at the beginning, and its slope is more dependent on the concentration of A), but it would not be easy to tell which was the better match for an unknown curve. Plots of concentration against time make it easy to recognize a reaction that is zeroth order in the substance being followed, but give little help in deciding among the other possibilities.

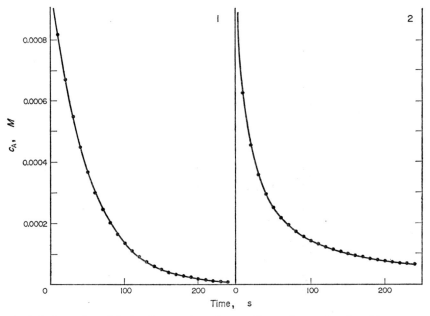

FIG. 3.3. Variations of c_A with time for reactions that are first and second order with respect to A. (1) First order, $k(c_B \ldots) = 0.02$ s^{-1}; (2) second order, $k(c_B \ldots) = 60$ dm^3 mol^{-1} s^{-1}.

SUMMARY: The order of a reaction is the sum of the exponents of the concentrations that appear in the rate equation. Its order with respect to A is the exponent of the concentration of A. Neither can be predicted from the overall equation; both must be obtained from experimental data.

3.5. Integrated rate equations

PREVIEW: This section shows how one can evaluate the order of a reaction with respect to a particular reactant.

Weapons that may be more powerful than simple plots of concentration against time are obtained by integrating common rate equations. If the concentrations of B and all the other substances that appear on the right-hand side of the rate equation are free to vary as the reaction proceeds, the integration is complicated and its result is not very useful. It helps enormously to keep all these concentrations constant, so that the product $(c_B \ldots)$ is constant. At first thought this may seem impossible. Suppose that our reaction is $EtOAc + OH^- = OAc^- + EtOH$ and that we have chosen to follow the concentration of ethyl acetate $(= A)$. If the reaction mixture contains 1 mM ethyl acetate and 1 mM hydroxide ion $(= B)$ initially, the concentration of hydroxide ion will decrease toward zero as the concentration of ethyl acetate decreases toward zero. It would then be silly to pretend that the concentration of hydroxide ion had remained constant. But what would happen if the two initial concentrations were different?

Initial concentrations		*Final concentrations*		
EtOAc	OH$^-$	EtOAc	OH$^-$	% decrease of c_{OH^-}
1 mM	5 mM	0	4 mM	25
1 mM	10 mM	0	9 mM	10
1 mM	50 mM	0	49 mM	2
1 mM	100 mM	0	99 mM	1

If the initial concentration of hydroxide ion is much larger than that of ethyl acetate, the fraction of the hydroxide ion that is consumed during the reaction becomes very small, and the concentration of hydroxide ion will be nearly constant throughout the reaction. The same thing would be true if the reaction mixture were buffered, but this works only with hydrogen and hydroxide ions.

We are now ready to integrate eqs. (3.9). In doing this we shall assume that the initial concentrations of all the other substances in the solution are much larger than the initial concentration of the reactant A that we are following. As we have just seen, all the other concentrations will thus be kept virtually constant and the concentration of A will be the only one that will change appreciably. Hence we can treat $(c_B \ldots)$ as a constant. The rate constant k depends on the temperature and pressure; if these are constant throughout the reaction the value of k will also be constant. We shall assume that both $(c_B \ldots)$ and k are constant, and shall use the symbol k' to represent their product $k(c_B \ldots)$ in each of the following integrations.

1. If the reaction is zeroth order in A,

$$dc_A/dt = -k' \tag{3.9a}$$

in which the variables may be separated to obtain

$$dc_A = -k' \, dt.$$

Integration now yields

$$c_A = -k't + C.$$

The constant of integration C is equal to the value of c_A at $t = 0$, which is the initial concentration of A and may be represented by the symbol c_A^0, so that

$$c_A = c_A^0 - k't. \tag{3.10a}$$

2. If the reaction is first order in A,

$$dc_A/dt = -k'c_A. \tag{3.9b}$$

Again the variables may be separated

$$dc_A/c_A = -k' \, dt$$

and the result may be integrated to obtain

$$\ln c_A = -k't + C.$$

Here the constant of integration C is equal to the value of $\ln c_A$ at $t = 0$, or $\ln c_A^0$, and

$$\ln c_A = \ln c_A^0 - k't. \tag{3.10b}$$

3. If the reaction is second order in A,

$$dc_A/dt = -k'c_A^2. \tag{3.9c}$$

On separating the variables

$$dc_A/c_A^2 = -k' \, dt$$

and integrating, we obtain

$$-1/c_A = -k't + C.$$

Now $C = -1/c_A^0$ and the final result may be written

$$1/c_A = 1/c_A^0 + k't. \tag{3.10c}$$

In each of these cases a reaction is made to behave as though its order were lower than it really is. The rate law may actually be something like $dc_A/dt = -kc_A^n c_B^o c_C^p \ldots$ so that the reaction is actually $(n+o+p+\ldots)$th order. By keeping all the concentrations constant except that of A the reaction is made to behave as though it were only nth order, and is then called a pseudo-nth-order reaction. In the presence of a large excess of hydroxide ion the reaction $EtOAc + OH^- = OAc^- + EtOH$ is a pseudo-first-order reaction, and the hydrolysis of an alkyl halide $RCl + H_2O(l) = ROH(aq) + H^+ + Cl^-$ in a dilute aqueous solution is also a pseudo-first-order reaction because there is a large excess of water present. A *pseudo-nth-order reaction* is a reaction that is nth order in one substance and is carried out in the presence of a large or constant concentration of every other substance appearing in its rate law. The rate constant k of a pseudo-nth-order reaction is called a *pseudo-nth-order rate constant*.

Equations (3.10) are called *integrated rate equations*. They enable us to find the order in A by finding which of the different plots they suggest gives a straight line. According to eq. (3.10a) a plot of c_A against time should be linear if the reaction is zeroth order in A. Figure 3.2 showed such plots for reactions having different orders in A, and you can see that it is easy to tell whether a reaction is zeroth order or not with respect to the reactant whose concentration was followed. According to eq. (3.10b) a plot of ln c_A against time should be linear if the reaction is first order in A. Such plots are shown in Fig. 3.4, and you can see that they make it easy to tell whether

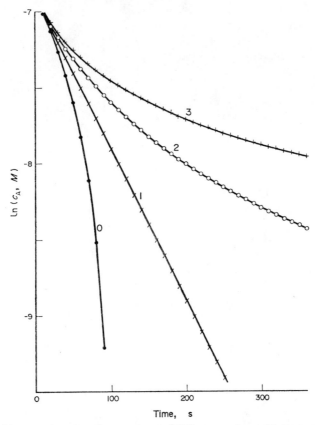

FIG. 3.4. Plots of ln c_A against time for reactions of different orders with respect to A. (0) Zeroth order, $k(c_B\ldots) = 1 \times 10^{-5}$ mol dm^{-3} s^{-1}; (1) first order, $k(c_B\ldots) = 1 \times 10^{-2}$ s^{-1}; (2) second order, $k(c_B\ldots) = 10$ dm^3 mol^{-1} s^{-1}; (3) third order, $k(c_B\ldots) = 1 \times 10^4$ dm^6 mol^{-2} s^{-1}. Each curve corresponds to the similarly numbered curve in Fig. 3.2.

or not a reaction is first order in A. Equation (3.10c) suggests plotting $1/c_A$ against time, and Fig. 3.5 shows that such a plot makes it easy to tell whether a reaction is second order in A.

In establishing the rate law for a reaction that has not been studied before, it is usually best to begin by plotting ln c_A against time. (Equation (3.10b) may be rewritten in the form

$$\log_{10} c_A = \log_{10} c_A^0 - 0.4343 k' t$$

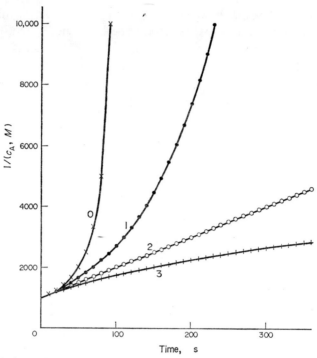

FIG. 3.5. Plots of $1/c_A$ against time for reactions of different orders with respect to A. (0) Zeroth order, $k(c_B \ldots) = 1 \times 10^{-5}$ mol dm^{-3} s^{-1}; (1) first order, $k(c_B \ldots) = 1 \times 10^{-2}$ s^{-1}; (2) second order, $k(c_B \ldots) = 10$ dm^3 mol^{-1} s^{-1}; (3) third order, $k(c_B \ldots) = 1 \times 10^4$ dm^6 mol^{-2} s^{-1}. Each curve corresponds to the similarly numbered curves in Figs. 3.2 and 3.4.

for convenience when values of $\log_{10} c_A$ are easier to obtain than those of $\ln c_A$, and $\log_{10} c_A$ may be plotted against time if the factor 0.4343 is remembered when k' is later evaluated from the slope of the linear plot.) There are two reasons why a plot of $\ln c_A$ against time is the best one to try first. One is that, of all the plots that might be constructed, it is the most likely to be a straight line, because first-order behavior is more common than anything else. The other is that it suggests what to try next if it is not a straight line. Figure 3.4 shows that it is concave downward if the reaction is zeroth order in A but is concave upward if the reaction is second order (or any higher order) in A; these shapes are easy to distinguish. If the first plot were one of c_A against time, Fig. 3.2 shows that failure to obtain a straight line would give you less information about what to try next.

Equations (3.10) must be slightly modified if it is a product, rather than a reactant, whose concentration is followed. Suppose that the concentration of acetate ion is followed while the reaction $EtOAc + OH^- = OAc^- + EtOH$ is carried out with a small initial concentration of ethyl acetate and a large (and therefore nearly constant) concentration of hydroxide ion. The reaction is pseudo-first-order and the integrated form of its rate equation is

$$\ln c_{EtOAc} = \ln c^0_{EtOAc} - k't \tag{3.10b}$$

although we would not know these things until after we had made the appropriate plot and found it to be linear. If there was no acetate ion present at the start of the reaction, each acetate ion that is present at any later time must have arisen from the disappearance of a molecule of ethyl acetate, so that

$$c_{OAc^-} = c_{EtOAc}^0 - c_{EtOAc}. \tag{3.11a}$$

By the same reasoning, the final concentration of acetate ion, $c_{OAc^-}^\infty$, will be the same as the initial concentration of ethyl acetate:

$$c_{OAc^-}^\infty = c_{EtOAc}^0. \tag{3.11b}$$

These three equations are easily combined to obtain

$$\ln (c_{OAc^-}^\infty - c_{OAc^-}) = \ln c_{OAc^-}^\infty - k't. \tag{3.12}$$

A plot of $\ln (c_{OAc^-}^\infty - c_{OAc^-})$ against time will be linear under these conditions, but would not be linear if the reaction were either zeroth or second order in ethyl acetate.

If we were interested only in the order with respect to A there would be easier ways to obtain it. In a first-order or pseudo-first-order reaction the time in which half of the reactant disappears is independent of the concentration of the reactant. This can be proved by solving eq. (3.10b) for t and setting c_A equal to $c_A^0/2$:

$$t = \frac{\ln c_A^0 - \ln c_A}{k'} = \frac{\left[\ln \dfrac{c_A^0}{c_A^0/2}\right]}{k'} = \frac{\ln 2}{k'} = \frac{0.693}{k'}. \tag{3.13}$$

The time described by eq. (3.13) is called the *half-time* for the reaction, and is given the symbol $t_{1/2}$. Half of the initial concentration or amount of reactant disappears during one half-time; half of the remaining half disappears during another one; of the quarter that remains after $2t_{1/2}$, half disappears during another half-time so that only an eighth is left after $3t_{1/2}$, and so on. If the reaction is zeroth order in A all of the A will be gone at $2t_{1/2}$; if it is second order in A, a third of the initial amount will still be left after $2t_{1/2}$, a quarter after $3t_{1/2}$, and a fifth after $4t_{1/2}$. These things are easily deduced from eqs. (3.10), and it can also be shown that the time required for the concentration of A to decrease to one-tenth of its initial value is equal to $1.8t_{1/2}$ if the reaction is zeroth order in A, $3.32t_{1/2}$ if it is first order in A, and $9t_{1/2}$ if it is second order in A. All these things are true regardless of the values of k' or c_A^0. You can often use them to find the order in a particular reactant by simply inspecting the data obtained under pseudo-nth-order conditions, and then construct a plot that you expect to be linear so that you can obtain the value of k' from its slope.

SUMMARY: The order of a reaction with respect to A can be evaluated by preparing a solution in which the concentration of every other reactant is much larger than that of A, and measuring the concentration of A at different times as the reaction proceeds. If a plot of c_A against time is a straight line, the reaction is zeroth order with respect to A; if a plot of $\ln c_A$ against time is a straight line, the reaction is first order with respect to A; and so on. The value of the pseudo-nth-order rate constant k' can be found from the slope of the straight line.

3.6. How rate laws are established

PREVIEW: The preceding section showed how one can evaluate the order n of a reaction with respect to one reactant A, using data on the variation of c_A with time under pseudo-nth-order conditions. To find the orders with respect to other reactants, one changes the concentrations of these one at a time, and evaluates the resulting effect on the slope of a plot of c_A, $\ln c_A$, $1/c_A$, or some other appropriate function of c_A, against time.

Curve a in Fig. 3.1 showed how the concentration of ethyl acetate changed when the reaction $EtOAc + OH^- = OAc^- + EtOH$ was carried out under pseudo-nth-order conditions. Figure 3.6 is a plot of $\ln c_{EtOAc}$ against time for the same data. It is a straight line; there is some curvature toward its end, but this could mean merely

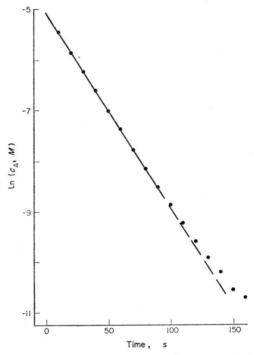

FIG. 3.6. Plot of $\ln c_{EtOAc}$ against time for the data shown in curve a of Fig. 3.1. The linearity of this plot shows that the reaction is first order with respect to ethyl acetate.

that our data are inexact if there is only a little ethyl acetate left. At the point, near the start of the reaction, where the concentration of ethyl acetate was 6.26 mM, an error of 0.01 mM in c_{EtOAc} would change $\ln c_{EtOAc}$ only from -5.0736 ($= \ln 0.00626$) to -5.0720 ($= \ln 0.0627$). The difference, 0.0016 unit, would be invisible on the scale of Fig. 3.6. When the reaction was 90% complete and the concentration of ethyl acetate had decreased to 0.626 mM, the same error would change $\ln c_{EtOAc}$ from -7.3762

$(= \ln 0.00626)$ to $-7.3603 \, (= \ln 0.00636)$, a difference of 0.016 unit. When the reaction was 99% complete and the concentration of ethyl acetate was only $0.0626 \, \text{m}M$, the difference would be 0.15 unit. We cannot expect the data to conform as well to a straight line when the concentration of ethyl acetate is low as they do when it is high. Unless we had made the measurements with such great care that we could not attribute the small deviations from linearity in Fig. 3.6 to experimental errors, we are certainly justified in concluding that the reaction is first order in ethyl acetate.

We have discovered part of the rate law: it is $dc_{\text{EtOAc}}/dt = -k'c_{\text{EtOAc}}$. The pseudo-first-order rate constant k' is the product of the actual rate constant k and a quantity $(c_B...)$ that might include the concentrations of various other substances. We must perform other experiments to find what these substances are and what exponents are attached to their concentrations in the term $(c_B...)$.

Seeing that hydroxide ion is consumed in the overall reaction, we may suspect that its concentration affects the rate. The suspicion can be tested by performing a second experiment in which the concentration of hydroxide ion is different from that in the first, and observing whether and how k' changes as a result. Since our value of k' may not be exactly correct, a very small change might be impossible to evaluate reliably or even to detect, and therefore the two concentrations of hydroxide ion should be considerably different. A factor of 2 might be appropriate. If the reaction was so fast that we could barely keep up with it in the preceding experiment, we would certainly prefer that it be slower in this one. It seems reasonable to expect that the rate of the reaction would be decreased by decreasing the concentration of hydroxide ion, and therefore we might divide the original concentration of hydroxide ion by 2 for the second experiment. The initial concentrations of ethyl acetate need not be the same in the two experiments, for the way in which the rate depends on c_{EtOAc} has already been discovered from Fig. 3.6. We must, however, make sure that the initial concentration of hydroxide ion is much larger than that of ethyl acetate, for eq. (3.10b) will not be correct if $c_{\text{OH}-}$ can vary appreciably during the reaction, and we must also make sure that there is no other difference between the two experiments. The solvent, the temperature, and the concentrations of everything except the hydroxide ion and ethyl acetate must be unchanged, for if two variables are changed at the the same time the result may be very difficult to interpret.

For this second experiment we would again plot $\ln c_{\text{EtOAc}}$ against time. Since this was a straight line for the first experiment it should also be a straight line for the second. Its slope yields a new value of k', which will differ from the original one unless the reaction is zeroth order in hydroxide ion. On comparing the values of k' obtained with a number of different concentrations of hydroxide ion, we would eventually conclude that k' is proportional to the concentration of hydroxide ion. The reaction must therefore be first order in hydroxide ion, and this knowledge makes it possible to write the rate law in a more complete form: $dc_{\text{EtOAc}}/dt = -k''c_{\text{EtOAc}}c_{\text{OH}-}$. Other concentrations might still remain concealed, and it might be premature to stop at this point.

However, there are no other reactants to try, and further experiments show that the slope of a plot of ln c_{EtOAc} against time is unaffected by adding a little acetate ion or ethanol to the initial solution. (Adding a comparatively large concentration of either will affect the rate, but only slightly and for reasons that cannot be explained here.) We would already have known that this was so unless our earlier experiments were made with solutions in which the initial concentrations of acetate ion and ethanol were a good deal larger than the initial concentration of ethyl acetate, for if the concentration of either product appeared in the rate law and varied appreciably during the reaction a straight line would not have been obtained in Fig. 3.6.

Having run out of concentrations that we can vary, we would conclude that eq. (3.4) is the rate law for this reaction under these conditions. The last three words provide for the possibility that changing the temperature or some other variable might provide new and different things to observe and persuade us that the rate law is really more complicated than this. For example, it might contain another term that is too small to be detected in strongly alkaline solutions but becomes prominent in weakly alkaline or neutral ones.

Once eq. (3.4) has been identified as the rate law, it is easy to calculate the value of k. The product kc_{OH^-} in eq. (3.4) is the quantity that is called k' in eq. (3.10b), and therefore the slope of the straight line in Fig. 3.6 is equal to $-kc_{OH^-}$. Dividing this by the known concentration of hydroxide ion gives a value of k, and the results obtained from all of the different experiments can be averaged to obtain the final result.

The procedure that has been described in this section and the preceding one consists of the following steps:

1. Find a reactant or product whose concentration can be followed in some convenient way.
2. Observe how its concentration changes with time in a reaction mixture that contains a relatively large (and therefore nearly constant) concentration of every substance but one that might affect the rate of the reaction.
3. Use the resulting data to identify the integrated rate equation that yields a linear plot. You now have the order of the reaction with respect to the substance whose concentration varied appreciably during the experiment.
4. Obtain a value of the pseudo-nth-order rate constant k' from the slope of this straight line.
5. Change the concentration of one of the other substances and repeat steps 2 and 4.
6. Find how the value of k' depends on the concentration that was changed in step 5. You now have the order of the reaction with respect to a second substance.
7. Repeat steps 5 and 6 for another substance, and keep on doing this until you have run out of substances.
8. Calculate the value of the rate constant k in the resulting rate law.

Each one of these steps is simple, but there are so many of them that the whole process seems very complicated until you understand the reasoning it involves and see what necessary bit of information each step provides. It will help you to go back through the foregoing discussion, identifying each of these steps as it arises, and there are some practice problems at the end of this chapter.

SUMMARY: In evaluating the order n of a reaction with respect to one reactant A whose concentration can be followed conveniently, one sets up pseudo-nth-order conditions and finds a function of c_A that changes linearly with time. The slope of the linear plot is equal to $-k(c_B^o c_C^p \ldots)$, where B and C are other reactants whose concentrations do not vary appreciably. The order o with respect to B is found by comparing the slope obtained in the original experiment with the slope obtained with a different constant concentration of B. Repeating this for all the reactants in turn yields the final form of the rate law and enables the chemist to calculate the value of k.

3.7. The use of rate laws

PREVIEW: This section deals with numerical calculations using rate laws.

Once the rate law for a reaction has been discovered, it can be used in either of two ways. One is to predict what the rate of the reaction will be under any particular conditions, or how long it will take for the reaction to proceed to any desired extent. Uses like these are discussed briefly in this section. Section 3.8 will show how a rate law can be used to obtain some understanding of the elementary steps by which the reaction occurs so that we can gain some insight into how and why it occurs and why it behaves as it does.

With the possible exception of the units in which rate constants are expressed, numerical calculations employing rate equations contain no surprises. The units of a rate constant depend on the order of the reaction. They can be deduced by solving the rate equation for k and inserting the units of the other quantities. Since concentrations are always expressed in mol dm^{-3} and times in s, the rate of change of a concentration must be expressed in mol dm^{-3} s^{-1}, and therefore

$k = -dc_A/dt$	$= \text{mol dm}^{-3}\text{s}^{-1}$	zeroth order
$k = -(dc_A/dt)/c_A$	$= \text{mol dm}^{-3}\text{s}^{-1}(\text{mol dm}^{-3})^{-1} = \text{s}^{-1}$	first order
$k = -(dc_A/dt)/c_A^2$	$= \text{mol dm}^{-3}\text{s}^{-1}(\text{mol dm}^{-3})^{-2} = \text{dm}^3 \text{mol}^{-1}\text{s}^{-1}$	second order
$k = -(dc_A/dt)/c_A^3$	$= \text{mol dm}^{-3}\text{s}^{-1}(\text{mol dm}^{-3})^{-3} = \text{dm}^6 \text{mol}^{-2}\text{s}^{-1}$	third order

The value of k' for a pseudo-nth-order reaction has the same units as k for an nth-order one.

Example 3.1. The decomposition of 1,1-dihydroxyethane is described by the equation

$CH_3CH(OH)_2(aq) = CH_3C\begin{smallmatrix}O\\\\H\end{smallmatrix}$ $(aq) + H_2O(l)$. In solutions that contain acetic acid at a concentration

of 0.1 mol dm^{-3} its rate equation is $dc_{CH_3CH(OH)_2}/dt = -kc_{CH_3CH(OH)_2}$ and its first-order rate constant is equal to 0.032 s^{-1}. What length of time is required for 90% of the 1,1-dihydroxyethane to disappear from a solution that originally contains 0.001 mol dm^{-3} of 1,1-dihydroxyethane and 0.1 mol dm^{-3} of acetic acid?

Answer. When 90% of it has disappeared, its concentration c will be equal to one-tenth of its initial concentration c^0. Using the integrated form of the rate equation,

$$\ln c = \ln (0.1\ c^0) = \ln c^0 - kt,$$

$$kt = \ln c^0 - \ln (0.1\ c^0) = \ln (c^0/0.1\ c^0) = \ln 10 = 2.303,$$

$$t = 2.303/k = 2.303/0.032\ \text{s}^{-1} = 72\ \text{s}.$$

Example 3.2. The reaction between peroxodisulfate and iodide ions is described by the equation $S_2O_8^{2-} + 3\ I^- = 2\ SO_4^{2-} + I_3^-$. Its rate law is $dc_{S_2O_8^{2-}}/dt = -kc_{S_2O_8^{2-}}c_{I^-}$, and its second-order rate constant is 2.22×10^{-3} dm^3 mol^{-1}s^{-1}. What length of time is required for 90% of the peroxodisulfate to disappear from a solution containing 1×10^{-4} mol dm^{-3} of $S_2O_8^{2-}$ and 0.01 mol dm^{-3} of I^-?

Answer. Although the reaction is second order, $c_{I^-}^0$ (the initial concentration of iodide ion) is so much larger than $c_{S_2O_8^{2-}}^0$ that the concentration of iodide ion will not change appreciably during the reaction. Combining its value with that of k in the rate law gives

$$dc_{S_2O_8^{2-}}/dt = -2.22 \times 10^{-5} c_{S_2O_8^{2-}},$$

a pseudo-first-order equation whose integrated form is

$$\ln c_{S_2O_8^{2-}} = \ln c_{S_2O_8^{2-}}^0 - 2.22 \times 10^{-5} t.$$

In the same way as in Example 3.1 we obtain

$$t = \ln(c_{S_2O_8^{2-}}^0/c_{S_2O_8^{2-}})/2.22 \times 10^{-5} = (\ln 10)/2.22 \times 10^{-5} = 2.303/2.22 \times 10^{-5} = 1.04 \times 10^5\ \text{s}$$

or

$$t = \frac{1.04 \times 10^5\ \text{s}}{3.6 \times 10^3\ \text{s h}^{-1}} = 28.8\ \text{h}.$$

As these examples suggest, such calculations are not difficult if the substance in which you are interested is the only one whose concentration changes appreciably during the reaction. This is not unusual, because large concentrations of reagents are often used to drive reactions to completion or cause them to occur at reasonable rates. However, if, in Example 3.2, the initial concentration of iodide ion had been 4×10^{-4} M, the concentration of iodide ion could not have been treated as a constant. It would have decreased from 4×10^{-4} M to 1×10^{-4} M as the concentration of peroxodisulfate ion decreased from 1×10^{-4} M to zero. The reaction consumes three moles of iodide ion for each mole of peroxodisulfate ion: when the concentration of peroxodisulfate ion has decreased from $c_{S_2O_8^{2-}}^0$ to $c_{S_2O_8^{2-}}$, the reaction has consumed $(c_{S_2O_8^{2-}}^0 - c_{S_2O_8^{2-}})$ moles of peroxodisulfate in each dm^3 of the solution, and therefore it has consumed three times as many, or $3\ (c_{S_2O_8^{2-}}^0 - c_{S_2O_8^{2-}})$, moles of iodide in each dm^3. Hence the concentration of iodide ion will be equal to $c_{I^-}^0 - 3(c_{S_2O_8^{2-}}^0 - c_{S_2O_8^{2-}})$ when that of peroxodisulfate ion is equal to $c_{S_2O_8^{2-}}$, and therefore the rate law becomes

$$dc_{S_2O_8^{2-}}/dt = -kc_{S_2O_8^{2-}}(c_{I^-}^0 - 3c_{S_2O_8^{2-}}^0 + 3c_{S_2O_8^{2-}}) \tag{3.14}$$

which would have to be integrated to obtain the necessary relation between $c_{S_2O_8^{2-}}$ and t. The integration is not very difficult, but it is more complicated than it was under pseudo-first-order conditions in Example 3.2.

It is also easy to calculate how much reactant will remain after a given length of time if every other concentration that appears in the rate law is constant, and again there are complications otherwise. Example 3.3 is typical of the former case, and you can guess at the possible complications on the basis of the preceding paragraph.

Example 3.3. Under the conditions described in Example 3.2, what concentration of peroxodisulfate ion would still be present after 48 hours?

Answer. $48 \text{ h} \times 3.6 \times 10^3 \text{ s h}^{-1} = 1.728 \times 10^5 \text{ s}$. When $t = 1.728 \times 10^5$ s and $k' = kc_{I^-} = 2.22 \times 10^{-5} \text{ s}^{-1}$,

$$\ln c_{S_2O_8^{2-}} = \ln c^0_{S_2O_8^{2-}} - k't = \ln c^0_{S_2O_8^{2-}} - (2.22 \times 10^{-5})(1.728 \times 10^5) = \ln c^0_{S_2O_8^{2-}} - 3.836,$$

$$\ln \left(c_{S_2O_8^{2-}} / c^0_{S_2O_8^{2-}} \right) = -3.836 \text{ or } c_{S_2O_8^{2-}} / c^0_{S_2O_8^{2-}} = e^{-3.836} = 2.16 \times 10^{-2},$$

$$c_{S_2O_8^{2-}} = 2.16 \times 10^{-2} \, c^0_{S_2O_8^{2-}} = (2.16 \times 10^{-2})(1 \times 10^{-4}) = 2.16 \times 10^{-6} \ M.$$

The value of $c_{S_2O_8^{2-}} / c^0_{S_2O_8^{2-}}$ means that $100 \times 2.16 \times 10^{-2} = 2.16$ per cent of the peroxodisulfate ion will remain unreacted after 48 h have elapsed.

SUMMARY: If either the extent or the duration of a reaction is known, the other can be calculated from the rate law.

3.8. The interpretation of rate laws

PREVIEW: The rate law for a reaction provides information about the rate-determining step and any fast steps that precede it. By combining this information with chemical and structural considerations and intuition, one can achieve much understanding of the details of the reaction.

A rate law is interpreted by imagining a *mechanism*, which is a sequence of chemical steps, that accounts for it and that makes chemical sense. The ingenuity and resourcefulness of the chemist are deeply involved and essential to success. Quite apart from the importance of the questions it seeks to answer, the process is therefore a fascinating and challenging one, and for these two reasons it has a continuing attraction for chemists of all kinds.

As an illustration we shall consider the reaction between vanadium(II) and vanadium(IV) in strongly acidic aqueous solutions. This was mentioned briefly in Section 3.4, where its overall equation $V^{2+} + VO^{2+} + 2 H^+ = 2 V^{3+} + H_2O$ and rate law $dc_{V^{2+}}/dt = -kc_{V^{2+}}c_{VO^{2+}}c_{H^+}$ were given. Its third-order rate constant, approximately 0.7 $\text{dm}^6 \text{ mol}^{-2} \text{ s}^{-1}$ at 25°, is neither very high nor very low as third-order rate constants go. In a reaction mixture that contains 3 M hydrogen ion, 0.025 M

5*

vanadium(IV), and a small concentration of vanadium(II), half of the vanadium(II) will disappear in about 13 s.

Although this is a long way from being extremely fast, it is too fast to enable us to suppose that the rate-determining step is a termolecular reaction in which a vanadium(II) ion, a vanadium(IV) ion, and a hydrogen ion combine to yield other substances that are consumed in subsequent fast steps. That is what the form of the rate law might suggest at first glance, but there is very little chance that these three ions, each going its own separate way through the solution, will all be present simultaneously in a volume of solution small enough to correspond to a collision among them. In part this is so because of their electrostatic repulsions; if they were all uncharged, such collisions would be much more frequent, though even then they would be rather uncommon. We must suppose that two of the ions collide first and react, forming a bond whose energy counterbalances the electrostatic repulsion. The product of this reaction may not be very stable, and either of two things can happen to it. It may simply dissociate into its constituents before an ion of the third kind collides with it. Then there will have been no net result. The other possibility is that an ion of the third kind will collide with it before it dissociates. If such an ion arrives from the right direction and if there is enough energy available, the last collision may result in the formation of vanadium(III).

The bare bones of this scheme were described by eqs. (3.8). Adding flesh and blood appropriate to the present reaction, and more or less arbitrarily picking V^{2+} and VO^{2+} as the reactants in its first step, we might write

$$\left.\begin{aligned}
V^{2+} + VO^{2+} &= V_2O^{4+} & &\text{fast prior equilibrium,}\\
V_2O^{4+} + H^+ &\rightarrow V^{III}OH^{2+} + V^{3+} & &\text{rate-determining step,}\\
V^{III}OH^{2+} + H^+ &= V^{3+} + H_2O & &\text{fast subsequent equilibrium.}
\end{aligned}\right\} \quad (3.15)$$

Adding these equations gives the equation for the overall reaction; they could not represent the correct mechanism if this were not true.

How can such a mechanism account for the observed rate law? Section 3.3 showed that the rate of the overall reaction is equal to that of its rate-determining step, which is given by

$$dc_{V_2O^{4+}}/dt = -kc_{V_2O^{4+}}c_{H^+}. \quad (3.16a)$$

If the prior reaction is at equilibrium, the equation

$$K = c_{V_2O^{4+}}/c_{V^{2+}}c_{VO^{2+}} \quad \text{or} \quad c_{V_2O^{4+}} = Kc_{V^{2+}}c_{VO^{2+}} \quad (3.16b)$$

must be satisfied. Substituting this description of $c_{V_2O^{4+}}$ into eq. (3.16a) yields

$$dc_{V_2O^{4+}}/dt = -kKc_{V^{2+}}c_{VO^{2+}}c_{H^+}. \quad (3.16c)$$

However, the consumption of V_2O^{4+} in the rate-determining step displaces the prior equilibrium, and therefore as one such ion is consumed it will be replaced by another, which must be formed at the expense of one V^{2+} ion and one VO^{2+} ion. Consequently,

$$dc_{V^{2+}}/dt = dc_{VO^{2+}}/dt = -kKc_{V^{2+}}c_{VO^{2+}}c_{H^+}. \tag{3.16d}$$

This is identical in form with the experimentally established rate law. If this mechanism is correct, the third-order rate constant obtained experimentally is the product of the second-order rate constant for the bimolecular rate-determining step and the equilibrium constant for the fast step that precedes it.

You can now see why third- and fourth- and even higher-order reactions are fairly common even though termolecular rate-determining steps are very rare and quadrimolecular ones are unknown. This rate-determining step involves neither V^{2+} nor VO^{2+}, but increasing the concentration of either one leads to an increase of the concentration of V_2O^{4+}, and thus to increases of the rates of the rate-determining step and the overall reaction. The overall rate is therefore proportional to each of these concentrations.

Since we can deduce the form of the experimental rate law from eqs. (3.15), that mechanism is consistent with that rate law, and we have already noted that it is also consistent with the equation for the overall reaction. If either of these things were untrue the mechanism could not possibly be correct, but even if both are true the mechanism might still not be correct. It must also make chemical sense, which is to say that it must be consistent with all the other information and understanding that we have.

The formula V_2O^{4+} looks a little odd. Certainly it could not represent a structure such as $(V^{II}-V^{IV}-O)^{4+}$, in which electrostatic repulsion would be enormous and any sort of bond between the vanadium atoms would be hard to envision. The structure $(V^{IV}-O-V^{II})^{4+}$ is much more reasonable; it resembles a molecule of water. Of course, it would not be linear. The two positively charged vanadium atoms repel each other, as do the two protons in a molecule of water. The covalent oxygen–hydrogen bonds are strong enough to hold a molecule of water together, and covalent oxygen–vanadium bonds might well be strong enough to hold $(V^{IV}-O-V^{II})^{4+}$ together. The VO^{2+} ion is a dipole, and a positively charged ion such as V^{2+} might well become attached to its negative end. Dimeric and polymeric species involving metal–oxygen–metal bonds are well known for iron(III), aluminum(III), and many other highly charged metal ions.

After we have satisfied ourselves that $(V^{IV}-O-V^{II})^{4+}$ might exist, we must imagine a way in which it might react with a hydrogen ion in the rate-determining step. With the analogy to a water molecule fresh in our minds, we might think of something like

$$\begin{bmatrix} V^{IV} \diagdown \diagup V^{II} \\ O \end{bmatrix}^{4+} + H^+ = \begin{bmatrix} V^{IV} \diagdown \diagup V^{II} \\ O \\ | \\ H \end{bmatrix}^{5+} \tag{3.17a}$$

This product would probably be tetrahedral. It resembles a hydronium ion or, from a different point of view, a hydrated vanadium ion in which one of the protons in a molecule of water of hydration has been replaced by another vanadium ion.

We have arrived at a crucial point in the reaction: the product in eq. (3.17a) is the *activated complex* or *transition state*. It is the result of tacking together all of the reactants in the proportions required by the rate law, and it is like a drop of water poised at the very peak of the Continental Divide. Exquisitely unstable and short-lived, it may simply dissociate into the components from which it was assembled. However, vanadium(IV) is a fairly strong electron-acceptor and vanadium(II) is a very strong electron-donor, and there is therefore a chance that an electron will be transferred between them during the brief moment of existence of the activated complex:

$$\left[\begin{array}{c} V^{IV} \quad V^{II} \\ O \\ | \\ H \end{array}\right]^{5+} \rightarrow \left[\begin{array}{c} V^{III} \quad V^{III} \\ O \\ | \\ H \end{array}\right]^{5+} \qquad (3.17b)$$

Now if one of the vanadium–oxygen bonds breaks, the result will be

$$\left[\begin{array}{c} V^{III} \quad V^{III} \\ O \\ | \\ H \end{array}\right]^{5+} \rightarrow \left[\begin{array}{c} V^{III} \\ O \\ | \\ H \end{array}\right]^{2+} + V^{3+} \qquad (3.17c)$$

Adding equations (3.17) yields the equation for the rate-determining step that appeared in eq. (3.15). It is pleasing to find that $V^{III}OH^{2+}$ is already well known and that, as the last step in eqs. (3.15) requires, it is known to react rapidly and, in strongly acidic solutions, almost completely with hydrogen ion to form V^{3+}.

Apart from all the objective arguments we have raised, this feels right. You will have to learn when something feels right, and you will enjoy the first time something does—and the thousandth. You are not likely to be so lucky with the first few mechanisms you invent yourself. Perhaps you will imagine a mechanism that does not account for the overall equation for the reaction, such as

$$\left.\begin{array}{ll} V^{2+} + VO^{2+} = V_2O^{4+} & \text{fast prior equilibrium,} \\ V_2O^{4+} \rightarrow V^{III}O^+ + V^{3+} & \text{rate-determining step,} \\ V^{III}O^+ + 2H^+ = V^{3+} + H_2O & \text{fast subsequent equilibrium} \end{array}\right\} \quad (3.18a)$$

which suffers from a fatal inability to account for the rate law obtained experimentally (the concentration of hydrogen ion would not appear in the rate law). You may imagine one that includes an unlikely or impossible step, such as $V^{2+} + H^+ = VH^{3+}$ (who could the vanadium and hydrogen atoms be bonded to each other?). When you have acquired enough experience to avoid such errors, and if you can succeed in keeping an open mind after you have found one mechanism that seems right, you will find that there is almost always at least one more that might also be right.

For instance, the mechanism

$$
\begin{array}{lll}
VO^{2+} + H^+ = VOH^{3+} & \text{fast prior equilibrium,} & \\
VOH^{3+} + V^{2+} \rightarrow V^{III}OH^{2+} + V^{3+} & \text{rate-determining step,} & \\
VOH^{3+} + H^+ = V^{3+} + H_2O & \text{fast subsequent equilibrium} &
\end{array} \Bigg\} \quad (3.18b)
$$

accounts for both the overall equation and the rate law. There are other reasons for supposing that a little VOH^{3+} may exist in strongly acidic solutions, the structures of $(V^{IV}-O-V^{II})^{4+}$ and $(V^{IV}-O-H)^{3+}$ are not importantly different, and the activated complex

$$
\left[\begin{array}{cc} V^{IV} & V^{II} \\ & \\ O & \\ | & \\ H & \end{array} \right]^{5+}
$$

is the same thing whether the last step in its assembly consists of adding a hydrogen ion to $(V^{IV}-O-V^{II})^{4+}$ or a V^{2+} ion to $(V^{IV}-O-H)^{3+}$. Arguments very similar to those that support eqs. (3.15) can be constructed to support eqs. (3.18b), and in many such situations it is impossible to get other information to help you decide among the acceptable possibilities. When you have excluded everything that is impossible, anything that remains might be right.[†] You can usually disprove most of the mechanisms that might be imagined, but you can never prove that a single one of them is correct, and you must always allow for the possibility of imagining other mechanisms that might account for everything that has been observed. If you know more than the overall equation and rate law, you can often disprove one or more of the mechanisms that survive the process described here, but you may still have two or more equally acceptable alternatives left when you have finished, and you may or may not be able to imagine some experimental measurement or theoretical calculation that would help you decide which was more probable.

This is much less discouraging than you might think. In the reaction that we have been considering, it seems very probable that the first step yields $(V^{IV}-O-X)^{p+}$ even if we cannot be sure whether X represents an atom of hydrogen or vanadium, and that the rate-determining step involves $(V^{IV}V^{II}OH)^{5+}$ as the activated complex. For other reactions the uncertainty may be quite different. Two mechanisms that account for the overall equation and rate law for the reaction between hypochlorite and iodide ions are

---

[†] This was originally stated, in a slightly different form and a very different context, by A. Conan Doyle.

$OCl^- + H_3O^+ =$ $HOCl(aq) + H_2O(l)$	fast prior equilibrium,	$OCl^- + H_3O^+ =$ $HOCl(aq) + H_2O(l)$
$HOCl(aq) + I^- \rightarrow HOI(aq) + Cl^-$	rate-deter-mining step,	$HOCl(aq) + I^- \rightarrow ICl(aq) + OH^-$
$HOI(aq) + H_2O(l) = H_3O^+ + OI^-$	fast subsequent reactions.	$\begin{cases} ICl(aq) + 2H_2O(l) = HOI(aq) \\ \qquad\qquad\qquad + H_3O^+ + Cl^- \\ HOI(aq) + OH^- = OI^- + H_2O(l) \end{cases}$
$OCl^- + I^- = Cl^- + OI^-$	overall reaction	$OCl^- + I^- = Cl^- + OI^-$

Here the prior equilibrium is common to the two schemes but the rate-determining steps are different, and they are different in such a way that different activated complexes must be involved: there must be an iodine-oxygen bond in one that decomposes to yield HOI, but an iodine–chlorine bond in one that decomposes to yield ICl. It might even be possible to imagine yet another mechanism, involving some quite different prior equilibrium, that would account for all the facts that are known about this reaction today, and would do so just as reasonably as these two mechanisms do. It would become necessary to imagine another mechanism if some new facts were observed that neither of these two served to explain.

Of course we could say similar things about all hypotheses. Even one that has been made to seem overwhelmingly probable, by the accumulation of an enormous amount of information that supports it, may be shown to be incomplete by the discovery of new information that it cannot explain, or to be wrong by the discovery of new information that contradicts it. Some hypotheses, such as the one that matter is composed of atoms, are consistent with so much information of so many different kinds that our intellectual reservations about them are crushed by the weight of the evidence. Others are sometimes retained, even in the face of knowledge that conflicts with them, because of inertia or the extreme difficulty of imagining alternatives that are as good. Other areas of chemistry involve the same kind of thinking that chemical kinetics does, but few illustrate it quite as clearly and explicitly.

SUMMARY: The mechanism of a reaction is a series of elementary steps, each consistent with chemical and structural principles, that accounts for both the overall equation for the reaction and its rate law. A crucial feature of a mechanism is the activated complex, an unstable substance that can decompose either to regenerate the starting materials or to form the products. The rate law for a reaction furnishes information about the steps that lead to the formation of the activated complex.

3.9. Catalysis, enzymes, and the prior equilibrium

PREVIEW: When a prior equilibrium is so nearly complete that it consumes most of one reactant, there are unexpected consequences.

Urea reacts with water and hydrogen ion as shown by the equation

$$\begin{array}{c} NH_2 \\ \diagup \\ C\!=\!\!O \quad (aq) + H^+ + 2H_2O(l) = 2NH_4^+ + HCO_3^- \\ \diagdown \\ NH_2 \end{array}$$

Important biochemical processes in our bodies would be severely impeded if this reaction did not occur at a reasonable rate. However, it involves a good deal of demolition and reconstruction, and we might therefore expect it to be rather slow; experiments reveal that it is very slow indeed. It can just be made to occur at a conveniently measurable rate by boiling an aqueous solution, but a decrease of temperature decreases the rate for reasons that we shall examine in Section 3.12. At body temperature (37° C, or 310 K) the half-time is something like 100 000 years.

Fortunately its rate is significantly increased by the enzyme urease, which is a protein having a molecular weight of about 5×10^5. Urease is not consumed when urea hydrolyzes in its presence, and therefore a very small amount of urease can serve to accelerate the hydrolysis of a very large amount of urea. The urease is called a *catalyst*. Many enzymes and other catalysts are extremely selective: of all the biochemical reactions that occur in the body, the hydrolysis of urea is the only one whose rate is affected by the presence of urease. A biochemist would call the urea the *substrate* in this reaction, and define a substrate as the particular substance on which an enzyme acts. Other catalysts are far less specific. Metallic platinum serves as a catalyst for the rearrangements and oxidations of hydrocarbons, the addition of hydrogen to unsaturated organic compounds, the oxidation of sulfur dioxide to sulfur trioxide, and a host of other reactions.

The rate of hydrolysis of urea in the presence of urease behaves in a curious way. It is proportional to the concentration of urease, and it is also proportional to the concentration of urea if that concentration is not too high. Reasoning as we did in Section 3.8, we envision a mechanism such as

$$\left. \begin{array}{ll} E + S = ES & \text{fast prior equilibrium,} \\ ES \rightarrow A + B + \ldots + E & \text{rate-determining step} \end{array} \right\} \quad (3.19)$$

where E represents the enzyme (urease), S represents the substrate (urea), and A, B, ... may be ammonium and hydrogen carbonate ions or intermediates that are converted into these ions in fast steps that follow the rate-determining one. The enzyme is released again so that it can catalyze the decomposition of another molecule of the substrate. Comparing this scheme with eqs. (3.15) and (3.16), we would write

$$\text{rate} = dc_{\text{products}}/dt = kc_{ES} = kKc_Ec_S. \quad (3.20)$$

This describes a second-order reaction, first order in urease and first order in urea. The data shown in Fig. 3.7 were obtained in experiments with solutions that contained different concentrations of urea but were otherwise exactly the same. It shows the rate of the reaction to be proportional to the concentration of urea only if that concentration is small; if it is very large the rate of the reaction no longer depends on it at all! The order with respect to urea varies from 1 at low concentrations of urea to 0 at higher ones.

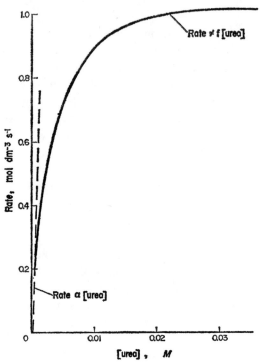

FIG. 3.7. Effect of the concentration of urea on the rate of its hydrolysis in the presence of a fixed concentration of urease. The data were obtained by G. B. Kistiakowsky and A. J. Rosenberg, *J. Amer. Chem. Soc*, **74**, 502 (1952).

Reactions that behave in this way are extremely common. Indeed, any reaction having the mechanism described by eqs. (3.8) will show such behavior if the initial concentration of one reactant is so high that the prior equilibrium consumes almost all of another reactant. Biochemists are especially interested in this behavior because the concentrations of enzymes are so small in biochemical systems. Two biochemists, Michaelis and Menten, discovered more than 60 years ago that the behavior can be accounted for by the mechanism of eqs. (3.19), which has been called the Michaelis–Menten mechanism ever since.

What happens is merely that the position of the initial equilibrium shifts as the concentration of the substrate S is increased. If that concentration is low, only a small fraction of the enzyme E is converted into the enzyme-substrate complex ES, and the concentration c_E of the unbound enzyme is only very slightly smaller than the concentration of enzyme that was put into the solution. As more and more of the substrate

S is added, the formation of the complex becomes more and more complete, and the concentration of unbound enzyme decreases. This accounts for the curved portion of Fig. 3.7. Adding more S to the solution increases c_S, which tends to increase the rate of the reaction, but at the same time it decreases c_E, which tends to decrease the rate of the reaction; the net effect is that the rate increases, but increases less rapidly than c_S does. Finally, when c_S is very high almost all of the enzyme is converted into the complex; then c_{ES} is almost equal to the original concentration of the enzyme, which is represented by c_E^0 in the equations below. Now a further increase of c_S can no longer increase c_{ES}, and the rate of the reaction becomes independent of c_S. This situation corresponds to the flat upper portion of Fig. 3.7.

To account for the curve quantitatively we need only recognize that the occurrence of the prior reaction does not affect the total amount of enzyme in the solution; it merely alters the way in which the enzyme is divided between the free and complexed forms. That is,

$$c_E^0 = c_E + c_{ES}. \tag{3.21}$$

By solving this for c_E $(= c_E^0 - c_{ES})$, inserting the resulting description of c_E into the equation that describes the equilibrium constant of the prior equilibrium $(K = c_{ES}/c_E c_S)$, and solving for c_{ES} we can obtain

$$c_{ES} = Kc_E^0 c_S/(1 + Kc_S) \tag{3.22}$$

which can be combined with eq. (3.20) to give

$$\text{rate} = \frac{kKc_E^0 c_S}{1 + Kc_S}. \tag{3.23}$$

There are two extreme situations. In one, c_S is so small that $Kc_S \ll 1$, and[†]

$$\text{rate} \approx kKc_E^0 c_S \tag{3.24a}$$

which agrees with the observed fact that the reaction is first order in S when c_S is small. In the other, c_S is so large that $Kc_S \gg 1$, and

$$\text{rate} \approx kc_E^0 \tag{3.24b}$$

which agrees with the observed fact that the reaction is zeroth order in S when c_S is large.

Whether any particular value of c_S is "small" or "large" or somewhere in between obviously depends on the value of K. For the reaction between hypochlorite ion and iodide ion, which is represented by the overall equation $OCl^- + I^- = Cl^- + OI^-$ and in which the prior equilibrium is $H^+ + OCl^- = HOCl(aq)$, Section 3.4 gave the rate equation as $dc_{OCl^-}/dt = -kc_{H^+}c_{OCl^-}c_{I^-}$. This applies to strongly alkaline solutions, where c_{H^+} is very low and where only a tiny fraction of the hypochlorite

---

[†] The sign "$\approx$" is used to mean "is approximately equal to".

ion is converted into HOCl. In strongly acidic solutions the formation of HOCl would be almost complete, and adding more hydrogen ion to such a solution would not lead to an appreciable increase of the concentration of HOCl. The concentration of hydrogen ion would not appear in the rate equation for strongly acidic solutions.

This sort of behavior provides some pitfalls and some opportunities. One of the pitfalls lies in wait for the chemist who jumps to conclusions that are based on just a little experimental work. Such a chemist is often unfortunate enough to hit upon the middle of the curved portion of a plot like the one in Fig. 3.7, where the order of the reaction with respect to urea is somewhere between 1 and 0. Two or three experiments that did not cover a very wide range of concentrations of urea might give rates that seemed to be fairly closely proportional to the square root of the concentration of urea, and much very foolish and useless discussion might result.

One of the opportunities is that the form of eq. (3.23) makes it possible to obtain the values of both k and K. A particularly simple way of doing this is suggested by rearranging the equation to give

$$\frac{\text{rate}}{c_S} = kKc_E^0 - K \times \text{rate}$$

which means that, if the ratio of the rate to the concentration of urea is plotted against the rate, a straight line should be obtained. Its slope will be equal to $-K$, and its intercept at zero rate will be equal to kKc_E^0. If c_E^0 is known, the value of k is easily obtained from the intercept and slope.

Another opportunity arises from the fact that the concentration of a substance does not disappear from a rate equation when it becomes large unless that substance is involved in a prior equilibrium. If the mechanism of the reaction between hypochlorite ion and iodide ion were something like

$$OCl^- + I^- = IOCl^{2-} \qquad \text{fast prior equilibrium,}$$
$$IOCl^{2-} + H^+ \rightarrow HOI + Cl^- \qquad \text{rate-determining step,}$$
$$HOI = H^+ + OI^- \qquad \text{fast subsequent equilibrium}$$

there would be no way in which the overall rate could become independent of the concentration of hydrogen ion. By changing the concentration of a reactant over a very wide range and observing whether the order with respect to that reactant remains constant or not, we can often tell whether it appears in the rate-determining step or in a prior equilibrium. This is an extremely powerful weapon. You should be able to see, for example, how it could help us to decide whether eqs. (3.15) or eqs. (3.18b) were correct for the reaction between vanadium(II) and vanadium(IV). It would not work if the equilibrium constant K for the prior equilibrium were so small that a term like Kc_S in the denominator of eq. (3.23) could not be made much larger than 1 even by saturating a reaction mixture with the reactant S, or if k were so small that the reaction would become too slow for convenience if Kc_S were made very small by employing an extremely low concentration of S.

SUMMARY: If one of the reactants is almost completely consumed in a prior equilibrium, the order with respect to that reactant decreases. Reactions catalyzed by enzymes often behave in this way because the concentration of the enzyme is so small that the prior equilibrium is driven nearly to completion by the excess of the substrate. Such behavior enables the chemist to decide whether a reactant takes part in a prior equilibrium or in the rate-determining step.

3.10. Reversible reactions

PREVIEW: This section deals with the rate laws for the forward and backward processes in a reaction that reaches equilibrium instead of proceeding to completion, with the additional information about the mechanism that can be obtained from the rate law for the reverse process, and with the connection between the rate law and the expression for the equilibrium constant.

So far in this chapter we have dealt only with reactions that are "irreversible" in the sense that they go nearly to completion, and only with their rates in the forward direction because their rates in the reverse direction are negligibly small. Many reactions can be appropriately described in this way under most experimentally feasible conditions; most reactions can be appropriately described in this way under some experimentally feasible conditions. Often, however, we must modify the treatment because a reaction is so far from complete that the rate of the reverse reaction cannot be ignored.

At the instant when the reaction between arsenic(V) acid and iodide ion in an acidic solution

$$H_3AsO_4(aq) + 2H^+ + 3I^- = H_3AsO_3(aq) + I_3^- + H_2O(l)$$

is initiated by mixing solutions of its reactants, the reverse reaction between arsenic(III) acid and triiodide ion clearly cannot occur at all unless these products have been deliberately added to the mixture. As the forward reaction proceeds, the concentrations of the products increase, and consequently the rate of the reverse reaction increases. While these things are happening, the rate of the forward reaction decreases as its reactants are consumed and their concentrations decrease. Eventually the rates of the two reactions become equal. Even after they have become equal, some molecules of H_3AsO_4 will be reduced by iodide ion, but others will be formed at the same rate by the reaction of H_3AsO_3 with triiodide ion, and so there will be no further change in the composition of the solution. This is, of course, what we call equilibrium. It would be quite proper to neglect the rate of the reverse reaction at the very start, where it is equal to zero, but quite impossible to neglect it at equilibrium, where it is equal to the rate of the forward reaction.

The equilibrium constant for this reaction is not very large; its value is approximately 3. If the concentrations of hydrogen, iodide, and triiodide ions are all $1\,M$, the ratio $[H_3AsO_3]/[H_3AsO_4]$ will become equal to 3 when the reaction has reached equilibrium. Only about three-quarters of the arsenic(V) acid will be reduced. Since the rates of the forward and reverse reactions are equal when three-quarters of the H_3AsO_4 is reduced and $[H_3AsO_3]/[H_3AsO_4] = 3$, they cannot be very widely different except near the very beginning of the reaction. When one-quarter of the H_3AsO_4 has been reduced, the concentration of H_3AsO_4 is only three times as large as its eventual value at equilibrium, while the concentration of H_3AsO_3 is already equal to one-third of its eventual value. You can confirm these figures by supposing that there were four molecules of H_3AsO_4 in the original solution: at equilibrium three of them will have been reduced to give three molecules of H_3AsO_3 and only one molecule of H_3AsO_4 will be left, but when the reaction is 25% complete one molecule of H_3AsO_3 will have been formed and three molecules of H_3AsO_4 will remain. Experiments show that the forward reaction is first order in H_3AsO_4 and that the reverse reaction is first order in H_3AsO_3. When 25% of the original H_3AsO_4 has reacted, the forward reaction is just three times as fast as it will be at equilibrium, while the reverse reaction is one-third as fast as it will be at equilibrium. Since the two rates are equal at equilibrium, the forward one is only nine times as fast as the reverse one when the reaction is 25 per cent complete. Depending on what is measured and how, it might be either just possible to detect the occurrence of the reverse reaction or just possible to overlook it.

This is not a favorable situation for using eqs. (3.10). If we are confined to studying the first 25% of a reaction because the rate of the reverse reaction becomes too large to ignore thereafter, any of the three plots corresponding to those equations might appear to give a fairly good straight line. This is illustrated by Fig. 3.8, in which two different sets of data, each covering just the first 25% of a reaction, are plotted according to eq. (3.10b). We might well decide that each of these plots is linear and therefore that both of these reactions were first order, but actually one of the two reactions is zeroth order while the other is second order. Figure 3.4 showed how easy it would be to tell that neither of these plots is linear if we could see more of them.

There are several things that we could do. One would be to evaluate the initial rates of the forward reaction with solutions that contained different concentrations of H_3AsO_4. Determining the concentration of H_3AsO_4 after each reaction had proceeded for only a short time Δt would enable us to calculate values of $\Delta c_{H_3AsO_4}/\Delta t$, which would be nearly equal to $dc_{H_3AsO_4}/dt$ if Δt were small enough, and by inspecting the way in which these depended on $c_{H_3AsO_4}$ we could find the order of the reaction with respect to H_3AsO_4. At the same time, of course, we could obtain a value of the pseudo-nth-order rate constant for the particular concentrations of hydrogen and iodide ions that were present in all these solutions. Then we could go on, in the general fashion described in Section 3.6, to change the concentration of one of these ions at a time, find how changing it affected the pseudo-nth-order rate constant, and discover the order with respect to that ion.

A second approach would be to increase the concentrations of hydrogen and iodide ions so that the reaction would be driven much more nearly to completion. If it is at

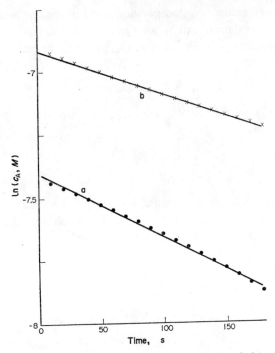

FIG. 3.8. Plots of ln c_A against time for the first 25% of a reaction that is (a) zeroth, and (b) second order with respect to A. Curve a is concave downward while curve b is concave upward, but the data cover so narrow a range that their departures from linearity are not easy to discern.

least 99% complete at equilibrium, the rate of the reverse reaction will not be as much as a tenth of the rate of the forward reaction until about 90% of the reaction has occurred. For the reason given in the discussion of Fig. 3.6, the data often deteriorate beyond that point even if the backward reaction does not occur at all, and consequently we could use eqs. (3.10) without danger if this were done. A possible disadvantage of this approach is that the forward reaction may become too fast to be studied conveniently.

A third approach would be to find the appropriate integrated rate equation and construct a plot based on it. Our search for that equation might begin with the assumptions that the forward reaction is first order with respect to H_3AsO_4, that the backward one is first order with respect to H_3AsO_3, and that the concentrations of all the other reactants and products are large and therefore constant. Then the rate of the forward reaction will be given by $k'c_{H_3AsO_4}$, where k' is its pseudo-first-order rate constant, and the rate of the reverse reaction will be given by $k'_-c_{H_3AsO_3}$, where k'_- is the pseudo-first-order rate constant for the reaction that is the reverse of the one to which k' belongs. Since the forward reaction consumes H_3AsO_4 while the reverse reaction produces it, the net rate of the reaction in the forward direction is given by

$$dc_{H_3AsO_4}/dt = -k'c_{H_3AsO_4} + k'_-c_{H_3AsO_3}. \tag{3.25a}$$

Atoms of arsenic are neither created nor destroyed in the reaction; they are merely converted from one oxidation state to another. At any stage of the reaction all of the arsenic that was present originally (as H_3AsO_4) must still be present in one of the two forms:

$$c_{H_3AsO_4}^0 = c_{H_3AsO_4} + c_{H_3AsO_3}. \qquad (3.25b)$$

This equation can be used to eliminate $c_{H_3AsO_3}$ from eq. (3.25a); after doing so and separating the variables the result is

$$\frac{dc_{H_3AsO_4}}{(k' + k'_-)c_{H_3AsO_4} - k'_- c_{H_3AsO_4}^0} = -dt.$$

This is not especially difficult to integrate, but we shall pursue it no farther. You can see that the final equation will be somewhat more complicated than eq. (3.10b), and also that it will contain both k' and k'_- and may therefore enable us to evaluate both if we handle it in the right way.

In one or another of these ways we can establish the following rate laws for the forward and reverse reactions separately:

$$\text{forward} \quad dc_{H_3AsO_4}/dt \ (= -dc_{H_3AsO_3}/dt) = -kc_{H_3AsO_4}c_{H^+}c_{I^-}, \qquad (3.26a)$$

$$\text{reverse} \quad dc_{H_3AsO_3}/dt \ (= -dc_{H_3AsO_4}/dt) = -k_-c_{H_3AsO_3}c_{I_3^-}/c_{H^+}c_{I^-}^2. \qquad (3.26b)$$

It is not difficult to imagine a mechanism for the forward reaction. Equation (3.26a) suggests a fast prior equilibrium between two of its reactants and some intermediate, which reacts with the third reactant in the next step. Representing the structure of a molecule of H_3AsO_4 as $O{=}As(OH)_3$, we might find it easy to see where a proton could attach itself but less easy to see how an iodide ion could do so, and then the equations

$$\begin{array}{lll} H_3AsO_4(aq) + H^+ = H_4AsO_4^+ & \text{fast prior equilibrium,} \\ H_4AsO_4^+ + I^- \xrightarrow{\ k\ } ? & \text{rate-determining step} \end{array} \Bigg\} \quad (3.27a)$$

are easy to find. We might be able to go farther if we had to, but let us leave this here and turn to eq. (3.26b).

Equation (3.26b) is the first rate equation we have seen that contains concentrations in the denominator of its right-hand side. This may be puzzling at first, but is not difficult to understand. The concentration of hydrogen ion appears in the numerator on the right-hand side of eq. (3.26a) because hydrogen ion is a reactant in a fast prior equilibrium. In eq. (3.26b) it appears in the denominator because hydrogen ion is a product in a fast prior equilibrium. Suppose that the reverse reaction has the mechanism

$$\begin{array}{lll} I_3^- + H_2O(l) = HOI(aq) + H^+ + 2I^- & \text{fast prior equilibrium,} \\ H_3AsO_3(aq) + HOI(aq) \xrightarrow{\ k_-\ } ? & \text{rate-determining step} \end{array} \Bigg\} \quad (3.27b)$$

Then the rate equation will be

$$dc_{H_3AsO_3}/dt = -k_- c_{H_3AsO_3} c_{HOI} \qquad (3.28a)$$

and the equilibrium constant for the reaction in which HOI is produced will be given by

$$K = c_{HOI} c_{H^+} c_{I^-}^2 / c_{I_3^-}. \qquad (3.28b)$$

Solving eq. (3.28b) for c_{HOI} and combining the result with eq. (3.28a) gives

$$dc_{H_3AsO_3}/dt = -k_- K c_{H_3AsO_3} c_{I_3^-} / c_{H^+} c_{I^-}^2 \qquad (3.29)$$

which has exactly the same form as eq. (3.26b).

The *principle of microscopic reversibility* asserts that every reaction follows the same path in both directions. To say the same thing at greater length:

1. The activated complex or transition state is the same for the forward and reverse reactions.
2. Every substance that is an intermediate in the forward reaction is also an intermediate in the reverse reaction.
3. The step that is rate-determining in the forward reaction is also rate-determining in the reverse reaction.
4. Any fast equilibrium that precedes the rate-determining step in the forward reaction must follow the rate-determining step in the reverse reaction.

By employing the principle of microscopic reversibility we can combine eqs. (3.27) into a single mechanism that applies to both the forward and the reverse reaction:

Forward reaction   *Reverse reaction*

fast prior equilibrium	$H_3AsO_4(aq) + H^+ = H_4AsO_4^+$	fast subsequent equilibrium
rate-determining step	$H_4AsO_4^+ + I^- \underset{k_-}{\overset{k}{\rightleftharpoons}} HOI(aq) + H_3AsO_3(aq)$	rate-determining step
fast subsequent equilibrium	$HOI(aq) + H^+ + 2I^- = I_3^- + H_2O$	fast prior equilibrium

$$(3.30)$$

On the basis of eqs. (3.30) it would be difficult to avoid the idea that the activated complex is

$$^+H—O—As(OH)_3$$
$$|$$
$$I^-$$

It is easy to see how this could decompose to yield either HOI and $As(OH)_3$

($= H_3AsO_3$) in the forward reaction or iodide ion and As(OH)$_4^+$ ($= H_4AsO_4^+$) in the reverse one; it would be difficult to imagine another substance that could do so as easily.

We have now examined reversible reactions from two points of view that have seemed to be quite different. Their equilibria were discussed in Chapter 2, and their rates have been discussed here. Are these connected in some way?

They certainly are; let us go back to eqs. (3.26) for an illustration. The rate of the forward reaction, which is described by eq. (3.26a), decreases as the reaction proceeds; the rate of the reverse reaction, which is described by eq. (3.26b), increases as the reaction proceeds. At equilibrium the two rates are equal:

$$kc_{H_3AsO_4}c_{H^+}c_{I^-} = k_-c_{H_3AsO_3}c_{I_3^-}/c_{H^+}c_{I^-}^2.$$

Collecting the concentrations onto one side and the rate constants onto the other, we obtain

$$\frac{k}{k_-} = \frac{c_{H_3AsO_3}c_{I_3^-}}{c_{H_3AsO_4}c_{H^+}^2 c_{I^-}^3}$$

of which the right-hand side is identical with the equilibrium constant of the reaction. Hence

$$K = k/k_-. \tag{3.31}$$

Equation (3.31) is valid for each elementary step in a reaction if k and k_- are the forward and reverse rate constants for the elementary step. It is also valid for any overall reaction if k and k_- are the overall forward and reverse rate constants and if there is just one term on the right-hand side of each rate equation, as is true for all the reactions discussed in this chapter. Some rate equations contain two or more terms, and then the equilibrium constant is not equal simply to the ratio of two rate constants, but the expression for it can always be obtained by equating the rates of the forward and reverse reactions.

You should conclude that equilibrium and kinetics are very closely related although they deal with quite different ideas and observations. Equilibrium is a dynamic state rather than a static one, and therefore it can be described by equations derived from kinetics. At the same time, the law of conservation of energy means that the energy changes involved in the successive steps of a reaction are inseparably connected with the tendency for the overall reaction to occur. In the next section we shall examine the most crucial and interesting of these energy changes.

SUMMARY: When a reaction does not proceed to completion, the rate laws for the forward and backward processes can be established separately. They are related by the principle of microscopic reversibility, which enables us to use the rate law for the backward reaction to obtain information about the decomposition of the activated complex, and the steps that follow it, in the forward reaction. The expression for the equilibrium constant can be obtained by equating the rates of the forward and backward reactions.

3.11. Diffusion-controlled reactions and activation energies

PREVIEW: A few reactions are so rapid that every collision between two particles of the reactants leads to reaction. Others are much slower because the unstable activated complex cannot be formed unless the colliding particles have exceptionally high kinetic energies.

Some reactions are almost, if not quite, instantaneous; others are so slow that they could not be observed to occur at all within a human lifetime. The reaction between hydrogen and hydroxide ions is at the first of these extremes. Its rate law is $dc_{H^+}/dt = -kc_{H^+}c_{OH^-}$, and the value of k is on the order of 10^{11} dm^3 mol^{-1} s^{-1}, which is so large that only 7×10^{-12} s is required to consume half of a small amount of hydrogen ion added to a solution containing 1 mol dm^{-3} of hydroxide ion.

This reaction is said to be *diffusion-controlled*, meaning that its rate is governed by the rate at which the ions collide with each other as they diffuse through the solution, and that nearly every collision leads to a reaction between the ions involved. There are a number of factors that influence the rate at which collisions occur. Of course they include the concentrations of the reactants, but as these appear explicitly in the rate equation it is more useful to focus attention on others that affect the value of the rate constant.

Increasing the temperature increases the kinetic energies of the reacting particles, and therefore causes them to move through the solution more rapidly and collide more frequently. Increasing the viscosity of the solution has the opposite effect: it slows their motions and makes collisions less frequent. Large particles encounter more viscous resistance than small ones do, and therefore move more slowly, but large particles collide even if their centers are so far apart that two small particles following the same paths could pass each other without colliding. The signs and magnitudes of the charges on the reacting particles are very important. If either of the particles is uncharged, there will be a collision whenever their separate random paths through the solution bring them so close together that their bonding orbitals can overlap. But if both are ionic and have charges of the same sign, collisions will be less frequent because their mutual electrostatic repulsion will often deflect two approaching ions and avert a collision that would have occurred in its absence, while if their charges have opposite signs collisions will be more frequent because of electrostatic attraction.

As a rough general guide, if at least one of the reacting particles is an uncharged molecule the frequency with which two particles collide at 298 K may be taken as 10^{11} s^{-1} in a solution containing 1 mol dm^{-3} of each kind of particle. The frequency is proportional to the concentration of each reactant, and this figure represents the upper limit of the second-order rate constant for a bimolecular elementary step in which every collision leads to reaction. For a reaction between two univalent (singly charged) ions the figure is about three times as large if they have opposite signs, or about a third as large if they have the same sign. The variation would be far larger if each reactant were a polyvalent ion.

A reaction may be slow even though its rate-determining step is diffusion-controlled. As an illustration we may consider the reaction between arsenic(V) acid and iodide ion, for which we imagined that the mechanism might be (ignoring the backward reaction for the sake of simplicity)

$$\left.\begin{array}{ll} H_3AsO_4 + H^+ = H_4AsO_4^+ & K = c_{H_4AsO_4^+}/c_{H_3AsO_4}c_{H^+} \\ H_4AsO_4^+ + I^- \xrightarrow{k_1} HOI + H_3AsO_3 & \text{rate} = k_1 c_{H_4AsO_4^+} c_{I^-} \\ HOI + H^+ + 2I^- = I_3^- + H_2O & \end{array}\right\} \quad (3.31)$$

If the equilibrium constant K for the formation of $H_4AsO_4^+$ is very small, there may be very little $H_4AsO_4^+$ in the solution, and then collisions of $H_4AsO_4^+$ and iodide ions cannot be very frequent. Exactly the same thing can be said in another way: since the overall rate constant k for the forward reaction is equal to the product Kk_1, if K is small k may also be small even though k_1 is large. Even if a reaction occurs in a single step, its rate depends (unless the reaction is zeroth order) on the concentrations of some or all of the reactants as well as on the value of its rate constant. What is of most practical interest is the time required to attain some particular degree of completion, such as 99%. Though this time decreases with decreasing concentration of reactant for a zeroth-order reaction and is independent of the concentration of reactant for a first-order reaction, it increases as the concentration of reactant decreases if the reaction is second or higher order. Of course this is implicit in the rate law, but you must not confuse the rate of a reaction with its rate constant and allow yourself to think that a reaction must be rapid because its rate constant is large.

Diffusion-controlled reactions are in the minority. For most reactions that have small overall rate constants the rate-determining steps are not diffusion-controlled. We can regard a rate-determining step as consisting of two half-steps, of which the first is always uphill and the second downhill. In the rate-determining step of the mechanism just given, these half-steps could be shown by writing

$$H_4AsO_4^+ + I^- = \left[\begin{array}{c} ^+H\text{—}O\text{—}As(OH)_3 \\ | \\ I^- \end{array}\right] = HOI + H_3AsO_3$$

The activated complex is the crest of the hill. Because it can decompose spontaneously in either of the two ways shown, it must have a higher energy than either $H_4AsO_4^+$ and iodide ion or HOI and arsenic(III) acid. In the forward reaction the formation of the activated complex from $H_4AsO_4^+$ and iodide ion is the uphill part of the rate-determining step. If an $H_4AsO_4^+$ ion collides with an iodide ion, and if their combined kinetic energies do not suffice to convert them into the activated complex, they will merely bounce without reacting. Only those collisions that involve highly energetic ions or molecules will result in the formation of the activated complex and make conversion into HOI and arsenic(III) acid possible.

Such collisions are infrequent because most ions and molecules are not highly energetic. Figure 3.9 shows how the fraction of the total number of $H_4AsO_4^+$ or iodide ions having any particular energy E depends on E under typical conditions. Most of the ions have relatively low energies; only a few have high energies, and their number drops off rapidly as E increases.

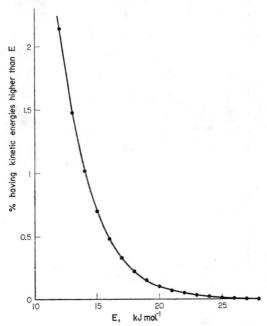

Fig. 3.9. Percentages of particles having kinetic energies exceeding E at 298 K.

In forming the activated complex the kinetic energies of the colliding molecules are transformed into chemical energy. The amount of energy that is required can be represented by the diagram shown in Fig. 3.10. The abscissa scale, labelled "reaction coordinate", would be simply the distance between the centers of the atoms if each reactant were monatomic, but since these reactants are not it is a complex function of all the interatomic distances involved. In the forward reaction we proceed from left to right in such a diagram. Initially the reacting ions have a certain amount of chemical energy between them. As they approach each other they also have a certain amount of kinetic energy between them. Kinetic energy is transformed into chemical energy when they collide, and if enough kinetic energy is available the resulting increase in the chemical energy of the system will suffice to form the activated complex, which corresponds to the peak of the curve. Then the activated complex may decompose to yield the products, liberating energy as it does so. The energy that is liberated appears as the kinetic energies of the molecules of HOI and arsenic(III) acid produced. What is important is the magnitude of the chemical energy that must be supplied in order to obtain the activated complex. This is called the *energy of activation* or *activation energy* and is represented by the symbol E_a.

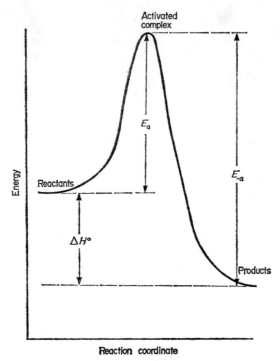

FIG. 3.10. Reaction-coordinate diagram showing the relations among E_a, E_{-a}, and ΔH^0.

Figure 3.10 also applies to the reverse reaction, but for this purpose it must be read from right to left. Two approaching molecules of HOI and arsenic(III) acid have the combined chemical energy shown at the right-hand side of the curve. If their combined kinetic energies equal or exceed the energy of activation for the reverse reaction, E_{-a}, the activated complex may form and then decompose to yield $H_4AsO_4^+$ and iodide ion. In this decomposition the excess of chemical energy is equal to E_a, and it appears in the form of the kinetic energies of the ions produced.

The value of the rate constant depends on the energy of activation. An increase of the energy of activation decreases the fraction of the total number of collisions in which the necessary amount of kinetic energy is available, and the resulting decrease of the frequency with which reaction occurs is expressed by a decrease in the value of the rate constant. This value also depends on the frequency with which collisions occur and on the geometry of the system: if an iodide ion collides with one of the oxygen atoms in $H_4AsO_4^+$ the activated complex may form, but if it strikes some other part of the ion it may simply recoil without reacting. The simplest description of all these effects is given by the Arrhenius equation

$$k = pZe^{-E_a/RT} \tag{3.32}$$

The *frequency factor* Z is equal to the total number of collisions that will occur during 1 s in a solution having a volume of 1 dm^3 and containing 1 mole of each reactant.

As was said above, the value of Z under typical conditions is roughly 10^{11} dm^3 mol^{-1} s^{-1} except for reactions between polyvalent ions. The factor $e^{-E_a/RT}$ gives the fraction of this total number in which the energy of activation E_a is available; E_a is expressed in J mol^{-1}, R is the gas constant (8.3143 J K^{-1} mol^{-1}), and T is the temperature (K). The factor p, called the *steric factor*, is the fraction of the collisions in which the reactants have orientations in space that make it possible for them to combine to form the activated complex. The value of p must lie between 0 and 1 and depends on the structures of the reacting particles. If these are very simple, as they are for hydronium and hydroxide ions, p is usually indistinguishable from 1, but for a reaction between two very complicated molecules p is much smaller, and values as low as 10^{-3} are not uncommon.

For a bimolecular elementary step for which p was equal to 1 and E_a was zero, the rate constant would be equal to the frequency factor Z. The following lines show how the value of the rate constant would decrease as that of E_a increased, assuming $p = 1$ and $Z = 1 \times 10^{11}$ dm^3 mol^{-1} s^{-1}:

E_a, kJ mol^{-1}	0	10	20	40	100	200	300
k, dm^3 mol^{-1} s^{-1}	1×10^{11}	2×10^9	3×10^7	1×10^4	3×10^{-7}	9×10^{-25}	3×10^{-42}

There is no exaggeration here: energies of activation even larger than 300 kJ mol^{-1} are not uncommon. Clearly a rate-determining step can be very slow indeed if the energy of activation is large. What is more, the rate constant for the rate-determining step often has to be multiplied by the equilibrium constant for a prior equilibrium to obtain the overall rate constant for a reaction, and if the prior equilibrium is unfavorable to the formation of the substance that takes part in the rate-determining step, the equilibrium constant may also have an extremely small value. It is no wonder that some reactions are too slow to be observed!

SUMMARY: The rate of a reaction depends on the frequency with which reacting particles collide, on the fraction of the collisions in which the spatial orientations of the particles permit reaction, and on the fraction in which the particles have kinetic energies high enough to provide the energy of activation needed to form the activated complex.

3.12. The effects of temperature on rate and equilibrium constants

PREVIEW: Equations are written and compared for the effects of temperature on rate and equilibrium constants.

The effect of temperature on a rate constant is most easily described by rewriting eq. (3.32) in logarithmic form:

$$\ln k = \ln pZ - E_a/RT. \tag{3.33}$$

For most reactions the steric factor p, the frequency factor Z, and the energy of activation E_a are all nearly independent of temperature, and therefore a plot of $\ln k$ against

$1/T$ is usually linear. Since its slope is equal to $-E_a/R$, it is a simple matter to evaluate E_a from such a plot. Once its value is known, the value of k at any desired temperature may be obtained from eq. (3.33), most easily by writing it for two different temperatures

$$\ln k_{T_2} = \ln pZ - E_a/RT_2$$
$$\ln k_{T_1} = \ln pZ - E_a/RT_1$$

and subtracting one of these from the other to secure

$$\ln k_{T_2} = \ln k_{T_1} - \frac{E_a}{R}\left(\frac{1}{T_2} - \frac{1}{T_1}\right). \tag{3.34}$$

Because different reactions have different energies of activation, their rates are affected to different extents by changes of the temperature. Some values calculated from eq. (3.34) for $T_1 = 298$ K $(= 25°$ C$)$ and $T_2 = 308$ K $(= 35°$ C$)$ are given in the following lines:

E_a, kJ mol^{-1}	10	20	40	100	200	300
$k_{35°}/k_{25°}$	1.14	1.30	1.69	3.7	13.7	51

Heating a reaction mixture from 298 K to 308 K increases the value of the rate constant only 14 per cent if the energy of activation is 10 kJ mol^{-1}, but almost quadruples it if the energy of activation is 100 kJ mol^{-1}, and increases it by a factor of 50 if the energy of activation is 300 kJ mol^{-1}. When the energy of activation is large the rate constant is small and increases rapidly as the temperature increases; when the energy of activation is small the rate constant is large and only slightly dependent on the temperature. Small rate constants therefore increase more rapidly with increasing temperature than large ones do.

We turn now to the effect of temperature on the equilibrium constant, which we shall discuss in two different ways so that the two results can be compared.

The first of these uses equations taken from Chapter 2. Rearranging eq. (2.12) gives

$$\ln K = -\Delta G^0/RT. \tag{3.35a}$$

Combining this with eq. (2.5) yields

$$\ln K = -(\Delta H^0 - T\,\Delta S^0)/RT = -\Delta H^0/RT + \Delta S^0/R. \tag{3.35b}$$

Since both ΔH^0 and ΔS^0 are nearly independent of temperature, the procedure that was used above to obtain eq. (3.34) can be used again. The result is

$$\ln K_{T_2} = \ln K_{T_1} - \frac{\Delta H^0}{R}\left(\frac{1}{T_2} - \frac{1}{T_1}\right). \tag{3.36}$$

Although this has exactly the same form as eq. (3.34), there is an important difference between the ways in which these two equations behave. In eq. (3.34) the value of E_a must be positive: in the rate-determining step of a reaction, kinetic energy must always be consumed in order to obtain the activated complex. Consequently, the value of a rate constant must increase as the temperature increases.[†] The same argument can be applied to eq. (3.36) if ΔH^0 is positive, and the equilibrium constants of endothermic reactions do become larger as the temperature rises. However, it is equally possible for ΔH^0 to have a negative value, and if it does the value of the equilibrium constant will decrease on increasing the temperature.

Let us set this aside for a moment and turn to eq. (3.34), which might represent the behavior of the rate constant for some forward reaction. An exactly similar equation can be written for the corresponding reverse reaction:

$$\ln k_{-,T_2} = \ln k_{-,T_1} - \frac{E_{-a}}{R}\left(\frac{1}{T_2} - \frac{1}{T_1}\right).$$

Subtracting this from eq. (3.34) gives

$$\ln\left(\frac{k}{k_-}\right)_{T_2} = \ln\left(\frac{k}{k_-}\right)_{T_1} - \frac{E_a - E_{-a}}{R}\left(\frac{1}{T_2} - \frac{1}{T_1}\right).$$

According to eq. (3.31), each of these ratios of rate constants is equal to the equilibrium constant at the same temperature, or

$$\ln K_{T_2} = \ln K_{T_1} - \frac{E_a - E_{-a}}{R}\left(\frac{1}{T_2} - \frac{1}{T_1}\right). \tag{3.37}$$

Comparing eqs. (3.36) and (3.37), we see that

$$\Delta H^0 = E_a - E_{-a}. \tag{3.38}$$

Figure 3.10 shows what this means. An amount of kinetic energy that is equal to E_a is consumed in the formation of the activated complex, whose decomposition into the products then liberates an amount of kinetic energy that is equal to E_a. Since kinetic energy is the manifestation of heat on the molecular scale, it should be no surprise that the difference between the two activation energies corresponds to an amount of heat.

In Fig. 3.10 the forward reaction is exothermic because E_{-a} exceeds E_a, but the diagram can also be read from right to left. When this is done, the energy of activation for the "forward" process, in which HOI and arsenic(III) acid combine to form the activated complex, is larger than the amount of energy liberated in the decomposition of the activated complex to give $H_4AsO_4^+$ and iodide ion. The net result is that energy is consumed, ΔH^0 is positive, and the reaction in this direction is endothermic.

SUMMARY: The effect of temperature on a rate constant depends on the value of the energy of activation; the effect of temperature on an equilibrium constant depends on the value of ΔH^0 for the reaction.

[†] The overall rate constant for the urease-catalyzed hydrolysis of urea has a maximum value at a temperature not much above 37° C; at higher temperatures, it decreases as the temperature increases. Can you imagine an explanation for such behavior?

3.13. Conclusion

Chapters 2 and 3 have dealt with some very fundamental matters. Most of the rest of this book will deal with some of the consequences of the ideas and relations that have been described. In doing so it will consider many different kinds of chemical situations. These will represent, but by no means exhaust, the variety of areas in which those ideas and relations help the chemist to correlate, organize, and understand the behaviors that chemical systems display.

Problems

Answers to some of these problems are given on page 520.

3.1. For each of the following reactions

Overall reaction	Rate law
(a) $In^+ + 2H^+ = In^{3+} + H_2(g)$	$dc_{In^+}/dt = -kc_{In^+}c_{H^+}$
(b) $2 Mo(V) + NH_3OH^+ = 2 Mo(VI) + NH_4^+$	$dc_{NH_4^+}/dt = kc_{Mo(V)}c_{NH_3OH^+}c_H^2 +$
(c) $I_2(aq) + 10 Ce(IV) + 6 H_2O(l) =$ $2 IO_3^- + 10 Ce(III) + 12 H^+$	$dc_{I_2}/dt = -kc_{I_2}^2 c_{Ce(IV)}$

(d) $CH_3\overset{\overset{\displaystyle O}{\|}}{C}CH_3(aq) + I_2(aq) = ICH_2\overset{\overset{\displaystyle O}{\|}}{C}CH_3(s) + H^+ + I^-$ $dc_{I_2}/dt = -kc_{(CH_3)_2CO}c_{H^+}$

what are
   (i) the order of the reaction?
   (ii) its order with respect to hydrogen ion?

3.2. The decomposition of dinitrogen pentoxide, N_2O_5, is a first-order reaction; an equation for the overall reaction and the rate law are given on page 39. In one experiment a solution of dinitrogen pentoxide in carbon tetrachloride was allowed to decompose, and the concentration of dinitrogen pentoxide was found to decrease to half of its original value in 900 s. What was the value of the rate constant under the conditions of this experiment?

3.3. The following data were obtained by following the reaction

$$(CH_3)_3CBr(aq) + H_2O(l) = (CH_3)_3COH(aq) + H^+ + Br^- \text{ at 298 K:}$$

t, s	5 000	20,000	40,000	60,000	100,000	150,000
$c_{(CH_3)_3CBr}$, mol dm^{-3}	0.0380	0.0308	0.0233	0.0176	0.0100	0.005 02

(a) Find the order n of the reaction with respect to $(CH_3)_3Br$.
(b) What is the value of the pseudo-nth-order rate constant at 298 K?
(c) What was the concentration of $(CH_3)_3$ CBr at the beginning of the reaction?

3.4. Alkaline aqueous solutions of hypochlorite ion, OCl^-, decompose in accordance with the equation $3 OCl^- = ClO_3^- + 2 Cl^-$. The following data were secured with a solution that originally contained 0.0127 mol dm^{-3} of hypochlorite ion and 0.260 mol dm^{-3} of hydroxide ion:

t, s	1 000	3 000	10,000	20,000	40,000	100,000
c_{OCl^-}, mol dm^{-3}	0.0122	0.0113	0.0089	0.0069	0.0047	0.0024

The following data were obtained with a second solution that originally contained 0.0271 mol dm^{-3} of hypochlorite ion and 0.495 mol dm^{-3} of hydroxide ion:

t, s	2 000	10,000	20,000	30,000	50,000	100,000
c_{OCl^-}, mol dm^{-3}	0.0230	0.0143	0.0097	0.0074	0.0050	0.0027

(a) What is the order of the reaction with respect to hypochlorite ion?
(b) What is its order with respect to hydroxide ion?
(c) Suggest a possible structure for the activated complex in this reaction.

3.5. The hydration of isobutene in an acidic aqueous solution follows the overall equation

$$\underset{CH_3}{\overset{CH_3}{\diagdown}}C=CH_2(aq) + H_2O(l) = \underset{CH_3}{\overset{CH_3}{\diagdown}}C\underset{OH}{\overset{CH_3}{\diagup}}(aq)$$

and its rate is given by $dc_{(CH_3)_2C=CH_2}/dt = -kc_{(CH_3)_2C=CH_2}$. At 298 K the value of k is equal to $2.2\times10^{-4}\,\text{s}^{-1}$.

(a) What is the half-time for this reaction at 298 K?
(b) How long a time would be required for the concentration of isobutene to decrease to 0.1% of its initial value at 298 K?
(c) If the concentration of isobutene in a solution is 1.00×10^{-4} mol dm^{-3} after the reaction has proceeded for 3600 s at 298 K, what was it when the reaction began?

3.6. The reaction $Co(OH_2)_4Cl_2^+ + OH^- = Co(OH_2)_4ClOH^+ + Cl^-$ obeys the rate law $dc_{Co(OH_2)_4Cl_2^+}$ $/dt = -kc_{Co(OH_2)_4Cl_2^+}$, and the value of k is equal to $1.0\times10^3\,\text{s}^{-1}$ at 297 K.

(a) What is the half-time for the reaction at 297 K?
(b) Suggest a possible mechanism for the reaction.

3.7. What would be the rate law for the reaction $V^{2+} + VO^{2+} + 2\,H^+ = 2\,V^{3+} + H_2O(l)$ if eqs. (3.18a) correctly described the mechanism?

3.8. In alkaline solutions at 298 K the rate law for the reaction $OCl^- + I^- = Cl^- + OI^-$ is $dc_I-/dt = -6\times10^{15}c_{H^+}c_{OCl^-}c_{I^-}$. The dissociation constant of hypochlorous acid, HOCl, is given by $K_a = c_H+c_{OCl^-}/c_{HOCl(aq)} = 3.4\times10^{-8}$ mol dm^{-3}. What would be the rate law in solutions so acidic that the extent of dissociation of hypochlorous acid was negligible?

3.9. If $c_{H_3AsO_4}$ is the concentration of $H_3AsO_4(aq)$ at equilibrium in the reaction $H_3AsO_4(aq) + 2\,H^+ + 3\,I^- = H_3AsO_3(aq) + I_3^- + H_2O(l)$, show by using eqs. (3.25) that

$$c_{H_3AsO_4} = \frac{k'_-c^0_{H_3AsO_4}}{k'+k'_-}.$$

3.10. The overall equation for the oxidation of chromium (III) by cerium (IV) is $Cr(III) + 3\,Ce(IV) = Cr(VI) + 3\,Ce(III)$. The rate law is $dc_{Cr(VI)} = kc_{Cr(III)}c^2_{Ce(IV)}/c_{Ce(III)}$. Suggest a possible mechanism for the reaction.

3.11. The rate constant for a certain reaction is equal to 8.93×10^{-7} dm^3 mol^{-1} s^{-1} at 650 K and to 8.12×10^{-5} dm^3 mol^{-1} s^{-1} at 750 K. What is the energy of activation for the reaction?

3.12. The rate constant for the hydration of isobutene (Problem 3.5) is doubled by increasing the temperature from 20° C to 30° C. How is it affected by increasing the temperature from 85° C to 95° C?

3.13. At 298 K the value of ΔH^0 for the reaction $2\,H_2O(l) = H_3O^+ + OH^-$ is 56.4 kJ mol^{-1} and the value of K_W is 1.01×10^{-14} mol^2 dm^{-6}. If ΔH^0 were independent of temperature, what would be the value of K_W at 373 K $(= 100° C)$? The literature value is 5×10^{-13} mol^2 dm^{-6}; what do you conclude?

CHAPTER 4

Equilibria Involving Non-electrolytes

4.1. Introduction

In this chapter and the four that follow it we shall examine the behaviors of six very common kinds of chemical processes from the points of view described in Chapters 2 and 3. Much of our attention will be devoted to the kinds of information about them that the expressions for, and the numerical values of, their equilibrium constants can be made to provide. The order in which we shall consider them was foreshadowed in Section 2.9. Because, as was shown there, expressions having very simple forms suffice to describe the solubility of a non-electrolyte and its distribution between two immiscible solvents, it is convenient to deal with these before going on to the more complicated kinds of equilibrium constants that are needed to describe other processes. Moreover, because solubility and distribution equilibria will turn out to be very closely related, it is convenient to consider them together rather than separately.

4.2. The solubility of a non-electrolyte

PREVIEW: This section discusses the factors that influence the solubility of a non-ionic solute.

According to Section 2.9 the dissolution of a solid or liquid non-electrolyte in water or any other solvent can be described by equations like

$$C_6H_6(l) = C_6H_6(aq); \quad K = [C_6H_6(aq)]. \tag{4.1}$$

An excess of the solid or liquid solute is needed to ensure that equilibrium has been reached, and therefore the activity of the undissolved material is taken to be equal to 1 and is omitted from the expression for the equilibrium constant. Equation (4.1) simply says that the concentration of such a solute is the same in every saturated solution prepared under the same conditions. The equilibrium constant is usually called the *solubility* and given the symbol S. Handbooks and tables generally give solubilities in units such as g dm^{-3} or grams per 100 cm^3 for convenience in practical work, but the fundamental units of solubility are mol dm^{-3}. Of course, the solubility varies with the temperature, in accordance with eq. (3.36). It also varies with the pressure, but so slightly (except for gaseous solutes) that we shall ignore this variation.

Different substances have solubilities that differ widely and depend on the temperature in different ways. The solubility of any particular solute in any particular solvent depends on the changes of both enthalpy and entropy that accompany its dissolution. Let us suppose that some benzene and some water are put into a container together, and consider what happens if one molecule of benzene leaves the benzene layer and dissolves in the water. The overall process can be divided into three steps. In the first, the molecule of benzene is detached from the pure liquid by overcoming the forces exerted on it by the molecules surrounding it. Some energy must be expended to do this. In the second, the forces between molecules of water in the aqueous layer are counteracted to an extent that enables them to be separated far enough to make room for the molecule of benzene. This step also requires the expenditure of energy, and it requires more energy than the first step because the forces between molecules of water are stronger than those between molecules of benzene. In the third step, the molecule of benzene is put into the hole that has been made to receive it. The energy that is involved in this step depends on the forces to which the molecule of benzene will be subjected in its new environment. If these are sufficiently strong, the amount of energy that is liberated in the third step may exceed the amount of energy that is consumed in the first two steps, and if it does the overall process will be exothermic. On the other hand, if these forces are very weak very little energy will be liberated in the third step and the overall process will then be endothermic because energy was consumed in the first two steps.

Molecules of water are capable of reacting with many kinds of dissolved molecules to form hydrated species. Similar reactions occur in other solvents as well, although the process must be called *solvation* rather than hydration if the solvent is non-aqueous. Most strongly solvated species are held together by hydrogen bonds. When acetic acid is dissolved in water, the hydrogen atom contained in its carboxyl (—COOH) group can be shared with a molecule of water to yield a species like

$$CH_3-C \overset{\textstyle O}{\underset{\textstyle OH---O-H}{}}$$
$$\underset{H}{|}$$

The molecule of water becomes polarized to some extent. Even before it attached itself to the molecule of acetic acid it had a dipole moment, with a positive charge localized on the oxygen atom and a positive charge centered between the two hydrogen atoms:

$$\delta^- O \overset{\textstyle H}{\underset{\textstyle H}{}} \delta^+$$

The positively charged end of this dipole, denoted by the symbol δ^+, is strengthened by the partial withdrawal of electrons into the hydrogen bond, and this facilitates the attachment of another molecule of water. In addition there is some possibility, though a smaller one, of hydrogen bonding to the oxygen atom of the carbonyl $\left(\!\!\begin{array}{c}\diagdown\\ \diagup\end{array}\!\!C\!=\!O\right)$ group, and the eventual result might look like

$$
\begin{array}{c}
\hspace{3.5cm} \text{H} \\
\hspace{3.5cm} | \\
\hspace{2.5cm} \text{O}\cdots\text{H}\!-\!\text{O} \\
\hspace{1.2cm}\diagup \\
\text{CH}_3\!-\!\text{C} \\
\hspace{1.2cm}\diagdown \\
\hspace{2.2cm} \text{OH}\cdots\text{O}\!-\!\text{H}\cdots\text{O}\!-\!\text{H} \\
\hspace{2.6cm} | \hspace{1.3cm} | \\
\hspace{2.6cm} \text{H} \hspace{1.3cm} \text{H}
\end{array}
$$

Water certainly provides a receptive environment for molecules of acetic acid. A less receptive one is provided by diethyl ether, in which solvation would produce something like

$$
\begin{array}{c}
\hspace{2.5cm} \text{O} \\
\hspace{1.2cm}\diagup \\
\text{CH}_3\!-\!\text{C} \\
\hspace{1.2cm}\diagdown \\
\hspace{2.2cm} \text{OH}\cdots\text{O}\!-\!\text{C}_2\text{H}_5 \\
\hspace{2.6cm} | \\
\hspace{2.6cm} \text{C}_2\text{H}_5
\end{array}
$$

but in which hydrogen bonding between the carbonyl group and a molecule of diethyl ether is nearly impossible.

Solvation is accompanied by the liberation of energy, and the liberated energy appears in the form of heat. There are very few non-ionic solutes that liberate spectacular amounts of heat when they dissolve, but many ionic ones do because their ions are very strongly solvated. For example, if you dissolve a mole of sodium hydroxide in a cubic decimeter of water, the mixture may boil locally, endangering the eyes of anyone nearby, unless it is very well stirred. Molecules of benzene can have no such strong interactions with molecules of water: the pi-electron cloud in a molecule of benzene is too diffuse to be shared with an atom of hydrogen, and it prevents the hydrogen atoms in the molecule from sharing electrons donated by the oxygen atom of water molecules. So little energy can be liberated in the third step of the dissolution that the overall process must be endothermic.

Neither pure water nor pure benzene has an orderly structure, although hydrogen bonding between its molecules causes water to have a more orderly one than benzene or any other liquid. Nevertheless, a solution of benzene in water is even more disorderly. In water there is some chance of predicting correctly whether there will be a molecule of water at any given distance and in any given direction from another one, although this chance decreases rapidly as the distance increases and becomes negligibly small beyond a few molecular diameters. If there is some benzene dissolved in the water the chance decreases, for the position in question may be occupied by a molecule of benzene rather than by one of water. Extensive hydration would tend to make the solution more orderly, for then the knowledge that there was a molecule of solute at a certain point would give some assurance that one or more molecules of water would be found in certain places nearby. However, this is unlikely to be the predominating effect with a non-ionic solute, for some of the molecules of water that are attached to molecules of the solute in the solution would be attached to other molecules of water if the solute were not there. The occurrence of solvation can affect the value of the change of entropy that accompanies the dissolution of a non-ionic solid or a liquid, but that change of entropy is almost certain to be positive. The solvation of a metal ion may be much more extensive, and the ordering of solvent molecules around the cation of an anhydrous salt can actually lead to a decrease of entropy on dissolution.

These are the fundamental considerations that govern the values of equilibrium constants such as the one in eq. (4.1). For the dissolution of benzene in water, ΔH^0 is positive and so is $T\Delta S^0$, but ΔH^0 is considerably the larger of the two. Hence the difference between them, which is equal to ΔG^0, is positive, and in accordance with eq. (2.12) ln K must be negative and K must be smaller than 1. Benzene is not very soluble in water at 25°. If the temperature is increased, the value of $T\Delta S^0$ increases, and ΔG^0 becomes less positive as a result. The solubility of benzene in water increases as the temperature rises. We might have used eq. (3.36) to obtain the same conclusion.

By reasoning in much the same way we can compare the solubilities of different solutes. A molecule of toluene, $C_6H_5CH_3$, in which there is a methyl group in the place of one of the hydrogen atoms in a molecule of benzene, is larger than a molecule of benzene. More energy is needed to detach a molecule from pure liquid toluene than to detach one from pure liquid benzene, and more energy is also needed to make room for the larger molecule of toluene in the aqueous phase. Since replacing the atom of hydrogen with a methyl group does not increase the possibility of hydration, ΔH^0 is even more positive for the dissolution of toluene in water than for that of benzene. This causes ΔG^0 to be more positive for toluene, and therefore toluene is less soluble in water than benzene is.

A molecule of aniline, $C_6H_5NH_2$, has about the same size as a molecule of toluene, but its amino ($-NH_2$) group can participate in hydrogen bonding. There is some hydrogen bonding in aniline itself, which makes it more difficult to free a molecule of aniline from the pure liquid than it is to free one of toluene, but it is much weaker than the hydrogen bonding that occurs in an aqueous solution of aniline because, although the amino group can accept protons readily, it is only a very weak proton donor. In an aqueous solution hydration leads to the formation of something like

$$C_6H_5N\underset{\displaystyle H}{\overset{\displaystyle H}{\diagup}}\cdots H-\underset{\displaystyle H}{O}$$

This looks very much like an activated complex, and indeed it can decompose in either of two ways. The nitrogen–hydrogen bond is more likely to break than the oxygen–hydrogen bond because it is the weaker of the two, but sometimes the oxygen–hydrogen bond breaks instead and then the reaction $C_6H_5NH_2(aq)+H_2O(l) = C_6H_5NH_3^+ +OH^-$ has occurred. More will be said about reactions like this in Chapter 6: here we need only note that energy is liberated in the formation of hydrogen-bonded species. Enough energy is liberated in the formation of this one to make ΔH^0 almost equal to $T\Delta S^0$ at 25°, so that ΔG^0 is nearly zero and the solubility of aniline in water at this temperature is very close to 1 mol dm^{-3}. However, the amount of energy that is liberated by hydration is not quite large enough to make ΔH^0 negative, and therefore the dissolution of aniline in water is endothermic and the solubility increases as the temperature rises.

We shall not be much concerned with the dissolution of gases, and it can be dismissed rather briefly. The dissolution of a gas is always exothermic, so that ΔH^0 is always negative. In effect, the gas is condensed as its molecules are transferred from the gaseous phase to the liquid solvent, and the heat of vaporization is evolved. Additional heat is evolved if there is a strong interaction between the molecules of the gas and those of the liquid in which it is dissolved. Because ΔH^0 is negative, the solubility of a gas decreases on increasing the temperature. On the other hand, ΔS^0 is also negative because there is less randomness in a liquid phase than in a gaseous one. The negative values of ΔH^0 and $T\Delta S^0$ counterbalance each other, and ΔG^0 may be either negative or positive. It is negative if the chemical interactions are so strong, and the heat of vaporization so large, that ΔH^0 is very negative, and a gas for which this is true has a high solubility (and one that changes rapidly as the temperature varies); the dissolution of ammonia in water is an example. If the chemical interactions are so weak, and the heat of vaporization so small, that ΔH^0 is less negative than $T\Delta S^0$, then ΔG^0 is positive, the gas is only slightly soluble, and its solubility decreases slowly as the temperature rises; the dissolution of hydrogen in water is an example.

Chemical bonding between molecules of the solvent and solute is only one of several ways in which dissolved molecules can be stabilized. It is much less important in most other solvents than it is in water, and even in water there are other processes that can occur, and that affect the values of ΔH^0 and ΔS^0 when they do. Electrostatic interaction is one such process. A polar molecule dissolved in water can be stabilized by the formation of a structure something like

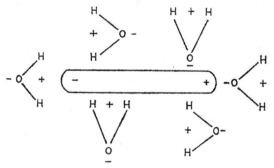

where the signs indicate the positive and negative ends of each dipole. Similar structures can be formed in other polar solvents, but a polar solute cannot be stabilized in this way in a non-polar solvent such as benzene or carbon tetrachloride. In general, substances having polar molecules are therefore more soluble in polar solvents than in non-polar ones. There are other forces that hold molecules together in the liquid state, and if the molecules of a solvent and solute are very similar these forces between them may be scarcely different from those between the molecules of the pure substances. Then the enthalpy change that accompanies dissolution will be very small. Because $T \Delta S$ is nearly certain to be positive, for reasons that were given earlier in this section, ΔG^0 is likely to have a negative value and the solubility may be very large.

There are many different kinds of processes that occur at a boundary between two different phases. These are called *heterogeneous processes*. They include the dissolution of ionic as well as non-ionic substances, precipitation, adsorption, electrolysis, and heterogeneous catalysis (such as the catalysis, at a platinum surface, of a reaction occurring in the gas phase). Regardless of whether the two phases are a solid and a liquid, two immiscible liquids, or a gas and either a solid or a liquid, every such process has two peculiarities from the kinetic viewpoint.

One is that its rate depends on the area of contact between the two phases. Representing this by A, the rate equation for the dissolution of benzene in water might be written

$$da_{C_6H_6(aq)}/dt = ka_{C_6H_6(l)}A - k_-a_{C_6H_6(aq)}A.$$

Provided that the organic phase was pure benzene (rather than, say, a mixture of benzene and toluene), the activity of benzene in that phase would be equal to 1, and consequently it could simply be dropped from the first term. Replacing the activity of benzene in the aqueous phase by its concentration, as is done in writing expressions for equilibrium constants involving dissolved species, yields the more usual form

$$dc_{C_6H_6(aq)}/dt = kA - k_-c_{C_6H_6(aq)}A. \tag{4.2}$$

The first term on the right-hand side gives the rate at which benzene enters the aqueous phase; the second gives the rate at which benzene leaves the aqueous phase and re-enters the benzene phase. Both rates increase in proportion to the area A.[†]

If some water and some benzene were simply placed in a flask and allowed to stand, equilibrium would be only very slowly attained. One reason is that the two liquids would be in contact over only a limited area. Another is that a molecule of benzene can neither enter nor leave the aqueous phase unless it is very close to the interface between the two liquids. The concentration of dissolved benzene that appears in eq. (4.2) is the concentration in a layer of solution that is no more than one or two molecular diameters thick and that is in contact with the liquid benzene. We are accustomed to thinking of the concentration of a solute as being uniform throughout a solution. This is true, except on a submicroscopic scale that need not concern us, at

---

[†] Another minor peculiarity of heterogeneous processes is that their rate constants have units quite different from those given in Section 3.7 for homogeneous processes. In eq. (4.2), k is a zeroth-order heterogeneous rate constant because it appears in a term that does not include a concentration, and since the product kA must have the units mol dm^{-3} s^{-1} the units of k must be mol dm^{-5} s^{-1}, while the first-order heterogeneous rate constant k_- must have the units dm^{-2} s^{-1}.

equilibrium, and it is also true during a homogeneous reaction unless the reactants were incompletely mixed at the start. However, it cannot be true during a heterogeneous process; this is the second of the two major peculiarities of heterogeneous processes. In the aqueous phase the concentration of benzene will be highest at the interface and it will decrease as the distance from the interface increases. The concentration gradient causes molecules of benzene to diffuse away from the interfacial layer, where their concentration is high, into the bulk of the aqueous phase, where their concentration is smaller. Eventually the concentration of benzene will become uniform throughout the aqueous phase, but a very long time will be required because diffusion is an extremely slow process. It is so slow that it would certainly limit the overall rate at which equilibrium would be attained.

Consequently the process is shockingly inefficient under these conditions. According to eq. (4.2), molecules of benzene enter the aqueous phase at a rate that is fixed in the sense that it does not depend on the length of time for which the liquids have been in contact, though of course it does depend on the interfacial area. Since diffusion carries them into the interior of the aqueous phase at a rate that would be grossly misrepresented by calling it a snail's pace,[†] most of these molecules remain near the interface. Their concentration approaches the equilibrium value near the interface, and the rate of the reverse process becomes almost equal to that of the forward one. Many of the molecules of benzene that get into the aqueous layer are lost again because they escape back into the benzene phase.

Clearly there are several different ways of speeding things along. The one that is the best, because it attacks the slowest step in the overall process, is to stir, shake, or agitate the aqueous phase so that the molecules of benzene are distributed throughout its volume as quickly as possible. By removing these molecules from the interfacial layer as soon as they enter it, this prevents them from accumulating there and greatly decreases the fraction of them that escapes back into the benzene. If this is done, the rate at which molecules enter the aqueous phase is the next thing to attack, and it can be increased by increasing the interfacial area A, in a way that depends on the nature of the solute. A solid can be ground to a fine powder. A liquid can be shaken with the solvent in a closed container, thereby breaking each of the two liquids up into droplets dispersed throughout the other liquid and having a combined surface area much larger than the area of a stationary interface. Shaking two liquids together in a separatory funnel achieves both of these ends at once. If the solute is a gas, it can be shaken with the solvent to throw droplets of liquid into the gas phase and form bubbles of gas in the liquid, or a stream of gas can be circulated through the solution in such a way as to obtain very small bubbles.

The least useful thing to do is to heat the mixture in an attempt to increase k, chiefly because k will not be increased very much by increasing the temperature un-

---

[†] Under typical conditions diffusion causes a molecule to move at a rate of approximately 1 cm day $^{-1}$.

less the energy of activation is large (Section 3.12). There is no reason why it should be large unless the dissolution is very strongly endothermic. Since the transfer of solute across the phase boundary is unlikely to be the rate-determining step in the overall process, energy expended on increasing its rate is simply wasted if nothing is done to accelerate the transfer of dissolved material away from the interfacial layer. When heating does increase the overall rate of dissolution it is almost always because convection currents are produced by localized differences of temperature and help to distribute the solute throughout the solvent.

SUMMARY: The solubility of a solute depends on the changes of enthalpy and entropy that accompany dissolution, and is increased by solvation and by electrostatic and other interactions between molecules of the solute and the solvent.

4.3. Distribution equilibria

PREVIEW: This section deals with the equilibrium constant for a process in which a solute is partitioned between two immiscible solvents, and shows how its value can be calculated from the solubilities of the solute in the two solvents.

When a solution of a non-ionic solute in one solvent is brought into contact with a second solvent, which is immiscible with the first one, some of the solute will escape from the original solvent and enter the one that has been added. The process is called *distribution* or *partition; liquid–liquid extraction*, which will be described in Sections 4.4 and 4.5, is a practical application of it. A typical distribution equilibrium is

$$I_2(aq) = I_2(CCl_4); \quad D = [I_2(CCl_4)]/[I_2(aq)]. \tag{4.3}$$

There are several ways in which this equilibrium might be reached experimentally One is to add carbon tetrachloride to an aqueous solution of iodine, another is to add water to a solution of iodine in carbon tetrachloride, and a third is to add solid iodine to a mixture of water and carbon tetrachloride. The equilibrium constant or *distribution coefficient D* is equal to 1×10^2. Because this is fairly large, the addition of carbon tetrachloride to an aqueous solution of iodine has some useful and important practical applications. The concentration of dissolved iodine becomes much larger in the carbon tetrachloride layer than it is in the aqueous layer. If the process is properly designed you can extract almost all of the iodine from a large volume of water into a much smaller volume of carbon tetrachloride. By doing so, you could separate the iodine from other substances that are present in the aqueous solution but that are not extracted into the carbon tetrachloride because they are insoluble in it. You would also obtain a solution in which the concentration of iodine is higher than it was in the original one, and this would make it easier to recover the iodine, to find how much of it is present, or to discover whether some of it is present or not. Section 4.4 will show you how to design the process properly. The addition of water

7*

to a solution of iodine in carbon tetrachloride is much less interesting. So little iodine would be extracted from the carbon tetrachloride layer into the aqueous one that nothing useful would be accomplished.

You would never encounter the third way of setting up the equilibrium anywhere except in a textbook. Nevertheless, it is the one that will be discussed here, because it shows most clearly what the equilibrium constant in eq. (4.3) means. If an excess of solid iodine is shaken with both water and carbon tetrachloride until its concentrations in the two liquid phases no longer change, three separate equilibria will be attained. The solid iodine will be in equilibrium with iodine dissolved in the carbon tetrachloride:

$$I_2(s) = I_2(CCl_4); \quad K_a = S_{CCl_4} = [I_2(CCl_4)] \qquad (4.4a)$$

where S_{CCl_4} denotes the solubility of iodine in carbon tetrachloride. This equilibrium constant is given the symbol K_a to distinguish it from the equilibrium constant that will appear in eq. (4.4b). There will also be an equilibrium between the solid iodine and the iodine that is dissolved in the aqueous phase:

$$I_2(s) = I_2(aq); \quad K_b = S_{aq} = [I_2(aq)]. \qquad (4.4b)$$

Finally, since each of the two phases is in equilibrium with solid iodine they must be in equilibrium with each other, and the equilibrium constant of eq. (4.3) must also be obeyed. Its numerator and denominator are described by eqs. (4.4a) and (4.4b) in this particular situation, and combining these three equations gives

$$D = [I_2(CCl_4)]/[I_2(aq)] = S_{CCl_4}/S_{aq} \quad (= K_a/K_b). \qquad (4.5)$$

Hence the distribution coefficient is equal to the ratio of the solubilities of the solute in the two solvents.

The considerations described in Section 4.2 can therefore be used to predict how distribution coefficients behave. Being non-polar, iodine is more soluble in a non-polar solvent, such as carbon tetrachloride, than in a strongly polar one, such as water, and so the value of D is large. Since, in general, low solubilities increase more rapidly with increasing temperature than high ones do (because lower solubilities tend to reflect more positive values of ΔH^0) the solubility of iodine in water is increased more by raising the temperature than the solubility of iodine in carbon tetrachloride is, and so the value of D decreases as the temperature rises.

Equation (4.5) describes one equilibrium constant (D) in terms of two others (K_a and K_b). There are so many such relationships among equilibrium constants that you should know how to obtain them. The basic principle is that if the equations for two chemical reactions can be added to obtain the equation for a third, the equilibrium constant for the third reaction is equal to the product of the equilibrium constants for the first two. Here are three reactions that are related in this way:

Reaction	Equation	$\Delta G^0 =$	$K =$
a	A = B	ΔG_a^0	K_a
b	B = C	ΔG_b^0	K_b
c	A = C	ΔG_c^0	K_c

The equation for reaction c is the sum of the equations for reactions a and b. The standard free-energy change for reaction c is equal to the difference between the standard free energies of formation of C and A (Section 2.7) and is independent of the path that is followed (Section 2.6). It does not matter whether A is converted directly into C or is first converted into B, after which is B converted into C:

$$\Delta G_c^0 = \Delta G_a^0 + \Delta G_b^0. \tag{4.6}$$

According to eq. (2.12), each of these standard changes of free energy is related to the equilibrium constant of the corresponding reaction:

$$\Delta G_c^0 = -RT \ln K_c = \Delta G_a^0 + \Delta G_b^0 = -RT \ln K_a + (-RT \ln K_b) = -RT \ln K_a K_b.$$

Taking away the scaffolding used to construct this,

$$K_c = K_a K_b. \tag{4.7}$$

Of course it would not take long in this simple case to see that

$$K_c = \frac{[C]}{[A]} = \frac{[B]}{[A]} \times \frac{[C]}{[B]} = K_a K_b$$

but for the stepwise and overall formation constants mentioned in part 6 of Section 2.9 (page 29) it would not be quite so easy to see that $\beta_3 = K_1 K_2 K_3$ without the aid of eq. (4.7).

The principle has a corollary, which might have been used to obtain eq. (4.5). If subtracting the equation for reaction b from the equation for reaction a yields the equation for reaction c, the equilibrium constant for reaction c is given by

$$K_c = K_a/K_b. \tag{4.8}$$

This can be proved in the same way as eq. (4.7). As applied to the distribution equilibrium discussed earlier in this section, it becomes

Reaction	Equation	$K =$
a	$I_2(s) = I_2(CCl_4)$	$K_a(= S_{CCl_4})$
b	$I_2(s) = I_2(aq)$	$K_b(= S_{aq})$
c	$I_2(aq) = I_2(CCl_4)$	$K_c(= D)$

so that eq. (4.8) gives $D = S_{CCl_4}/S_{aq}$ without further ado. Tables of equilibrium constants would be incredibly long and difficult to use if eqs. (4.7) and (4.8) were not valid.

> SUMMARY: The distribution coefficient is equal to the ratio of the solubilities of a solute in two immiscible solvents.

4.4. Principles of liquid–liquid extraction

> PREVIEW: The chemist can often remove a solute from a solution by extracting it repeatedly with an immiscible solvent. This section discusses the factors that affect the completeness of such a procedure.

In this section we shall imagine that, for one of the purposes described in Section 4.3, a chemist wishes to extract the iodine that is dissolved in V_w cm³ of an aqueous solution by shaking that solution with V_o cm³ of carbon tetrachloride until equilibrium is reached. We shall use the symbols $_0c_w$ to denote the concentration of iodine in the original aqueous solution, $_1c_w$ to denote the concentration of iodine remaining in the aqueous solution at equilibrium, and $_1c_o$ to denote the concentration of iodine in the organic solvent at equilibrium. The subscripts "w" and "o" stand for "water" and "organic", respectively, and the number gives the number of extractions that have been performed.

The ratio of the concentrations at equilibrium can be described by using eq. (4.5):

$$D = {_1c_o}/{_1c_w} .$$
(4.9)

As neither of the concentrations is known, another relation between them is needed. One can be obtained by the same reasoning that led to eqs. (3.21) and (3.25b). The total number of moles of iodine in the system is not changed by dividing the iodine between the water and the carbon tetrachloride. Equations that represent such statements are called *conservation equations* or *mass-balance equations*.

Unfortunately the mole is an absurdly large unit for most chemical work. No beaker is large enough to contain the volume of water that would be needed to dissolve anything like a mole of iodine, which weighs almost 257 g. The millimole is much more practical and convenient, and it will be used instead of the mole throughout this book. There are only a few simple things that you need to know in order to handle statements about millimoles easily and comfortably:

1. The equation for a chemical reaction can be read as giving information about the numbers of moles of reactants consumed and the numbers of moles of products formed. Millimoles can be substituted for moles without changing the numbers. The equation $2 H_2 + O_2 = 2 H_2O$ means that 2 millimoles of hydrogen can react with 1 millimole of oxygen to yield 2 millimoles of water.

2. If a mole of some substance weighs M g, a millimole of it weighs M mg, which is exactly a thousandth as much. The number is the same; only the unit is different.

3. If the concentration of some solute in a solution is c mol dm^3, one cubic decimeter of the solution contains c moles of the solute. One cubic centimeter of the solution, which is exactly a thousandth of a cubic decimeter, contains exactly a thousandth as much of the solute, or c millimoles. The concentration is c millimol cm^{-3}, abbreviated c mmol cm^{-3}. Again the number is the same and only the unit is different. There is an added bonus here because the cm^3 is a much more convenient unit of volume than the dm^3.

Applying these ideas to the distribution equilibrium with which this section began, the number of millimoles of iodine in the original aqueous solution must be equal to the sum of the numbers of millimoles of iodine in the two separate solutions at equilibrium. The total number of millimoles of iodine in the system was originally equal to $V_w \times {}_0c_w$, and is still equal to this at equilibrium. Note that the product of a volume in cm^3 and a concentration in mmol cm^{-3} is equal to a number of millimoles. The conservation equation is

$$V_w \times {}_0c_w = V_w \times {}_1c_w + V_o \times {}_1c_o. \tag{4.10}$$

Using eq. (4.9) to eliminate ${}_1c_o$, and solving for ${}_1c_w$,

$$_1c_w = \frac{V_w \times {}_0c_w}{V_w + DV_o} = \frac{{}_0c_w}{1 + D(V_o/V_w)}. \tag{4.11}$$

The second expression, obtained by dividing each term in the first one by V_w, is slightly more compact and has the advantage of emphasizing that it is the ratio of volumes that is important.

Typical values might be $V_w = 100$ cm^3 and $V_o = 10$ cm^3. With $V_o/V_w = 0.1$ and $D = 1 \times 10^2$, eq. (4.11) becomes ${}_1c_w = {}_0c_w/11$. One-eleventh of the original amount of iodine is left in the aqueous phase; the other ten-elevenths has been extracted into the carbon tetrachloride. The separation is by no means complete, but it becomes more so if it is repeated. If the two solutions have been brought to equilibrium in a separatory funnel, the carbon tetrachloride can be drained off through the stopcock because it is denser than water, and then a fresh portion of carbon tetrachloride can be added. In the second extraction the initial concentration of iodine in the aqueous phase is ${}_1c_w$, and a repetition of the argument used to obtain eq. (4.11) gives

$$_2c_w = \frac{{}_1c_w}{1 + D(V_o/V_w)} \tag{4.12}$$

where ${}_2c_w$ is the concentration of iodine remaining in the water after two extractions have been completed. The volume of water V_w does not change from one extraction to the next, and if the successive portions of carbon tetrachloride all have the same volume V_o eqs. (4.11) and (4.12) can be combined to yield

$$_2c_w = \frac{_0c_w}{[1+D(V_o/V_w)]^2}$$

or, after n extractions have been performed,

$$_nc_w = \frac{_0c_w}{[1+D(V_o/V_w)]^n}. \tag{4.13}$$

Equation (4.13) can be used for either of two purposes. One is to find the result of any given number of extractions under any given conditions; the other is to find how many extractions must be performed to remove any given fraction of the solute that was present initially. Both are illustrated by the examples that follow.

Example 4.1. Under the conditions given in the text, what fraction of the iodine will have been removed from the aqueous phase after four extractions have been completed?

Answer. The fraction *remaining* in the aqueous phase is $_4c_w/_0c_w$ and is given by

$$_4c_w/_0c_w = 1/[1+1\times 10^2\times 0.1]^4 = 1/11^4 = 7\times 10^{-5}.$$

The fraction *removed* is equal to $1-{_4c_w}/{_0c_w} = 1-7\times 10^{-5} = 0.999\ 93 = 99.993\%$.

Example 4.2. Under the same conditions, how many extractions would be required to decrease the concentration of iodine in the aqueous phase to 0.1% of its initial value?

Answer. The easiest way to solve for n is to rearrange eq. (4.13) to give

$$[1+D(V_o/V_w)]^n = {_0c_w}/{_nc_w}\left(= \frac{1}{_nc_w/_0c_w} = \frac{1}{0.001} = 1\times 10^3\right)$$

and take the logarithm of each side. The result is

$$n = \frac{\log(_0c_w/_nc_w)}{\log[1+D(V_o/V_w)]} = \frac{\log(1\times 10^3)}{\log(11)} = \frac{3}{1.041} = 2.88.$$

It is not possible to perform a fraction of an extraction, and this number means that two extractions will not suffice. Three will remove a little more than the necessary fraction of the iodine but must be performed to achieve the desired result.

The idea of a *quantitative separation* is useful in discussing the results of such calculations. Whether the purpose is to recover the iodine that is present or to find out how much there is, unavoidable losses and errors of measurement make it very difficult to tell the difference between a separation that is perfectly complete and one that is only 99.9% complete. In a separation that is much less than 99.9% complete it is usually easy to tell that something has been lost. A separation, or any other process, that is at least 99.9% complete is said to be quantitative. The figure is an arbitrary dividing line, and you would not want to be content with a barely quantitative separation of a very toxic substance from a solution that you expected to dump into the river when you were finished with it.

Example 4.2 shows that three extractions with 10-cm³ portions of carbon tetra-chloride are needed to obtain a quantitative separation of iodine from the original 100 cm³ of water. Mixing the separate portions of carbon tetrachloride gives 30 cm³ of a solution in which the concentration of iodine will be equal to 3.3 times the original concentration of iodine in the aqueous phase.

All these figures depend on the value of V_o/V_w. Suppose that the same original solution of iodine in water is extracted with only 2 cm³ of carbon tetrachloride. Now $[1+D(V_o/V_w)]$ is equal to only 3, and one-third of the iodine is left after one extraction. One-third of that, or one-ninth of the original amount, is left after the first portion of carbon tetrachloride has been drawn off and a second extraction has been performed with another 2-cm³ portion. Only 4 cm³ of carbon tetrachloride has been used in these two extractions together and yet the fraction of the original amount of iodine remaining in the aqueous phase is not very different from the fraction that remained after one extraction with 10 cm³ of carbon tetrachloride. In the fashion illustrated by Example 4.2 it can be shown that seven extractions with 2-cm³ portions of carbon tetrachloride are needed to achieve a quantitative separation. The total volume of carbon tetrachloride that is needed will be only 14 cm³, and the concentra-tion of iodine in the solution obtained by mixing these portions will be just over seven times as large as it was in the aqueous solution initially. Going in the other direction, two extractions with 50-cm³ portions of carbon tetrachloride are needed to obtain a quantitative separation. A total of 100 cm³ of carbon tetrachloride will be used, and the concentration of iodine will not be increased at all.

You should certainly be willing to do three extractions with 10-cm³ portions instead of two with 50-cm³ portions. It is well worth while to do 50% more work in order to save 70% of the material used and increase the concentration of iodine by a factor of 3. Should you be willing to do seven extractions with 2-cm³ portions instead? That involves over twice as much work, saves barely half of the carbon tetrachloride, and increases the concentration of iodine by a factor barely exceeding 2. You often have to face this sort of trade-off between the excellence of a final result and the amount of work that has to be done to achieve it. How you respond to it depends on the kind of person you are, and also on the extent of your interest in the result. One reason why chemical education is so diversified is to enable you to discover the kinds of problems on which you are most willing to do a little more work to get a result that is a little better, because those are the ones on which you can base the most enjoyable and rewarding career.

SUMMARY: This section introduced conservation (or mass-balance) equations and described the algebraic relationships between the number of extractions performed and the degree of separation attained.

4.5. Applications of liquid–liquid extraction

PREVIEW: This section discusses the practical uses of liquid–liquid extraction and the chemical considerations that govern separations by liquid–liquid extraction.

Practical chemists use liquid–liquid extraction in several different ways. Synthetic chemists carry out reactions designed to yield certain products, and obtain mixtures of those products with left-over starting materials, by-products, and other substances. Liquid–liquid extraction is often an easy way of separating the desired product from the things that accompany it.

Ionic and non-ionic substances are the easiest to separate. Suppose, for example, that some reaction carried out in an aqueous solution of sodium hydroxide has yielded benzaldehyde, C_6H_5CHO, as the desired product. It might be isolated by extracting it into a solvent, such as diethyl ether or benzene, in which both water and sodium hydroxide are virtually insoluble.

Mixtures of non-ionic substances can be separated in either of two ways. One takes advantage of differences between their solubilities in different solvents. Phenol, C_6H_5OH, and 2-aminophenol

are both quite soluble in water, but phenol is extremely soluble in benzene while 2-aminophenol is almost insoluble in it. The distribution coefficient of phenol between water and benzene is fairly high, whereas that of 2-aminophenol is very small. If an aqueous solution containing both phenol and 2-aminophenol were shaken with benzene, a substantial fraction of the phenol would be extracted into the benzene layer but only a little 2-aminophenol would accompany it. Several extractions would yield a benzene solution containing almost all of the phenol contaminated by just a little 2-aminophenol, and leave an aqueous solution containing almost all of the 2-aminophenol along with just a little phenol.

The other way of separating two non-ionic compounds is to convert one of them into an ionic form. This is easy if they have different acid–base properties. The solubilities of benzaldehyde and benzoic acid, C_6H_5COOH, in different solvents are too nearly equal to permit them to be separated cleanly and easily by the simple procedure just described. However, benzoic acid is acidic while benzaldehyde is not. If an aqueous solution containing these two compounds is made slightly alkaline, the reaction $C_6H_5COOH(aq) + OH^- = C_6H_5COO^- + H_2O(l)$ will occur. Its product, benzoate ion, cannot be extracted into an organic solvent such as benzene, ether, or chloroform because it is ionic, and extraction of the aqueous solution with such a solvent will therefore remove the benzaldehyde and leave the benzoate ion behind.

With two ionic substances, one must be converted into a non-ionic form so that it can be extracted. There are two somewhat different ways of doing this. In one, called *chelate extraction*, an aqueous solution of a metal ion is shaken with a solution of some organic compound, with which it can react to form an uncharged *chelate*, in an organic solvent that is immiscible with water. The organic compound is called a *chelating agent*. A chelating agent is a substance that can form two or more non-ionic bonds with a metal ion. Ethylenediamine, $H_2NCH_2CH_2NH_2$, is a very simple chelating agent; it can react with many metal ions in the following way:

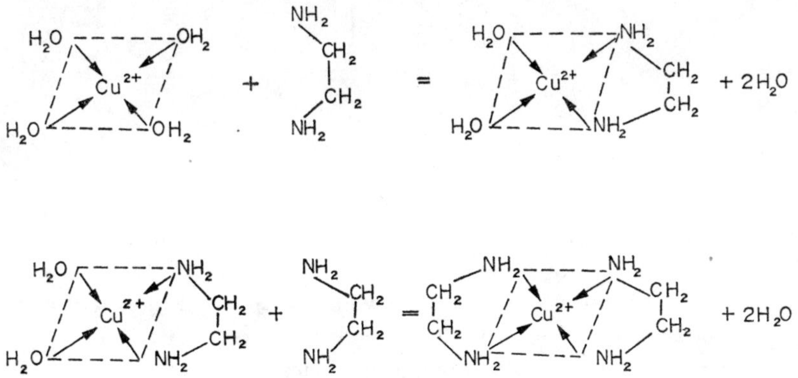

Each molecule of ethylenediamine can donate two pairs of electrons to the copper(II) ion. The final product contains two five-membered rings, each consisting of the copper(II) ion and the two nitrogen atoms and two carbon atoms of a molecule of ethylenediamine. Five- and six-membered rings are the most stable because they involve the least distortion of the normal bond angles. A ring with four or fewer members is less stable because the electron-donating atoms of the chelating agent are not far enough apart to meet the bonding orbitals of the metal ion, so that much strain is needed to form the chelate. One with seven or more members is also less stable, but for the opposite reason: here the ends of the free chelating agent are too far apart, and therefore the molecule must be bent and puckered into an unnatural shape in forming the chelate.

The copper(II)–ethylenediamine chelate could not be extracted from water into an organic solvent because, like copper(II) ion itself, it is ionic. Chelate-extraction systems entail the use of chelating agents that contain acidic functional groups from which protons are lost during coordination with metal ions. A typical chelating agent of this sort is 8-quinolinol:

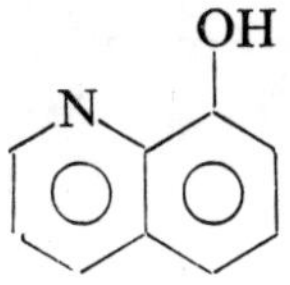

whose reaction with copper(II) ion is very much like that of ethylenediamine except for the loss of protons that accompanies it. This is crucial because it leads to the formation of an uncharged product rather than an ionic one, and the overall result when an aqueous solution of copper(II) ion is shaken with a solution of 8-quinolinol in chloroform, benzene, amyl acetate, or any of various other organic solvents is

$$+2H^+ + 4H_2O\,(l)$$

where the "(o)" means that the compound is dissolved in an organic solvent, which must of course be immiscible with water.

This chelate is stable both because it has two five-membered rings that include the metal ion and because the copper–nitrogen bonds are strong ones, it has a fairly high molecular weight and a good deal of organic character, and the reaction can be driven to completion by the presence of a high concentration of 8-quinolinol in the organic layer. For all of these reasons the extraction of copper(II) into the organic solvent is virtually complete—unless the aqueous layer is sufficiently acidic to repress the reaction.

There are a great many possibilities hidden in the last clause. The overall equilibrium constant for the reaction

$$M^{n+} + n \text{[quinolinol]}\,(o) = [M^{n+}(\text{quinolinolate})_n]\,(o) + nH^+$$

is much higher when the metal ion represented by M^{n+} is nickel(II) ion than when it is manganese(II) ion. It is so low for manganese(II) ion that an appreciable fraction of the manganese(II) cannot be extracted unless the pH of the aqueous solution is at least as high as 10, whereas nickel(II) can be quantitatively extracted even if the pH is as low as 5. Consequently it is possible to extract nickel(II) from an aqueous solution having a pH anywhere between 5 and 10, while the manganese(II) will be left behind. The hydrogen ions that are present in such a solution conceal the manganese(II) from the reagent; the scheme is called *masking*.

With the aid of some chemical facts you can often make a situation like this stand on its head. The fact that nickel(II) is more readily extracted than manganese(II) would not be very helpful if you were searching for a trace of manganese(II) that might or might not be present in a solution containing a great deal of nickel(II) ion. If you went about this by a procedure in which you extracted the nickel(II) and left the manganese(II) behind, you might have to do a lot of work to remove the nickel(II) very completely, and when you had done it you would still have to cope with a very dilute solution of manganese(II). It would be easier to remove most of the manganese(II) while leaving the nickel(II) behind, and you could do that by extracting the manganese(II) into a solution of 8-quinolinol in chloroform after adjusting the pH of the aqueous solution to 12–13 and adding an excess of potassium hexacyanoferrate(II) to it. The hexacyanoferrate(II) ion masks the nickel(II) ion by reacting with it to form a precipitate, $K_2Ni[Fe(CN)_6]$, that is insoluble in both water and chloroform.

The second general kind of system from which an ionic substance can be extracted is called an *ion-association* system. In the simplest kind of ion-association system, the ion that is to be extracted is made to react with an oppositely charged ion to form an uncharged molecule that can be extracted into another solvent. Mercury(II) ion can be extracted into diethyl ether if bromide ion is added to form $HgBr_2$. The process may be more roundabout. In an aqueous solution that contains a high concentration of hydrochloric acid, the molecules of water that are coordinated with an iron(III) ion can be successively replaced by chloride ions, and a negatively charged complex ion having the formula $FeCl_4(OH_2)_2^-$ is formed. In the presence of an organic solvent, such as an ether, that is immiscible with water and that has some basic character, this complex ion can associate with a proton to give an uncharged species that is extracted:

$$H^+ + FeCl_4(OH_2)_2^- = HFeCl_4(OH_2)_2(o).$$

In an ether the product can be stabilized by the formation of a hydrogen bond between its proton and the atom of oxygen in a molecule of ether, and one or both of the molecules of water remaining in the coordination shell may be replaced by molecules of ether; if the solvent could not donate electron pairs to hydrogen or metal atoms a species like this would not be soluble in it. Ions of cobalt, zinc, gold, plutonium, and a great many other elements can be extracted in these general ways from aqueous solutions containing halide or thiocyanate ions.

These approaches can often be combined by adding both an uncharged organic ligand and a suitable anion to the metal ion that is to be extracted. The organic ligand combines with the metal ion to yield a charged complex or chelate containing organic groups that facilitate extraction into an organic solvent, and this combines with the anion to yield an uncharged extractable species. For example, copper(II) can be extracted into chloroform from an aqueous solution containing pyridine and bromide ion; the overall reaction is

All these possibilities make liquid–liquid extraction widely useful for separating and purifying both inorganic and organic substances.

SUMMARY: Separations by liquid–liquid extraction can be based on differences between the ionic forms, solubilities or distribution coefficients, or acid-base or other chemical properties, of the substances involved. Chelation, ion association, and masking are widely employed in liquid–liquid extraction procedures.

Problems

Answers to some of these problems are given on page 52.

4.1. At 293 K the solubility of aspirin in water is 2.5 g dm^{-3}, and its solubility in diethyl ether is 35.7 g dm^{-3}.

(a) What is the value of the distribution coefficient for aspirin between water and diethyl ether at 293 K?

(b) If 100 cm^3 of an aqueous solution containing 50 mg of aspirin is shaken with 10 cm^3 of diethyl ether at 293 K, what weight of aspirin will remain dissolved in the aqueous layer when equilibrium is reached?

(c) How many such extractions, each with 10 cm^3 of diethyl ether, would be needed to remove at least 49.9 mg of aspirin from the aqueous layer?

4.2. At 289 K the solubility of caffeine in water is 13.5 g dm^{-3}, and its solubility in diethyl ether is 0.44 g dm^{-3}.

(a) What is the value of the distribution coefficient for caffeine between water and diethyl ether at 289 K?

(b) If 100 cm^3 of an aqueous solution containing 10 mg of caffeine is shaken with 10 cm^3 of diethyl ether at 289 K, what weight of caffeine will be dissolved in the ether layer when equilibrium is reached?

(c) What weight of caffeine would remain dissolved in the aqueous layer after four such extractions?

4.3. Twenty cm^3 of a solution of iodine in water is shaken with 10 cm^3 of carbon disulfide. When equilibrium is reached, the concentration of iodine in the carbon disulfide layer is found to be 175 times as large as the concentration of iodine in the aqueous layer.

(a) What is the value of the distribution coefficient for iodine between water and carbon disulfide at the temperature at which this experiment was performed?

(b) Compare the amounts of iodine in the aqueous and carbon disulfide layers at equilibrium.

(c) What percentage of the iodine originally present in the aqueous solution would be removed by two such extractions?

(d) How many extractions, each with 2 cm^3 of carbon disulfide instead of 10 cm^3, would be required to remove the same percentage of the iodine?

4.4. The solubility of yellow phosphorus in water at 283 K is 0.003 g dm^{-3}. At the same temperature its solubility in carbon disulfide is 8.8×10^3 g dm^{-3}. What volume of carbon disulfide would be needed to remove 99.9% of the yellow phosphorus dissolved in 100 cm^3 of an aqueous solution in a single extraction?

CHAPTER 5

The Solubilities of Ionic Compounds

5.1. Introduction

This chapter deals with the dissolution and precipitation of ionic substances. It emphasizes the behaviors of slightly soluble substances in aqueous solutions, and stresses the relationships among several different kinds of equilibrium constants. In later chapters we shall have frequent need for some of the kinds of equilibrium constants that are introduced in this chapter and for some of the algebraic ways of handling equilibrium-constant expressions that are discussed here.

5.2. Solubilities and solubility products of strong electrolytes

PREVIEW: The solubility of a sparingly soluble electrolyte in a pure solvent is easy to calculate from its solubility product if it is completely dissociated.

In Section 2.9 the solubility product was defined as the equilibrium constant for a reaction in which an electrolyte dissolves and yields its ions. In this section it is assumed that the dissolved ions are the only products obtained, so that the solubility product provides a complete description of the process. Unfortunately this is not true for the great majority of sparingly soluble electrolytes. Later sections will explain why it is not, and will describe what happens when it is not.

The solubility product of silver iodate is described by the equations

$$AgIO_3(s) = Ag^+ + IO_3^-; \quad K_{AgIO_3} = [Ag^+][IO_3^-]. \quad (5.1)$$

Its value in water at 298 K is 3.0×10^{-8} mol^2 dm^{-6}. The chemical equation says that 1 mole of silver ion and 1 mole of iodate ion are obtained when 1 mole of silver iodate is consumed in this reaction. Suppose that the solubility of silver iodate in water is equal to S mol dm^{-3}. Then in each cubic decimeter of the saturated solution S mole of solid silver iodate will have dissolved, and S mole of silver ion and S mole of iodate ion will have been formed. The concentration of silver ion in the saturated solution will be equal to SM, and so will that of iodate ion. We are assuming that neither ion reacts with water; if one did, its concentration would be smaller than S.

Combining these statements with eq. (5.1) and the numerical value of K_{AgIO_3},

$$K_{AgIO_3} = [Ag^+][IO_3^-] = (S)(S) = S^2 = 3.0 \times 10^{-8} \text{ mol}^2 \text{ dm}^{-6}$$

whence $S = (3.0 \times 10^{-8})^{1/2} = 1.7 \times 10^{-4} \text{ mol dm}^{-3}$.

If the solubility products of two different compounds have the same units, you can easily tell which of the compounds is the more soluble by comparing the numerical values of their solubility products. Knowing that the solubility product of thallium(I) iodate, $TlIO_3$, is equal to $3.1 \times 10^{-6} \text{ mol}^2 \text{ dm}^{-6}$, you can tell at once that thallium(I) iodate is more soluble in water than silver iodate because its solubility product has a larger value than the solubility product of silver iodate. It is not necessary to go to the trouble of calculating that the solubility of thallium(I) iodate is $1.7_6 \times 10^{-3}$ mol dm^{-3} while that of silver iodate is only $1.7 \times 10^{-4} \text{ mol dm}^{-3}$.

Such decisions are harder to make if the products have different units. The solubility product of barium iodate is given by

$$Ba(IO_3)_2 (s) = Ba^{2+} + 2 IO_3^-; \quad K_{Ba(IO_3)_2} = [Ba^{2+}][IO_3^-]^2 \qquad (5.2)$$

and its numerical value in water at 25^0 is $1.5 \times 10^{-9} \text{ mol}^3 \text{ dm}^{-9}$. This is less than a tenth as large as the value of K_{AgIO_3}, but you cannot conclude that barium iodate is ess soluble in water than silver iodate. If S mole of barium iodate dissolves in water to give a cubic decimeter of a saturated solution, S mole of barium ion will be formed along with $2S$ moles of iodate ion. There will be twice as many moles of iodate ion as of barium ion because the reaction produces two moles of iodate ion for each mole of barium ion. In the saturated solution the concentration of barium ion will be S mol dm^{-3} (M), while that of iodate ion will be $2S M$. From these figures and eq. (5.2),

$$K_{Ba(IO_3)_2} = [Ba^{2+}][IO_3^-]^2 = (S)(2S)^2 = 4S^3 = 1.5 \times 10^{-9} \text{ mol}^3 \text{ dm}^{-9}$$

so that

$$S = (1.5 \times 10^{-9}/4)^{1/3} = (0.375 \times 10^{-9})^{1/3} = (375 \times 10^{-12})^{1/3} = 7._2 + 10^{-4} \text{ mol dm}^{-3}.$$

Barium iodate is several times as soluble as silver iodate although its solubility product is only a twentieth as large as the solubility product of silver iodate, and it is almost as soluble as thallium(I) iodate even though its solubility product is less than 1/2000 as large as the solubility product of thallium(I) iodate. One advantage of giving the units of equilibrium constants is that it helps you to avoid making comparisons that you should not make. The solubility products of silver and thallium(I) iodates can be compared because both have the units $\text{mol}^2 \text{ dm}^{-6}$, but neither can be compared directly with the solubility product of barium iodate, of which the units are $\text{mol}^3 \text{ dm}^{-9}$.

Of course it is also possible to carry out the same sort of calculation in the reverse direction. If you know that the dissolution of copper(I) tetraphenylborate is described by the equation $Cu[B(C_6H_5)_4](s) = Cu^+ + B(C_6H_5)_4^-$ and that its solubility is equal to 1.0×10^{-4} mol dm^{-3} at 25°, you need only recognize that the concentration of each ion must be equal to the solubility to obtain $K_{\text{Cu}[\text{B}(\text{C}_6\text{H}_5)_4]} = (1.0 \times 10^{-4})^2 = = 1.0 \times 10^{-8}$ mol^2 dm^{-6}.

SUMMARY: When a completely dissociated electrolyte having the formula $C_c A_a$ is dissolved in a pure solvent, the concentrations of the ions in the saturated solution are given by $[C^{n+}] = cS$ and $[A^{m-}] = aS$, where S is the solubility in mol dm^{-3}. The value of S can be calculated by combining these equations with the one for $K_{C_c A_a}$.

5.3. The common-ion effect

PREVIEW: In a solution that already contains one of its ions, the solubility of a sparingly soluble compound is smaller than in the pure solvent.

Section 5.2 dealt only with the solubility in the pure solvent. Silver iodate is less soluble in a dilute solution of either potassium iodate or silver nitrate, each of which has an ion in common with silver iodate, than it is in water alone. The decrease of solubility is called the *common-ion effect*, and is nothing more than an illustration of le Chatelier's principle.

Calculations involving the common-ion effect may be either extremely simple or fairly complicated. It depends on how much of the common ion there is in the solution to begin with and how much more of it enters the solution as the sparingly soluble electrolyte dissolves. At the first extreme, the solubility of silver iodate in a solution containing 0.01 M potassium iodate can be calculated very easily. If the solubility of silver iodate is S mol dm^{-3}, then S mole of silver iodate will be dissolved in each cubic decimeter of the solution when equilibrium is reached, and S mole of silver ion and S mole of iodate ion will have been added to the solution as it dissolved. Exactly the same thing was said in Section 5.2, for this is true no matter what, if anything, was present in the solution originally. The only effect of the potassium iodate is to increase the concentration of iodate ion. If the potassium iodate were not present, the concentration of iodate ion at equilibrium would be equal to S M. With 0.01 M iodate ion in the solution before any silver iodate dissolves, the concentration of iodate ion at equilibrium will be equal to $(0.01 + S)$ M. Since there is no other source of silver ion, its concentration will be equal to S M. These facts lead to the equation

$$K_{\text{AgIO}_3} = [\text{Ag}^+][\text{IO}_3^-] = S(0.01 + S).$$

Silver iodate is not very soluble: even in pure water its solubility is only 1.7×10^{-4} mol dm^{-3}, and S will certainly be smaller than this in the potassium iodate solution. Without having any idea how much smaller it may turn out to be, we can see that it cannot be very far from the truth to say that the total concentration of iodate ion at equilibrium will be just 0.01 M. Then we can find the solubility S by writing

$$S = [\text{Ag}^+] = \frac{K_{\text{AgIO}_3}}{[\text{IO}_3^-]} = \frac{3.0 \times 10^{-8}}{0.01} = 3.0 \times 10^{-6} \text{ mol dm}^{-3}. \tag{5.3}$$

Whenever you make an assumption you should check your answer to see whether it justifies the assumption. Here the assumption was that S would be much smaller than 0.01 M, and on finding that S was only 3.0×10^{-6} M you could go away content.

The solubility of barium iodate in a solution containing 0.005 M barium chloride could be calculated in much the same way. If the solubility is S mol dm^{-3}, the solution will contain $(0.005 + S) M$ barium ion and $2S$ M iodate ion after it has become saturated with barium iodate. Since S would be only 7.2×10^{-4} mol dm^{-3} in the absence of the barium chloride and must be less than this in its presence, the total concentration of barium ion cannot be much above 0.005 M, and

$$[IO_3^-]^2 = (2S)^2 = 4S^2 = \frac{K_{Ba(IO_2)_3}}{[Ba^{2+}]} = \frac{1.5 \times 10^{-9}}{5 \times 10^{-3}} = 3 \times 10^{-10}$$

whence

$$S = (3 \times 10^{-10}/4)^{1/2} = (0.75 \times 10^{-10})^{1/2} = (75 \times 10^{-12})^{1/2} = 8.7 \times 10^{-6} \text{ mol dm}^{-3}.$$

Here the approximation consisted of assuming that S would be much smaller than 5×10^{-3} M, and it is.

You should not expect to be as lucky as this with every assumption you make. If silver iodate is dissolved in a solution initially containing 1×10^{-4} mol dm^{-3} of potassium iodate, the concentrations of silver and iodate ions will be equal to S and $(1 \times 10^{-4} + S)$, respectively, when equilibrium is reached. If you decide to neglect S in comparison with 1×10^{-4}, you will write

$$S = [Ag^+] = \frac{K_{AgIO_3}}{[IO_3^-] = 1 \times 10^{-4}} \frac{3.0 \times 10^{-8}}{} = 3 \times 10^{-4} \text{ mol dm}^{-3}$$

which is just like eq. (5.3) except that the concentration of iodate ion is different. You would be in trouble if you did not make sure that the assumption is justified, for this result is not only wrong but absurd: it is larger than the solubility of silver iodate in pure water, and that is impossible. You obtain protection against such catastrophes by observing that if S is equal to 3×10^{-4} M it is *not* very much smaller than 1×10^{-4} M. Now you can conclude that you should not have neglected S in comparison with 1×10^{-4}, and are well on the way toward writing

$$[Ag^+][IO_3^-] = (S)(1 \times 10^{-4} + S) = S^2 + 1 \times 10^{-4}S$$
$$= K_{AgIO_3} = 3.0 \times 10^{-8} \text{ mol}^2 \text{ dm}^{-6}. \tag{5.4}$$

You must solve a quadratic equation, but there is hardly any other way out. Its solution is $S = 1.3 \times 10^{-4}$ mol dm^{-3}, which is a little smaller than the solubility in pure water, as it should be. Unless you know that you never make mistakes in arithmetic, you should take the trouble to make sure at this point that S is *not* much smaller than 1×10^{-4} M; if it is, your original assumption should have succeeded and there must be an error somewhere.

You may have recognized that the word "hardly", in the line just below eq. (5.4), is a weasel word. Often there are different approximations that you can make in solving a chemical problem, and it is worth your while, when one is unsatisfactory, to try another one in the hope of saving work in the long run. In the problem just considered, there are certain limits between which the total concentration of iodate ion must lie. It is sure to be greater than 1×10^{-4} M because it was equal to 1×10^{-4} M before silver iodate was added, and the dissolution of silver iodate must increase it to some extent. It is sure to be less than 2.7×10^{-4} M because the solubility of silver iodate cannot be as high in the potassium iodate solution as it is in pure water, and even if it were it would add only 1.7×10^{-4} M iodate ion to the 1×10^{-4} M that was present originally. The average of these two limits, which is 1.9×10^{-4} M, cannot possibly be wrong by more than 40%. Replacing the concentration of iodate ion in eq. (5.3) by this value gives $S = 3.0 \times 10^{-8}/1.9 \times 10^{-4} = 1.6 \times 10^{-4}$ M. The correct value, obtained by solving the quadratic equation (5.4) at the cost of much more time, is 1.3×10^{-4} M; the error is not 40% but only about 25% $(= 100 \times (1.6 \times 10^{-4} - 1.3 \times 10^{-4})/1.3 \times 10^{-4})$. If you were surer of your insight into situations like this you might have noticed that the concentration of potassium iodate is smaller than the solubility of silver iodate in pure water, decided that the presence of the potassium iodate would not decrease the solubility of the silver iodate very much, and guessed that S would be closer to its value in pure water $(1.7 \times 10^{-4}$ $M)$ than to zero. This train of thought would have led you to conclude that the total concentration of iodate ion in the saturated solution would be closer to 2.7×10^{-4} M than to 1×10^{-4} M, and you might have picked a value like 2.2×10^{-4} M. This gives $S = 1.3_6 \times 10^{-4}$ M, which is only 5% away from the truth.

Whether or not you can get away with a result that might be in error by 5%, or 40%, or an order or two of magnitude, depends chiefly on what you plan to do with the result when you have it. Sometimes it will be obvious that an approximation is good enough; sometimes you will have to think in order to decide. In the first problem in this section there is no remotely conceivable purpose for which the approximation $S \ll 0.01$ M might not be good enough: the equilibrium constant in that problem is simply not known closely enough to make it worth saying that the concentration of iodate ion is 0.010003 M rather than 0.01 M. The last problem in this section is one that might arise if you had obtained a quantity of silver iodate that was wet with a solution containing various impurities and were trying to wash away those impurities without dissolving any significant amount of the silver iodate. Even if you knew the solubility of the silver iodate with an accuracy no better than 40% you could almost certainly decide whether or not a significant amount of it would dissolve. You cannot afford to be content with crude guesses when accurate results are needed, but neither should you be willing to waste your time on obtaining an accurate result when a crude guess will suffice.

It is often convenient to represent the results of equilibrium calculations in graphical form, and the most convenient graphical form is the one shown in Fig. 5.1. This

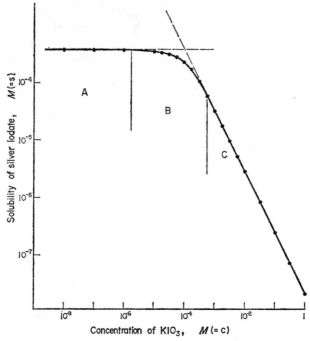

FIG. 5.1. Effect of the concentration c of potassium iodate on the solubility S of silver iodate. In region A $c \ll S$ and the solubility is indistinguishable from its value in pure water. In region C $c \gg S$ and the solubility is indistinguishable from the value given by $S = K_{AgIO_3}/c$. In region B the values of c and S are too nearly equal to permit neglecting either in comparison with the other.

plot shows how the concentration of silver ion in a saturated solution of silver iodate, which is equal to the solubility of silver iodate, depends on the concentration of potassium iodate that is present in the solutions. Logarithmic scales are used on both axes so that wide ranges of concentration can be represented, and also because they cause certain portions of such plots to become linear. The plot is a horizontal straight line in region A because extremely small concentrations of potassium iodate have no detectable effect on the solubility of silver iodate, and the value of the ordinate for this line is merely the logarithm of the solubility of silver iodate in pure water. In region C, the concentration of potassium iodate is so high that the concentration of iodate ion is not significantly increased by the addition of a few more iodate ions to the solution as silver iodate dissolves in it. This is the situation for which eq. (5.3) was written. Rewriting that equation in logarithmic form gives

$$\log S = \log K_{AgIO_3} - \log [IO_3^-] \approx \log K_{AgIO_3} - \log c \qquad (5.5)$$

where c is the concentration of potassium iodate. Plotting $\log S$ on the ordinate axis and $\log c$ on the abscissa axis makes this the equation of a straight line with a slope equal to -1. Between these two linear portions, in region B, the plot is curved. This intermediate region corresponds to the situation in which eq. (5.4) is needed because the silver iodate and potassium iodate each contribute too much iodate ion to permit ignoring the presence of either. From a curve like this one you can often read a concentration as accurately as it would be worth your while to calculate it, and with much less trouble. Such curves are especially valuable in more complex situations, where two or more different equilibria may be involved, because they make it easy to get a clear understanding of how these are affected by changing the independent variable.

The common-ion effect appears in a somewhat different way if two slightly soluble compounds having an ion in common are dissolved in the same solution. Suppose that both an excess of silver iodate and an excess of thallium(I) iodate are added to some water, and that the mixture is stirred until it has become saturated with both. The iodate ions contributed by the thallium(I) iodate will repress the solubility of the silver iodate, and vice versa: each compound will be less soluble than it is in pure water. If the solubility of silver iodate is S_{AgIO_3} M, the dissolution of silver iodate will contribute S_{AgIO_3} mole of silver ion and S_{AgIO_3} mole of iodate ion to each cubic decimeter of the solution. If the solubility of thallium(I) iodate is S_{TlIO_3} mol dm^{-3}, the dissolution of thallium(I) iodate will add S_{TlIO_3} mole of thallium(I) ion and S_{TlIO_3} mole of iodate ion. The total concentration of iodate ion will be $(S_{AgIO_3}+S_{TlIO_3})$ mol dm^{-3} (M) at equilibrium, while the concentrations of silver ion and thallium(I) ion will be simply S_{AgIO_3} M and S_{TlIO_3} M, respectively. All this is conveyed by the equation

$$[IO_3^-] = [Ag^+]+[Tl^+]   (= S_{AgIO_3}+S_{TlIO_3}). \tag{5.6}$$

Both solubility products must be satisfied because the solution is saturated with both of these compounds, and therefore

$$[Ag^+] = K_{AgIO_3}/[IO_3^-]   \text{and}   [Tl^+] = K_{TlIO_3}/[IO_3^-].$$

Combining these with eq. (5.6),

$$[IO_3^-] = \frac{K_{AgIO_3}}{[IO_3^-]} + \frac{K_{TlIO_3}}{[IO_3^-]}$$

from which

$$[IO_3^-] = (K_{AgIO_3}+K_{TlIO_3})^{1/2} = (3.0\times10^{-8}+3.1\times10^{-6})^{1/2} = (3.13\times10^{-6})^{1/2}$$
$$= 1.7_7\times10^{-3} \ M.$$

The solubility of silver iodate is equal to the concentration of silver ion:

$$S_{AgIO_3} = [Ag^+] = \frac{K_{AgIO_3}}{[IO_3^-]} = \frac{3.0\times10^{-8}}{1.7_7\times10^{-3}} = 1.7\times10^{-3} \ M$$

and the solubility of thallium(I) iodate is equal to the concentration of thallium(I) ion:

$$S_{TlIO_3} = [Tl^+] = \frac{K_{TlIO_3}}{[IO_3^-]} = \frac{3.1\times10^{-6}}{1.7_7\times10^{-3}} = 1.7_5\times10^{-3} \ M.$$

In Section 5.2 the solubilities of these compounds in pure water were found to be 1.7×10^{-4} M for silver iodate and $1.7_6 \times 10^{-3}$ M for thallium(I) iodate. Thallium(I) iodate is more soluble than silver iodate, and it contributes many more iodate ions to the mixed solution than the silver iodate does. Because the concentration of iodate ion is much larger in the mixed solution than in a saturated solution of silver iodate alone, the concentration of silver ion and the solubility of silver iodate are much smaller in the mixed solution. On the other hand, silver iodate is less soluble than thallium(I) iodate, and therefore the addition of silver iodate has scarcely any effect on the concentration of iodate ion in a solution that is already saturated with thallium(I) iodate. Consequently the concentration of thallium(I) ion and the solubility of thallium(I) iodate are almost the same in the mixed solution as in pure water.

SUMMARY: If a sparingly soluble compound is dissolved in a solution containing one of its ions, the concentration of the common ion at equilibrium is the sum of the concentration originally present and the additional concentration resulting from the dissolution. Often the solubility will be so small that its contribution can be neglected.

5.4. Solubility products of weak electrolytes

PREVIEW: Sections 5.2 and 5.3 dealt with substances that are completely dissociated in their solutions. For a substance that is only partially dissociated, neither the solubility in a pure solvent nor the effect of adding a common ion is correctly described by the solubility product alone. The behavior of a sparingly soluble weak acid is described to illustrate the ideas involved.

Benzoic acid has a solubility of 0.025 mol dm^{-3} in water at 298 K. In a saturated solution of benzoic acid there is an equilibrium between the solid and the dissolved molecules of the acid:

$$C_6H_5COOH(s) = C_6H_5COOH(aq); \quad K = K_0 = [C_6H_5COOH(aq)]. \quad (5.7a)$$

If benzoic acid were a non-electrolyte the equilibrium constant for this reaction would be equal to its solubility S (Section 4.2). A different symbol, K_0, is used here, however, because benzoic acid is not a non-electrolyte. It dissociates to some extent:

$$C_6H_5COOH(aq) + H_2O(l) = C_6H_5COO^- + H_3O^+;$$

$$K = K_a = [(C_6H_5COO^-][H_3O^+]/[C_6H_5COOH(aq)] \quad (5.7b)$$

and its acidic dissociation constant K_a is equal to 6.1×10^{-5} mol dm^{-3}. Adding the equations for these two reactions gives

$$C_6H_5COOH(s) + H_2O(l) = C_6H_5COO^- + H_3O^+;$$

$$K = K_0K_a = [C_6H_5COO^-][H_3O^+]. \quad (5.8a)$$

The product K_0K_a is the solubility product of benzoic acid. The chemical equation is just like the ones in eqs. (5.1) and (5.2); to be sure, there is a molecule of water on the left-hand side here, but only because we chose to recognize the fact that the hydrogen ion is very strongly hydrated in aqueous solutions. We might write

$$C_6H_5COOH(s) = C_6H_5COO^- + H^+;$$

$$K = K_{C_6H_5COOH} = K_0 K_a = [C_6H_5COO^-][H^+] \qquad (5.8b)$$

without qualms if it is not the acid-base behavior that is foremost in our minds. The product of the concentrations of hydronium (or hydrogen) and benzoate ions in eqs. (5.8) is just like the product of the concentrations of silver and iodate ions in eq. (5.1), or the product of the concentrations of silver and chloride ions in the expression for the solubility product of silver chloride.

Let us try to calculate the solubility of benzoic acid from this solubility product in the same way that we calculated the solubility of silver iodate from its solubility product in Section 5.2. The reasoning is exactly the same. We suppose that S mole of benzoic acid dissolves in water, giving a cubic decimeter of the solution, and that S mole of hydrogen ion has been produced along with S mole of benzoate ion. On this basis we write

$$[C_6H_5COO^-][H^+] = (S)(S) = S^2 = K_{C_6H_5COOH} = 1.43 \times 10^{-6} \, \text{mol}^2 \, \text{dm}^{-6}$$

and obtain $S = 1.2 \times 10^{-3} \, \text{mol dm}^{-3}$.

This is disastrous. The solubility of benzoic acid in water at 298 K is nothing like $1.2 \times 10^{-3} \, \text{mol dm}^{-3}$; it is $2.5 \times 10^{-2} \, \text{mol dm}^{-3}$, which is 21 times as large as the calculated value.

The calculated value is wrong because there is nowhere near S mole of either hydrogen ion or benzoate ion in a cubic decimeter of a solution in which S mole of benzoic acid has been dissolved. If you look at eqs. (5.7) instead of at eqs. (5.8) you can see why this is true. When K_0 mole of solid benzoic acid dissolves, K_0 mole of dissolved benzoic acid is produced. If none of it dissociated, the whole situation would be just like the ones discussed in Chapter 4, and the solubility of benzoic acid and the concentration of dissolved molecules of benzoic acid would both be equal to K_0 mol dm^{-3}. However, some of the benzoic acid does dissociate. The dissociation consumes a little of the dissolved acid, and a little more of the solid must dissolve to make up for this loss because the concentration of the undissociated molecules must be equal to K_0 (mol dm^{-3}) at equilibrium. This requirement is expressed by eq. (5.7a). If one benzoate ion (or one hydrogen ion) is produced by dissociation, one additional molecule of benzoic acid must dissolve to replace the one that dissociated. If the solution contains c M benzoate ion (or hydrogen ion) at equilibrium, then c mole of benzoic acid must have dissolved in each cubic decimeter of the solution to provide the molecules of benzoic acid that dissociated to produce the benzoate ion, and in addition K_0 mole of benzoic acid must have dissolved in each cubic decimeter of the solution to provide the concentration of undissociated acid that is needed to satisfy eq. (5.7a). The total solubility S of benzoic acid is given by

$$S = [C_6H_5COOH(aq)] + [C_6H_5COO^-] = K_0 + c. \qquad (5.9)$$

The first equality is a conservation equation; the second is a translation into the symbols being used here. In the calculation that gave rise to this discussion, K_0 was ignored and the solubility S was equated to the concentration of benzoate ion. The result was wrong because some of the benzoic acid remained undissociated: K_0 was not equal to zero. The result was grossly wrong because most of the benzoic acid remained undissociated: K_0 is about 20 times as large as c.

With the aid of this understanding the problem becomes simple. The value of K_0 for benzoic acid in water at 25° is $2.3_5 \times 10^{-2}$ mol dm^{-3}, and the earlier calculation gave the concentration c of benzoate ion as 1.2×10^{-3} M. Substituting these figures into eq. (5.9) gives $S = 2.3_5 \times 10^{-2} + 1.2 \times 10^{-3} = 2.5 \times 10^{-2}$ mol dm^{-3}, which of course is the correct result.

Exactly the same problem arises in calculations involving the common-ion effect, but here it causes errors that are even more severe. Consider the solubility of benzoic acid in an aqueous solution containing 0.01 M hydrochloric acid at 298 K. Following the procedure described in Section 5.3, we begin by saying that the concentration of hydrogen ion in this solution is 0.01 M before any benzoic acid dissolves in it, because hydrochloric acid is a strong acid. Then, because benzoic acid is a weak acid that does not dissociate extensively even when it is present alone, and because its dissociation is repressed by the presence of the hydrochloric acid, it would be reasonable to make the approximation that $[H^+] = 0.01$ M at equilibrium, neglecting the few additional hydrogen ions obtained from the benzoic acid. As the concentration c of dissolved benzoate ion in the saturated solution can be obtained from eq. (5.8b)

$$c = [C_6H_5COO^-] = K_{C_6H_5COOH}/[H^+] = 1.43 \times 10^{-6}/0.01 = 1.43 \times 10^{-4}\ M$$

one would say that the solubility of benzoic acid is 1.43×10^{-4} M if the concentration of the undissociated acid were small enough to neglect. However, that concentration is still equal to 2.35×10^{-2} M, as it is in a saturated solution of benzoic acid in pure water. The presence of the hydrochloric acid does not affect the position of the equilibrium represented by eq. (5.7a). The solubility of benzoic acid is equal to the sum of the concentration of benzoate ion (1.43×10^{-4} M) and the concentration of undissociated benzoic acid (2.35×10^{-2} M); it is 2.36×10^{-2} M. The presence of the hydrochloric acid causes the ionic concentration c to be very small because the dissociation of benzoic acid is repressed, but no matter how much hydrochloric acid is present the total solubility of benzoic acid cannot become smaller than 2.35×10^{-2} M, the equilibrium concentration of undissociated benzoic acid. The effect of the concentration of hydrochloric acid on the solubility of benzoic acid is shown in Fig. 5.2 to aid you in comparing this situation with the one depicted by Fig. 5.1.

SUMMARY: The solubility product of a substance describes the concentrations of its ions in a saturated solution. If the substance is incompletely dissociated, the concentration of its undissociated form must be taken into account separately.

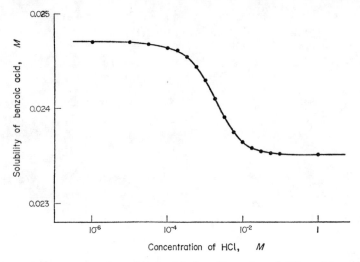

FIG. 5.2. Effect of the concentration of hydrochloric acid on the solubility of benzoic acid. The difference between the shape of this curve and the shape of the one in Fig. 5.1 reflects the fact that silver iodate is completely dissociated in solution whereas benzoic acid is not.

5.5. Solubilities and solubility products

PREVIEW: The ideas outlined in the preceding section are discussed further and applied to other kinds of sparingly soluble substances.

The dissolution of a substance CA can always be considered to occur in two steps:

$$CA(s) = CA(aq); \quad K = K_0 = [CA(aq)], \tag{5.10a}$$

$$CA(aq) = C^+ + A^-; \quad K = K_c = [C^+][A^-]/[CA(aq)]. \tag{5.10b}$$

The symbol K_c is used to represent the overall dissociation constant of the dissolved solute. There will be a formal definition of the overall dissociation constant in Section 5.6. For the purposes of this section it suffices to say that K_c would be equal to zero for a non-electrolyte, and that for benzoic acid in water it would be equal to the familiar acidic dissociation constant K_a. There will be no C^+ or A^- in the solution unless the solute is an electrolyte and K_c has some finite value, but if the ions are formed they are assumed to have equal but opposite charges (so that the solute is CA rather than, say, CA_2) because this simplifies the algebra.

The solubility product K_{CA} is given by

$$K_{CA} = K_0 K_c = [C^+][A^-] \tag{5.11}$$

which can be compared with eqs. (5.8). The solubility product is uninteresting if the solute is a non-electrolyte, for K_{CA} is equal to zero if K_c is equal to zero. Nevertheless eq. (5.11) is valid for non-electrolytes and electrolytes alike.

The solubility of CA in the pure solvent is given by

$$S = [CA(aq)]+[C^+] \ (= [CA(aq)]+[A^-]) = K_0+c. \tag{5.12}$$

We continue to assume that C^+ and A^- are stable; if either is consumed by hydrolysis or some other reaction the picture becomes more complicated. In the absence of any such reaction the concentrations of C^+ and A^- must be equal as long as there is no other source of either, and they will be given by

$$[C^+] = [A^-] = c = K_{CA}^{1/2} = (K_0 K_c)^{1/2}. \tag{5.13}$$

Of course c will be equal to zero if CA is a non-electrolyte.

Equations (5.12) and (5.13) can be combined to obtain

$$S = K_0+(K_0 K_c)^{1/2} = K_0^{1/2}(K_0^{1/2}+K_c^{1/2}) \tag{5.14}$$

in which the second equality is produced by separating the factor $K_0^{1/2}$, which is common to both terms. Equation (5.14) is the fundamental algebraic description of the solubility in the pure solvent.

There are two extreme cases. In one of them, K_c is equal to zero, CA is a non-electrolyte, and its solubility S is equal to K_0. In the other, $K_c^{1/2}$ is very much larger than $K_0^{1/2}$,[†] the dissolved CA is almost completely dissociated into its ions, and eq. (5.14) becomes

$$S = K_0^{1/2}(K_0^{1/2}+K_c^{1/2}) \approx K_0^{1/2}(K_c^{1/2}) = K_{CA}^{1/2}.$$

Many substances fall somewhere between these extremes. For calcium sulfate the solubility product is equal to 1.2×10^{-6} mol² dm⁻⁶ and K_c, which is the overall dissociation constant of dissolved $CaSO_4(aq)$, is equal to 5×10^{-3} mol dm⁻³. Consequently

$$K_0 = K_0 K_c/K_c = K_{CaSO_4}/K_c = 1.2\times10^{-6}/5\times10^{-3} = 2.4\times10^{-4} \text{ mol dm}^{-3}. \tag{5.15}$$

The solubility of calcium sulfate in pure water is therefore given by

$$S = K_0+(K_0 K_c)^{1/2} = 2.4\times10^{-4}+(1.2\times10^{-6})^{1/2}$$
$$= 2.4\times10^{-4}+1.1\times10^{-3} = 1.3\times10^{-3} \text{ mol dm}^{-3}.$$

It is about 20% larger than the value that would be obtained by neglecting the existence of the undissociated $CaSO_4(aq)$ and writing simply $S = K_{CaSO_4}^{1/2}$ as Section 5.2 might suggest.

In the presence of a common ion the undissociated solute becomes a much larger fraction of the total. For silver chloride the solubility product is equal to 1.77×10^{-10}

[†] Since both K_0 and K_c are smaller than 1 for any sparingly soluble substance, $K_c^{1/2}$ will not be very much larger than $K_0^{1/2}$ unless K_c is *very* much larger than K_0. For example, if $K_c = 1\times10^{-2}$ and $K_0 = 1\times10^{-4}$, $K_c^{1/2} (= 0.1)$ will be only 10 times as large as $K_0^{1/2} (= 0.01)$.

$mol^2 dm^{-6}$ and K_c is equal to $2 \times 10^{-3} mol dm^{-3}$, so that $K_0 = 1.77 \times 10^{-10}/2 \times 10^{-3} = 9 \times 10^{-8} mol dm^{-3}$. In pure water the solubility is equal to $9 \times 10^{-8} + (1.77 \times \times 10^{-10})^{1/2} = 9 \times 10^{-8} + 1.33 \times 10^{-5} = 1.34 \times 10^{-5} mol dm^{-3}$. Only for some very special purpose would the presence of so little undissociated AgCl(aq) in the saturated solution be worth taking into account. If silver chloride is dissolved in a solution containing 0.001 M silver ion the concentration of chloride ion in the saturated solution will be only $1.77 \times 10^{-7} M (= K_{AgCl}/[Ag^+])$. For each cubic decimeter of this solution, 1.77×10^{-7} mole of silver chloride must dissolve to furnish the chloride ions that it contains, and an additional 9×10^{-8} mole of silver chloride must dissolve to furnish the undissociated AgCl(aq). The undissociated material now accounts for a third of the total solubility. If the concentration of silver ion is increased to 0.01 M the concentration of chloride ion drops to $1.77 \times 10^{-8} M$, the total solubility becomes $1.08 \times 10^{-7} M (= 1.77 \times 10^{-8} + 9 \times 10^{-8})$, and five-sixths of this is due to the presence of the undissociated material.

Calculations like those in Sections 5.2, 5.3, and 5.4 yield values of the concentrations of the ions in solutions like these. Sometimes the amount of material that dissolves but remains undissociated is so much larger—or smaller—than the amount that dissolves and dissociates that you can neglect one or the other. Sometimes it is necessary to include both. Whether you can neglect either in any particular problem depends on the conditions of that problem, the accuracy you demand, and the values of K_0 and K_c. Since the values of the solubility product, K_0, and K_c are related by eq. (5.11), any one of them can be calculated if the other two are known, and values of K_0 are always obtained by calculating them from the solubility product and dissociation constant. You can be fairly sure that the undissociated solute should be taken into account if its ions are polyvalent or if the common-ion effect is involved. If you cannot decide whether the undissociated material is significant, you should calculate the ionic solubility from the expression for the solubility product, using procedures like those described in Sections 5.2 to 5.4, and compare the result with the value of K_0 calculated from the solubility product and the dissociation constant. It will then be easy for you to decide what to do next.

SUMMARY: The solubilities of non-electrolytes and strong and weak electrolytes can all be described by a single equation; the differences among them arise from differences among their dissociation constants.

5.6. Weak electrolytes and ion pairs

PREVIEW: This section explains why most electrolytes are not completely dissociated in solution.

There are two different ways in which two ions—say C^{n+} and A^{m-}—may be held together in a solution. There may be chemical bonding between them, as there is between hydrogen and benzoate ions, and there is also an electrostatic attraction between them because they are oppositely charged. For either or both of these reasons, it may be difficult to tear them apart once they have collided, and the concentrations of the free ions decrease as a result. The process may be represented by the equation

$$C^{n+} + A^{m-} = \text{adduct.}$$

The word "adduct" means merely "the product obtained on adding C^{n+} to A^{m-}". The adduct may be an uncharged molecule: if C^{n+} is hydrogen ion and A^{m-} is benzoate ion the adduct will be a molecule of benzoic acid. It may be an ion: if C^{n+} is iron(III) ion and A^{m-} is sulfate ion the adduct will be the $FeSO_4^+$ ion. Whether it is electrically charged or not, it may be held together by the electrostatic force alone, or by chemical bonding as well, and if there is a chemical bond it may be very much stronger than the electrostatic one. In silver nitrate chemical bonding is very weak and electrostatic attraction predominates, while in silver chloride the chemical force is much stronger than the electrostatic one. By using the non-committal term "adduct" we merely avoid emphasizing any particular one of these different possibilities.

If chemical bonding is known to predominate, the adduct is usually called either a *weak electrolyte*, if it is electrically neutral, or a *complex ion*, if it does have an electric charge. If A^{m-} is a sufficiently strong electron donor (Lewis base) and C^{n+} is a sufficiently strong electron acceptor (Lewis acid), an A^{m-} ion can donate a pair of electrons to a C^{n+} ion, and a covalent bond will be formed. The strength of that bond will depend both on the ability of C^{n+} to accept electron pairs and on the ability of A^{m-} to donate them.

If C^{n+} is a metal ion, its ability to accept electron pairs depends on the electronic structures of C^{n+} and the neutral C atom, and also on the charge carried by the C^{n+} ion. A useful generalization for our purposes is that the ability of C^{n+} to accept electron pairs increases as its charge increases: univalent ions such as sodium and potassium ions are very weak electron acceptors, divalent ions such as lead and calcium ions are stronger ones, and trivalent ions such as iron(III) and aluminum ions are stronger still. In aqueous solutions these differences are largely counteracted by the occurrence of hydration. Because an iron(III) ion accepts electron pairs more readily than a lead ion, it forms stronger bonds to either a chloride ion or a molecule of water than a lead ion does. The reaction between either of these metal ions and chloride ion in an aqueous solution is essentially

$$C^{n+}{-}OH_2 + Cl^- = C^{n+}{-}Cl^- + H_2O(l)$$

Both the bond that is formed and the bond that is broken are stronger when C^{n+} is iron(III) ion than when it is lead ion. Here the result is almost a stand-off: the ability of a chloride ion to replace a molecule of water in the coordinate shell is almost the same with iron(III) ion as it is with lead ion. In other cases, however, there are differences arising from the particular chemical natures and electronic structures of C^{n+} and A^{m-}: $HgCl^+$ is much more stable than $PbCl^+$ even though Hg^{2+} and Pb^{2+} have equal charges, and FeF^{2+} is much more stable than PbF^+ even though the stabilities of $FeCl^{2+}$ and $PbCl^+$ are not very different.

The ability of A^{m-} to donate electron pairs depends on its structure and on the identities of the electron-rich atoms that it contains. Inorganic oxyanions such as perchlorate, nitrate, sulfate, and carbonate are weak electron donors, and so are complex anions such as hexacyanoferrate(III), $Fe(CN)_6^{3-}$, and tetrafluoroborate, BF_4^-. Halide and thiocyanate ions are stronger electron donors, and cyanide ion is stronger still. The value of the overall dissociation constant of $HgNO_3^+$ is probably about 0.1 mol dm^{-3}; this adduct owes its existence to electrostatic attraction. For $HgCl^+$, $HgBr^+$, and HgI^+ the values are 5×10^{-6}, 1×10^{-9}, and 1.3×10^{-13} mol dm^{-3}; for $HgCN^+$ it is 1×10^{-18} mol dm^{-3}. Since the metal ion is the same in all of these cases, the energies that are consumed in breaking the bond between a mercury(II) ion and a molecule of water are also the same, and therefore the differences among the stabilities of these complexes reflect the differing strengths of the bonds that are formed. Ions that contain atoms of oxygen or sulfur with localized negative charges, such as OH^- and SH^-, are strong electron donors, although their strength is greatly decreased if the charge is delocalized, as it is in acetate ion by resonance between the extreme structures

or in phenoxide ion, $-O^-$, by interaction with the π-electron cloud of the aromatic ring. Ligands that contain two or more electron-donating groups, such as

oxalate ion, , or glycinate ion, H_2NCH_2C ,

may form very strong adducts if the sizes of the rings in the resulting chelates are conducive to stability (Section 4.5). The value of the dissociation constant of $CuOAc^+$ is 1.6×10^{-3} mol dm^{-3}, while that of CuC_2O_4 is 1.6×10^{-7} mol dm^{-3}. The bidentate ("two-toothed") ligand has the advantage of being able to form two bonds at once, but the advantage disappears if the ligand must be tortured into an unnatural shape to achieve it: succinate ion, $^-OOCCH_2CH_2COO^-$, is no more strongly bonded to copper(II) ion than acetate ion is, because the copper(II)–succinate chelate would contain a seven-membered ring if it formed at all.

The electrostatic force between C^{n+} and A^{m-} is described by Coulomb's law:

$$F \propto |z_1 z_2|/\varepsilon r^2 \tag{5.16}$$

where z_1 and z_2 are the charges on the ions, r is the distance between them, and ε is the dielectric constant of the solution. If each ion were a rigid sphere, the smallest possible value of r would be equal to the distance between their centers. For any particular pair of ions there is a minimum value of r, which is called the *distance of closest approach* and is represented by the symbol $å$. The value of $å$ would be equal to the sum of the ionic radii if the ions were rigid spheres, but that is an oversimplified picture. Not all ions are spherical, and the shape of an ion may be altered by the electric field surrounding another ion that is very close to it. Moreover, molecules of the solvent may be so firmly attached to the ions that actual contact is impossible. For these reasons it is difficult to predict or interpret the value of $å$ for any given pair of ions, and $å$ must therefore be regarded as an empirical parameter.

The force between the ions increases as r decreases, and reaches its maximum value when $r = å$. If the ions are large, even this maximum force may be so small that the ions can easily be knocked apart if another ion or a molecule of solvent collides with either of them. This is especially likely if the charges on both ions are small (as they are if both ions are univalent, so that $z_1 = 1$ and $z_2 = -1$) and if the dielectric constant is high (as it is in dilute aqueous solutions, for which ε is approximately equal to 80).

However, if $å$ is small, if either ion is polyvalent, or if ε is small, the electrostatic force between two oppositely charged ions can become so large that an average molecule of the solvent does not have enough kinetic energy to knock them apart. Two such ions will then remain together until an unusually energetic molecule of the solvent comes along. As was said in Section 3.11, unusually energetic molecules are rare, and therefore the ions may remain together for a long time. In this state they are said to constitute an *ion pair* and are represented by a formula like $C^{n+}A^{m-}$. Ion pairs are in equilibrium with free ions:

$$C^{n+}A^{m-} = C^{n+} + A^{m-}; \quad K_d = [C^{n+}][A^{m-}]/[C^{n+}A^{m-}]. \tag{5.17}$$

The equilibrium constant K_d is the *dissociation constant* of the ion pair.

There are some ions, especially polymeric ones, that have sizes quite out of the ordinary range, but the great majority of common ions are all about the same size and the values of $å$ for different ion pairs do not differ very much. A rough, but useful, guide for aqueous solutions at 298 K is

$$\log K_d = -|z_1 z_2|/2 \tag{5.18}$$

where K_d is understood to have the units mol dm^{-3}. For calcium sulfate, where $z_1 = 2$ and $z_2 = -2$, this gives $\log K_d = -2$ or $K_d = 0.01$ mol dm^{-3}; for the ion pair $Fe^{3+}SO_4^{2-}$, where $z_1 = 3$ and $z_2 = -2$, it gives $\log K_d = -3$ or $K_d = 10^{-3}$ mol dm^{-3}. These and other values obtained from eq. (5.18) will be discussed in the following section. A comparable guide for solutions in other solvents at 298 K is

$$\log K_c = -40|z_1 z_2|/\varepsilon \tag{5.19}$$

where K_d again has the units mol dm^{-3}. Since ε is approximately equal to 80 for water at 298 K, eq. (5.19) includes eq. (5.18) as a special case. Whereas the dissociation constant of $Ca^{2+}SO_4^{2-}$ is approximately equal to 0.01 mol dm^{-3} in water, its value is roughly 10^{-5} mol dm^{-3} in methanol ($\varepsilon = 32.6$), 10^{-7} mol dm^{-3} in ethanol ($\varepsilon = 24.3$), and 10^{-26} mol dm^{-3} in glacial acetic acid ($\varepsilon = 6.15$), all at 298 K.

SUMMARY: Two dissolved ions may be bound by both chemical and electrostatic forces. If the chemical force predominates, the substance is called a weak electrolyte or a complex ion; if the electrostatic force predominates, it is called an ion pair. The magnitude of the electrostatic force, and hence the stability of the ion pair, depends on the charges and sizes of the ions and on the dielectric constant of the solvent.

5.7. Ionization and dissociation constants

PREVIEW: The "dissociation" of a dissolved electrolyte can be regarded as occurring in two steps. In the first, called *ionization*, the chemical bond between the ions is broken and an ion pair is formed. In the second, called *dissociation*, the electrostatic bond is broken and the free ions are formed.

Equation (5.11) showed that the solubility product for an electrolyte depends on its overall dissociation constant, and eq. (5.14) showed that the same thing is true of the solubility. With the aid of the ideas presented in Section 5.6 it is now possible to define the overall dissociation constant and discuss the factors that affect it.

We consider an undissociated molecule, CA, that is held together by chemical or electrostatic forces or both, and imagine that its transformation into free C^+ and A^- ions takes place in two steps. In the first, any chemical bond is broken and the product is an ion pair that is held together by electrostatic attraction alone:

$$CA(aq) = C^+A^-; \quad K_i = [C^+A^-]/[CA(aq)]. \tag{5.20}$$

This process is called *ionization*, and its equilibrium constant K_i is called an *ionization constant*. The value of K_i for any particular substance depends on the strength of the chemical bond that is broken: if the bond is very weak the value of K_i is very large, but if the bond is very strong very little of the CA is ionized and the value of K_i is very small. A non-electrolyte is simply a substance in which even the weakest bond is so strong that the value of K_i is indistinguishable from zero.

In the second step, the ion pair dissociates to give the free ions. The equations

$$C^+A^- = C^+ + A^-; \quad K_d = [C^+][A^-]/[C^+A^-] \tag{5.17}$$

that describe this *dissociation* were given in Section 5.6.

What is interesting in practical applications is the relationship between the concentrations of the ions and the total concentration of dissolved but undissociated CA, regardless of whether that CA is ionized or not. You should not allow eqs. (5.20) and (5.17) to mislead you into thinking that there are two different species, CA and C^+A^-, for there is really only one, and splitting the overall process into two steps is simply an aid to understanding the factors that govern its extent. If chemical bonding in the one species that does exist is much stronger than electrostatic attraction, then it is chiefly the chemical bonding that must be overcome in order to obtain the free ions, and saying that K_i is small and K_d large for such a substance is merely a short and convenient way of describing the situation. At the other extreme there might be an equally stable species in which chemical bonding is weak or even non-existent, so that it is chiefly or entirely the electrostatic force that must be overcome in order to obtain the free ions; here the same ideas can be conveyed more briefly by saying that K_i is large and K_d small.

Regardless of the relative strengths of the chemical and electrostatic forces, the useful expression is that for the *overall dissociation constant K_c* defined by

$$K_c = \frac{[C^+][A^-]}{\text{concentration of undissociated CA}}.$$ (5.21)

In the symbols used in eqs. (5.20) and (5.17), the denominator of this expression is the sum of the concentrations of the unionized molecule $CA(aq)$ and the ion pair C^+A^-. From eq. (5.20),

$$[CA(aq)] = [C^+A^-]/K_i$$

so that

$$\text{concentration of undissociated CA} = [CA(aq)] + [C^+A^-]$$

$$= [C^+A^-]/K_i + [C^+A^-] = [C^+A^-](1 + 1/K_i).$$

Combining this with eq. (5.21), with an eye on eq. (5.17) in writing the second equality

$$K_c = \frac{[C^+][A^-]}{[C^+A^-](1 + 1/K_i)} = \frac{K_d}{1 + 1/K_i} = \frac{K_i K_d}{1 + K_i}.$$ (5.22)

You can see that a careful distinction must be drawn between the dissociation constant K_d and the overall dissociation constant K_c.

As usual, there are two extreme situations. In one of them, chemical bonding is very weak and the stability of the undissociated CA is almost entirely due to the electrostatic attraction of its ions. Here K_i is very large, $(1 + K_i)$ is nearly equal to K_i, and

$$K_c \approx K_d.$$ (5.23a)

In the other, chemical bonding is very strong and K_i is very small. Here $(1 + K_i)$ is nearly equal to 1, and

$$K_c \approx K_i K_d.$$ (5.23b)

In eq. (5.22) K_d is multiplied by a factor $[K_i/(1 + K_i)]$ whose value cannot exceed 1. Consequently the value of K_d that is crudely approximated by eqs. (5.18) and (5.19) is the maximum value that the overall dissociation constant can have for an electrolyte of any particular charge type in any particular solvent. That value decreases as the dielectric constant decreases, and therefore every electrolyte is a weak electrolyte in a solvent of low dielectric constant. It also decreases as the ionic charges increase, and therefore every electrolyte is a weak electrolyte even in water unless both of its ions are univalent. (According to eq. (5.18) even a uni-univalent electrolyte should have a finite dissociation constant in water, and there are many such electrolytes that do. However, there are a few ions that are larger than the average assumed in eq. (5.18); these include hydrogen, lithium, perchlorate, and iodate ions, and some compounds containing these ions do appear to be completely dissociated in water at 298 K. This is why silver and thallium(I) iodates were so prominent in Sections 5.2 and 5.3.) There are a few solvents—

$$\text{formamide} \quad \left(\text{HC} \underset{\text{NH}_2}{\overset{\text{O}}{\diagup}} \right)$$

and some compounds related to it, liquid hydrogen cyanide, and anhydrous sulfuric acid among them—that have dielectric constants higher than that of water, but the resulting tendency for K_d to be larger in these solvents than in water is usually counteracted by the fact that ionic species are less stabilized by solvation in these solvents. There is no way of getting around it: in the overwhelming majority of solvents, including water, the overwhelming majority of electrolytes are weak electrolytes.

Table 5.1 illustrates these ideas with the aid of some numerical values pertaining to aqueous solutions at 298 K. Its first column gives the ionic charges z_1 and z_2,

TABLE 5.1. *Overall dissociation constants for some ion pairs and complexes*

Charge type	Value of K_d (mol dm^{-3}) estimated from eq. (5.18)	Overall dissociation constants, mol dm^{-3}	
		Ion pairs	Complexes or undissociated molecules
1–1	0.3	5 (NaOH), 2 (AgNO$_3$), 1.7 (KNO$_3$), 1.5 (AgF)	5×10^{-3} (AgOH), 6×10^{-5} (AgBr), 5×10^{-10} (HCN)
2–1	0.1	0.8 (CuCl$^+$), 0.6 (AgSO$_4^-$), 0.08 (PbNO$_3^+$ or NH$_4$SO$_4^-$), 0.054 (NaCO$_3^-$), 0.04 (CaOH$^+$)	1.2×10^{-2} (HSO$_4^-$), 2×10^{-3} (PbOAc$^+$), 5×10^{-11} (HCO$_3^-$), 10^{-18} (HgCN$^+$)
3–1	0.03	0.1 (FeNO$_3^{2+}$), 0.04 (K[Co^{III}(CN)$_6$]$^{2-}$), 0.012 (Ce^{III}ClO$_4^{2+}$)	5×10^{-5} (BiBr^{2+}), 8×10^{-7} (AlF^{2+}), 8×10^{-12} (FeOH^{2+})
2–2 } 4–1 }	0.01	$\begin{cases} 5\times10^{-3} \text{ (BaC}_2\text{O}_4 \text{ or CaSO}_4\text{)}, \\ 6\times10^{-4} \text{ (CaCO}_3\text{)} \\ 5\times10^{-3} \text{ (K[Fe}^{II}\text{(CN)}_6\text{]}^{3-}\text{)}, \\ 6\times10^{-4} \text{ (Tl}^I\text{[Fe}^{II}\text{(CN)}_6\text{]}^{3-}\text{)} \end{cases}$	1×10^{-3} (Cu[tartrate]), 5×10^{-6} (NiC$_2$O$_4$) 8×10^{-9} (HP$_2$O$_7^{3-}$)
3–2	10^{-3}	1.4×10^{-2} ([Co^{III} (H$_2$NCH$_2$CH$_2$NH$_2$)$_3$]CO$_3^+$), 1.5×10^{-3} (Ca[Fe^{III}(CN)$_6$]$^-$), 1×10^{-3} (Fe^{III}SO$_4^+$)	6×10^{-4} (Mg[citrate]$^-$), 4×10^{-10} (Fe^{III}C$_2$O$_4^+$), 6×10^{-15} (Cu^{II} [citrate]$^-$)
4–2	10^{-4}	1.8×10^{-4} (Ca[Fe^{II}(CN)$_6$]$^{2-}$), 2×10^{-5} (BaP$_2$O$_7^{2-}$)	2×10^{-7} (CuP$_2$O$_7^{2-}$)
3–3	3×10^{-5}	1.8×10^{-4} (La[Fe^{III}(CN)$_6$])	1.6×10^{-12} (Fe^{III} [citrate])

without regard to sign because they must always have opposite signs and because the signs disappear from eq. (5.18). With the exception indicated by the brace, the value of $|z_1 z_2|$ increases from each line to the next. The second column gives the value that is predicted by eq. (5.18) for the dissociation constant of an ion pair formed from ions of the charge type given in the first column. The third column gives the experimentally determined values of the overall dissociation constants for some species in which chemical bonding is weak or non-existent, and includes the formulas of these species. The values in this column are directly comparable with those in the second column because both refer to ion pairs. The fourth column gives similar information for some other species of the same charge type in which there is more or less strong chemical bonding in addition to electrostatic attraction.

Several conclusions can be drawn. Comparing the second and third columns indicates that eq. (5.18) may be in error by a factor of 3 or more for any particular ion pair. This is because it ignores the specific properties of the individual ions; since these are not all the same, the dissociation constants of ion pairs of the same charge type can range over an order of magnitude or more. Comparing the third and fourth columns shows that chemical bonding can cause the overall dissociation constant to be many orders of magnitude smaller than it would be if electrostatic attraction alone were involved, but it also shows that some cases are not at all clearcut. Some of the classifications are necessarily arbitrary. Few chemists would be willing to believe that there is no chemical bonding between cupric and tartrate ions, and fewer yet would accept the idea that there is appreciable bonding between calcium and carbonate ions, but it would be very difficult to draw a distinction between these ion pairs on the basis of their overall dissociation constants.

> SUMMARY: Most electrolytes are incompletely dissociated in solution. The overall dissociation constant of any particular electrolyte depends on both its ionization constant and its dissociation constant. The ionization constant reflects the strength of the chemical bond that is broken in forming the ion pair; the dissociation constant reflects the strength of the electrostatic bond.

5.8. Conclusion

Far from covering everything that needs to be said about the solubilities of ionic substances, this chapter has left three important topics to be discussed in later ones. The first is the behavior of electrolytes—such as hydroxides, sulfides, phosphates, and carbonates, among a great many others—that contain basic (or acidic) ions. The dissolution of such an electrolyte in water (or any other amphoteric solvent) affects, and is affected by, the autoprotolytic process, and to such an extent that nothing said so far in this book would enable us to calculate the solubility of iron(III) hydroxide or copper(II) sulfide in pure water within a factor of 10^6. The relations between solubility and acid-base processes, including autoprotolysis, are discussed in Section 6.11. The second deals with the effects of stepwise complex formation on solubility: whereas the solubility of silver chloride in water is decreased by adding a little hydrochloric acid, it increases again on adding more, and it increases so much that silver chloride is far more soluble in concentrated hydrochloric acid than in water. Phenomena like this are discussed in Section 7.4. The third includes the rate and mechanism of precipitation, the sizes of the particles of precipitate obtained under different conditions, the nature and extent of adsorption onto the surfaces of those particles, and the effects of all these on the purity of the precipitate. These subjects are discussed in Chapter 10. Dissolution and precipitation are so common and so important that you should not be surprised to find them reappearing in many different contexts.

Problems

Neglect possible ion-pair formation in solving Problems 5.1 through 5.10. Answers to some of these problems are given on page 520.

5.1. The concentration of thallium(I) ion, Tl^+, in a saturated aqueous solution of thallium(I) chloride at 298 K is 0.013 M. What is the solubility product of thallium(I) chloride in water at 298 K?

5.2. A saturated aqueous solution of silver oxide, Ag_2O, contains 1.6×10^{-4} M hydroxide ion at 298 K. What is the solubility product of silver oxide in water at 298 K? The reaction that occurs when silver oxide dissolves in water is $\frac{1}{2} AgO(s) + \frac{1}{2} H_2O(l) = Ag^+ + OH^-$.

5.3. Excess solid silver oxide is added to an aqueous solution containing 1.00×10^{-4} M sodium hydroxide at 298 K. What will be the concentration of hydroxide ion in the solution when equilibrium is reached?

5.4. At 298 K the solubility product of mercury(I) iodide, Hg_2I_2, in water is equal to 4.5×10^{-29} mol^3 dm^{-9}. What is the concentration of mercury(I) ion, Hg_2^{2+}, in a saturated aqueous solution of mercury(I) iodide at 298 K? Mercury(I) iodide dissolves in water according to the equation $Hg_2I_2(s) = Hg_2^{2+} + 2 I^-$.

5.5. A solution containing 0.100 M strontium nitrate, $Sr(NO_3)_2$, was added dropwise to 100 cm^3 of a solution containing 0.0100 M sodium iodate. The first permanent precipitate of strontium iodate appeared when 3.6 cm^3 of strontium nitrate solution had been added. What was the solubility product of strontium iodate under the conditions of this experiment?

5.6. An excess of solid lead fluoride is added to a solution containing 1.00×10^{-3} M potassium iodide. The solubility products of lead fluoride and lead iodide under the conditions employed are $K_{PbF_2} = 2.7 \times 10^{-8}$ mol^3 dm^{-9} and $K_{PbI_2} = 7.1 \times 10^{-9}$ mol^3 dm^{-9}, respectively. What will be the concentration of iodide ion in the solution when equilibrium is reached?

5.7. The solubility product of silver bromate, $AgBrO_3$, in water at 298 K is equal to 5.2×10^{-5} mol^2 dm^{-6}. Calculate the solubility of silver bromate at 298 K

(a) in pure water,
(b) in an aqueous solution containing 0.100 M sodium bromate,
(c) in one containing 1.00×10^{-3} M sodium bromate,
(d) in one containing 1.00×10^{-2} M sodium bromate.

5.8. The solubility product of calcium fluoride, CaF_2, in water at 298 K is equal to 4.9×10^{-11} mol^3 dm^{-9}. Calculate the solubility of calcium fluoride at 298 K

(a) in pure water,
(b) in an aqueous solution containing 0.0100 M sodium fluoride,
(c) in one containing 1.00×10^{-4} M sodium fluoride,
(d) in one containing 1.00×10^{-4} M calcium perchlorate.

5.9. The solubility of barium sulfate in water at 298 K is 2.46 mg dm^{-3}, and the formula weight of $BaSO_4$ is 233.40. Calculate the solubility product of barium sulfate in water at 298 K.

5.10. At 298 K the solubility product of thallium(I) bromate, $TlBrO_3$, is equal to 8.5×10^{-5} mol^2 dm^{-6}, while that of silver bromate is equal to 5.2×10^{-5} mol^2 dm^{-6}, both in water. Calculate the solubility of each of these compounds in water that is saturated with both of them at 298 K.

5.11. The solubility of 3-methylbenzoic acid,

in water at 298 K is 0.855 g dm^{-3}. The formula weight of the acid is 136.14, and its (overall) dissociation constant is equal to 5.3×10^{-5} mol dm^{-3}. What is the solubility of the acid in an aqueous solution containing 0.0100 M hydrochloric acid at 298 K?

5.12. At 298 K the solubility product of copper(II) oxalate, CuC_2O_4, in water is equal to 2.3×10^{-8} mol^2 dm^{-6} and the dissociation constant of the ion pair $Cu^{2+}C_2O_4^{2-}$ is equal to 6×10^{-7} mol dm^{-3}.

(a) What is the solubility of copper(II) oxalate in water at 298 K?
(b) What is its solubility in a solution of sodium oxalate containing 0.0100 M oxalate ion?

5.13. How many moles of sodium oxalate, $Na_2C_2O_4$, would have to be dissolved in water at 298 K to give 1 dm^3 of a solution containing 0.100 M oxalate ion? Use eq. (5.18) to estimate K_d for the ion pair $NaC_2O_4^-$.

5.14. The solubilities, in g dm^{-3}, of cesium perchlorate ($CsClO_4$) and thallium(I) iodide (TlI) in water at various temperatures are

T, K	283	293	303	313	323	333	343	353	363	373
S_{CsClO_4}	10	16	26	40	54	73	98	144	205	300
S_{TlI}	0.036	0.06	0.08	0.15	...	0.35	...	0.70	...	1.20

The formula weight of cesium perchlorate is 232.27 and that of thallium(I) iodide is 331.31. What are the values of ΔH for the dissolution of cesium perchlorate and thallium(I) iodide in water over this range of temperatures?

CHAPTER 6

Acid-base Reactions

6.1. Introduction

Acid-base reactions are common and important in every branch of chemistry. The synthesis and analysis of many organic and organometallic compounds depend on their acid-base behaviors and on those of the substances from which they are made and with which they can react. Acid-base reactions play prominent roles in electro-chemistry, colloid chemistry, corrosion, water purification, and a host of other chemical fields, as well as in biochemistry: none of the functions of the body can continue unimpeded in the face of even quite a small change of acidity in the organ that effects it. Much of chemistry is therefore concerned with acid-base reactions. This chapter describes some of the fundamental ideas about such reactions, some of the consequen. ces of those ideas, and some of the ways in which they are expressed quantitatively-

6.2. Some fundamental ideas

PREVIEW: This section deals with the vocabulary of acid-base chemistry.

There are several different ways in which an acid and a base can be defined. One, originated by G. N. Lewis, defines a base as a substance that can donate a pair of electrons to another substance, called an acid, with the formation of a covalent bond. Conversely, an acid is defined as a substance that can form a covalent bond with a base by accepting a pair of electrons from it. According to this definition the reaction of pyridine with tin(IV) chloride

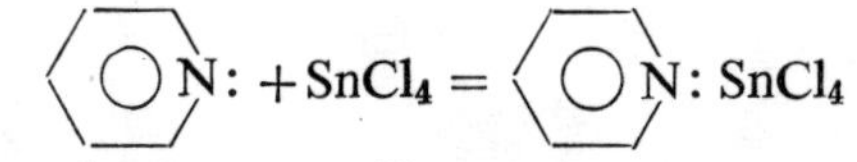

$$\langle \bigcirc N: + SnCl_4 = \langle \bigcirc N: SnCl_4$$

base   acid

in benzene or another inert solvent is an acid-base reaction, and so are the formation of the activated complex in the reaction between iodine and arsenic(III) acid in aqueous solution

$$H{-}O: + As(OH)_3 = H{-}O: As(OH)_3$$
$$\qquad | \qquad\qquad\qquad\qquad |$$
$$\qquad I \qquad\qquad\qquad\qquad\ \ I$$

base   acid

and the reaction of acetate ion with hydronium ion

$$CH_3C \overset{\nearrow O}{\underset{\searrow O \overline{:}}{}} + H_3O^+ = CH_3C \overset{\nearrow O}{\underset{\searrow O:H-OH_2}{}}$$

$$\text{base} \qquad\qquad \text{acid}$$

The products of these and other Lewis acid-base reactions are usually called *adducts* (Section 5.6). In each of these equations the formulas of the original base and the adduct show the electron pair that is donated by the base and accepted by the acid.

The Lewis definitions are widely useful because they help us to correlate observations about many different kinds of reactions that, like the three just mentioned, might seem to have very little in common. Much of our understanding of coordination chemistry and of chemical behavior in inert solvents would have been much more difficult to secure without the help of these definitions.

For other purposes, however, their broad generality is something of a disadvantage, and this is especially true in amphoteric solvents such as water. We are accustomed to thinking of acid-base reactions in water as being reactions that involve protons. This makes it useful to draw a distinction between a reaction such as the hydration of copper(II) sulfate

$$H_2O:+CuSO_4(s) = H_2O:Cu^{2+}+SO_4^{2-}$$

which does not affect the pH, and a reaction such as the one between the fully hydrated copper(II) ion and water

$$H_2O:+ \overset{H\searrow}{\underset{H\nearrow}{}} OCu(OH_2)_3^{2+} = H_3O^+ +HOCu(OH_2)_3^+$$

which does affect it. According to the Lewis definitions, however, each of these is an acid-base reaction, and a distinction between them is hard to draw.

The Lowry–Brønsted[†] definitions are often more convenient. These define an acid as a substance that can donate a proton, and a base as a substance that can accept one. Of course a Lowry–Brønsted base accepts a proton by donating a pair of electrons to it and forming a covalent bond with it: there is no difference between a Lowry–Brønsted base and a Lewis base. The difference between the two definitions is that many substances—such as the anhydrous copper(II) ion in the first example of the preceding paragraph—are Lewis acids because they can accept electron pairs but are not Lowry–Brønsted acids because they have no protons to donate. Most of this book is devoted to the behaviors of solutions in water, and it emphasizes proton-transfer reactions and measurements of pH. Consequently the Lowry–Brønsted definitions will be the ones most often used here.

[†] The Danish letter "ø" is pronounced like the "u" in "hurt"

It follows from these definitions that an acid-base reaction is a reaction that consists of the transfer of a proton from one substance (an acid) to another (a base). Moreover, an acid-base reaction always consists of the reaction of one acid with one base to form a different acid and a different base. In the reaction

$$H_2O(l) + CH_3COOH(aq) = H_3O^+ + CH_3COO^-$$
$$\text{base} \qquad \text{acid} \qquad\qquad \text{acid} \qquad \text{base}$$

the molecule of water is a base because it can accept a proton from the molecule fo acetic acid, and the acetate ion is also a base because it can accept a proton from a hydronium ion. Similarly, both the acetic acid and the hydronium ion are acids because each can donate a proton to a base. Adding a proton to a base gives an acid that is called the *conjugate acid* of the original base, and removing a proton from an acid gives a base that is called the *conjugate base* of the original acid. Hydronium ion is the conjugate acid of water, and acetic acid is the conjugate acid of acetate ion; water is the conjugate base of hydronium ion, and acetate ion is the conjugate base of acetic acid. An acid and its conjugate base together are called a *conjugate acid-base pair*.

An acid may be uncharged, or it may be either a cation or an anion, and the same thing is true of a base. Acetic acid, ammonium ion, and hydrogen sulfate ion (HSO_4^-) are all acids; and ammonia, the triaquahydroxocopper(II) complex $Cu(OH)(OH_2)_3^+$, and acetate ion are all bases. Of course the charge on a base is one unit less positive (or one unit more negative) than the charge on its conjugate acid. Hence a base is uncharged if its conjugate acid is a univalent or singly charged cation (such as ammonium ion), while an acid that is uncharged (such as acetic acid) has a univalent anion as its conjugate base.

A substance is said to be *amphoteric* if it can behave as either an acid or a base. Water is amphoteric: it can behave as an acid, donating a proton to a base such as oxalate ion

$$H_2O(l) + {}^-OOCCOO^- = HOOCCOO^- + OH^-$$
$$\text{acid} \qquad \text{base} \qquad\qquad \text{acid} \qquad \text{base}$$

and it can also behave as a base, accepting a proton from an acid such as oxalic acid

$$H_2O(l) + HOOCCOOH(aq) = HOOCCOO^- + H_3O^+$$
$$\text{base} \qquad \text{acid} \qquad\qquad \text{base} \qquad \text{acid}$$

These two equations show that the hydrogen oxalate ion, $HOOCCOO^-$, is also amphoteric. The first equation, read from right to left, shows it donating a proton to a base (hydroxide ion), and thus behaving as an acid; the second equation, also read from right to left, shows it accepting a proton from an acid (hydronium ion), and thus behaving as a base.

Being both an acid and a base, an amphoteric substance can react with itself:

$$H_2O(l) + H_2O(l) = H_3O^+ + OH^-$$
$$HOOCCOO^- + HOOCCOO^- = HOOCCOOH(aq) + {}^-OOCCOO^-$$
$$\text{acid} \qquad\quad \text{base} \qquad\qquad \text{acid} \qquad\qquad \text{base}$$

When such a reaction occurs between two molecules of a solvent, it is called *auto-protolysis*, and its equilibrium constant is called the *autoprotolysis constant* of the solvent.

Both acids and bases have widely varying strengths, and the strength of each acid is closely related to that of its conjugate base. Perchloric acid is a very strong acid: its ability to donate protons is very high. Because this is true, its conjugate base (perchlorate ion) has very little ability to accept protons and is an extremely weak base. Conversely, water is a very weak acid, having only very little ability to donate protons; its conjugate base, hydroxide ion, accepts protons very readily and is a very strong base. A conjugate acid-base pair consists of a base B^{n+} and its conjugate acid $BH^{(n+1)+}$. The strength of the base reflects its ability to accept protons, and this ability depends on the strength of the covalent bond that is formed when a proton is added to it. Increasing the strength of that bond increases the strength of the base. Increasing the strength of that bond also decreases the ease with which a proton can be removed from the conjugate acid $BH^{(n+1)+}$ and thus decreases its strength as an acid. Hydroxide ion is a stronger base than ammonia because the oxygen–hydrogen bond that is formed when a proton is added to a hydroxide ion is stronger than the nitrogen–hydrogen bond that is formed when a proton is added to a molecule of ammonia. Saying exactly the same thing in different words, the oxygen–hydrogen bonds in water (the conjugate acid of hydroxide ion) are stronger than the nitrogen–hydrogen bonds in ammonium ion (the conjugate acid of ammonia), and because this is true water has less ability to lose protons, and is a weaker acid, than ammonium ion. As the strength of a base increases the strength of its conjugate acid decreases; as the strength of an acid increases the strength of its conjugate base decreases.

The strengths of acids in aqueous solutions are traditionally expressed by their overall dissociation constants, which are given the symbol K_a. The overall dissociation constant of acetic acid is given by

$$HOAc(aq) + H_2O(l) = H_3O^+ + OAc^-; \quad K_a = [H_3O^+][OAc^-]/[HOAc(aq)]$$
$$= 1.8 \times 10^{-5} \, \text{mol dm}^{-3} \, (\text{at 298 K}) \quad (6.1)$$

while that of ammonium ion is given by

$$NH_4^+ + H_2O(l) = H_3O^+ + NH_3(aq); \quad K_a = [H_3O^+][NH_3(aq)]/[NH_4^+]$$
$$= 5.5 \times 10^{-10} \, \text{mol dm}^{-3} \, (\text{at 298 K}). \quad (6.2)$$

The fact that the value of K_a is larger for acetic acid than for ammonium ion means that acetic acid loses protons more easily, and is a stronger acid, than ammonium ion. It also means that acetate ion accepts protons less easily, and is a weaker base, than ammonia.

Once upon a time the strength of a base was expressed by an overall dissociation constant K_b, which for ammonia was given by

$$NH_3(aq) + H_2O(l) = NH_4^+ + OH^-; \quad K_b = [NH_4^+][OH^-]/[NH_3(aq)]$$
$$= 1.8 \times 10^{-5} \, \text{mol dm}^{-3} \, (\text{at 298 K}). \quad (6.3)$$

Values of K_b are still encountered occasionally but are disappearing from the chemical literature: all the most recent compilations of overall dissociation constants give only the value of K_a for the acidic member of each conjugate acid-base pair. Nothing is lost by this, because eq. (6.3) contains no information that is not also contained in eq. (6.2), but there are two advantages. One is that all of our data on the strengths of acids and bases can be collected into one table instead of being divided between two tables. The other, and more important, one is that comparisons are made much easier. From eqs. (6.1) and (6.2) we can tell immediately that acetic acid is a stronger acid than ammonium ion, and also that acetate ion is a weaker base than ammonia. These things would be much more difficult to judge on the basis of eqs. (6.1) and (6.3).

Expressions for K_b are not used in this book. If you ever encounter a value of K_b for a base, you can convert it into one of K_a for the conjugate acid of that base by using the relation

$$K_a = K_s/K_b \qquad (6.4)$$

where K_s is the autoprotolysis constant of the solvent. Since the autoprotolysis constant of water is represented by the symbol K_w, eq. (6.4) may be written as

$$K_a = K_w/K_b \qquad (6.5)$$

for aqueous solutions. The last few paragraphs of Section 4.3 indicate how these equations can be obtained from eqs. (6.2) and (6.3).

Because each of the reactions in eqs. (6.1) and (6.2) involves the solvent, expressing the strength of an acid in terms of its overall dissociation constant—or the strength of a base in terms of the overall dissociation constant of its conjugate acid—makes that strength dependent on the properties of the solvent. The overall dissociation constant of an acid depends on the ability of the solvent to accept protons as well as on the ability of the acid to donate them. Acetic acid is a weak acid in aqueous solutions, not only because acetic acid has only a limited ability to donate protons, but also because water is an extremely weak base and has very little ability to accept protons. Perchloric acid is a strong acid in aqueous solutions because it loses protons so easily that the reaction

$$HClO_4(aq) + H_2O(l) = H_3O^+ + ClO_4^-$$

proceeds to completion even though water is only a very weak base.

These ideas will be pursued further in Section 6.12, which is a brief introduction to acid-base chemistry in non-aqueous solvents. The intervening sections are limited to aqueous solutions, and will provide a basis for comparing and contrasting the behaviors of non-aqueous solutions with those of aqueous ones.

SUMMARY: A Lowry–Brønsted acid is a substance that can donate a proton, and a Lowry–Brønsted base is a substance that can accept one. An acid-base reaction involves two conjugate acid-base pairs, and it consumes one acid and one base and produces another acid and another base. An amphoteric substance is one that can either accept or donate a proton, and the reaction between two molecules of an amphoteric solvent is called autoprotolysis. The overall dissociation constant K_a of an acid depends on both the ability of the acid to donate protons and the ability of the solvent to accept them.

6.3. pH scales

PREVIEW: This section describes the ways in which the acidity of a solution can be expressed and introduces the electroneutrality principle, which is a powerful weapon in dealing with ionic equilibria.

There are two ways in which the acidity of a dilute aqueous solution can be described. One is to speak of the concentration of hydronium ion; the other is to speak of its activity.

More will be said about the activity in Chapter 13. For the time being it will suffice to regard the concentration as expressing how much hydronium ion there really is in a solution, and the activity as expressing how much there appears to be. There are many properties of a solution of, say, perchloric acid that change as the concentration of the acid changes. A few of these are the freezing-point depression, the boiling-point elevation, the vapor pressure of water that is in equilibrium with it, its osmotic pressure, and its effects on the rates and equilibria of many reactions, including the gain or loss of protons by acid-base indicators added to it. Measurements of all these properties give closely agreeing results. They indicate, for example, that there are about 0.4 M hydronium ion and 0.4 M perchlorate ion in a solution prepared by combining 0.5 mole of perchloric acid with enough water to yield a total of 1 dm³ of solution. Chemists once interpreted these figures as meaning that perchloric acid is a weak acid in aqueous solutions. If that were true, the fraction of the perchloric acid that is dissociated should decrease as the concentration of the solution increases. However, the same kinds of measurements indicate that there are almost exactly 2 M hydronium ion and 2 M perchlorate ion in a solution prepared by combining 2 moles of perchloric acid with enough water to yield a total of 1 dm³ (and that there are very roughly 50 M hydronium ion and 50 M perchlorate ion in a solution prepared by combining 7 moles of perchloric acid with enough water to yield a total of 1 dm³!). Moreover, many attempts have been made to detect the presence of unionized or undissociated perchloric acid in such solutions, and all have failed. Their failure forces us to conclude that perchloric acid is a strong and completely dissociated acid in aqueous solutions, and this in turn forces us to believe that the concentrations of hydronium ion in the three solutions mentioned are 0.5, 2, and 7 M, respectively. We say that the corresponding activities are 0.4, 2, and 50 M. The activity and the concentration are nearly equal in very dilute solutions (and would be exactly equal in an infinitely dilute one), but may be very different in concentrated solutions.

Since it is clearly essential to distinguish between the activity and the concentration, we define two different quantities by the equations

$$\text{pcH} = -\log_{10} c_{H_3O^+} \quad \text{or} \quad 10^{-\text{pcH}} = c_{H_3O^+} \tag{6.6}$$

and

$$\text{paH} = -\log_{10} a_{H_3O^+} \quad \text{or} \quad 10^{-\text{paH}} = a_{H_3O^+} \tag{6.7}$$

for the concentration and activity, respectively. The pcH will be used in most of the rest of this book, but the paH will reappear in a later chapter. Neither should be confused with the pH, which is defined operationally as the number that is obtained by subjecting a solution to a measurement made with a pH meter or some similar device, following some prescribed procedure. The pH is a reproducible and useful property of a solution, but in general it is not equal to either the pcH or the paH.

Using the pcH enables us to express the concentration of hydronium ion in a short convenient form. It is so much easier to write pcH = 7.00 than to write $[H_3O^+]$ = 1.00×10^{-7} M that the latter form is rarely seen anywhere except in the course of a numerical calculation. Many other quantities are treated in the same way, and you should be prepared to encounter the negative logarithms of equilibrium constants ($\text{p}K = -\log_{10} K$) and of the concentrations ($\text{pcAg} = -\log_{10} [Ag^+]$) and activities ($\text{paCl} = -\log_{10} a_{Cl^-}$) of many different substances. In all of them the "p" represents the initial letter of the German word *Potenz*, meaning "power" or "exponent"; the second half of eq. (6.6) or (6.7) shows what this signifies. If the "p" is followed immediately by the symbol of a quantity, as in "pK", it denotes the negative decadic logarithm of that quantity; if it is followed by the formula of a chemical substance (from which any electrical charge is always omitted, as in "pH", "pCO$_2$", or "pCa") it denotes a measured quantity. The "c" or "a", if one appears, denotes the concentration or activity of the substance whose formula follows it.

The universe contains so many different things that claim our attention, and there are so few letters in the English alphabet, that there are many pairs of symbols that look very similar but have completely different meanings. You can easily make yourself incomprehensible by being careless with them. Whereas "pK" signifies the negative logarithm of an equilibrium constant, "pK" conveys information about the potassium ion contained in a solution, or in someone's blood. The symbol "pO$_2$" means the result of an experimental measurement of the dissolved oxygen content; the symbol "p_{O_2}" means the partial pressure of oxygen in a gas. The "pH" is a quantity that is measured with a pH meter, but "Ph" is an abbreviation for the phenyl (C_6H_5-) group and "PH" is a phosphorus hydride.

The "p" is so widely used that you should establish some reflexes for dealing with it. In the following examples the symbols X and pX are used for the sake of generality. If X is an equilibrium constant K, then pX = pK; if X is the concentration of hydronium ion $c_{H_3O^+}$, then pX = pcH; if X is the activity of fluoride ion a_{F^-}, then pX = paF.

1. *To translate a value of X into one of pX*, write the value of X in the form

$$X = a \times 10^b$$

where b is an integer and a is a number between 1 and 9.99.... . In dealing with the pcH, the value of b will usually be negative because the concentration of hydronium ion is usually less than 1 ($= 10^0$) M. However, it is also possible for the value of b to be zero or positive, as in Example 6.2 below. Taking the logarithm of each side and changing its sign,

$$pX = -\log_{10} X = -\log_{10}(a \times 10^b) = -(\log_{10} a + \log_{10} 10^b) = -(\log_{10} a + b). \quad (6.8)$$

2. *To translate a value of pX into one of X*, begin by writing

$$-pX = c + d$$

where c is a positive decimal number between 0 and 0.99 ... and d is an integer, which must be equal to the largest (most positive) integer contained in $-pX$. In dealing with the pcH the value of d will usually be negative, but it may also be zero or positive, as in Example 6.4. Now

$$X = 10^{-pX} = 10^{c+d} = 10^c \times 10^d = (\text{antilog}_{10}\ c) \times 10^d. \quad (6.9)$$

The following four examples illustrate all of the possibilities.

Example 6.1. A solution contains 3.00×10^{-4} M hydronium ion. What is its pcH?

Answer. pcH $= -\log(3.00 \times 10^{-4}) = -(\log 3.00 + \log 10^{-4}) = -(0.48 + (-4)) = 4 - 0.48 = 3.52$.

Example 6.2. According to nuclear magnetic resonance measurements the overall dissociation. constant K_a of nitric acid in an aqueous solution at 298 K is equal to 21 mol dm^{-3}. What is the value of pK_a?

Answer. $21 = 2.1 \times 10^1$,

$$pK_a = -\log(2.1 \times 10^1) = -(\log 2.1 + \log 10^1) = -(0.32 + 1) = -1.32.$$

Example 6.3. The pcH of a solution is 7.60. What is the concentration of hydronium ion?

Answer. $-7.60 = c + d$. The largest integer contained in -7.60 is -8, so that $d = -8$. Since $-7.60 = c + (-8), c = 8 - 7.60 = 0.40$. Consequently

$$[H_3O^+] = 10^{-pcH} = 10^{-7.60} = 10^{0.40 + (-8)} = 10^{0.40} \times 10^{-8} = (\text{antilog}_{10}\ 0.40) \times 10^{-8} = 2.5_0 \times 10^{-8}\ M$$

Example 6.4. For a certain solution of sodium hydroxide the pcOH is equal to -1.10. What is the concentration of hydroxide ion?

Answer. $-(-1.10) = +1.10 = c + d$. The largest integer contained in $+1.10$ is $+1$, so that $d = 1$ and $c = 0.10$, and

$$[OH^-] = 10^{-pcOH} = 10^{-(-1.10)} = 10^{0.10} \times 10^1 = (\text{antilog}_{10}\ 0.10) \times 10^1 = 1.26 \times 10^1\ M\ (= 12.6\ M).$$

Appendix I contains a discussion of how the number of figures that appear to the right of the decimal point in a value of $\log_{10} X$ is related to the certainty with which X is known. Giving a value of pcX to two decimal places means that the concentration of X has a *relative precision* of 2%; the term is defined and discussed in Appendix I. It is often possible to determine a concentration with a relative precision that is smaller than 2%, but a relative precision as small as 0.2% is not easy to obtain, and this is what would be needed to justify giving a value of pcX to three decimal places. You are usually entitled to give a value of pcX to two decimal places, but you should never use more unless you make the deliberate decision that more are justified by some unusual circumstance. Values of paX are much more difficult to obtain, chiefly because of imperfections in the theory involved, and almost never deserve three decimal places. There are a very few values of pK—including those of pK_w, of pK_a for formic and acetic acids and a few other acids and bases in aqueous solutions, and a handful of others—that have been evaluated with enough painstaking care to deserve three decimal places, but most deserve one rather than two, and some deserve none at all. Handbooks currently being printed contain two different values of pK_{CuS}, of which one is close to 35 while the other is approximately 44; and the literature contains two estimates of the value of pK_a for hydroxide ion (corresponding to the reaction $H_2O(l) + OH^- = H_3O^+ + O^{2-}$): they are 130 and 28.

A *neutral* solution is one that contains no acid or base except the solvent itself, or one that has the same acidity as the pure solvent. In a neutral aqueous solution small concentrations of hydronium and hydroxide ions are produced by the autoprotolysis of water:

$$2 H_2O(l) = H_3O^+ + OH^-; \quad K_w = [H_3O^+][OH^-] = 1.008 \times 10^{-14} \text{ mol}^2 \text{ dm}^{-6}$$

$$\text{(at 298 K)}. \tag{6.10}$$

Two lines of reasoning lead to the conclusion that the concentrations of hydronium and hydroxide ions must be equal in such a solution. One is stoichiometric and the other is electrostatic. The stoichiometric one says that the autoprotolysis yields equal numbers of hydronium and hydroxide ions. There is no other source of either ion, and no other reaction by which either can be consumed. The solution must contain equal numbers, and therefore equal concentrations, of the two ions. The electrostatic one says that the solution must contain equal numbers of positive and negative charges, for if it did not it would have a very large electrostatic potential that would act to restore the balance of charge. To remove a single ion from an electrically neutral solution requires the expenditure of enough work to overcome the electrostatic attraction between the ion that is removed and the oppositely charged solution that is left behind. To remove a second ion having the same charge requires twice as much work because the charge that is left behind is twice as large as it was while the first ion was being removed. Long before enough ions have been removed to constitute any measurable fraction of a mole, impossibly large amounts of work will have had to be expended.

This electrostatic argument leads to what is called the *electroneutrality principle*, which is that any substantial volume† of any solution must be electrically neutral or uncharged. An algebraic statement of the electroneutrality principle is

$$\sum_+ |z_+| c_+ = \sum_- |z_-| c_- \tag{6.11}$$

where z_+ is the charge on a particular kind of cation present at the concentration c_+ (M), z_- is the charge on a particular kind of anion present at the concentration c_- (M), and the symbols $\sum_+$ and $\sum_-$ denote summations for all of the different kinds of cations and anions contained in the solution. The electroneutrality equations describing pure water and aqueous solutions of sodium chloride and oxalic acid are, respectively,

$$[H_3O^+] = [OH^-]$$
$$[Na^+] + [H_3O^+] = [Cl^-] + [OH^-]$$
$$[H_3O^+] = [HOOCCOO^-] + 2[^-OOCCOO^-] + [OH^-]$$

The first two of these are straightforward because all the ions are univalent. In the third, the concentration of oxalate ion is multiplied by two because oxalate ion is divalent and the solution must therefore contain two hydronium ions for each oxalate ion.

No matter how it is reached, the conclusion that $[H_3O^+] = [OH^-]$ may be combined with eq. (6.10) to obtain

$$[H_3O^+] = [OH^-] = \sqrt{K_w} = \sqrt{1.008 \times 10^{-14}} = 1.00 \times 10^{-7} \, M \tag{6.12}$$

for a neutral dilute aqueous solution at 25°. The same result could be obtained in a different way by rewriting eq. (6.10) in logarithmic form and changing all the signs:

$$\log_{10}[H_3O^+] + \log_{10}[OH^-] = \log_{10} K_w$$

and/or

$$-\log_{10}[H_3O^+] + (-\log_{10}[OH^-]) = -\log_{10} K_w$$

---

† In an aqueous solution the volume of a single potassium or chloride ion is about 1×10^{-23} cm³. The electroneutrality principle would not apply to any such volume of a potassium chloride solution: if 1×10^{-23} cm³ of such a solution contains a potassium ion it cannot also contain a chloride ion at the same time, and even a volume a hundred times as large might contain twice as many ions of one kind as of the other at a particular instant. Here the principle describes the average composition over some sufficiently long time: the probability of finding a potassium ion in any particular 1×10^{-23} cm³ of the solution is no different from the probability of finding a chloride ion there, and so in the long run even a very small volume of a homogeneous solution is also electrically neutral. Definite local violations of the electroneutrality principle do occur in a very thin layer of solution that is in immediate contact with the surface of a precipitate, of an electrode, or of any second phase. Powerful short-range chemical, electrical, or other forces are at work at the boundaries between different phases and may cause ions of one charge to accumulate at an interface while repelling ions that are oppositely charged. Even in such circumstances, however, the electroneutrality principle may be safely applied to the system as a whole.

or

$$pcH + pcOH = pK_w. \tag{6.13}$$

Since the pcH and pcOH must be equal if the concentrations of hydronium and hydroxide ions are equal, eq. (6.13) becomes

$$pcH = pcOH = pK_w/2 = 13.996/2 = 7.00 \tag{6.14}$$

for a neutral solution at 298 K. This does not mean that distilled water has a pcH of 7.00: the last traces of acidic and basic impurities (such as carbon dioxide, ammonia, and dissolved glass) are so difficult to remove that the measured pH may lie anywhere between about 5.5 and 8.5.

In an acidic solution the concentration of hydronium ion is larger than 1.00×10^{-7} M and the pcH is smaller than 7.00; since increasing the concentration of hydronium ion causes the concentration of hydroxide ion to decrease, the concentration of hydroxide ion is smaller than 1.00×10^{-7} M in an acidic solution and the pcOH is larger than 7.00. In an alkaline solution $[H_3O^+] < 1.00 \times 10^{-7}$ M and pcH > 7.00, while $[OH^-] > 1.00 \times 10^{-7}$ M and pcOH < 7.00. The concentration of hydronium ion in a solution may be as high as 10 or 20 M (as it is in concentrated solutions of hydrochloric, nitric, perchloric, and sulfuric acids), and so may that of hydroxide ion (as it is in concentrated solutions of sodium or potassium hydroxide). Hence the pcH (or pcOH) may be as low as about -1.3, or as high as about 15.3, in an aqueous solution. Both much higher and much lower values can be attained in non-aqueous solutions.

SUMMARY: The pH (an experimentally measured number), pcH ($= -\log_{10} c_{H_3O^+}$), and paH ($= -\log_{10} a_{H_3O^+}$) are three different measures of the acidity of a solution. The electroneutrality principle states that any substantial volume of any solution is electrically neutral, and is much used in calculations based on ionic equilibria.

6.4. The acidities of solutions of acids and bases

PREVIEW: This section describes calculations of the pcH of solutions that contain a single strong or weak acid or base.

By employing equations already given in this chapter and the preceding ones, it is possible to calculate the pcH of an aqueous solution of any acid or base. There are a number of different but overlapping possibilities. The acid or base may be so strong that it is almost completely ionized and dissociated, or it may be so weak that the extents of ionization and dissociation are too small to have any detectable effect on the concentration of the unionized acid or base. Every aqueous solution contains hydronium and hydroxide ions, which are produced by the autoprotolysis of water. The extent of the autoprotolytic reaction is affected by adding an acid or a base, and it may or may not be small enough to ignore in any particular solution. If the acid or base is extremely dilute or extremely weak, its effect on the extent of autoprotolysis is small, and the pcH has nearly the same value that it would have if the acid or base were not present at all. This section describes the possibilities that can arise in a solution. containing only a single dissolved acid or base. Mixtures of acids, mixtures of bases, and mixtures containing both acids and bases will be considered in subsequent sections.

There are two important quantities. For an acid these are its concentration and its overall dissociation constant; for a base they are its concentration and the overall dissociation constant of its conjugate acid. The concentration is represented by the symbol c_a (for an acid) or c_b (for a base), is expressed in mol dm^{-3}, and is the number of moles of the acid (or base) that were used in preparing each cubic decimeter of the solution. This is called the *total concentration*, the *analytical concentration*, or the *formal concentration* of the acid (or base).

It is vitally important to distinguish between the total (or analytical or formal) concentration of a solute and its *equilibrium concentration*, which is the number of moles that are present in each cubic decimeter of the solution when equilibrium has been achieved. If one mole of hydrogen chloride is dissolved in enough water to yield one cubic decimeter of solution, the total concentration of hydrochloric acid in the solution will be 1 mol dm^{-3}, or 1 M. However, the reactions

$$HCl(aq) + H_2O(l) = H_3O^+Cl^-(aq) = H_3O^+ + Cl^-$$

proceed so nearly to completion that no detectable trace of either HCl or $H_3O^+Cl^-$ remains when equilibrium is reached. The equilibrium concentration of hydrochloric acid in such a solution is indistinguishable from zero, even though its total concentration is 1 M. In a solution prepared by dissolving 0.10 mole of iodic acid (HIO_3) in enough water to give a volume of 1 dm^3, the total concentration of iodic acid is 0.10 M, but because 70% of the dissolved iodic acid is consumed by ionization and dissociation the equilibrium concentration of iodic acid is only 0.03 M. In a solution prepared by dissolving 0.01 mole of ammonia and 0.01 mole of ammonium chloride in enough water to give a volume of 1 dm^3, the reaction

$$NH_3(aq) + H_2O(l) = NH_4^+ + OH^-$$

occurs to some extent, consuming a little of the ammonia and producing a little more ammonium ion. The equilibrium concentration of ammonia is a little less than 0.01 M, and the equilibrium concentration of ammonium ion is a little higher than 0.01 M. Here we would speak of the analytical or formal concentrations of ammonia and ammonium ion, rather than of their total concentrations, because it would seem peculiar to say that the equilibrium concentration of ammonium ion was higher than its total concentration. We might, for example, say that the analytical concentration of ammonia is 0.01 M, and that the analytical concentration of ammonium ion is also 0.01 M.

Many chemists consider the distinction between total (or analytical or formal) concentrations and equilibrium concentrations to be so important that they use the symbol F ("formal") to denote the former and reserve the symbol M for equilibrium concentrations. Since both are expressed in mol dm^{-3}, this book will use the symbol M for both, but will always specify which is meant.

The simplest situation is the one that arises in fairly concentrated solutions of strong acids, such as hydrochloric acid. If the total concentration of hydrochloric acid in a solution is $c_a\,M$, and if the hydrochloric acid that was put into the solution is completely consumed by the overall reaction

$$HCl(aq) + H_2O(l) = H_3O^+ + Cl^-$$

then the equilibrium concentration of hydronium ion will be given by the simple equation

$$[H_3O^+] = c_a \qquad (6.15a)$$

—provided that c_a is not too small. The exactly similar equation

$$[OH^-] = c_b \qquad (6.15b)$$

applies to a solution in which the total concentration of a completely dissociated ("strong") base was equal to $c_b\,M$—provided that c_b is not too small. Having obtained the concentration of hydroxide ion from eq. (6.15b), we could calculate the concentration of hydronium ion by rewriting eq. (6.10) in the form

$$[H_3O^+] = K_w/[OH^-] \quad (= 1.008 \times 10^{-14}/[OH^-] \quad \text{at } 298\ \text{K}) \qquad (6.16a)$$

and then obtain the pcH from the concentration of hydronium ion. Or we could calculate the pcOH from the concentration of hydroxide ion, and then obtain the pcH from eq. (6.13), rewritten for the purpose as

$$pcH = pK_w - pcOH \quad (= 13.996 - pcOH \quad \text{at } 298\ \text{K}). \qquad (6.16b)$$

Equations (6.15) become inaccurate, and eventually ridiculous, as the total concentration of the acid or base decreases toward zero. Consider a solution in which the total concentration of hydrochloric acid is equal to $1.00 \times 10^{-7}\ M$. From eq. (6.15a) you would calculate $[H_3O^+] = 1.00 \times 10^{-7}\ M$. This is an impossible result, for the solution must be acidic: the concentration of hydronium ion cannot be the same as it would be in pure water. Writing the electroneutrality equation

$$[H_3O^+] = [Cl^-] + [OH^-] \qquad (6.17)$$

makes it clear what has happened: it is the concentration of chloride ion that is equal to $1.00 \times 10^{-7}\ M$, and the concentration of hydroxide ion has simply been ignored. It would be quite safe to ignore the concentration of hydroxide ion if the total concentration of hydrochloric acid were 0.1 M, for then the acid would furnish such a high concentration of hydronium ion that the autoprotolytic reaction

$$2\,H_2O(l) = H_3O^+ + OH^-$$

would be almost completely suppressed; since this reaction is the only source of hydroxide ion in the solution, the concentration of hydroxide ion would be negligibly small. But if c_a is only 1.00×10^{-7} M the concentration of hydronium ion cannot be very large, and the autoprotolytic reaction proceeds too far, and provides too much additional hydronium ion, to be ignored. Combining eq. (6.17) with eq. (6.10) and the value $[Cl^-] = 1.00 \times 10^{-7} M$,

$$[H_3O^+] = 1.00 \times 10^{-7} + \frac{1.008 \times 10^{-14}}{[H_3O^+]} .$$

This is a quadratic equation, and its solution is $[H_3O^+] = 1.62 \times 10^{-7}$ M, so that pcH = 6.79.

You can always substitute numbers into an equation and get some sort of numerical result by solving it, but because it is human to err you will sometimes get the wrong result, either because you have used the wrong equation or because you have made an algebraic or arithmetic mistake. You can often insure yourself against errors by looking for limits between which the result should lie. Here the concentration of hydronium ion would be equal to 1.00×10^{-7} M if the autoprotolysis of water were completely suppressed (so that all of the hydronium ion came from the hydrochloric acid), and would be equal to 2.00×10^{-7} M if the autoprotolysis were not suppressed at all (so that the hydrochloric acid and the water each contributed 1.00×10^{-7} M hydronium ion). After you have thought this through you can tell that a result like $[H_3O^+] = 2.62 \times 10^{-7}$ M must be wrong because it does not fall between these limits. You could not be sure that the result, $[H_3O^+] = 1.62 \times 10^{-7}$ M, obtained in the preceding paragraph was exactly right, but at least you could tell that it was reasonable and that the error in it could not be very large.

This discussion should have made it clear that eqs. (6.15) are only approximately correct. More accurate equations are

$$[H_3O^+] = c_a + K_w/[H_3O^+] \qquad (6.18a)$$

for a solution in which the total concentration of a strong acid HB is equal to c_a M and

$$[OH^-] = c_b + K_w/[OH^-] \qquad (6.18b)$$

for a solution in which the total concentration of a strong base B is equal to c_b M. Equations (6.18) are written for aqueous solutions, but they can be applied to solutions in other amphoteric solvents if $[H_3O^+]$ is replaced by the concentration of the conjugate acid of the solvent, if $[OH^-]$ is replaced by the concentration of the conjugate base of the solvent, and if K_w is replaced by K_s.

In practical calculations the thing to do, as usual, is to use the simpler equations (eqs. (6.15)) unless you are sure that they cannot be right, and to examine the answer to make sure that it was justifiable to neglect the term that you neglected in using those equations. With $c_a = 1.00 \times 10^{-7}$ M, eq. (6.15a) gives $[H_3O^+] = 1.00 \times 10^{-7} M$, which would be correct if the concentration of hydroxide ion were small enough to ignore. However, if $[H_3O^+] = 1.00 \times 10^{-7}$ M then $[OH^-] = 1.00 \times 10^{-7}$ M, which is equal to c_a and is therefore not so much smaller than c_a that it can be neglected. In this situation you have to solve the quadratic equation, as was done in an earlier paragraph. If c_a were 1.00×10^{-6} M, eq. (6.15a) would give $[H_3O^+] = 1.00 \times 10^{-6}$ M, which would correspond to $[OH^-] = 1.00 \times 10^{-8}$ M. You could accept this result because $[OH^-]$ is much smaller than c_a, as you assumed it to be in using eq. (6.15a).

If you experiment with several different values of c_a (or c_b), you will be able to convince yourself that eqs. (6.18) are hardly worth using unless c_a (or c_b) is less than about 10^{-6} M. Hence these equations are not often needed for aqueous solutions, because such extremely dilute solutions are not very common. In another solvent having a higher autoprotolysis constant, the error in eqs. (6.15) would be perceptible for more concentrated solutions. For example, $K_s = 1.4 \times 10^{-4}$ mol^2 dm^{-6} for concentrated sulfuric acid, and eqs. (6.15) would not give an acceptable result either for a solution in which the total concentration of perchloric acid (which is a strong acid in this solvent) is 0.05 M or for a solution in which the total concentration of sodium hydrogen sulfate (which dissociates to give the hydrogen sulfate ion, a strong base in this solvent) is 0.05 M.

Equations (6.15) are also incorrect for weak acids and bases because these are not completely ionized and dissociated. For a solution of a weak acid HB in water we might write

$$[H_3O^+] = [B^-] \tag{6.19a}$$

and

$$[HB(aq)] = c_a. \tag{6.19b}$$

The first of these equations says that the reaction

$$HB(aq) + H_2O(l) = H_3O^+ + B^-$$

produces equal numbers (and therefore equal concentrations) of hydronium and B$^-$ ions, and that there is no other source of an appreciable amount of either, and nothing else in the solution with which either can react. The second says that only an insignificant fraction of the acid is consumed by this reaction. Combining eqs. (6.19) with the expression for the overall dissociation constant gives

$$K_a = \frac{[H_3O^+][B^-]}{[HB(aq)]} = \frac{[H_3O^+]^2}{c_a} \quad \text{or} \quad [H_3O^+] = \sqrt{c_a K_a}. \tag{6.20}$$

The reaction $B(aq) + H_2O(l) = BH^+ + OH^-$ occurs in an aqueous solution of a weak base B. Reasoning in exactly the same way that led to eqs. (6.19) gives

$$[BH^+] = [OH^-], \tag{6.21a}$$

$$[B(aq)] = c_b. \tag{6.21b}$$

10*

Combining these with the expression for the overall acidic dissociation constant of BH^+, the conjugate acid of B, we obtain

$$K_a = \frac{[H_3O^+][B(aq)]}{[BH^+]} = \frac{[H_3O^+]c_b}{[OH^-]} = \frac{[H_3O^+]c_b}{K_w/[H_3O^+]} = \frac{[H_3O^+]^2 c_b}{K_w}$$

in which the third equality is written by invoking eq. (6.10). Solving the result for the concentration of hydronium ion,

$$[H_3O^+] = \sqrt{K_w K_a / c_b}. \tag{6.22}$$

Example 6.5. What is the pcH of a solution in which the total concentration of acetic acid (for which $K_a = 1.75 \times 10^{-5}\,\text{mol dm}^{-3}$) is 0.100 M?

Answer. Assuming that $[H_3O^+] = [OAc^-]$ and that $[HOAc(aq)] = 0.100\ M$,

$$[H_3O^+] = \sqrt{0.100 \times 1.75 \times 10^{-5}} = \sqrt{1.75 \times 10^{-6}} = 1.32 \times 10^{-3}\,M; \text{pcH} = 2.88.$$

Example 6.6. What is the pcH of a solution in which the total concentration of sodium acetate is 0.100 M?

Answer. We assume that $[HOAc(aq)] = [OH^-]$ and that $[OAc^-] = 0.100\ M$. The value of K_a that is needed is the one for the conjugate acid of acetate ion, which is acetic acid; according to Example 6.5 it is equal to $1.75 \times 10^{-5}\,\text{mol dm}^{-3}$. Hence

$$[H_3O^+] = \sqrt{1.008 \times 10^{-14} \times 1.75 \times 10^{-5}/0.100} = \sqrt{1.76 \times 10^{-18}} = 1.33 \times 10^{-9}\ M; \text{pcH} = 8.88.$$

Equations (6.19) and (6.21) contain two independent assumptions, and either or both of them may be unjustified. One is that only an insignificant fraction of the acid (or base) is consumed in the reaction that takes place. This is justifiable if the acid (or base) is sufficiently weak and if its total concentration is sufficiently high. The concentration is important because the extent of the reaction increases as the concentration decreases. If the acid (or base) is fairly strong, or if its total concentration is fairly small, this assumption becomes incorrect. Example 6.5 is just on the safe side of the borderline. If the equilibrium concentration of hydronium ion (or acetate ion) is $1.32 \times 10^{-3}\ M$, the dissociation must have consumed $1.32 \times 10^{-3}\ M$ acetic acid, and therefore the equilibrium concentration of acetic acid is not 0.100 M but $0.100 - 1.32 \times 10^{-3} = 0.098_7\ M$. From this figure we can obtain a more accurate result by writing

$$[H_3O^+] = \sqrt{0.098_7 \times 1.75 \times 10^{-5}} = 1.31 \times 10^{-3}\ M,$$

but this is hardly worth the trouble that it costs: to two decimal places it gives pcH = 2.88, which is identical with the result obtained in Example 6.5.

Chloroacetic acid, $ClCH_2COOH$, is a stronger acid than acetic acid: its overall dissociation constant is equal to $1.40 \times 10^{-3}\,\text{mol dm}^{-3}$ in water at 298 K. Because this is true, eq. (6.19b) is less satisfactory for a solution of chloroacetic acid than it is for a solution of acetic acid at the same total concentration. For a solution in which the total concentration of chloroacetic acid is 0.100 M, eq. (6.20) gives $[H_3O^+] = 1.18 \times 10^{-2}$ M. If this result is correct, then almost 12% of the acid that was put into the solution must have dissociated in reaching equilibrium, and the equilibrium concentration of the undissociated acid is almost 12% less than the total concentration employed in eqs. (6.19b) and (6.20). Matters become even worse if the total concentration of chloroacetic acid is only 1.00×10^{-3} M, for then eq. (6.20) gives $[H_3O^+] = 1.18 \times 10^{-3}$ M, which corresponds to the dissociation of more acid than the solution contained. You can easily recognize such a situation by comparing the concentration of hydronium ion calculated from eq. (6.20) with the value of c_a (or, for a base, comparing the concentration of hydroxide ion with the value of c_b). If the former is more than a few per cent as large as the latter, the extent of dissociation is too large to ignore.

Example 6.7. What is the pcH of a solution in which the total concentration of ammonia is 1.00×10^{-5} M? The overall acidic dissociation constant of ammonium ion, which is the conjugate acid of ammonia, is equal to 5.76×10^{-10} mol dm^{-3} in water at 298 K.

Answer. Equation (6.22) yields

$$[H_3O^+] = \sqrt{1.008 \times 10^{-14} \times 5.76 \times 10^{-10}/1.00 \times 10^{-5}} = 7.62 \times 10^{-10}\ M.$$

To decide whether it is justifiable to apply eqs. (6.21b) and (6.22) to this solution, we must calculate the concentration of hydroxide ion. We can do this by using eq. (6.10):

$$[OH^-] = K_w/[H_3O^+] = 1.008 \times 10^{-14}/7.62 \times 10^{-10} = 1.32 \times 10^{-5}\ M.$$

The result is impossible because the reaction $NH_3(aq) + H_2O(l) = NH_4^+ + OH^-$ could not furnish more than 1.00×10^{-5} M hydroxide ion even if it proceeded to completion. The answer to this problem is continued on page 136.

When the dissociation of an acid is too extensive to be ignored, as it is in the solutions of chloroacetic acid described in the preceding paragraph, eq. (6.19b) must be replaced by the conservation equation

$$[HB(aq)] = c_a - [B^-] \tag{6.23}$$

which says that the total amount of B^- in any given volume of the solution is not affected by removing some of the protons that were originally combined with the B^-. Combining eq. (6.23) with eq. (6.19a) and the expression for K_a gives

$$K_a = \frac{[H_3O^+][B^-]}{[HB(aq)]} = \frac{[H_3O^+]^2}{c_a - [H_3O^+]}.$$

This is a quadratic equation, and its solution is

$$[H_3O^+] = \frac{-K_a + \sqrt{K_a^2 + 4c_aK_a}}{2}. \tag{6.24}$$

Pure mathematics says that the radical should be preceded by a "$\pm$" sign, but pure mathematics must take a back seat to our unwillingness to accept a negative concentration. There are few chemical problems in which both signs yield answers that make sense. Equation (6.24) yields $[H_3O^+] = 1.12 \times 10^{-2}$ M for a solution in which the total concentration of chloroacetic acid is 0.100 M, and $[H_3O^+] = 6.7_5 \times 10^{-4}$ M for one in which the total concentration of chloroacetic acid is 1.00×10^{-3} M. These results can be compared with the incorrect ones obtained from eq. (6.20) and given in the preceding paragraph.

For a solution of a weak base B that is dissociated to an appreciable extent, the corresponding equations are

$$[B(aq)] = c_b - [BH^+] = c_b - [OH^-] \tag{6.25}$$

(whose second equality is obtained from eq. (6.21a)) and

$$K_a = \frac{[H_3O^+][B(aq)]}{[BH^+]} = \frac{[H_3O^+](c_b - K_w/[H_3O^+])}{K_w/[H_3O^+]}.$$

Solving this quadratic equation for the concentration of hydronium ion gives

$$[H_3O^+] = \frac{K_w + \sqrt{K_w^2 + 4c_b K_a K_w}}{2c_b}. \tag{6.26}$$

Example 6.7 *(continued)*. On page 135 it was concluded that eq. (6.22) yields an impossible value for the concentration of hydronium ion in a solution in which the total concentration of ammonia is 1.00×10^{-5} M. Equation (6.26) yields

$$[H_3O^+] = \frac{1.008 \times 10^{-14} + \sqrt{(1.008 \times 10^{-14})^2 + (4)(1.00 \times 10^{-5})(5.76 \times 10^{-10})(1.008 \times 10^{-14})}}{(2)(1.00 \times 10^{-5})}$$

$$= \frac{1.008 \times 10^{-14} + \sqrt{1.016 \times 10^{-28} + 2.322 \times 10^{-28}}}{2.00 \times 10^{-5}} = 1.42 \times 10^{-9} \ M.$$

The corresponding value of $[OH^-]$ is 7.11×10^{-6} M, which is a possible result because it is smaller than c_b.

The other assumption contained in eqs. (6.19) and (6.21) is that the dissociation of the acid (or base) furnishes enough hydronium (or hydroxide) ion to suppress the autoprotolysis of water completely. This is justifiable if the acid (or base) is sufficiently strong and if its total concentration is sufficiently high, but it becomes incorrect if the acid (or base) is fairly weak or if its total concentration is fairly small. The paragraphs following eqs. (6.16) (page 131) showed that a similar assumption fails even for solutions of strong acids (or bases) if these are extremely dilute, and the danger increases as the acid (or base) becomes weaker.

Let us consider a solution in which the total concentration of sodium iodate is 0.100 M. Iodate ion is an extremely weak base: the overall dissociation constant of its conjugate acid, iodic acid, is equal to 0.16_6 mol dm^{-3} in water at 298 K. If eq. (6.22) is employed to calculate the concentration of hydronium ion, the result is $[H_3O^+] = 1.29 \times 10^{-7}$ M, which corresponds to pcH = 6.89. This is absurd and impossible: although iodate ion is a very weak base, it is a base, and a solution of it in pure water must be alkaline and must have a pcH higher than 7.00. The reaction $IO_3^- + H_2O(l) = HIO_3(aq) + OH^-$ produces a few hydroxide ions, and the autoprotolysis of water also produces some hydroxide ions. Because the solution must be more alkaline than eq. (6.22) indicates, we cannot neglect the hydroxide ions obtained from the autoprotolytic reaction. One way of taking them into account is based on a stoichiometric argument like the one described on page 127, a few lines below eq. (6.10). There are two reactions that occur in an aqueous solution of iodate ion, and they are described by the equations

$$2 H_2O(l) = H_3O^+ + OH^-, \tag{6.27a}$$

$$IO_3^- + H_2O(l) = HIO_3(aq) + OH^-. \tag{6.27b}$$

The number of hydroxide ions produced by the first reaction is equal to the number of hydronium ions present in the solution, for there is no way of producing a hydronium ion in this solution without producing a hydroxide ion at the same time. The number of hydroxide ions produced by reaction (6.27b) is equal to the number of molecules of iodic acid present in the solution, for a similar reason. The total number of hydroxide ions is the sum of the numbers produced by these two reactions, so that

$$[OH^-] = [H_3O^+] + [HIO_3(aq)] \quad \text{or} \quad [HIO_3(aq)] = [OH^-] - [H_3O^+].$$

In general, and with the symbols employed in eq. (6.21a),

$$[BH^+] = [OH^-] - [H_3O^+]. \tag{6.28a}$$

Although eq. (6.21a) must be replaced by eq. (6.28a) for an extremely weak base, eq. (6.21b) can often be retained. Iodate ion is so weak a base that only a very tiny fraction of it can have been consumed by reaction (6.27b), and therefore its equilibrium concentration must be indistinguishable from its total concentration c_b. Combining these descriptions of $[HIO_3(aq)]$ and $[IO_3^-]$ with the expression for the overall dissociation constant of iodic acid, we obtain

$$K_a = \frac{[H_3O^+][IO_3^-]}{[HIO_3(aq)]} = \frac{[H_3O^+]c_b}{[OH^-] - [H_3O^+]} = \frac{[H_3O^+]c_b}{K_w/[H_3O^+] - [H_3O^+]}.$$

Rearranging this equation and solving it for the concentration of hydronium ion yields

$$[H_3O^+] = \sqrt{\frac{K_w K_a}{c_b + K_a}} \tag{6.28b}$$

from which, if $c_b = 0.100$ M and $K_a = 0.16_6$ mol dm^{-3}, we obtain $[H_3O^+] = 7.93 \times 10^{-8}$ M or pcH $= 7.10$.

With a solution of an extremely weak acid it may be necessary to take into account the hydronium ions produced by the autoprotolysis of water as well as those produced by the dissociation of the acid. The resulting expression is

$$[H_3O^+] = \sqrt{c_a K_a + K_w}. \tag{6.29}$$

It is easy to tell when you need to use eq. (6.29) or eq. (6.28b) instead of a simpler equation. If eq. (6.20) gives a pcH that is above about 6 for a solution of a weak acid, or if eq. (6.22) gives one that is below about 8 for a solution of a weak base, the autoprotolysis of water cannot have been completely suppressed and should be taken into account.

If you are trying to compute the pcH of a solution of a weak acid, the sensible thing to do is to use eq. (6.20) first. In using it you are making the two assumptions described by eqs. (6.19), and your result cannot be correct unless both of them are satisfied. You should compare your value of $[H_3O^+]$ with c_a, and you should also compare your value of the pcH with 6. If $[H_3O^+]$ is no larger than a few per cent of c_a, eq. (6.19b) is satisfactory for the purpose; if the pcH is below 6, eq. (6.19a) is satisfactory. If eq. (6.19b) is not satisfactory, you should replace it with eq. (6.23); if eq. (6.19a) is not satisfactory, you should replace it with an electroneutrality equation, which will be $[H_3O^+] = [B^-] + [OH^-]$ if the acid is uncharged. It is possible for both eqs. (6.19a) and (6.19b) to be incorrect, but for aqueous solutions this is so unlikely that you can safely ignore it for the present.

There have been a good many equations in this section, and there are more to come in the sections that follow. Do not make the mistake of trying to memorize them, for some of them look so much alike that you will never be able to get them right. Worse yet, you will never be able to remember what situation any particular equation was intended for, and this will make them all useless to you. It is far better, and very much longer lasting, to understand how the equations are obtained, how an incorrect or impossible result is recognized, how its cause is identified, and how the trouble can be cured. The same thing is true in most other branches of chemistry. Practice on the problems at the end of this chapter, but do not turn the pages of the book while you are doing so. If you solve them by looking up equations and substituting numbers into these, you will probably get the right answers, but you will learn very little in the process.

It is important for you to realize that the discussion and the equations in this section apply only to solutions that contain just one dissolved acid, or just one dissolved base. They are valid for solutions of acetic acid in water, but they are not valid for solutions that contain hydrochloric acid, or sodium acetate, as well as acetic acid. They are valid for solutions of ammonia in water, but they do not apply to solutions that contain both sodium hydroxide and ammonia, or both ammonium chloride and ammonia. Solutions to which two or more acids have been added (such as a solution containing both hydrochloric acid and acetic acid), or to which two or more bases have been added (such as a solution containing both sodium hydroxide and ammonia, or one containing sodium acetate and ammonia), are discussed in Section 6.5. Solutions to which both an acid and a base have been added are discussed in Section 6.8.

SUMMARY: The equilibrium concentration of a dissolved substance will differ from its total or analytical concentration if the substance is consumed or produced by a chemical reaction. In calculating the equilibrium concentration of hydronium ion in a solution of an acid or base, you must consider the nature of the substance that is present in the solution, the natures and extents of the reactions that occur, and the relations among the concentrations of their reactants and products. Approximations are often possible but should always be tested to see whether they are justified by the answers obtained.

6.5. Competing equilibria: mixtures of acids or bases

PREVIEW: When a solution contains two acids, or two bases, calculations of its pcH involve the same principles as those already discussed, but there are more factors that must be considered.

A dissolved ion or molecule may be involved in two (or more) different reactions. Its concentration will then appear in two (or more) equilibrium-constant expressions that are satisfied simultaneously. In such a situation there are said to be *competing equilibria*. Two examples of competing equilibria have already appeared in this book. One was in the last few paragraphs of Section 5.3, where iodate ion was produced by each of the reactions $AgIO_3(s) = Ag^+ + IO_3^-$ and $TlIO_3(s) = Tl^+ + IO_3^-$, and the concentration of iodate ion appeared in each of the solubility-product expressions. The iodate ions obtained from the silver iodate suppressed the dissolution of thallium(I) iodate, while the iodate ions obtained from the thallium(I) iodate suppressed the dissolution of silver iodate. The other example was near the end of Section 6.4, where hydroxide ion was produced by the reaction $IO_3^- + H_2O(l) = HIO_3(aq) + OH^-$ and also by the reaction $2 H_2O(l) = H_3O^- + OH^-$, and again the occurrence of each reaction affected the extent of the other. Competing equilibria are very common. This section, and several others that follow it, will deal with some of the most common situations and will show you how to handle others.

Suppose that a solution contains two weak acids: HB_1, which has an overall dissociation constant equal to $K_{a,1}$ and a total concentration of $c_1 M$, and HB_2, which has an overall dissociation constant equal to $K_{a,2}$ and a total concentration of $c_2 M$. What is the equilibrium concentration of hydronium ion in the solution?

From the chemical point of view this is really a quite complicated question. Each of the acids suppresses the dissociation of the other; the autoprotolysis of water suppresses the dissociations of both, and in return both suppress the autoprotolysis of water; and either or both of the acids may be dissociated to an appreciable extent. The principles that are involved are no different from the ones described in Section 6.4, but their application is more complicated because there are more chemical substances to be considered. Fortunately, it is not often that all of these problems have to be confronted at once. Nearly always one of the two acids will furnish enough hydronium ion to suppress the autoprotolysis of water almost completely; often the dissociation of the other acid will be almost completely suppressed as well.

The fundamental equation that describes a mixture of two acids is the electroneutrality equation

$$[H_3O^+] = [B_1^-] + [B_2^-] + [OH^-] \tag{6.30}$$

and the concentration of each anion is related to the equilibrium concentration of the corresponding acid by an equation of the form

$$[B_i^-] = K_{a,i}[HB_i(aq)]/[H_3O^+] \tag{6.31}$$

where i may be either 1 or 2. Combining this with eq. (6.30), replacing $[OH^-]$ by $K_w/[H_3O^+]$, and multiplying each term by $[H_3O^+]$

$$[H_3O^+]^2 = [HB_1]K_{a,1} + [HB_2]K_{a,2} + K_w.$$

Provided that neither acid has dissociated to an appreciable extent—which is an assumption that will not always be justified—the equilibrium concentration $[HB_i]$ of each will be nearly equal to its total concentration c_i, and then

$$[H_3O^+] = \sqrt{c_1 K_{a,1} + c_2 K_{a,2} + K_w}. \tag{6.32}$$

This has almost the same form as eq. (6.29), and you can handle it in the same general way. Usually both of the cK_a products will be so much larger than K_w that you can ignore K_w, and often one of those products will be so much larger than the other that you can neglect the smaller one as well. For a solution in which the total concentration of acetic acid ($K_a = 1.75 \times 10^{-5}$ mol dm^{-3}) is 0.100 M and the total concentration of phenol ($K_a = 1.00 \times 10^{-10}$ mol dm^{-3}) is 0.200 M, the numerical values of the three terms under the radical are 1.75×10^{-6} for acetic acid, 2.00×10^{-11} for phenol, and 1.008×10^{-14} for the autoprotolysis of water. The second and third are so much smaller than the first that you should ignore them and write simply $[H_3O^+] = \sqrt{c_{HOAc}K_{a,HOAc}}$. This gives $[H_3O^+] = 1.322\,876 \times 10^{-3}$ M and pcH $= 2.878\,481$; including all three terms would give $[H_3O^+] = 1.322\,883 \times 10^{-3}$ M and pcH $= 2.878\,478$. Nowhere near this number of significant figures is justified [chiefly because activity coefficients (Chapter 13) are being neglected, and also because of the uncertainty in the value of K_a for acetic acid], and the difference between the two results is entirely negligible. Both the autoprotolysis of water and the dissociation of phenol are almost completely suppressed by the hydronium ions obtained from the dissociation of acetic acid, and the dissociation of acetic acid is the only source of hydronium ion that is worth taking into account. However, there is a danger involved in calculations of this sort. It would be easy to recognize with a mixture in which the total concentration of chloroacetic acid ($K_a = 1.40 \times 10^{-3}$ mol dm^{-3}) was 1.00×10^{-3} M and the total concentration of acetic acid ($K_a = 1.75 \times 10^{-5}$ mol dm^{-3}) was also 1.00×10^{-3} M. The numerical values of the three terms under the radical in eq.

(6.32) would be 1.40×10^{-6} for chloroacetic acid, 1.75×10^{-8} for acetic acid, and 1.008×10^{-14} for the autoprotolysis of water. Again the first of these is the only one worth taking into account: almost all of the hydronium ion in this mixture is obtained from the dissociation of the chloroacetic acid. Equation (6.32) gives $[H_3O^+] = 1.18 \times 10^{-3}\ M$ (as did eq. (6.20) in Section 6.4). This is an impossible result (as it was in Section 6.4), because you cannot obtain $1.18 \times 10^{-3}\ M$ hydronium ion from $1.00 \times 10^{-3}\ M$ chloroacetic acid. You can easily tell that it is impossible, because you have already decided that there is no other important source of hydronium ion in this solution. The dissociation of the chloroacetic acid must be fairly extensive, and its equilibrium concentration must be much smaller than its total concentration. At this point you might resort to eqs. (6.23) and (6.24).

The danger would be much more difficult to recognize with a mixture in which the total concentration of chloroacetic acid was $1.00 \times 10^{-3}\ M$ while the total concentration of acetic acid was $0.100\ M$. The third term (K_w) under the radical in eq. (6.32) is still negligible, but the first two have numerical values of 1.40×10^{-6} for chloroacetic acid and 1.75×10^{-6} for acetic acid. They are so nearly the same that you cannot ignore either of them, and you therefore write $[H_3O^+] = \sqrt{1.40 \times 10^{-6} + 1.75 \times 10^{-6}} = 1.77 \times 10^{-3}\ M$. Again the result exceeds the total concentration of chloroacetic acid (as it did in the preceding paragraph), but here this is quite possible because acetic acid is an important source of hydronium ion. You can tell whether it is reasonable by using eq. (6.31) to calculate the concentrations of the acetate and chloroacetate ions. In using eq. (6.32) you assumed that the equilibrium concentration of each acid is equal to its total concentration; if you make the same assumption in eq. (6.31), you obtain

$$[ClCH_2COO^-] = \frac{1.40 \times 10^{-3} \times 1.00 \times 10^{-3}}{1.77 \times 10^{-3}} = 7.9 \times 10^{-4}\ M$$

and

$$[CH_3COO^-] = \frac{1.75 \times 10^{-5} \times 0.100}{1.77 \times 10^{-3}} = 9.9 \times 10^{-4}\ M.$$

The first of these means that 79% of the chloroacetic acid appears to have dissociated: if this is correct, the equilibrium concentration of this acid is only about a fifth of its total concentration, and you cannot have been justified in assuming that they are equal. On the other hand, it appears that only about 1% of the acetic acid has dissociated, and this is so small that it is quite justifiable to say that the equilibrium concentration of acetic acid is equal to its total concentration.

On this basis you might write a conservation equation like eq. (6.23):

$$[ClCH_2COO^-] = 1.00 \times 10^{-3} - [ClCH_2COOH(aq)]$$

and combine it with the expression for the dissociation constant of chloroacetic acid to obtain

$$[ClCH_2COO^-] = 1.00 \times 10^{-3} - \frac{[H_3O^+][ClCH_2COO^-]}{K_{a,ClCH_2COOH}} = \frac{1.00 \times 10^{-3} K_a}{K_a + [H_3O^+]}$$

where the second equality is obtained by solving the first one. This makes no assumption about the equilibrium concentration of chloroacetic acid and therefore takes the dissociation of this acid into account. Since the extent of dissociation of acetic acid is negligible, you can ignore it and write simply

$$[CH_3COO^-] = \frac{[HOAc(aq)]K_{a,HOAc}}{[H_3O^+]} = \frac{c_{a,HOAc}K_{a,HOAc}}{[H_3O^+]}.$$

Combining these with eq. (6.30) (and neglecting $[OH^-]$ because you already know that it is negligible), you would obtain

$$[H_3O^+] = \frac{1.00 \times 10^{-3}K_{a,ClCH_2COOH}}{K_{a,ClCH_2COOH} + [H_3O^+]} + \frac{0.10K_{a,CH_3COOH}}{[H_3O^+]}$$

which is a cubic equation. Cubic equations used to be so difficult to solve that most chemists will still go to great lengths to avoid them, but nowadays it takes only a few minutes to write a computer program that will solve one within a fraction of a second. The solution to this one is $[H_3O^+] = 1.58 \times 10^{-3} M$, about 10% less than the approximate result obtained from eq. (6.32). Whether the difference is large enough to justify the expenditure of the time and trouble required to obtain it is a question that you must decide for each problem. In deciding it for the present one, you might note that the concentration of hydronium ion would be equal to 1.32×10^{-3} M $\left(= \sqrt{1.75 \times 10^{-6}}\right)$ if there were no chloroacetic acid present at all. In the mixture it must be larger than this. If the chloroacetic acid dissociated completely, it would furnish 1.00×10^{-3} mol dm^{-3} of hydronium ion, but the concentration of hydronium ion in the mixture cannot be as large as 2.32×10^{-3} M $(= 1.32 \times 10^{-3} + 1.00 \times 10^{-3})$ because the dissociation of the chloroacetic acid would suppress that of the acetic acid to some extent. You might guess that the hydronium-ion concentration might be at most 2×10^{-3} M in the mixture. If the difference between a hydronium-ion concentration of 1.32×10^{-3} M (pcH $= 2.88$) and one of 2×10^{-3} M (pcH $= 2.70$) is unimportant for the problem at hand it would certainly be a waste of time to solve the cubic equation even if you had the computer program already written and a terminal at your fingertips.

For a mixture of two bases B_1 and B_2 the equation corresponding to eq. (6.32) is

$$[H_3O^+] = \sqrt{\frac{K_w}{1 + (c_1/K_{a,1}) + (c_2/K_{a,2})}} \tag{6.33}$$

where c_1 is the total concentration of the base B_1, $K_{a,1}$ is the overall dissociation constant of its conjugate acid B_1H^+, c_2 is the total concentration of the base B_2, and $K_{a,2}$ is the overall dissociation constant of its conjugate base B_2H^+. Equation (6.33) can be handled in exactly the same way as eq. (6.32).

SUMMARY: In a solution that contains two acids, or two bases, the dissociation of one may predominate to such an extent that the other need not be considered. Such situations are indistinguishable from those discussed in Section 6.4. However, both may dissociate to an appreciable extent, and special care must then be taken to ensure that unjustifiable approximations do not pass unnoticed.

6.6. Solutions of polyfunctional acids or bases

PREVIEW: This section defines polyfunctional acids and bases, discusses the factors that govern the value of the important ratio K_1/K_2 of the successive overall dissociation constants of a difunctional acid, and explains how the pcH of a solution of a polyfunctional acid or base can be calculated.

A polyfunctional acid is an acid of which each molecule or ion can donate two or more protons to a base. A polyfunctional base is a base of which each molecule or ion can accept two or more protons from an acid. Sulfuric acid and oxalic acid are familiar difunctional acids, and the ion $^+H_3NCH_2COOH$ (which is obtained by adding a proton to the H_2N- group of the amino acid glycine, H_2NCH_2COOH) is also a difunctional acid. Oxalate ion, ethylenediamine ($H_2NCH_2CH_2NH_2$), and the ion $H_2NCH_2COO^-$ (which is obtained by removing a proton from the $-COOH$ group of glycine) are difunctional bases, and phosphate ion is a trifunctional base.

A difunctional acid has two overall dissociation constants. For the acid H_2B they are described by the equations

$$H_2B(aq) + H_2O(l) = H_3O^+ + HB^-; \quad K_1 = [H_3O^+][HB^-]/[H_2B(aq)], \quad (6.34a)$$

$$HB^- + H_2O(l) = H_2O^+ + B^{2-}; \quad K_2 = [H_3O^+][B^{2-}]/[HB^-]. \quad (6.34b)$$

The value of the ratio K_1/K_2 is an important property of a difunctional acid. The ratio can be regarded as the equilibrium constant of the reaction

$$H_2B(aq) + B^{2-} = 2HB^-; \quad K = K_1/K_2 = [HB^-]^2/[H_2B(aq)][B^{2-}]. \quad (6.35)$$

If its value is large, the reaction will occur to such an extent that no solution can contain high concentrations of both H_2B and B^{2-}. A solution of H_2B must contain some HB^- produced by the first dissociation step; if K_1/K_2 is large, it cannot also contain an appreciable concentration of B^{2-}, and therefore the second dissociation can occur to only a very small extent. Conversely, a solution of the difunctional base B^{2-} must contain some HB^- produced by the reaction $B^{2-} + H_2O(l) = HB^- + OH^-$, but if K_1/K_2 is large it cannot also contain an appreciable concentration of H_2B. On the other hand, if K_1/K_2 is small, there is only a limited tendency for H_2B and B^{2-} to react with each other, and it is possible for a solution to contain substantial concentrations of both.

The value of K_1/K_2 depends on a number of factors. If the two acidic groups in the acid H_2B are different, their acidic strengths are of the first importance. The sulfonic acid group ($-SO_3H$) is a strong proton donor, the carboxyl ($-COOH$) and ammonium ($-NH_3^+$) groups are much weaker ones, and the phenolic group ($-OH$ in compounds of the form R—OH, where R is an aromatic group such as a phenyl group) is much weaker still. These differences are reflected in the overall dissociation constants of some simple compounds. For benzenesulfonic acid ($C_6H_5SO_3H$) K_a is equal to 2.8×10^{-3} mol dm^{-3}; for benzoic acid (C_6H_5COOH) $K_a = 6.1 \times 10^{-5}$ mol dm^{-3}; for anilinium ion ($C_6H_5NH_3^+$) $K_a = 2.5 \times 10^{-5}$ mol dm^{-3}; and for phenol $K_a = 1.00 \times 10^{-10}$ mol dm^{-3}. The same acidic groups appear in the difunctional acids

The value of K_1/K_2 is very large for the first of these because the sulfonic acid group is very much more strongly acidic than the phenolic hydroxyl group. It is also large for the second because the carboxyl group is much more strongly acidic than the phenolic group. If the two functional groups did not interact at all, the values of K_1 and K_2 for the second compound would be the same as the values of K_a for benzoic acid and phenol, respectively, and K_1/K_2 would be approximately 6×10^5. Experimentally it is found to be 8×10^4. For the third compound, if there were no interaction between the two functional groups, K_1 would be equal to K_a for benzoic acid while K_2 would be equal to K_a for anilinium ion, and K_1/K_2 would be about 2.5; the experimental value is about 250.

Predictions made in this very simple way are crude: they do indicate correctly that K_1/K_2 is larger for the second of the above acids than for the third, but they do not give good ideas of the numerical values. They are crude because they ignore interactions of the acidic groups, not only with each other but also with other functional groups that are present in the molecule.

These interactions also affect the behaviors of difunctional acids, such as oxalic acid and *o*-phthalic acid

in which the two acidic groups are identical. If interactions did not exist in such a compound H_2B and its conjugate bases HB^- and B^{2-}, the rate of loss of protons from H_2B molecules would be twice as large as for HB^- ions because the former have twice as many sites from which protons can be lost, while the B^{2-} ions would accept protons twice as rapidly as HB^- ions because the B^{2-} ions have twice as many sites to which protons can be attached. The result would be

$$H_2B(aq) + H_2O(l) \underset{k_{-1}}{\overset{k_1}{\rightleftharpoons}} H_3O^+ + HB^-; \quad K_1 = k_1/k_{-1},$$

$$HB^- + H_2O(l) \underset{k_{-2}}{\overset{k_2}{\rightleftharpoons}} H_3O^+ + B^{2-}; \quad K_2 = k_2/k_{-2},$$

$$\frac{K_1}{K_2} = \frac{k_1/k_{-1}}{k_2/k_{-2}} = \frac{k_1/k_2}{k_{-1}/k_{-2}} = \frac{2}{1/2} = 4.$$

This limit is approached for the compounds $HOOC$—$(CH_2)_n$—$COOH$ as n becomes large: if $n = 7$ the value of K_1/K_2 is only about 6.5. In the monoanion $HOOC$—$(CH_2)_7$—COO^- the negatively charged carboxylate group is far away from the carboxyl group that remains, and the long saturated aliphatic chain between them does not transmit electrostatic effects. Consequently the acidic strength of the carboxyl group is hardly at all affected by the presence of the carboxylate group, and K_1/K_2 is scarcely different from the "statistical" value, 4. As n decreases the interaction becomes stronger. If $n = 0$ (oxalic acid) there are two possibilities. One is that a hydrogen-bonded monoanion such as

can be formed; the other is that the second proton will be more difficult to remove from the nearby negative charge than the first one was. The value of K_1/K_2 is much larger for oxalic acid than for $HOOC$—$(CH_2)_7$—$COOH$. The same possibilities arise with inorganic acids, where the acidic groups are usually —OH groups attached to a single atom and therefore not very far apart, as in sulfuric acid $[O_2S(OH)_2]$ or arsenic(V) acid $[OAs(OH)_3]$. The magnitude of the electrostatic effect depends on the dielectric constant of the solvent and the charge carried by H_2B, which is not necessarily an uncharged molecule. If it is, its successive ionizations in an amphoteric solvent HSo will yield the ion pairs $H_2So^+HB^-$ and $H_2So^+B^{2-}$. More work must be expended in the dissociation of the second. If the dielectric constant of the solvent is high, each of these amounts of work may be small enough to be provided by collisions with solvent molecules of average energy. As the dielectric constant ε decreases, both dissociation constants will decrease, but that for $H_2So^+B^{2-}$ will decrease much more rapidly than that for $H_2So^+HB^-$, and the value of K_1/K_2 will increase correspondingly. For sulfuric acid the value of K_1/K_2 is many orders of magnitude larger in anhydrous acetic acid ($\varepsilon = 6.15$) than in water.

There are electronic effects as well. Crude pictorial representations of acetic acid and chloroacetic acid are

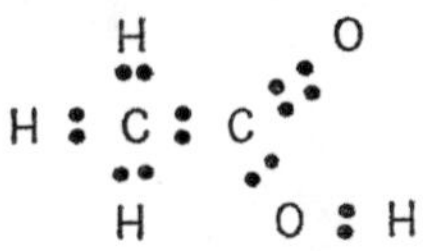

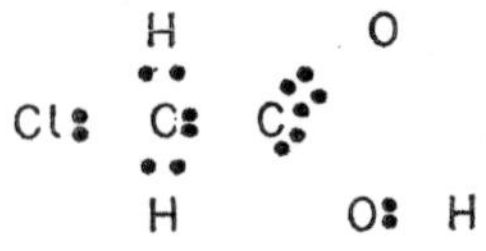

$$
\begin{array}{cc}
\text{H} & \text{O} \\
\text{H : C : C} & \\
\text{H} & \text{O : H}
\end{array}
\qquad
\begin{array}{cc}
\text{H} & \text{O} \\
\text{Cl : C : C} & \\
\text{H} & \text{O : H}
\end{array}
$$

The chlorine atom attracts electrons to itself, and in doing so weakens the bond between the oxygen and hydrogen atoms in the carboxyl group: although K_a is only 1.75×10^{-5} mol dm^{-3} for acetic acid, it is 1.40×10^{-3} mol dm^{-3} for chloroacetic acid. Such effects are cumulative: if another of the hydrogen atoms in the original methyl group is replaced with a chlorine atom to give dichloroacetic acid, the oxygen–hydrogen bond is further weakened and K_a increases to 3.3×10^{-2} mol dm^{-3}, while for trichloroacetic acid K_a is 0.2 mol dm^{-3}. They depend on the position of the substituent atom or group as well as on its identity. For propionic acid, CH_3CH_2COOH, K_a is 1.34×10^{-5} mol dm^{-3}; for 2-chloropropionic acid, $CH_3CHClCOOH$, it is 1.47×10^{-3} mol dm^{-3}; and for 3-chloropropionic acid, $ClCH_2CH_2COOH$, it is 1.04×10^{-4} mol dm^{-3}; the effect of the chlorine atom decreases as it moves farther away from the carboxyl group. The value of K_a is slightly decreased by substituting a methyl group for one of the hydrogen atoms in acetic acid because the methyl group repels electrons and strengthens the attachment of the electron pair to the proton in the carboxyl group.

Acidic substituent groups have similar effects. One of the three isomeric hydroxybenzoic acids was shown above; the structures and values of K_1, K_2 (of which both have the units mol dm^{-3}), and K_1/K_2 for all three are

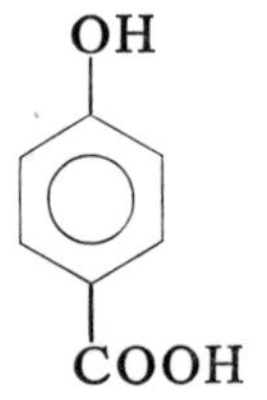

para-hydroxybenzoic acid	*meta*-hydroxybenzoic acid	*ortho*-hydroxybenzoic acid
$K_1 = 3.3 \times 10^{-5}$	$K_1 = 8.7 \times 10^{-5}$	$K_1 = 1.1 \times 10^{-2}$
$K_2 = 4.8 \times 10^{-10}$	$K_2 = 1.2 \times 10^{-10}$	$K_2 = 2.8 \times 10^{-13}$
$K_1/K_2 = 6.9 \times 10^4$	$K_1/K_2 = 7.2 \times 10^5$	$K_1/K_2 = 4.0 \times 10^{10}$

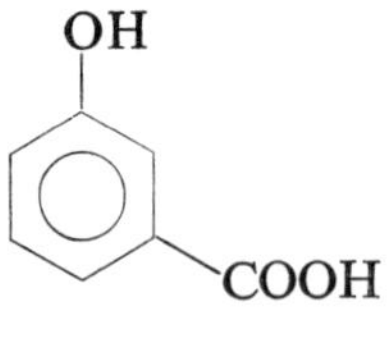

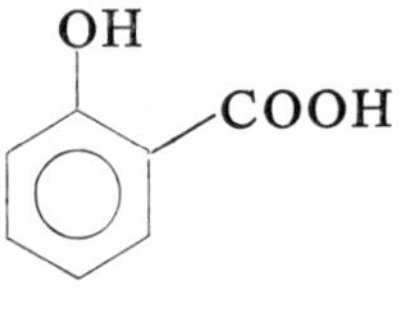

Two effects are involved: the values of K_1 are different because the presence and location of the phenolic group affect the acidity of the carboxylic group, and the values of K_2 are different because the presence and location of the carboxylate group resulting from the first dissociation affect the acidity of the phenolic group. Organic chemistry is largely concerned with the study and understanding of such effects.

A solution of a difunctional acid H_2B in water is really a mixture of two acids. One of the acids is H_2B itself; the other is the HB^- produced by the dissociation of H_2B. Calculating the pcH of such a solution involves the same sort of thinking that was described in Section 6.7, but is simplified by two facts. One is that the overall dissociation constant K_1 for H_2B is always larger, and is often much larger, than the overall dissociation constant K_2 for HB^-. The other is that the concentration of H_2B is almost always much larger than that of HB^-, for there is no HB^- in the solution except that produced by the dissociation of the H_2B. It would be a very unusual situation in which you could not ignore the dissociation of the HB^- and write

$$[H_3O^+] = c_a \qquad (6.36a)$$

(compare eq. (6.15a)) if K_1 is so large that the H_2B is almost completely dissociated into hydronium and HB^- ions, as is true for sulfuric acid; or

$$[H_3O^+] = \sqrt{c_a K_1} \qquad (6.36b)$$

(compare eq. (6.20)) if K_1 and the extent of dissociation are small, which is the commonest situation; or one of the other equations from Section 6.4. In any event you will take c_a as the total concentration of H_2B and use K_1 as the overall acidic dissociation constant of the only acid (H_2B) whose dissociation is worth taking into account.

Very similar things are true for solutions of polyfunctional bases. Suppose that you wish to calculate the pcH of a solution in which the total concentration of disodium oxalate is 0.100 M. From reasoning just like that described in the preceding paragraph you should conclude that the oxalate ion is the only base that is worth taking into account. Only a little hydrogen oxalate ion will be produced by the first of the reactions

$$C_2O_4^{2-} + H_2O(l) = HC_2O_4^- + OH^-,$$

$$HC_2O_4^- + H_2O(l) = H_2C_2O_4(aq) + OH^-,$$

and hydrogen oxalate ion is a much weaker base than oxalate ion. Consequently $[HC_2O_4^-] = [OH^-]$ while $[C_2O_4^{2-}] = 0.100$ M. The equilibrium constants that describe the behaviors of oxalate solutions are the overall dissociation constants of oxalic acid:

$$H_2C_2O_4(aq) + H_2O(l) = H_3O^+ + HC_2O_4^-;$$
$$K_1 = [H_3O^+][HC_2O_4^-]/[H_2C_2O_4(aq)]$$
$$HC_2O_4^- + H_2O(l) = H_3O^+ + C_2O_4^{2-};$$
$$K_2 = [H_3O^+][C_2O_4^{2-}]/[HC_2O_4^-].$$

Since you know nothing about the concentration of oxalic acid except that it must be very small, the expression for K_1 is useless to you, but the one for K_2 can be manip-

11

ulated to give

$$K_2 = \frac{[H_3O^+][C_2O_4^{2-}]}{[HC_2O_4^-]} = \frac{[H_3O^+](0.100)}{[OH^-]}$$

$$= \frac{[H_3O^+](0.100)}{K_w/[H_3O^+]} = \frac{[H_3O^+]^2(0.100)}{K_w}$$

so that

$$[H_3O^+] = \sqrt{K_w K_2/0.100} = \sqrt{(1.01 \times 10^{-14})(5.4 \times 10^{-5})/0.100}$$
$$= \sqrt{5.4 \times 10^{-18}} = 2.3 \times 10^{-9}\ M$$

or pcH $= 8.63$.

SUMMARY: The value of the ratio K_1/K_2 for a difunctional acid depends on the intrinsic acidities of the two acidic groups, on the framework to which they are attached, and on their relative positions. Since K_1/K_2 is usually much larger than the "statistical" value of 4 for the acid H_2B, you can usually ignore the trace of B^{2-} present in a solution of H_2B, or the trace of H_2B present in a solution of B^{2-}.

6.7. Solutions of amphoteric substances

PREVIEW: This section discusses solutions that contain both an acid and a base at equal concentrations. The acid and the base may be identical, as in a solution of sodium hydrogen oxalate, or they may not, as in a solution of ammonium benzoate.

Hydrogen oxalate ion, like water and many other substances, is amphoteric. It can either donate a proton to a base, yielding an oxalate ion, or accept a proton from an acid, yielding a molecule of oxalic acid. In a solution of sodium hydrogen oxalate in water the following reaction can occur:

$$2\,HC_2O_4^- = H_2C_2O_4(aq) + C_2O_4^{2-};$$
$$K = [H_2C_2O_4(aq)][C_2O_4^{2-}]/[HC_2O_4^-]^2 = K_2/K_1. \tag{6.37}$$

The equilibrium constant of this reaction is equal to K_2/K_1 because it is the reciprocal of the one in eq. (6.35).

Equation (6.37) does not contain the concentration of hydronium ion explicitly, but it does provide information from which that concentration can be calculated. The chemical equation tells us that molecules of oxalic acid and oxalate ions are produced in equal numbers, so that

$$[H_2C_2O_4(aq)] = [C_2O_4^{2-}] = (K_2/K_1)^{1/2}[HC_2O_4^-].$$

This can be combined with the expression for K_1 to obtain

$$[H_3O^+] = K_1\frac{[H_2C_2O_4(aq)]}{[HC_2O_4^-]} = K_1\frac{(K_2/K_1)^{1/2}[HC_2O_4^-]}{[HC_2O_4^-]} = \sqrt{K_1K_2} \tag{6.38}$$

or with the expression for K_2, which yields the same result. Identical reasoning can be used to show that

$$[H_3O^+] = \sqrt{K_2 K_3} \tag{6.39}$$

in a pure solution of monohydrogen phosphate ion, HPO_4^{2-}, where the reaction that takes place is

$$2\,HPO_4^{2-} = H_2PO_4^-{}_{;}+PO_4^{3-}; \quad K = [H_2PO_4^-][PO_4^{3-}]/[HPO_4^{2-}]^2 = K_3/K_2$$

and where K_2 and K_3 are the second and third overall acidic dissociation constants of phosphoric acid.

These are special cases because K_1 and K_2 in eq. (6.38) are properties of the same substance, as are K_2 and K_3 in eq. (6.39). For other amphoteric substances the reasoning is similar but the result has a different look. A solution of ammonium benzoate, $C_6H_5COONH_4$, also contains both an acid (ammonium ion) and a base (benzoate ion), and naturally these can react:

$$NH_4^+ + C_6H_5COO^- = NH_3(aq) + C_6H_5COOH(aq);$$

$$K = \frac{[NH_3(aq)]\,[C_6H_5COOH(aq)]}{[NH_4^+]\,[C_6H_5COO^-]}. \tag{6.40}$$

Here again the concentrations of the products are equal because one molecule of ammonia is formed along with each molecule of benzoic acid. The concentrations of the reactants are also equal because ammonium benzoate contains equal numbers of ammonium and benzoate ions and because these are consumed in equal numbers by the reaction. The first of these things would not be true if any ammonia or benzoic acid were present in the solution along with the ammonium benzoate, and the second would not be true if another ammonium salt or another benzoate were added. They are true only for a solution of pure ammonium benzoate that contains no other acid or base.

When they are true, so that $[NH_3(aq)] = [C_6H_5COOH(aq)]$ and $[NH_4^+] = [C_6H_5COO^-]$, eq. (6.40) can be rewritten in either of two ways:

$$[NH_3(aq)]/[NH_4^+] = K^{1/2}$$

or

$$[C_6H_5COOH(aq)]/[C_6H_5COO^-] = K^{1/2}.$$

Arbitrarily choosing the second, and combining it with the expression for the overall acidic dissociation constant of benzoic acid (for which the symbol $K_{a,\,C_6H_5COOH}$ will be used here to avoid possible confusion with the overall acidic dissociation constant of ammonium ion, $K_{a,\,NH_4^+}$):

$$[H_3O^+] = K_{a,\,C_6H_5COOH} \frac{[C_6H_5COOH(aq)]}{[C_6H_5COO^-]} = K_{a,\,C_6H_5COOH}\,K^{1/2}, \tag{6.41}$$

11*

but a value of K will be needed before we can obtain a numerical result. Expressions like the one for K in eq. (6.40) are fairly common. They have to be transformed into combinations of other equilibrium constants whose values are given in tables, or can be calculated from other values that are given in tables. You do this by recognizing portions of the expressions for other equilibrium constants in them, and supplying terms needed to complete those expressions in the fashion illustrated below.

Looking at the expression for K in eq. (6.40), you might notice that the first term in the numerator and the first term in the denominator together bear a strong resemblance to the expression $K_{a, NH_4^+} = [NH_3(aq)] [H_3O^+]/[NH_4^+]$ for the overall acidic dissociation constant of ammonium ion. The concentration of hydronium ion is missing from eq. (6.40), but it can be supplied without disturbing the equality if the right-hand side of the equation is multiplied by $[H_3O^+]/[H_3O^+]$. (If you needed to supply a concentration of hydroxide ion you could do so by multiplying by $OH^-]/[OH^-]$.) The result is

$$K = \frac{[NH_3(aq)] [H_3O^+]}{[NH_4^+]} \times \frac{[C_6H_5COOH(aq)]}{[H_3O^+] [C_6H_5COO^-]}$$

$$= K_{a, NH_4^+} \times \frac{[C_6H_5COOH(aq)]}{[H_3O^+] [C_6H_5COO^-]}.$$

Now you might recognize the remaining fraction as being the reciprocal of the overall acidic dissociation constant of benzoic acid, so that

$$K = K_{a, NH_4^+}/K_{a, C_6H_5COOH}. \tag{6.42}$$

By substituting this expression for K into eq. (6.41) you can easily obtain

$$[H_3O^+] = \sqrt{K_{a, C_6H_5COOH} K_{a, NH_4^+}}. \tag{6.43}$$

Introducing the appropriate numerical values gives $[H_3O^+] = 1.87 \times 10^{-7}$ M and pcH = 6.73.

Now compare eq. (6.43) with eq. (6.38). In each of them the second equilibrium constant under the radical is the acidic dissociation constant of the acid that the solution contains; in eq. (6.43) this acid is ammonium ion, and in eq. (6.38) it is hydrogen oxalate ion. In each of them the first equilibrium constant under the radical is the overall acidic dissociation constant of the conjugate acid of the base that the solution contains; in eq. (6.43) this base is benzoate ion and its conjugate acid is benzoic acid, while in eq. (6.38) the base is hydrogen oxalate ion and its conjugate acid is oxalic acid. The two equations are really identical, because the two chemical situations are really just the same.

SUMMARY: Values of the equilibrium constants for the reactions that take place in solutions of amphoteric substances (and in many other common chemical situations) do not appear in tables but may be found by combining the values of other equilibrium constants. This section showed how a useful combination may be discovered, and also discussed the equilibrium compositions of solutions that contain equal concentrations of an acid and a base.

6.8. Mixtures of acids with bases

PREVIEW: This section introduces and discusses the idea of the principal acid-base equilibrium and shows how it can be applied to solutions containing an acid and a base, regardless of whether their concentrations are equal (as in the preceding section) or unequal.

The preceding section discussed solutions of substances, such as hydrogen oxalate ion or ammonium benzoate, that contain both an acid and a base at equal total concentrations. Almost always the acid (hydrogen oxalate ion or ammonium ion) is a stronger acid than water and the base (hydrogen oxalate ion or benzoate ion) is a stronger base than water. When these things are true, the composition of the solution at equilibrium can be described by relatively simple equations, such as eqs. (6.38) and (6.43). The reactions that take place are described by eqs. (6.37) and (6.40), respectively. Each of these contains the equation for the acid-base reaction that proceeds more nearly to completion than any other in the same solution, that yields larger concentrations of its products than any other, and that governs the pcH of the solution. The acid-base equilibrium of which these things are true is called the *principal acid-base equilibrium.*

To see why eq. (6.40) represents the principal acid-base equilibrium in a solution of ammonium benzoate, let us inspect the equations for all four of the acid-base reactions that can occur in such a solution. Listed in the order of the numerical values of their equilibrium constants, and with indications of how these values might have to be calculated from the ones readily available in common tables, they are:

1. $NH_4^+ + C_6H_5COO^- = NH_3(aq) + C_6H_5COOH(aq)$;
$$K = K_{a,\,NH_4^+}/K_{a,\,C_6H_5COOH} = 9.4 \times 10^{-6}.$$

2. $NH_4^+ + H_2O(l) = NH_3(aq) + H_3O^+$;
$$K = K_{a,\,NH_4^+} = 5.8 \times 10^{-10} \text{ mol dm}^{-3}.$$

3. $H_2O(l) + C_6H_5COO^- = OH^- + C_6H_5COOH(aq)$;
$$K = K_w/K_{a,\,C_6H_5COOH} = 1.65 \times 10^{-10} \text{ mol dm}^{-3}.$$

4. $2H_2O(l) = OH^- + H_3O^+$;
$$K = K_w = 1.008 \times 10^{-14} \text{ mol}^2 \text{ dm}^{-6}.$$

The numerical value of the equilibrium constant is much larger for reaction 1 than for reaction 2 because benzoate ion is a stronger base than water, is much larger for reaction 1 than for reaction 3 because ammonium ion is a stronger acid than water, and is very much larger for reaction 1 than for reaction 4 for both of these reasons. Consequently reaction 1 produces so much ammonia that reaction 2 is almost completely suppressed, and it also produces so much benzoic acid that reaction 3 is almost completely suppressed. The pcH of the equilibrium mixture is fixed by the relatively high concentrations of ammonia and benzoic acid that are produced in reaction 1 and by the much larger concentrations of ammonium and benzoate ions that remain unreacted, and therefore reaction 4 is suppressed as well. The first equilibrium is the principal acid-base equilibrium because it is the only one that can produce appreciable concentrations of its products.

All these things were tacitly assumed to be true in Section 6.7. They are indeed true in many cases, but they are not always true. They are not true if either the acid or the base is extremely weak, and they are not true in very dilute solutions. In either of these circumstances it may be necessary to take two or more acid-base reactions into account because they proceed to comparable extents. Moreover, some modification of the equation is needed if the concentrations of the acid and base are different.

Taking these three things in turn, let us first consider a solution containing sodium hydrogen sulfate at a total concentration of 0.100 M. At first glance the acid-base processes that occur in this solution may seem to be just like those that occur in a solution of sodium hydrogen oxalate, and it may therefore seem appropriate to use eq. (6.38). For sulfuric acid $K_2 = 1.20 \times 10^{-2}$ mol dm^{-3}; a value for K_1 is less easy to obtain, but it has been estimated to be 50 mol dm^{-3} and this estimate will serve our purpose. Hence

$$[H_3O^+] = \sqrt{K_1 K_2} = \sqrt{50 \times 1.20 + 10^{-2}} = \sqrt{0.60} = 0.77 \ M?$$

The question mark is appropriate: it is absolutely impossible to obtain so much hydronium ion from so little hydrogen sulfate ion. Let us look at the equilibria involved. The chemical equations are arranged in the same order as those for ammonium benzoate above, and K_1 and K_2 are the overall dissociation constants of sulfuric acid:

1. $HSO_4^- + HSO_4^- = SO_4^{2-} + H_2SO_4(aq)$; $K = K_2/K_1 = 2.4 \times 10^{-4}$.

2. $HSO_4^- + H_2O(l) = SO_4^{2-} + H_3O^+$; $K = K_2 = 1.2 \times 10^{-2}$ mol dm^{-3}.

3. $H_2O(l) + HSO_4^- = OH^- + H_2SO_4(aq)$; $K = K_w/K_1 = 2.0 \times 10^{-16}$ mol dm^{-3}.

4. $2 H_2O(l) = OH^- + H_3O^+$; $K = K_w = 1.008 \times 10^{-14}$ mol^2 dm^{-6}.

Reaction 1 is not the principal equilibrium! Water is a stronger base than hydrogen sulfate ion, and therefore it is reaction 2 that has the largest numerical value of the equilibrium constant and that proceeds to the greatest extent. It provides sulfate ions that suppress reaction 1, and therefore the extent of this reaction can be ignored, as can the extents of reactions 3 and 4. The value of K for reaction 2 is so high that a fairly large fraction of the hydrogen sulfate ion will be consumed, and the proper description is

$$K_2 = \frac{[H_3O^+]^2}{0.100 - [H_3O]^+} = 1.20 \times 10^{-2} \text{ mol dm}^{-3}$$

which yields $[H_3O^+] = 2.92 \times 10^{-2}\ M$ (and $[HSO_4^-] = 7.08 \times 10^{-2}\ M$). Combining these values with

$$[H_2SO_4(aq)] = \frac{[H_3O^+][HSO_4^-]}{K_1}$$

gives $[H_2SO_4(aq)] = 4.1 \times 10^{-5}\ M$, which is so much smaller than $[H_3O^+]$ that it is certainly justifiable to neglect reaction 1.

You should check every such answer in some way, for the values of the equilibrium constants do not tell the whole story about a situation like this. Part of the difficulty arises from the fact that the different equilibrium constants often have different units. The extents of the reactions are much more important than the numerical values of the equilibrium constants, and usually depend on the concentration of the original solute. Calculations just like the one above for solutions containing different total concentrations of hydrogen sulfate ion give the following results:

$c_{HSO_4^-}$, M	0.100	0.300	1.00	3.00	10.0
$[H_3O^+]$, M	2.92×10^{-2}	5.4×10^{-2}	0.104	0.184	0.341
$[H_2SO_4]$, M	4.1×10^{-5}	2.7×10^{-4}	1.9×10^{-3}	1.0×10^{-2}	6.6×10^{-2}
$[H_3O^+]/[H_2SO_4]$	7.1×10^2	2×10^2	55	18	5.2

The concentration of sulfuric acid is negligible in dilute solutions but becomes appreciable in concentrated ones. This is because the expressions for different equilibrium constants have different algebraic forms. To illustrate this, let us assume that three different reactions occur in the same solution:

$$2\,A = B; \qquad K_1 = [B]/[A]^2.$$

$$2\,A = C + D; \qquad K_2 = [C][D]/[A]^2.$$

$$A = E + F; \qquad K_3 = [E][F]/[A].$$

The second and third of these have the same forms as reactions 1 and 2 for hydrogen sulfate ion. Let us assume for simplicity that none of them proceeds to a large extent, so that the concentration of A at equilibrium will be equal to its total concentration c_A. Setting [C] equal to [D] and [E] equal to [F], $[B] = K_1 c_A^2$, $[C] = [D] = \sqrt{K_2} c_A$, and $[E] = [F] = \sqrt{K_3 c_A}$. If the value of c_A is multiplied (or divided) by a factor of 10, the concentration of B is multiplied (or divided) by a factor of 100, while the concentrations of C and D are multiplied (or divided) by a factor of only 10 and the concentrations of E and F are multiplied (or divided) by a factor of only $\sqrt{10} = 3.16$. The extent of the second reaction—that is, the fraction of the A that is converted into C or D—does not depend on the concentration of A, but the extent of the first reaction increases as the concentration of A increases (and decreases as the concentration of A decreases) and the extent of the third reaction decreases as the concentration of A increases (and increases as the concentration of A decreases).

Let us consider a solution containing sodium hydrogen succinate at a total concentration of 1.00×10^{-5} M from this point of view. Succinic acid, which is represented here as H_2Su to save space, has the formula $HOOC$—CH_2CH_2—$COOH$, and its overall dissociation constants are $K_1 = 6.2 \times 10^{-5}$ mol dm^{-3} and $K_2 = 2.32 \times 10^{-6}$ mol dm^{-3}. According to eq. (6.38) the concentration of hydronium ion in a pure solution of hydrogen succinate ion is given by $[H_3O^+] = \sqrt{K_1 K_2} = 1.2 \times 10^{-5}$ M. This is impossible, for there is no way in which 1.2×10^{-5} M hydronium ion could be obtained from 1.00×10^{-5} M hydrogen succinate ion. Arranged in the same order as before, the equilibria are:

1. $HSu^- + HSu^- = Su^{2-} + H_2Su(aq)$; $K = 3.7 \times 10^{-2}$.

2. $HSu^- + H_2O(l) = Su^{2-} + H_3O^+$; $K = 2.32 \times 10^{-6}$ mol dm^{-3}.

3. $H_2O(l) + HSu^- = OH^- + H_2Su(aq)$; $K = 1.63 \times 10^{-10}$ mol dm^{-3}.

4. $2H_2O(l) = H_3O^+ + OH^-$; $K = 1.008 \times 10^{-14}$ mol^2 dm^{-6}.

The values of K for reactions 3 and 4 are so small that we focus our attention on reactions 1 and 2. The numerical value of the equilibrium constant for reaction 1 is larger than that for reaction 2, which means that hydrogen succinate ion is a stronger base than water. Acids (including the hydrogen succinate ion itself) will therefore react more strongly with hydrogen succinate ion than they do with water. In a solution as dilute as this one, however, there are so few hydrogen succinate ions to accept protons that many of the available protons will be accepted by molecules of water instead. Because of the form of its equilibrium constant, reaction 2 proceeds to a larger and larger extent as the solution is diluted, whereas reaction 1 does not, and in this very dilute solution the extent of reaction 2 is comparable with that of reaction 1 even though its equilibrium constant has a much smaller value.

In either a very concentrated solution of hydrogen sulfate ion or a very dilute solution of hydrogen succinate ion there are two equilibria that must be taken into account. In each case these are the first two of the four reactions that occur. Correct results can be obtained by applying the same sort of argument that was applied to eqs. (6.27) in Section 6.4, but the algebra is too messy to be described here.

A more important situation is the one in which the concentrations of the acid and base are unrelated. This would arise with a solution in which the total concentration of ammonium chloride was 0.100 M ($= c_a$) while that of sodium benzoate was 0.0100 M ($= c_b$). The principal reaction is represented by eq. (6.40), which may be combined with eq. (6.42) to give

$$NH_4^+ + C_6H_5COO^- = NH_3(aq) + C_6H_5COOH(aq);$$
$$K = K_{a,NH_4^+}/K_{a,C_6H_5COOH} = 9.4 \times 10^{-6}. \qquad (6.44)$$

This reaction produces equal concentrations of ammonia and benzoic acid, no matter whether the concentrations of ammonium ion and benzoate ion are equal or not.

Since the value of K is very small, very little of the ammonium and benzoate ions will be consumed, and their concentrations at equilibrium will be virtually equal to c_a and c_b, respectively. Consequently

$$K = \frac{[NH_3(aq)]\,[C_6H_5COOH(aq)]}{[NH_4^+]\,[C_6H_5COO^-]} = \left(\frac{[NH_3(aq)]^2}{c_a c_b} = \frac{[C_6H_5COOH(aq)]^2}{c_a c_b} \right)$$

$$= K_{a,\,NH_4^+}/K_{a,\,C_6H_5COOH}.$$

You can choose either of the two expressions between the parentheses and use it to obtain a value for the one unknown concentration that it contains. The second one gives

$$[C_6H_5COOH(aq)] = \left(\frac{c_a c_b K_{a,\,NH_4^+}}{K_{a,\,C_6H_5COOH}} \right)^{1/2}$$

which can be combined with the expression for the overall acidic dissociation constant of benzoic acid to give

$$[H_3O^+] = \frac{[C_6H_5COOH(aq)]}{[C_6H_5COO^-]} K_{a,\,C_6H_5COOH} = \left(\frac{c_a c_b K_{a,\,NH_4^+}}{K_{a,\,C_6H_5COOH}} \right)^{1/2} \frac{K_{a,\,C_6H_5COOH}}{c_b}$$

$$= \sqrt{\frac{c_a}{c_b} K_{a,\,C_6H_5COOH} K_{a,\,NH_4^+}} \tag{6.45}$$

of which eq. (6.43) is merely the special case that corresponds to $c_a = c_b$. With $c_a = 0.100\ M$, $c_b = 0.0100\ M$, $K_{a,\,C_6H_5COOH} = 6.1 \times 10^{-5}$ mol dm^{-3}, and $K_{a,\,NH_4^+} = 5.8 \times 10^{-10}$ mol dm^{-3}, eq. (6.45) gives $[H_3O^+] = 5.9 \times 10^{-7}\ M$ and pcH $= 6.23$.

> SUMMARY: There is often one acid-base reaction that proceeds to a much larger extent than any other, and that governs the pcH of the solution: that reaction is called the principal acid-base equilibrium. Identifying the principal acid-base equilibrium makes it easy to calculate the compositions of many solutions by enabling the chemist to ignore other reactions whose extents are negligible. Sometimes, however, there are two (or more) reactions that proceed to comparable extents, and then both (or all) of these must be taken into account.

6.9. Graphical representations of acid-base equilibria: formation functions

> PREVIEW: Many equilibrium problems are easier to solve by graphical approaches than by algebraic ones. This section describes the use and construction of graphs that are useful for this purpose, and defines the formation function, on which they are based.

It is the second of the three situations described in Section 6.8—in which it is impossible to deal with a single principal equilibrium while ignoring the extents of all the other reactions—that is the most difficult one. As the number of acids and bases in a solution increases, the number of different possible acid-base reactions increases, and so does the probability that two or more of them will occur to comparable extents. Graphical diagrams are often extremely helpful in dealing with such problems.

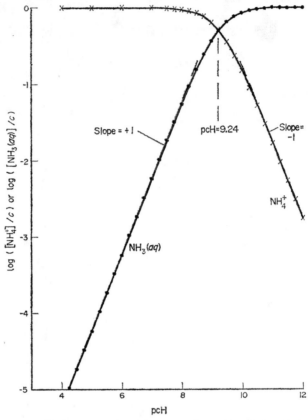

FIG. 6.1. Plots of $\log{([NH_4^+]/c)}$ and $\log{([NH_3(aq)/c)}$ against pcH, where c is the fixed sum of the concentrations of ammonium ion and ammonia.

Figure 6.1 shows how the concentrations of ammonium ion and ammonia depend on the pcH if their sum is kept constant and equal to some value c. Toward the left-hand side of the figure, in fairly strongly acidic solutions, the concentration of ammonia is very small and the concentration of ammonium ion is indistinguishable from c. In this region the concentration of ammonia is given by

$$[NH_3(aq)] = K_{a, NH_4^+}[NH_4^+]/[H_3O^+]$$

or

$$\log_{10}[NH_3(aq)] = \log_{10} K_{a, NH_4^+} c + pcH$$

so that a plot of its logarithm against the pcH is a straight line with a slope equal to $+1$. Toward the right-hand side, in fairly strongly alkaline solutions, the concentration of ammonium ion is very small and the concentration of ammonia is indistinguishable from c. Here

$$[NH_4^+] = [NH_3(aq)] [H_3O^+]/K_{a, NH_4^+}$$

or

$$\log_{10} [NH_4^+] = \log_{10} (c/K_{a, NH_4^+}) - pcH$$

so that a plot of the logarithm of the concentration of ammonia against the pcH is a straight line with a slope equal to -1. Substantial concentrations of ammonia and ammonium ion can exist together only over a fairly narrow range of pcH-values. The two curves cross at the point where these concentrations are equal and

$$[H_3O^+] = K_{a, NH_4^+}[NH_4^+]/[NH_3(aq)] = K_{a, NH_4^+} = 5.7 \times 10^{-10}\,M \qquad (6.46a)$$

or, in logarithmic form,

$$pcH = pK_{a, NH_4^+} - \log ([NH_4^+]/[NH_3(aq)]) = pK_{a, NH_4^+} = 9.24. \qquad (6.46b)$$

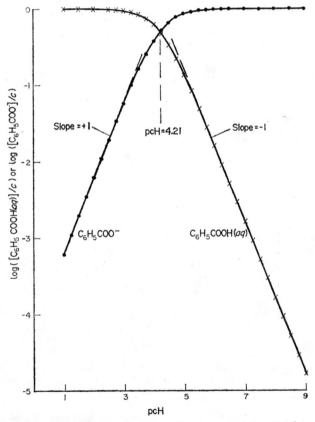

FIG. 6.2. Plots of $\log ([C_6H_5COOH(aq)]/c)$ and $\log ([C_6H_5COO^-]/c)$ against pcH, where c is the fixed sum of the concentrations of benzoic acid and benzoate ion, and is smaller than $2.3_5 \times 10^{-2}\,M$ for the reason explained in the text.

Figure 6.2 is a similar plot for benzoic acid and benzoate ion. The curves have exactly the same shapes as those in Fig. 6.1;[†] the only difference is that they cross at the point where $pcH = pK_{a, C_6H_5COOH} = 4.21$.

Let us superimpose the curves of Figs. 6.1 and 6.2 on the same graph in such a way that c is the same for both. We shall arbitrarily take c to be 0.01 M. The result corresponds to a solution of ammonium benzoate, which was discussed in Section 6.7, and is shown in Fig. 6.3. Since the concentrations of ammonia and benzoic acid will be equal in such a solution if no other acid or base is added to it, we seek the point at which the curves for ammonia and benzoic acid cross. This lies at $pcH = 6.72$, as closely as it can be read from the graph; numerical calculation from eq. (6.43) gave $pcH = 6.73$.

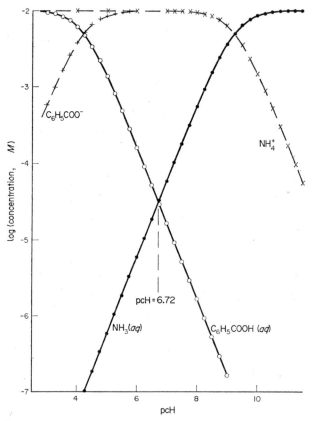

FIG. 6.3. Plots of $\log[NH_4^+]$, $\log[NH_3(aq)]$, $\log[C_6H_5COOH(aq)]$, and $\log[C_6H_5COO^-]$ against pcH for a solution in which $[NH_4^+]+[NH_3] = [C_6H_5COOH]+[C_6H_5COO^-] = 0.01\ M$.

[†] It is assumed that c is smaller than $2.3_5 \times 10^{-2}\ M$, which is equal to the solubility of undissociated benzoic acid in water at 25°. If c exceeds this value there will be a discontinuity at the point where benzoic acid begins to precipitate. Except for complications like this, the curves for every conjugate acid-base pair are just like the ones in Figs. 6.1 and 6.2.

To find the pcH of a solution that contains 0.100 M ammonium ion and 0.0100 M benzoate ion, which was the last problem discussed in Section 6.8, we need only shift the curves for ammonium ion and ammonia in a direction parallel to the ordinate axis so as to make c equal to 0.100 M for this conjugate acid-base pair. The result is shown in Fig. 6.4. Again it is the point where the curves for ammonia and benzoic acid intersect that is of interest, because the concentrations of these substances must be equal even though those of ammonium and benzoate ions are not. This point lies at pcH = 6.22; eq. (6.45) gave pcH = 6.23.

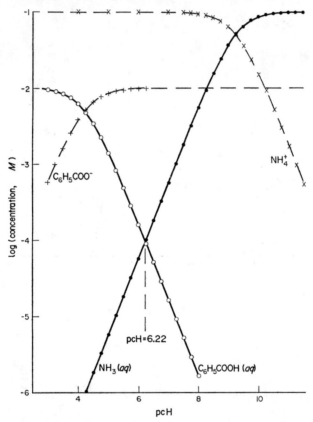

FIG. 6.4. Plots of log $[NH_4^+]$, log $[NH_3(aq)]$, log $[C_6H_5COOH(aq)]$, and log $[C_6H_5COO^-]$ against pcH for a solution in which $[NH_4^+]+[NH_3] = 0.1$ M and $[C_6H_5COOH]+[C_6H_5COO^-] = 0.01$ M.

If the system contains a difunctional acid H_2B, the plot consists of three curves: one each for H_2B, HB^-, and B^{2-}. Figure 6.5 shows such a plot for oxalic acid, for which $pK_1 = 1.27$ and $pK_2 = 4.27$. Since

$$pcH = pK_1 - \log([H_2C_2O_4(aq)]/[HC_2O_4^-]) = pK_2 - \log([HC_2O_4^-]/[C_2O_4^{2-}])$$

the curves for oxalic acid and hydrogen oxalate ion cross at pcH = pK_1 = 1.27 (point A), while those for hydrogen oxalate and oxalate ions cross at pcH = pK_2 =

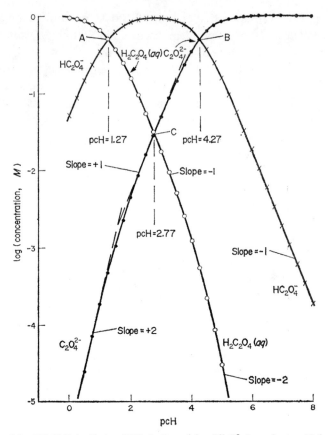

FIG. 6.5. Plots of $\log [H_2C_2O_4(aq)]$, $\log [HC_2O_4^-]$, and $\log [C_2O_4^{2-}]$ against pcH for a solution in which $[H_2C_2O_4(aq)] + [HC_2O_4^-] + [C_2O_4^{2-}] = 1\ M$.

4.27 (point B). In more alkaline solutions oxalate ion predominates and its concentration is indistinguishable from the total or formal concentration c; in this region

$$[H_2C_2O_4(aq)] = \frac{[H_3O^+][HC_2O_4^-]}{K_1} = \frac{[H_3O^+]^2[C_2O_4^{2-}]}{K_1K_2}$$

or

$$\log_{10}[H_2C_2O_4(aq)] = \log_{10} c/K_1K_2 - 2\,\mathrm{pcH}$$

so that the curve for oxalic acid is a straight line with a slope equal to -2. Point C, where the concentrations of oxalic acid and oxalate ion are equal, corresponds to a solution that contains hydrogen oxalate ion but no other added acid or base (see Section 6.9). It lies at pcH $= 2.77\ [= (pK_1 + pK_2)/2$, which is the logarithmic form of eq. (6.38)], and the maximum on the curve for hydrogen oxalate ion occurs at the same pcH-value. The maximum concentration of hydrogen oxalate ion is just slightly smaller than c because the equilibrium constant K for the reaction $2\,HC_2O_4^- =$

$H_2C_2O_4(aq) + C_2O_4^{2-}$ is not very small: it is described by eq. (6.37) and is equal to 1.00×10^{-3}. About 3% of the hydrogen oxalate ion is consumed by this reaction.

Figure 6.6 shows a similar plot for succinic acid, for which $pK_1 = 4.21$ and $pK_2 = 5.63$. Its general shape is similar to that of Fig. 6.5, but the range over which the monohydrogen anion predominates is much narrower because pK_1 and pK_2 are much more nearly equal than they are for oxalic acid, and in addition the maximum concentration of this anion is a distinctly smaller fraction of the total concentration c than it was in Fig. 6.5.

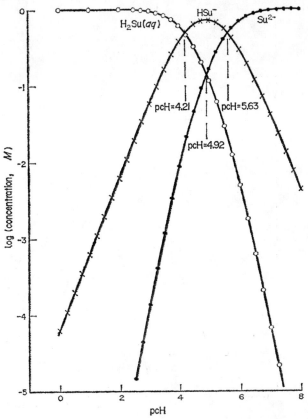

FIG. 6.6. Plots of $\log[H_2Su(aq)]$, $\log[HSu^-]$, and $\log[Su^{2-}]$ against pcH for a solution in which $[H_2Su(aq)] + [HSu^-] + [Su^{2-}] = 1\,M$.

In Section 6.10 it was said that the algebra involved in computing the pcH of a solution containing sodium hydrogen succinate at a total concentration of $1.00 \times 10^{-5}\,M$ was too complicated to be given here. There are two reactions that have to be considered:

1. $HSu^- + HSu^- = Su^{2-} + H_2Su(aq);$ $K = 3.7 \times 10^{-2}.$

2. $HSu^- + H_2O(l) = Su^{2-} + H_3O^+;$ $K = 2.32 \times 10^{-6}$ mol dm^{-3}.

The total number of divalent succinate ions in the solution is the sum of the numbers produced in these two reactions. The first yields a molecule of succinic acid along with each succinate ion; the second yields a hydronium ion along with each succinate ion. The fundamental equation is

$$[Su^{2-}] = [H_2Su(aq)] + [H_3O^+] \qquad (6.47)$$

Figure 6.7 is identical with Fig. 6.6 except that c has been set to 1.00×10^{-5} M, a line having the equation $\log_{10} [H_3O^+] = -pcH$ has been added to represent the

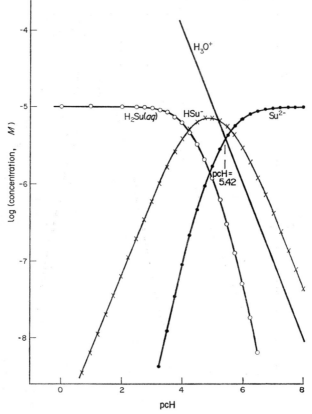

FIG. 6.7. Plots of log [H₂Su(aq), log [HSu⁻], log [Su²⁻], and log [H₃O⁺] against pcH for a solution in which [H₂Su(aq)]+[HSu⁻]+[Su²⁻] = 1.00×10⁻⁵ M.

concentration of hydronium ion, and the curve for hydrogen succinate ion has been drawn very lightly because this ion does not appear in eq. (6.47). The curves for succinate and hydronium ions cross at pcH = 5.42; at this pcH the right-hand side of eq. (6.47) is already larger than the left-hand side, because the concentration of succinic acid is positive. Therefore the equilibrium mixture must have a pcH that is above 5.42. Since the curve for succinate ion is fairly steep in this region, let us try pcH = 5.50.

Reading the coordinates of the curves at this point from the graph gives $\log[\text{Su}^{2-}] =$
$= -5.39$, $\log[\text{H}_2\text{Su}(aq)] = -6.54$, and of course $\log[\text{H}_3\text{O}^+] = -5.50$. Consequently
$[\text{Su}^{2-}] = 4.1\times10^{-6}$ M, $[\text{H}_2\text{Su}(aq)] = 2.9\times10^{-7}$ M, and $[\text{H}_3\text{O}^+] = 3.2\times10^{-6}$ M.
Since 4.1×10^{-6} is larger than $2.9\times10^{-7}+3.2\times10^{-6}$ ($= 3.5\times10^{-6}$), a pcH of 5.50 is
too high. If 5.42 is too low and 5.50 is too high, the truth must be very close to 5.46.
Much algebra would be needed to achieve the same result, and a quartic equation
would have to be solved in the process.

Plots like these are easy to construct, especially with the aid of a plotter driven by a
computer or programmable desk calculator to perform the mechanical labor. For a
single conjugate acid-base pair, say HB and B⁻, the conservation equation is

$$[\text{HB}]+[\text{B}^-] = c$$

and the expression for K_a can be rearranged to give either

$$[\text{HB}] = [\text{H}_3\text{O}^+][\text{B}^-]/K_a \quad \text{or} \quad [\text{B}^-] = K_a[\text{HB}]/[\text{H}_3\text{O}^+]$$

which can be used to eliminate one of the two concentrations that appear in the conservation equation. The resulting expression

$$\frac{[\text{H}_3\text{O}^+][\text{B}^-]}{K_a}+[\text{B}^-] = c \quad \text{or} \quad [\text{HB}]+\frac{K_a[\text{HB}]}{[\text{H}_3\text{O}^+]} = c$$

relates the other concentration to the independent variable whose effect is being investigated. That variable is called the *master variable*; in acid-base problems the master
variable is always the concentration of hydronium ion, but other choices are possible
in other areas. Solving for the concentration of B⁻ or HB yields

$$[\text{B}^-] = \left(\frac{K_a}{K_a+[\text{H}_3\text{O}^+]}\right) c \quad \text{or} \quad [\text{HB}] = \left(\frac{[\text{H}_3\text{O}^+]}{K_a+[\text{H}_3\text{O}^+]}\right) c.$$

The quantities within parentheses are called *formation functions* and given the symbol
α with an appropriate subscript:

$$[\text{B}^-] = \alpha_{\text{B}^-}c \qquad\qquad [\text{HB}] = \alpha_{\text{HB}}c$$

$$\alpha_{\text{B}^-} = \frac{K_a}{K_a+[\text{H}_3\text{O}^+]} \qquad \alpha_{\text{HB}} = \frac{[\text{H}_3\text{O}^+]}{K_a+[\text{H}_3\text{O}^+]} \qquad\left.\right\} \quad (6.48)$$

The formation function is the fraction of the total concentration that is present in
the specified form when the master variable has any particular value.

For a difunctional acid H_2B the conservation equation is

$$[\text{H}_2\text{B}]+[\text{HB}^-]+[\text{B}^{2-}] = c. \qquad (6.49)$$

To eliminate the concentrations of HB⁻ and B²⁻ we must write

$$[\text{HB}^-] = K_1[\text{H}_2\text{B}]/[\text{H}_3\text{O}^+] \quad \text{and} \quad [\text{B}^{2-}] = K_1K_2[\text{H}_2\text{B}]/[\text{H}_3\text{O}^+]^2.$$

12

Combining these with eq. (6.49) and solving for [H_2B] gives

$$[H_2B] = \alpha_{H_2B} c; \quad \alpha_{H_2B} = \frac{[H_3O^+]^2}{[H_3O^+]^2 + K_1[H_3O^+] + K_1K_2} . \tag{6.50a}$$

In much the same fashion one can obtain

$$[HB^-] = \alpha_{HB^-} c; \quad \alpha_{HB^-} = \frac{K_1[H_3O^+]}{[H_3O^+]^2 + K_1[H_3O^+] + K_1K_2} \tag{6.50b}$$

and

$$[B^{2-}] = \alpha_{B^{2-}} c; \quad \alpha_{B^{2-}} = \frac{K_1K_2}{[H_3O^+]^2 + K_1[H_3O^+] + K_1K_2} . \tag{6.50c}$$

Formation functions will reappear in Chapter 7 and it will help you to recognize, write, and handle them easily if you note several things about eqs. (6.50):

1. All of the denominators are the same.
2. The individual terms in the denominators appear one after another in the successive numerators.
3. In general, for a j-functional acid H_jB, the denominator is equal to $[H_3O^+]^j + K_1[H_3O^+]^{j-1} + K_1K_2[H_3O^+]^{j-2} + \ldots + K_1K_2 \ldots K_{j-1}[H_3O^+] + K_1K_2 \ldots K_j$.

> SUMMARY: The formation function is the ratio of the equilibrium concentration of a particular species to the total concentration of the solute from which that species is formed. A graph showing how the concentration of each species depends on some independent variable, called the master variable, can greatly simplify the analysis of situations in which there are many competing equilibria.

6.10. Buffer solutions and buffer capacities

> PREVIEW: A buffer solution is a solution that has the ability to resist changes of pcH when an acid or a base is added to it. This section explains how buffer solutions are prepared and why they behave as they do, and defines the buffer capacity and describes the factors that affect it.

A *buffer solution* is conventionally defined as a solution that contains substantial concentrations of both a weak acid and its conjugate base. Such a solution has the useful property of being able to react with either an added base or an added acid, and thus to resist changes of pcH resulting from such additions. The use of buffer solutions greatly simplifies the study of chemical reactions that consume or liberate or are catalyzed by hydronium ion, because it nearly eliminates the effects that variations of pcH would exert on the equilibria and rates of the reactions.

The action of a buffer solution can be illustrated in the following way. Suppose that a very small amount—say, 1.00×10^{-6} mole—of a completely dissociated acid such as hydrochloric acid is added to 1 dm³ of pure water, and that the volume is unchanged by the addition. Before the acid is added the pcH is 6.998 ($= pK_w/2$); after it is added the pcH is 5.996 (from eq. (6.18a) with $c_a = 1.00 \times 10^{-6} M$). There is a decrease of 1.00 unit. Now suppose that the same amount of hydrochloric acid is added to the same volume of a solution in which the equilibrium concentrations of a weak acid HB and its conjugate base B^- are equal to 1.00×10^{-3} M, and that the overall dissociation constant of HB is equal to 1.005×10^{-7} mol dm^{-3}. Before the hydrochloric acid is added, the concentration of hydronium ion will be given by

$$[H_3O^+] = K_a \frac{[HB(aq)]}{[B^-]} = 1.005 \times 10^{-7} \times \frac{1.00 \times 10^{-3}}{1.00 \times 10^{-3}} = 1.005 \times 10^{-7}\ M.$$

The pcH is 6.998, exactly the same as in pure water. When the hydrochloric acid is added, the reaction

$$B^- + H_3O^+ = HB(aq) + H_2O(l); \quad K = 1/K_a = 9.95 \times 10^6\ \text{mol}^{-1}\ \text{dm}^3$$

will occur, and will consume almost all of the added acid because its equilibrium constant is large and because there is an excess of B^-. The concentration of B^- will decrease from 1.00×10^{-3} M to $(1.00 \times 10^{-3} - 1.00 \times 10^{-6}) = 0.999 \times 10^{-3}$ M, and the concentration of HB will increase from 1.00×10^{-3} M to 1.001×10^{-3} M. The concentration of hydronium ion at equilibrium will be given by

$$[H_3O^+] = K_a \frac{[HB(aq)]}{[B^-]} = 1.005 \times 10^{-7} \times \frac{1.001 \times 10^{-3}}{0.999 \times 10^{-3}} = 1.007 \times 10^{-7}\ M$$

and the pcH will be 6.997. Not all of the figures used in this calculation are significant, but three decimal places are needed to show that any change has occurred at all. The pcH decreases only 0.001 unit, instead of 1.00 unit as it did with pure water. A similar calculation would show that the pcH of 1 dm^3 of the same buffer solution would increase only from 6.998 to 6.999 if 1.00×10^{-6} mole of sodium hydroxide were added to it, whereas adding the same amount of base to 1 dm^3 of pure water would increase the pcH from 6.998 to 8.001.

There are two factors that govern the ability of a buffer solution to resist changes of pcH by reacting with a base or an acid that is added to it. One is the total concentration of the buffer solution—that is, the sum of the concentrations of the weak acid and its conjugate base. The other is the ratio of these concentrations.

A solution containing 1.00×10^{-4} M HB ($pK_a = 6.998$) and 1.00×10^{-4} M B^- would have a pcH of 6.998, just as did the solution in which the concentration of each was 1.00×10^{-3} M. The pcH of a buffer solution is most conveniently described by the Henderson–Hasselbalch equation, which is merely eq. (6.46b) in its most general form:

$$pcH = pK_a - \log \frac{[HB]}{[B^-]}. \tag{6.51}$$

It is the value of the ratio [HB]/[B$^-$] that is important, not the individual values of [HB] and [B$^-$]. If 1.00×10^{-6} mole of hydrochloric acid is added to 1 dm^3 of this more dilute solution, the concentration of B$^-$ will decrease from 1.00×10^{-4} M to $(1.00 \times 10^{-4} - 1.00 \times 10^{-6}) = 0.99 \times 10^{-4}$ M, while the concentration of HB will increase from 1.00×10^{-4} M to 1.01×10^{-4} M. The change of pcH will be given by

$$\Delta\text{pcH} = \left(-\log\frac{[\text{HB}]}{[\text{B}^-]}\right)_{\text{final}} - \left(-\log\frac{[\text{HB}]}{[\text{B}^-]}\right)_{\text{initial}}$$

$$= \left(\log\frac{[\text{HB}]}{[\text{B}^-]}\right)_{\text{initial}} - \left(\log\frac{[\text{HB}]}{[\text{B}^-]}\right)_{\text{final}}. \tag{6.52}$$

Since the initial concentrations of HB and B$^-$ are the same, $(\log [\text{HB}]/[\text{B}^-])_{\text{initial}} = \log 1 = 0$, but $(\log [\text{HB}]/[\text{B}^-]_{\text{final}}) = \log (1.01 \times 10^{-4}/0.99 \times 10^{-4}) = \log 1.02 = 0.009$. The change of pcH is -0.009 unit. The negative sign merely means that the pcH decreases when acid is added, which is not surprising; what is important is that the change is about ten times as large as for the solution in which each of the initial concentrations was 1.00×10^{-3} M instead of 1.00×10^{-4} M. If each concentration were 1.00×10^{-5} M, the addition of the same amount of hydrochloric acid would produce a change of 0.087 unit in the pcH, which is ten times as large again. Concentrated buffer solutions are more effective than weak ones.

To investigate the effect of the ratio of concentrations in a buffer solution, let us imagine that still another buffer with pcH = 6.998 is made from another weak acid that has a different dissociation constant. Acetic acid, for which $pK_a = 4.759$, might be used. Consider a buffer solution in which the sum of the concentrations of acetic acid and acetate ion is equal to 2.00×10^{-3} M, as was the sum of the concentrations of HB and B$^-$ in the most concentrated of the three buffers above. To obtain a pcH of 6.998 with acetic acid, we need

$$\log\frac{[\text{HOAc}]}{[\text{OAc}^-]} = pK_a - \text{pcH} = 4.759 - 6.998 = -2.239$$

so that [HOAc]/[OAc$^-$] = 5.77×10^{-3}. Combining this with the requirement that [HOAc]+[OAc$^-$] = 2.00×10^{-3} M, we obtain [HOAc] = 1.15×10^{-5} M and [OAc$^-$] = 1.989×10^{-3} M. Let us prepare 1 dm^3 of such a solution and add 1.00×10^{-6} mole of hydrochloric acid to it. The concentration of acetic acid will increase to $(1.15 \times 10^{-5} + 1.00 \times 10^{-6}) = 1.25 \times 10^{-5}$ M, while that of acetate ion will decrease to $(1.989 \times 10^{-3} - 1.00 \times 10^{-6}) = 1.988 \times 10^{-3}$ M. Equation (6.52) gives

$$\Delta\text{pcH} = \log 5.77 \times 10^{-3} - \log\frac{1.25 \times 10^{-5}}{1.988 \times 10^{-3}} = \log\frac{5.77 \times 10^{-3} \times 1.988 \times 10^{-3}}{1.25 \times 10^{-5}}$$

$$= \log 0.918 = -0.037.$$

For the buffer that contained equal concentrations of HB and B$^-$ the change was only -0.001 unit. It is much larger here because the concentrations of acetic acid and acetate ion are widely different: one of them is very small and is therefore considerably affected by the addition of acid or base. Solutions in which the ratio of concentrations is close to 1 are better buffers than those in which it is different from 1.

There is an arbitrary rule of thumb that says that this ratio should be between 0.1 and 10—or, in other words, that the pcH should be between $pK_a - 1$ and $pK_a + 1$. Solutions for which this is not true are not very effective buffers. The basis of the rule will be examined more closely in a later paragraph.

These considerations govern what you should do if you want to prepare a buffer solution. Suppose that you want to obtain a pcH of 3.0. You begin with a table listing the values of pK_a for different acids and look for those having values of pK_a between 2 and 4. A few that might catch your eye are *ortho*-bromobenzoic acid ($pK_a = 2.85$), chloroacetic acid ($pK_a = 2.87$), citric acid ($pK_1 = 3.13$), hydrofluoric acid ($pK_a = 3.14$), and trifluoropropionic acid ($CF_3CH_2COOH, pK_a = 3.02$). The bromobenzoic acid will not be very soluble in water and a reasonably effective buffer will be impossible to prepare from it. Hydrofluoric acid attacks glass, trifluoropropionic acid is expensive, and you would probably choose citric acid because it is more easily available and cheaper than chloroacetic acid. To obtain pcH = 3.0 you would want

$$\log\frac{[HB]}{[B^-]} = \log\frac{[H_3\,citrate]}{[H_2\,citrate^-]} = pK_1 - pcH = 3.13 - 3.0 = 0.13$$

or $[H_3\,citrate]/[H_2\,citrate^-] = 1.35$. Buffers having total concentrations less than 0.01 M or more than 1 M are rarely used: the former are too ineffective and the latter too expensive. You might select a total concentration of 0.10 M as a compromise between these extremes. If $[H_3\,citrate]/[H_2\,citrate^-] = 1.35$ and $[H_3\,citrate] + [H_2\,citrate^-] = 0.10$, then the concentration of citric acid must be 0.057_5 M and that of dihydrogen citrate ion must be 0.042_5 M.

These are equilibrium concentrations. They are not the same as the concentrations that you must use in preparing the solution, and the difference may be appreciable. Here you want the solution to contain 1.0×10^{-3} M hydronium ion at equilibrium, and it must come from the dissociation of the citric acid. To provide this concentration of hydronium ion and leave 0.057_5 M citric acid, you must begin by putting 0.058_5 mole of citric acid into each cubic decimeter of the solution. Since the dissociation will produce 1.0×10^{-3} M hydrogen citrate ion along with the hydronium ion, you must start with only 0.041_5 M sodium dihydrogen citrate. The correction is not large here; it is never large unless the concentration of either hydronium ion or hydroxide ion is appreciable in comparison with the concentration of the acid or its conjugate base. Usually you can ignore it, but Examples 6.6 and 6.7 (which are taken from real life) show that ignoring it may not always be safe.

Example 6.6. The value of pK_a for trichloroacetic acid is 0.70. A chemist, who owned 0.100 mole of trichloroacetic acid and wished to make a buffer solution having a pcH of 0.70, dissolved the acid in water, added 0.050 mole of sodium hydroxide to it (with the idea of neutralizing half of the acid and giving $[Cl_3CCOOH(aq)] = [Cl_3CCOO^-]$), and diluted the mixture to 1 dm^3. What was its pcH?

Answer. Neglecting the concentration of hydroxide ion, which cannot be significant, the electroneutrality equation is

$$[Na^+] + [H_3O^+] = [Cl_3CCOO^-].$$

The concentration of sodium ion was 0.050 M and the concentration of trichloroacetate ion was (eq. (6.48))

$$[Cl_3CCOO^-] = \alpha_{Cl_3CCOO^-} \times 0.100 = \frac{K_a}{K_a + [H_3O^+]} \times 0.100 = \frac{0.20}{0.20 + [H_3O^+]} \times 0.100$$

so that

$$0.050 + [H_3O^+] = \frac{0.020}{0.20 + [H_3O^+]}$$

or

$$[H_3O^+]^2 + 0.25[H_3O^+] - 0.010 = 0$$

whose solution is $[H_3O^+] = 0.035$ M. The pcH was 1.45 instead of 0.70.

Example 7.7. In the preceding example, what should have been done to obtain the desired pcH?

Answer. That pcH corresponds to $[H_3O^+] = 0.20$ M. Since even an 0.100 M solution of trichloroacetic acid would not contain this much hydronium ion, an acid must be added to it. If hydrochloric acid is used, the electroneutrality equation becomes

$$[H_3O^+] = 0.20 = [Cl_3CCOO^-] + [Cl^-] = \alpha_{Cl_3CCOO^-} \times 0.100 + [Cl^-]$$

$$= \frac{0.20}{0.20 + 0.20} \times 0.100 + [Cl^-] = 0.050 + [Cl^-]$$

so that the required concentration of hydrochloric acid is 0.15 M. Hence 0.15 mole of hydrochloric acid should have been added instead of the sodium hydroxide. The desired result could be achieved more economically by making an 0.20 M solution of hydrochloric acid, saving the trichloroacetic acid for some worthwhile purpose.

Earlier in this section the abilities of different solutions to resist changes of pcH were compared by calculating their pcH-values before and after some acid or base had been added to them. This is clumsy and time-consuming, and it often fails to reveal important and interesting aspects of the behavior being investigated. A more elegant and rewarding approach is to obtain a general description of the phenomenon and then to inspect the information that it provides.

The ability of a solution to resist changes of pcH is described by its *buffer capacity* or *buffer index* β, which is defined by the equation

$$\beta = \left| \frac{dx}{d(pcH)} \right| \tag{6.53}$$

where $d(pcH)$ is the change of pcH that results when an infinitesimal number of moles, dx, of a strong acid or base is added to 1 dm^3 of the solution without changing its volume. The units of β are mol dm^{-3}. As the buffer capacity increases, the change of pcH decreases and the buffering action becomes more effective. Good buffers have high buffer capacities; poor buffers have low ones.

If a solution does not contain a weak acid and its conjugate base, its pcH can be described by an equation that can be obtained by rearranging eq. (6.18a):

$$x = [H_3O^+] - K_w/[H_3O^+]. \tag{6.54}$$

In an acidic solution x is positive and represents the concentration of strong acid that has been added to the solution; in an alkaline solution x is negative and represents the concentration of strong base that has been added; in pure water x is zero. To obtain an expression for β we shall write

$$\beta = \left| \frac{dx}{d(pcH)} \right| = \left| \frac{dx}{d[H_3O^+]} \times \frac{d[H_3O^+]}{d(pcH)} \right| = \left| \frac{dx}{d[H_3O^+]} \middle/ \frac{d(pcH)}{d[H_3O^+]} \right|.$$

Since pcH $= -\log_{10}[H_3O^+] = -0.434 \ln[H_3O^+]$,

$$\frac{d(pcH)}{d[H_3O^+]} = -\frac{0.434}{[H_3O^+]}$$

and, by differentiating eq. (6.54),

$$\frac{dx}{d[H_3O^+]} = 1 + \frac{K_w}{[H_3O^+]^2} = \frac{[H_3O^+]^2 + K_w}{[H_3O^+]^2}.$$

The result is

$$\beta = \left| \left(\frac{[H_3O^+]^2 + K_w}{[H_3O^+]^2} \right) \middle/ \left(-\frac{0.434}{[H_3O^+]} \right) \right| = \frac{2.303\,([H_3O^+]^2 + K_w)}{[H_3O^+]}. \tag{6.55}$$

In pure water this gives $\beta = 4.6 \times 10^{-7}$ mol dm^{-3}, which is extremely small. Very large changes of pcH result from the addition of small amounts of acid or base to pure water. If a significant concentration of either a strong acid or a strong base is present, one of the two terms in the numerator will be very much smaller than the other and

$$\beta \approx 2.303\,[H_3O^+] \quad \text{(in acidic solutions)}$$

or

$$\beta \approx 2.303\,[OH^-] \quad \text{(in alkaline solutions)}. \tag{6.56}$$

The buffer capacity increases dramatically as the pcH moves away from 7 in either direction.

A solution that does contain a weak acid and its conjugate base has a pcH that is described by eq. (6.51), which can be rewritten as

$$pcH = pK_a - \log_{10}\frac{c_a}{c_b} = pK_a - 0.434 \ln\frac{c_a}{c_b}$$

so that

$$\frac{d(\text{pcH})}{dx} = -\frac{0.434}{(c_a/c_b)} \times \frac{c_b(dc_a/dx) - c_a(dc_b/dx)}{c_b^2}.$$

If the base is sufficiently strong to react completely with the infinitesimal amount of strong acid that is added, its concentration will decrease from c_b before the addition to $c_b - dx$ after it, and at the same time the concentration of the weak acid will increase from c_a to $c_a + dx$. Hence $dc_b/dx = -1$ and $dc_a/dx = +1$, and therefore

$$\beta = \left| \frac{dx}{d(\text{pcH})} \right| = \left| \frac{1}{d(\text{pcH})/dx} \right|$$

$$= \left| \frac{1}{-\dfrac{0.434}{c_a c_b}(c_b + c_a)} \right| = \frac{2.303 c_a c_b}{c_a + c_b}. \tag{6.57a}$$

This can be further simplified by letting the total concentration of the buffer, $c_a + c_b$, equal c_T, and by defining a quantity f as the fraction of this total concentration that is in the acidic form:

$$c_a = f c_T \quad \text{and} \quad c_b = (1 - f)c_T$$

Combining these with eq. (6.57a) yields

$$\beta = 2.303 f(1 - f)c_T. \tag{6.57b}$$

Equation (6.57b) shows that the buffer capacity is proportional to c_T. Because it can be differentiated to give

$$d\beta/df = 2.303\, c_T(1 - 2f) \quad [= 0 \text{ when } f = 1/2]$$

it also shows that the buffer capacity is greatest when $f = 1/2$, so that $c_a = c_b$. Finally, it shows that the value of β will be very small if either f or $(1 - f)$ is close to zero, as will be true if the pcH is smaller than $(pK_a - 1)$ or larger than $(pK_a + 1)$.

This section began with the words "A buffer solution is conventionally defined as a solution that contains substantial concentrations of both a weak acid and its conjugate base". Let us imagine two solutions, each having a pcH of 2.00. One contains 1.00×10^{-2} M hydrochloric acid. According to eq. (6.56) its buffer capacity is equal to 2.3×10^{-2} mol dm^{-3}. The other solution contains some weak acid HB, for which $pK_a = 2.00$, at an equilibrium concentration of 2.00×10^{-2} M, and it also contains 2.00×10^{-2} M B$^-$. For this solution $c_T = 4.00 \times 10^{-2}$ and $f = 1/2$, and according to eq. (6.57b) its buffer capacity is equal to 2.3×10^{-2} mol dm^{-3}. According to the conventional definition the second solution is a buffer solution and the first one is not, and yet their buffer capacities are exactly the same: they have identical abilities to resist changes of pcH. Draw your own conclusions.

SUMMARY: The buffer capacity or buffer index β reflects the ability of a solution to resist changes of pcH when an acid or base is added to it. In a mixture of a weak acid and its conjugate base, the buffer capacity increases as the sum of their concentrations increases. It also increases as the ratio of their concentrations approaches 1, and as the pcH approaches the value of pK_a for the acid. An important rule of thumb asserts that a buffer system is useful only within the range of pcH-values from $(pK_a - 1)$ to $(pK_a + 1)$.

6.11. Effects of pcH on the solubilities of weak bases

PREVIEW: Many sparingly soluble compounds contain ions that are acidic or basic, or both. This section combines ideas presented in this chapter and the preceding one [to describe the solubilities of such compounds.

The solubility product of cadmium sulfide, CdS, is equal to 1.6×10^{-28} mol^2 dm^{-6} in water at 25°; what is its solubility? Questions like this were avoided in Chapter 5. The basic properties of sulfide ion, phosphate ion, and many other common ions have profound effects on the solubilities of their compounds. Now that we have considered acid-base reactions at some length we are able to take these effects into account.

The chemist who ignored them would write $[Cd^{2+}] = [S^{2-}] = S_{CdS} = K_{CdS}^{1/2} = 1.3 \times 10^{-14}$ M. However, it is not at all difficult to see that this result cannot be even approximately correct. Supposing that it were, the pcH of the saturated solution would be 7.00, for although sulfide ion is as strong a base as hydroxide ion, no acid or base that is present at such a low concentration could possibly have any effect on the pcH. Since the overall acidic dissociation constants of hydrogen sulfide are $K_1 = 6.0 \times 10^{-8}$ mol dm^{-3} and $K_2 = 1 \times 10^{-14}$ mol dm^{-3}, the formation function for sulfide ion at pcH $= 7.00$ is

$$\alpha_{S^{2-}} = \frac{K_1 K_2}{[H_3O^+]^2 + K_1[H_3O^+] + K_1 K_2}$$

$$= \frac{6 \times 10^{-22}}{(1.00 \times 10^{-7})^2 + (6.0 \times 10^{-8})(1.0 \times 10^{-7}) + (6.0 \times 10^{-8})(1 \times 10^{-14})}$$

$$= \frac{6 \times 10^{-22}}{1.6 \times 10^{-14}} = 4 \times 10^{-8}. \tag{6.58}$$

If a total of 1.3×10^{-14} mole of sulfide dissolves in 1 dm^3 of the solution, the concentration of sulfide ion is $[S^{2-}] = \alpha_{S^{2-}}(1.3 \times 10^{-14}) = 5 \times 10^{-22}$ M. The product of the concentrations of cadmium and sulfide ions is $[Cd^{2+}][S^{2-}] = (1.3 \times 10^{-14})(5 \times 10^{-22})$ $= 6 \times 10^{-36}$ mol^2 dm^{-6}. This is many orders of magnitude smaller than the solubility product; a 1.3×10^{-14} M solution of cadmium sulfide is unsaturated, and the solubility must be very much larger than 1.3×10^{-14} M.

Again we are confronted with competing equilibria: the solubility product must be satisfied, and so must the equilibrium constants for at least *two* acid-base reactions.

One of these is the reaction in which the anion behaves as a base and accepts a proton from the solvent; the other is the autoprotolysis of water. If the anion B^- is a monofunctional base we can describe these three equilibrium constants by the equations

1. $MB(s) = M^+ + B^-$; $K_{MB} = [M^+][B^-]$

2. $B^- + H_2O(l) = HB(aq) + OH^-$; $K = [HB(aq)][OH^-]/[B^-] = K_w/K_{a,HB}$

3. $2 H_2O(l) = H_3O^+ + OH^-$; $K_w = [H_3O^+][OH^-]$

With sulfide ion a small additional complication arises from the fact that HB ($=$ hydrogen sulfide ion) is also basic. The further reaction $HS^- + H_2O(l) = H_2S(aq) + OH^-$, for which $K = K_w/K_{1,H_2S}$, can and does occur as well. Indeed, if you inspect the three terms in the denominator in eq. (6.58) you will see that there is more hydrogen sulfide than hydrogen sulfide ion in a solution having pcH $= 7.00$.

There are three possible extremes in such a situation:

1. B^- is so weakly basic, and the solubility of MB is so small, that reaction 2 does not consume an appreciable fraction of the dissolved B^- and does not produce enough hydroxide ion to affect the pcH. Here reaction 2 can be neglected and the situation is very simple because reactions 1 and 3 are independent: their equilibrium constants have no concentration in common. There were many examples in Chapter 5.
2. B^- is so strongly basic, and the solubility of MB is so high, that reaction 2 produces enough hydroxide ion to suppress the autoprotolysis of water almost completely. Reaction 3 can be neglected, but the situation is less simple than the preceding one because the concentration of B^- appears in the expressions for both K_{MB} and the equilibrium constant of the second reaction.
3. Regardless of the basic strength of B^- and the solubility of MB, the solution is so well buffered that its pcH does not change when it is saturated with MB. This is really the same as the first of our three extremes: in either one the pcH at equilibrium is known in advance.

Let us assume that a saturated solution of cadmium sulfide in pure water represents the first of these possibilities. If the solubility of cadmium sulfide is $S\ M$, then $[Cd^{2+}] = S$ and $[H_2S(aq)] + [HS^-] + [S^{2-}] = S$. The second of these is the conservation equation for sulfide. It may be rewritten as $[S^{2-}] = \alpha_{S^{2-}}S$, so that

$$K_{CdS} = [Cd^{2+}][S^{2-}] = S(\alpha_{S^{2-}}S) = \alpha_{S^{2-}}S^2 = 1.6 \times 10^{-28}\ mol^2\ dm^{-6}. \quad (6.59)$$

At pcH $= 7.00$ the value of $\alpha_{S^{2-}}$ is equal to 4×10^{-8} according to eq. (6.58), and

$$S = \sqrt{K_{CdS}/\alpha_{S^{2-}}} = \sqrt{1.6 \times 10^{-28}/4 \times 10^{-8}} = 6 \times 10^{-11}\ M$$

which is approximately 5000 times as large as the incorrect value obtained in the first paragraph of this section. The assumption that the pcH is 7.00 at equilibrium can be checked by noting that 6×10^{-11} M sulfide ion could not alter the pcH of pure water even if both of the reactions $S^{2-} + H_2O(l) = HS^- + OH^-$ and $HS^- + H_2O(l) = H_2S(aq) + OH^-$ went to completion.

The third possibility may be represented by the solubility of cadmium sulfide in a buffer solution having pcH = 5.00. We begin by assuming that the dissolution of cadmium sulfide does not affect the pcH. The formation function for sulfide ion at pcH = 5.00 is

$$\alpha_{S^{2-}} = \frac{6 \times 10^{-22}}{(1.00 \times 10^{-5})^2 + (6.0 \times 10^{-8})(1.00 \times 10^{-5}) + 6 \times 10^{-22}}$$

$$= \frac{6 \times 10^{-22}}{1.00 \times 10^{-10}} = 6 \times 10^{-12}.$$

Again $K_{CdS} = \alpha_{S^{2-}} S^2$, as in eq. (6.59), so that $S = 5 \times 10^{-9}$ M, and again the assumption can be checked by noting that this solubility is too low to affect the pcH.

The second of the three possibilities is the most complicated and may be illustrated by considering the solubility of calcium carbonate in pure water. The solubility product of calcium carbonate is equal to 4.8×10^{-9} mol^2 dm^{-6}, and the chemist who ignored acid-base reactions would write $[Ca^{2+}] = [CO_3^{2-}] = S = K_{CaCO_3}^{1/2} = 6.9 \times 10^{-5}$ M. The behavior of carbonate ion is described by the equations

$$CO_3^{2-} + H_2O(l) = HCO_3^- + OH^- ; \quad K_1 = [HCO_3^-][OH^-]/[CO_3^{2-}]$$
$$= K_w/K_{a,2} = 2.14 \times 10^{-4} \text{ mol dm}^{-3}$$

$$HCO_3^- = CO_2(aq) + OH^- ; \quad K_2 = [CO_2(aq)][OH^-]/[HCO_3^-]$$
$$= K_w/K_{a,1} = 2.25 \times 10^{-8} \text{ mol dm}^{-3}$$

where $K_{a,1}$ and $K_{a,2}$ are the first and second overall acidic dissociation constants of carbon dioxide. The first of these reactions will consume a substantial fraction of the carbonate ion present in such a dilute solution, and we assume that it will produce enough hydroxide ion to suppress both the second reaction and the autoprotolysis of water. Since the first reaction produces equal numbers of hydroxide and hydrogen carbonate ions, we can write

$$[HCO_3^-] = [OH^-] = (K_1[CO_3^{2-}])^{1/2}.$$

The dissolution of calcium carbonate yields equal numbers of calcium and carbonate ions, but some of the carbonate ions are transformed into hydrogen carbonate ions by the basic dissociation, and therefore

$$[Ca^{2+}] = S = [CO_3^{2-}][HCO_3^-]$$
$$= [CO_3^{2-}] + (K_1[CO_3^{2-}])^{1/2} = K_{CaCO_3}[CO_3^{2-}].$$

Introducing the numerical values of K_1 and K_{CaCO_3} into the last equality and rearranging gives

$$[CO_3^{2-}]^2 + 1.46 \times 10^{-2}[CO_3^{2-}]^{3/2} - 4.8 \times 10^{-9} = 0$$

which can be solved by successive approximation to obtain $[CO_3^{2-}] = 3.7_7 \times 10^{-5}\ M$. Hence

$$[Ca^{2+}] = S = 4.8 \times 10^{-9}/3.7_7 \times 10^{-5} = 1.27 \times 10^{-4}\ M,$$

and the solubility is accordingly $1.27 \times 10^{-4}\ M$. The assumptions may be checked by noting that they give

$$[OH^-] = [HCO_3^-] = \sqrt{2.14 \times 10^{-4} \times 3.7_7 \times 10^{-5}}$$
$$= 9.0 \times 10^{-5}\ M\ (\ = [Ca^{2+}] - [CO_3^{2-}]).$$

This is certainly high enough to suppress the autoprotolysis of water. Moreover, if $[OH^-] = [HCO_3^-]$ then $[CO_2(aq)] = K_2 = 2.25 \times 10^{-8}\ M$; this is much smaller than $9.0 \times 10^{-5}\ M$ and signifies that the concentrations of carbon dioxide and hydroxide ion formed in the second dissociation step are negligible, as they were assumed to be. The calculated solubility is nearly twice as large as the value obtained by neglecting the fact that carbonate ion is basic. The difference would be even larger for an equally soluble phosphate because phosphate ion is a stronger base than carbonate ion, and it would also be larger for a less soluble carbonate because a larger fraction of the carbonate ion is consumed if its total concentration decreases.

Danger lurks amidst these relatively simple extremes, and may not be conspicuous at first glance. You would not at this point say that the solubility of tin(II) sulfide, SnS, in pure water is equal to the square root of its solubility product ($K_{SnS} = 1 \times 10^{-25}\ mol^2\ dm^{-6}$), but neither could you calculate it in the same way that we calculated the solubility of cadmium sulfide. Tin(II) ion, unlike cadmium ion, is strongly acidic. Two different chemists might describe its acidity by two different but equivalent equations:

$$Sn(OH_2)_6^{2+} + H_2O(l) = Sn(OH)(OH_2)_5^+ + H_3O^+;$$
$$K_{a,1} = [SnOH^+][H_3O^+]/[Sn^{2+}] = 8 \times 10^{-3}\ mol\ dm^{-3}$$

or

$$Sn(OH_2)_6^{2+} + OH^- = Sn(OH)(OH_2)_5^+ + H_2O(l);$$
$$K_1 = [SnOH^+]/[Sn^{2+}][OH^-] = 8 \times 10^{11}\ mol^{-1}\ dm^3.$$

Molecules of coordinated water are omitted from the expressions for the equilibrium constants to simplify them. The product of this reaction is also acidic (which is to say that tin(II) ion is a polyfunctional acid):

$$Sn(OH)(OH_2)_5^+ + H_2O(l) = Sn(OH)_2(OH_2)_4(aq) + H_3O^+;$$
$$K_{a,2} = [Sn(OH)_2(aq)][H_3O^+]/[SnOH^+] = 1 \times 10^{-5}\ mol\ dm^{-3}$$

or

$$Sn(OH)(OH_2)_5^+ + OH^- = Sn(OH)_2(OH_2)_4(aq) + H_2O(l);$$
$$K_2 = [Sn(OH)_2(aq)]/[SnOH^+][OH^-] = 1 \times 10^9\ mol^{-1}\ dm^3.$$

Assuming that the saturated solution has a pcH of 7.00, as we did for cadmium sulfide above, we can again write $[S^{2-}] = \alpha_{S^{2-}}S = 4 \times 10^{-8} S$, but now the acidity of the tin(II) ion must be taken into account as well. Guided by eqs. (6.48) and continuing to assume that the pcH will be 7.00, we write

$$[Sn^{2+}] = \alpha_{Sn^{2+}}S = \frac{[H_3O^+]^2}{[H_3O^+]^2 + K_{a,1}[H_3O^+] + K_{a,1}K_{a,2}} S = \frac{1.00 \times 10^{-14}}{8.08 \times 10^{-8}} S$$

$$= 1.2_4 \times 10^{-7} S \tag{6.60}$$

so that

$$[Sn^{2+}][S^{2-}] = (\alpha_{Sn^{2+}}S)(\alpha_{S^{2-}}S) = 1.2_4 \times 10^{-7} \times 4 \times 10^{-8} S^2$$
$$= K_{SnS} = 1 \times 10^{-25} \, mol^2 \, dm^{-6}$$

whence $S = 4._5 \times 10^{-6} \, M$. Neglecting the acidity of tin(II) ion would give a value about 1/3500 as large.

You should not be content with this result until you have proved to your own satisfaction that it justifies the assumption (pcH = 7.00) you made to obtain it. Unfortunately, it may be very difficult to decide whether it does or not. Different chemists adopt different approaches. One is to say that the solution cannot contain an appreciable concentration of either tin(II) ion or sulfide ion at equilibrium: the former is so strong an acid, and the latter is so strong a base, that the reaction $Sn^{2+} + S^{2-} + H_2O = SnOH^+ + HS^-$ goes practically to completion. The dissolution of SnS yields tin(II) ions and sulfide ions in equal numbers, and this reaction converts them into equal numbers of $SnOH^+$ and HS^- ions. Hence the solution may be regarded as having contained equal initial concentrations of $SnOH^+$ and HS^- ions. According to the value of $K_{a,2}$ for tin(II) ion, the first of these is a weak acid, but not a very weak one; according to the value of $K_{a,1}$ for hydrogen sulfide, the hydrogen sulfide ion is quite a weak base. Comparing the strength of an acid with the strength of a base is rather like comparing apples with oranges, but it is not too difficult to conclude that the reaction $Sn(OH)(OH_2)_5^+ + H_2O(l) = Sn(OH)_2(OH_2)_4(aq) + H_3O^+$ will furnish more hydronium ions than the reaction $HS^- + H_3O^+ = H_2S(aq) + H_2O(l)$ consumes. At the worst the first one might produce $4._5 \times 10^{-6} \, M$ hydronium ion while the second consumed only a tiny fraction of that. Probably things are not really quite that bad, but if you know that the pcH is below 7.00 and can imagine a way in which it might be as small as 5.35 $(= -\log 4._5 \times 10^{-6})$ you can be fairly confident that the assumption pcH = 7.00 is a bad one.

Other chemists, more numerically minded, might test the assumption by seeing whether it agrees with other things that are known about the solution. One important and useful fact is that the solution must be electrically neutral; it is useful because it was not employed in arriving at the above answer and can therefore be used to test it without danger of going in circles. If we write

$$[H_3O^+] + 2[Sn^{2+}] + [SnOH^+] = [OH^-] + [HS^-] + 2[S^{2-}] \tag{6.61}$$

and the formation functions, at the assumed pcH of 7.00, for the tin(II) and sulfide species, we obtain

$$[H_3O^+]-[OH^-] = [HS^-]+2[S^{2-}]-2[Sn^{2+}]-2[SnOH^+]$$
$$= \frac{6.0\times 10^{-15}}{1.6\times 10^{-4}} S + \frac{1.2\times 10^{-21}}{1.6\times 10^{-14}} S - \frac{2\times 10^{-14}}{8.1\times 10^{-8}} S - \frac{8\times 10^{-10}}{8.1\times 10^{-8}} S. \quad (6.62)$$

Some of the numerical values come from eqs. (6.58) and (6.60), and the others may be obtained similarly. Combining the terms on the right-hand side, and taking $S = 4._5\times 10^{-6}$ mol dm^{-3} to go along with the assumption that the pcH is 7.00, gives $[H_3O^+]-[OH^-]= 1.7\times 10^{-6} M$, or $[H_3O^+]= 1.7\times 10^{-6} M$ and $[OH^-]= 6\times 10^{-9} M$. Chemically, this means that the solution is acidic, which is the same thing that the qualitatively minded chemist would conclude. Logically, it means that if the solution is neutral (on the right-hand side of eq. (6.62)), then it is acidic (on the left-hand side): if the pcH is 7.00 then the pcH is 5.77 $(=-\log 1.7\times 10^{-6})$. This is deplorable logic and the discrepancy is too large to shrug off. The assumption was wrong and so is the answer obtained by making it.

You should realize that the difficulty arises because the buffer capacity of pure water is so extremely low. Dissolving the same amount of tin(II) sulfide in 1 dm^3 of any reasonably well-buffered solution at pcH $= 7.00$ would have no significant effect on its pcH, and the value of S that has been calculated would be correct.

There is a brute-force approach that will produce the correct solution to any problem of this sort, no matter how complicated it may be and no matter how little insight into it you may be able to obtain. Only its general outlines will be sketched here. By combining the conservation equation

$$[Sn^{2+}]+[SnOH^+]+[Sn(OH)_2(aq)] = S$$

with the expressions for $K_{a,1}$ and $K_{a,2}$—or by writing the expression for $\alpha_{Sn^{2+}}$, which comes to the same thing—you can obtain

$$[Sn^{2+}] = \frac{[H_3O^+]^2}{[H_3O^+]^2+K_{a,1,Sn^{2+}}[H_3O^+]+K_{a,1,Sn^{2+}}K_{a,2,Sn^{2+}}} S. \quad (6.63a)$$

From either the conservation equation for sulfide ion

$$[H_2S]+[HS^-]+[S^{2-}] = S$$

or the expression for $\alpha_{S^{2-}}$ you can similarly obtain

$$[S^{2-}] = \frac{K_{a,1,H_2S}K_{a,2,H_2S}}{[H_3O^+]^2+K_{a,1,H_2S}[H_3O^+]+K_{a,1,H_2S}K_{a,2,H_2S}} S. \quad (6.63b)$$

In eq. (6.63a) $K_{a, 1, Sn^{2+}}$ and $K_{a, 2, Sn^{2+}}$ are the acidic dissociation constants of tin(II) ion; in eq. (6.63b) $K_{a, 1, H_2S}$ and $K_{a, 2, H_2S}$ are the acidic dissociation constants of hydrogen sulfide. Multiplying these two equations together and replacing the product $[Sn^{2+}]$ $[S^{2-}]$ by K_{SnS} yields an equation containing only two unknowns: S and the concentration of hydronium ion. Another such equation can be obtained by starting with the electroneutrality equation, eq. (6.61), and combining it with eqs. (6.63) and the expressions for α_{SnOH^+} and α_{HS^-}, again eliminating everything except S and $[H_3O^+]$. Having thus obtained two equations containing these as the only unknowns, you need only solve them for S—which is easier said than done. You can simplify the task by noting that $[S^{2-}]$ is negligible, according to eq. (6.62), even at pcH $= 7.00$ and that it will be still smaller at a lower pcH: this is equivalent to dropping the product $K_{a, 1, H_2S} K_{a, 2, H_2S}$ in the denominator of eq. (6.63b). Usually there are several such ways of simplifying the algebra: there are others in this problem. You can sometimes discern enough simplifications or assumptions to make the final expression quite tractable; otherwise there is nothing to do except set a computer to the task of finding a number of values that satisfy an equal number of very complicated equations.

The details of such calculations will not be discussed here because they are of very little practical value. Most solutions of this kind are extremely dilute, their buffer capacities are extremely low, and even a trace of an acidic or basic impurity (such as carbon dioxide or ammonia) would affect their pcH-values appreciably and alter the solubilities accordingly. Even if the solubility products were highly reliable, which they are not because they are calculated from the results of other experiments that were similarly affected by traces of impurities, it would be very difficult to obtain an experimental value that agreed with a calculated one. The chief value of being able to solve problems like those considered in this section is that it guarantees you against incurring the very large errors that may result from incomplete understanding of the complications involved.

SUMMARY: This section has considered the effects of acid-base reactions involving one or both of the ions of a sparingly soluble compound. Such reactions tend to increase the solubility by consuming one or both of the ions, and the increase may be very large. Exact descriptions are likely to be very complicated, but they can often be simplified by ignoring processes that occur to only very small extents.

6.12. Non-aqueous solvents

PREVIEW: The behaviors of acids and bases in a non-aqueous solvent depend on the strengths of the solvent as an acid and a base and on its dielectric constant. This section defines amphoteric, basic, and inert solvents and describes their behaviors, and it also defines and discusses the levelling effect.

In other solvents that are more strongly acidic or basic than water, the strengths of acids and bases are very different from what they are in water. Acetic acid is a weak acid when it is dissolved in water, but it is a strong acid when it is dissolved in liquid ammonia. Ammonia is a much stronger base than water, and its ability to accept protons is so high that it removes them easily from dissolved molecules of acetic acid. Strongly basic solvents, such as ammonia, increase the strengths of weak acids dissolved in them. On the other hand, acetate ion is an extremely weak base in liquid ammonia, for ammonia is so weak an acid that acetate ions cannot remove protons from it. Conversely, acetate ion is a much stronger base in a solvent such as anhydrous formic acid than it is in water, for formic acid is much more strongly acidic than water and its readiness to donate protons causes acetate ion to react with it almost completely.

These ideas are obscured by using the overall dissociation constant to express the strength of an acid or a base. Although acetic acid is a strong acid in liquid ammonia —although the transfer of protons from molecules of acetic acid to molecules of ammonia is virtually complete—the overall dissociation constant of acetic acid

$$CH_3COOH + NH_3(l) = CH_3COO^- + NH_4^+ ;$$
$$K_a = [CH_3COO^-][NH_4^+]/[CH_3COOH] \qquad (6.64)$$

is fairly small, largely because the dielectric constant of anhydrous ammonia, which is only 22 at 223 K as compared with 80 for water at 298 K, is too small to permit the existence of large concentrations of free ions. The distinction between ionization and dissociation, which was explained in Sections 5.6 and 5.7, is essential. Acetic acid is a strong acid when dissolved in anhydrous ammonia because its ionization

$$CH_3COOH + NH_3(l) = CH_3COO^- NH_4^+ ;$$
$$K_i = [CH_3COO^- NH_4^+]/[CH_3COOH] \qquad (6.65a)$$

is almost complete and the value of its ionization constant K_i is very high. However, the ion pairs are only slightly dissociated

$$CH_3COO^- NH_4^+ = CH_3COO^- + NH_4^+ ;$$
$$K_d = [CH_3COO^-][NH_4^+]/[CH_3COO^- NH_4^+] \qquad (6.65b)$$

and the value of the dissociation constant K_d is fairly small. In these circumstances the overall dissociation constant is virtually equal to K_d (eq. (5.23a)). Clearly it is the value of the ionization constant, rather than that of the overall dissociation constant, that best depicts the strength of a base or an acid.

Many familiar features of aqueous acid-base chemistry appear almost unchanged in anhydrous acetic acid and other amphoteric solvents. An autoprotolytic reaction such as

$$2\,CH_3COOH(l) = CH_3COOH_2^+ + CH_3COO^- ;$$
$$K_s = [CH_3COOH_2^+][CH_3COO^-] \qquad (6.6)$$

is exactly analogous to the one described by eq. (6.2) for aqueous solutions. The numerical value of the autoprotolysis constant K_s varies, of course, from one solvent to another, for it depends on the ability of the solvent both to donate protons and to accept them, and it also depends on the dielectric constant. For acetic acid K_s is equal to 3.6×10^{-15} mol^2 dm^{-6} at 25°. It is a coincidence that this is so nearly the same as the value for water, which is 1×10^{-14} mol^2 dm^{-6}: acetic acid is a very much stronger acid than water, and its much greater ability to donate protons counterbalances both its lesser ability to accept them and the lesser tendency for dissociation to occur in a solvent whose dielectric constant is low. The autoprotolysis constants of other common amphoteric solvents range from about 1×10^{-4} mol^2 dm^{-6} for anhydrous sulfuric acid through 9×10^{-20} mol^2 dm^{-6} for ethanol to 10^{-33} mol^2 dm^{-6} for liquid ammonia (at 223 K). The autoprotolytic reactions in these solvents are

$$2\,H_2SO_4(l) = H_3SO_4^+ + HSO_4^-; \qquad K_s = [H_3SO_4^+]\,[HSO_4^-],$$

$$2\,C_2H_5OH(l) = C_2H_5OH_2^+ + C_2H_5O^-; \quad K_s = [C_2H_5OH_2^+]\,[C_2H_5O^-],$$

$$2\,NH_3(l) = NH_4^+ + NH_2^-; \qquad K_s = [NH_4^+]\,[NH_2^-].$$

Each involves the transfer of a proton from one molecule of the solvent, acting as an acid, to another, which acts as a base. Any amphoteric solvent may be represented by the general formula HSo. It can not only undergo autoprotolysis

$$2\,HSo(l) = H_2So^+ + So^-; \quad K_s = [H_2So^+]\,[So^-]$$

but can also accept a proton from a dissolved molecule (or ion) of an acid HB

$$HB + HSo(l) = H_2So^+ + B^-; \qquad K_a = [H_2So^+]\,[B^-]/[HB]$$

or can donate a proton to a dissolved molecule (or ion) of a base B

$$B + HSo(l) = BH^+ + So^-; \quad K = [BH^+]\,[So^-]/[B] = K_s/K_{a,BH^+}.$$

The last two of these equations show that B$^-$ (or B) becomes a weaker base as HB (or BH$^+$) becomes a stronger acid, just as in aqueous solutions.

Not all solvents are amphoteric. Some, including pyridine $\left(\langle\bigcirc N\rangle\right)$, other amines such as aniline $\left(\langle\bigcirc\rangle\text{—}NH_2\right)$, and others such as di-*n*-butyl ether (CH$_3$CH$_2$CH$_2$CH$_2$—O—CH$_2$CH$_2$CH$_2$CH$_3$)   and   1,4-dioxane $\left(\bigcirc\begin{smallmatrix}O\\O\end{smallmatrix}\right)$ are more or less strongly basic but have no detectable ability to donate protons to bases dissolved in them. These are called *basic solvents*. Others, the so-called *inert solvents*, are neither appreciably acidic nor appreciably basic. These include hydrocarbons such as benzene and *n*-hexane, and also chlorinated compounds such as carbon tetrachloride and chloroform, which are all extremely unreactive. They also include solvents, such as methyl isobutyl ketone (4-methyl-2-pentanone)

$$\overset{\displaystyle O}{\underset{\displaystyle \|}{}}$$
$$CH_3C\text{—}CH_2CH(CH_3)_2$$

that must have both a conjugate acid and a conjugate base[†] and must really be amphoteric, but that have so little tendency to gain or lose protons that they do not react to appreciable extents with common acids or bases dissolved in them.

These distinctions among different kinds of solvents are associated with the so-called *levelling effect*. There are many acids that are much stronger than hydronium ion: perchloric, hydrochloric, and sulfuric acids are three familiar ones, while the protonated forms or conjugate acids of acetic acid, acetone $[(CH_3)_2C\!\!=\!\!OH^+]$, and many other organic compounds are less familiar but are comparably strong. When any such acid is dissolved in water it ionizes in accordance with the equation $HB(aq) + H_2O(l) = H_3O^+B^-$, and this reaction proceeds nearly to completion because the acid is a very powerful proton donor and the water is a willing proton acceptor. Once it has done so the HB has disappeared, and the result is that an aqueous solution cannot contain an appreciable concentration of any acid that is much stronger than hydronium ion. This is expressed by saying that, in aqueous solutions, strong acids are levelled down to the strength of hydronium ion.

Because water is acidic as well as basic, there is also a levelling effect for strong bases in aqueous solutions. Oxide ion (O^{2-}), amide ion (NH_2^-), and hydride ion (H^-), among many others, are stronger bases than hydroxide ion, but because water is acidic no appreciable concentration of any of these can survive the reaction $B + H_2O(l) = BH^+OH^-$ in an aqueous solution. In water there can be no substantial concentration of any base that is much stronger than hydroxide ion; strong bases are levelled down to the strength of hydroxide ion.

These two levelling effects occur together in any amphoteric solvent. They may be summarized by saying that neither extremely strong acids nor extremely strong bases can exist in an amphoteric solvent because they react with it. They have two consequences. One is that two very strong acids or bases cannot be distinguished in an amphoteric solvent by measurements of acidity. An aqueous solution of hydrochloric acid has exactly the same acidity as an equally concentrated one of nitric acid. The other is that there is only a limited range of acidities in any amphoteric solvent. No solution can be prepared that is more acidic than one containing a high concentration of the conjugate acid of the solvent, and none can be prepared that is more basic than one containing a high concentration of the conjugate base of the solvent. In water this range extends roughly from solutions containing 20 M hydronium ion to those containing 20 M hydroxide ion.

In a basic solvent such as aniline, however, acids are levelled but bases are not. When hydrochloric acid or perchloric acid is dissolved in aniline it is levelled down to the strength of anilinium ion, $C_6H_5NH_3^+$, just as it is levelled down to the strength of hydronium ion in water. Moreover, acids that are not levelled in water, because they

---

[†] For methyl isobutyl ketone these are $CH_3\overset{\overset{\textstyle OH^+}{\|}}{C}\!\!-\!\!CH_2CH(CH_3)_2$ and $CH_3\overset{\overset{\textstyle O^-}{|}}{C}\!\!=\!\!CHCH(CH_3)_2$, respectively.

are weaker than hydronium ion, may be levelled in a basic solvent such as aniline. The reaction $C_6H_5NH_2(l) + HIO_3 = C_6H_5NH_3^+IO_3^-$ is virtually complete, whereas the corresponding one $H_2O(l) + HIO_3 = H_3O^+IO_3^-$ is not. In an aqueous solution there may be an appreciable concentration of unionized iodic acid, and therefore a moderately concentrated solution of iodic acid is less acidic, and has a higher pH-value, than a solution of hydrochloric acid or perchloric acid at the same concentration. In aniline, however, the ionization of iodic acid is driven almost to completion by the basic strength of the solvent. A solution of iodic acid in aniline therefore contains the same concentration of anilinium ion, and has the same acidity, as an equally concentrated solution of hydrochloric acid or perchloric acid. Minor differences of acidity would doubtless arise from differences among the dissociation constants of the ion pairs $C_6H_5NH_3^+IO_3^-$, $C_6H_5NH_3^+Cl^-$, and $C_6H_5NH_3^+ClO_4^-$. Of these three anions chloride ion is the smallest and iodate ion the largest, and there will consequently be more ion-pair formation, and less dissociation into free anilinium ion, in a solution of hydrochloric acid than in one of iodic acid. This means that, in aniline, a solution of hydrochloric acid is slightly less acidic than a solution of iodic acid even though, in water, hydrochloric acid is a stronger acid than iodic acid!

Bases are not levelled in basic solvents, such as aniline, because such solvents are not appreciably acidic. Solutions in basic solvents can therefore contain bases that are much too strong to exist in water and other amphoteric solvents. In a basic solvent the scale of acidity is limited at one end but not at the other: there can be no acid that is much stronger than the conjugate acid of the solvent, but very strong bases can exist.

In inert solvents neither acids nor bases are levelled because no common acid is strong enough to donate protons to molecules of such a solvent, and because no common base is strong enough to remove protons from them. Consequently both very strong acids and very strong bases can exist in inert solvents, and the range of measured pH-values may be as wide as 35 or 40 units. The distinction between an inert solvent and a basic or amphoteric one is not really a structural one: if it were, we would have to classify sulfolane as a basic solvent because its structure does enable us to see how it could accept protons from extremely strong acids

and to classify methyl isobutyl ketone as an amphoteric solvent because both a conjugate acid and a conjugate base can be imagined for it. By classifying both as inert solvents we express the fact that neither is a sufficiently strong acid or base to react appreciably with any ordinary base or acid dissolved in it.

SUMMARY: The strength of a dissolved acid or base is best expressed by the value of its ionization constant. The strength of an acid is increased by dissolving it in a more basic solvent, while that of a base is increased by dissolving it in a more acidic solvent. There are three kinds of solvents: amphoteric ones that can both accept and donate protons, basic ones that can accept protons from dissolved acids but not donate them to dissolved bases, and inert ones that react with neither acids nor bases. The levelling effect limits the strength of any acid that is dissolved in a basic solvent, and limits the strength of either an acid or a base that is dissolved in an amphoteric solvent. The dielectric constant affects the overall dissociation constant by influencing the dissociation constants of ion pairs.

6.13. Conclusion

Acid-base chemistry is such a wide and important field that this has had to be a very long chapter. It has dealt with a large number of fundamental ideas about the strengths of acids and bases and the correlation of these with molecular constitution and structure, about the natures and extents of acid-base reactions in solvents of different kinds, about the algebraic and numerical consequences of the expressions for equilibrium constants of various kinds, and about the ways in which one chemical equilibrium can interact with others that are established at the same time. Many of the topics introduced in this chapter will reappear frequently in later ones; some will be explored much more thoroughly in other courses. They play vital roles in our knowledge and understanding of chemical behavior and our ability to manipulate that behavior in ways that are useful or that, by leading to further knowledge and understanding, may become useful in the future.

Problems

Answers to some of these problems are given on page 520.

6.1. A solution containing 0.0100 mol dm^{-3} of ammonia and 1.90 mol dm^{-3} of ammonium chloride has a hydronium-ion concentration equal to 1.00×10^{-7} M. What is the value of the overall acidic dissociation constant K_a of ammonium ion?

6.2. A solution of acetic acid is prepared by adding water to 0.0100 mole of acetic acid, using enough water to give a total volume of 1000 cm^3. The equilibrium concentration of acetate ion in this solution is 1.33×10^{-3} M. What is the value of the overall acidic dissociation constant K_a of acetic acid?

6.3. Exactly 0.01 mole of tribromoacetic acid, Br_3CCOOH, is dissolved in enough water to yield 1000 cm^3 of solution. The concentration of hydronium ion in the solution is 9.5×10^{-3} M. What is the value of the overall acidic dissociation constant of the acid?

6.4. The concentration of hydronium ion in pure water at 373 K is 7.25×10^{-7} M. What is the value of the autoprotolysis constant of water at that temperature?

6.5. The autoprotolysis constant of liquid ammonia at 223 K is equal to 1×10^{-33} mol^2 dm^{-6}. What would be the concentration of amide ion, NH_2^-, in pure liquid ammonia at this temperature?

6.6. Convert each of the following hydronium-ion concentrations to a pcH-value:

$\qquad$ (a) 0.206 M $\quad$ (b) 3.00×10^{-6} M. $\quad$ (c) 4.5×10^{-9} M.

6.7. Convert each of the following values to exponential form:

(a) pcH = 3.50. $\qquad\qquad$ (d) pK_a = 9.60.
(b) pcH = −0.40. $\qquad\qquad$ (e) pK_s = 14.47.
(c) paH = 5.10. $\qquad\qquad$ (f) pcI = 6.70.

6.8. Perchloric acid, $HClO_4$, is completely ionized and dissociated in aqueous solutions. What is the pcH of an aqueous solution in which the total concentration of perchloric acid is

(a) $2.00 \times 10^{-3} M$?   (b) $2.00 \times 10^{-8} M$?

6.9. What is the pcH of an aqueous solution containing iodic acid (HIO_3) at a total concentration of 0.100 mol dm^{-3}? The overall acidic dissociation constant of iodic acid is equal to 0.166 mol dm^{-3}.

6.10. At $25°$ the value of K_a for benzoic acid is 6.14×10^{-5} mol dm^{-3}. What is the pcH of an aqueous solution in which the total concentration of benzoic acid is

(a) $0.04 M$?   (b) $1.00 \times 10^{-4} M$?   (c) $2.00 \times 10^{-7} M$?

6.11. Trimethylammonium ion, $(CH_3)_3NH^+$, is the conjugate acid of trimethylamine, $(CH_3)_3N$. In water at 298 K the overall acidic dissociation constant of trimethylammonium ion is equal to 1.60×10^{-10} mol dm^{-3}. What is the pcH of a solution in which the total concentration of solute is

(a) 0.100 mol dm^{-3} of trimethylammonium chloride?
(b) 0.100 mol dm^{-3} of trimethylamine?
(c) 5.00×10^{-5} mol dm^{-3} of trimethylammonium chloride?

6.12. What is the pcH of a solution in which the total concentrations of hydrochloric acid and iodic acid are $0.01 M$ and $0.10 M$, respectively? The value of K_a for iodic acid is 0.166 mol dm^{-3}.

6.13. What is the pcH of a solution in which the total concentration of trichloroacetic acid $(Cl_3CCOOH, K_a = 0.20$ mol $dm^{-3})$ is $0.50 M$ and the total concentration of picric acid

$(K_a = 0.42$ mol $dm^{-3})$ is $0.10 M$?

6.14. In water at 298 K the overall acidic dissociation constants of citric acid,

are $K_1 = 7.5 \times 10^{-4}$ mol dm^{-3}, $K_2 = 1.7 \times 10^{-5}$ mol dm^{-3}, and $K_3 = 4.0 \times 10^{-7}$ mol dm^{-3}. The pcH of a sample of lime juice is 1.90. Assuming that citric acid is the only acid present, what is its total concentration?

6.15. What are the concentrations of hydronium ion, hydrogen oxalate ion $(HC_2O_4^-)$, and oxalate ion $(C_2O_4^{2-})$ in an aqueous solution at 298 K in which the total concentration of oxalic acid is $0.100 M$? The overall acidic dissociation constants of oxalic acid at that temperature are $K_1 = 5.4 \times 10^{-2}$ mol dm^{-3} and $K_2 = 5.4 \times 10^{-5}$ mol dm^{-3}.

6.16. In aqueous solutions at 298 K the overall acidic dissociation constants of phosphoric acid, H_3PO_4, are $K_1 = 7.1 \times 10^{-3}$ mol dm^{-3}, $K_2 = 6.3 \times 10^{-8}$ mol dm^{-3}, and $K_3 = 4.2 \times 10^{-13}$ mol dm^{-3}. What is the pcH of a solution in which the total concentration of solute is

(a) 0.0100 mol dm^{-3} of trisodium phosphate?
(b) 0.0100 mol dm^{-3} of disodium hydrogen phosphate?
(c) 0.0100 mol dm^{-3} of sodium dihydrogen phosphate?
(d) 0.0100 mol dm^{-3} of phosphoric acid?

6.17. Pyridinium ion ($C_5H_5NH^+$, often abbreviated "pyH$^+$") is the conjugate acid of pyridine (C_5H_5N, abbreviated "py"). The overall acidic dissociation constant of pyridinium ion in water at 298 K corresponds to $pK_{a, pyH^+} = 5.22$. For acetic acid in water at the same temperature $pK_{a, HOAc} = 4.76$. Calculate the pcH of a solution containing

(a) 0.100 mol dm^{-3} of pyridinium acetate.
(b) 0.0100 mol dm^{-3} of pyridinium chloride and 0.100 mol dm^{-3} of sodium acetate.

6.18. Use eq. (6.43) to find the pcH of an aqueous solution at 298 K in which the total concentrations of acetic acid and sodium acetate are equal. The overall acidic dissociation constant of acetic acid in water at 298 K is equal to 1.75×10^{-5} mol dm^{-3}.

6.19. How much sodium hydrogen sulfate must be dissolved in water to make 1.00 dm^3 of a solution having pcH = 2.00 at 298 K? The overall acidic dissociation constants of sulfuric acid at this temperature are $K_1 = 50$ mol dm^{-3}, $K_2 = 1.20 \times 10^{-2}$ mol dm^{-3}.

6.20. The value of K_a for ammonium ion in water at 298 K is 5.7×10^{-10} mol dm^{-3}.

(a) How much solid sodium hydroxide would have to be added to 1 dm^3 of an 0.10 M solution of ammonium chloride to yield a buffer solution having pcH = 9.70?
(b) What would be the buffer capacity of the resulting solution?

6.21. If the solubility product of copper(II) sulfide, CuS, is equal to 6×10^{-36} mol^2 dm^{-6} and if the acidic dissociation constants of hydrogen sulfide are $K_1 = 6.0 \times 10^{-8}$ mol dm^{-3} and $K_2 = 1.0 \times 10^{-14}$ mol dm^{-3}, what is the solubility of CuS in

(a) pure water?   (b) 0.10 M perchloric acid?

Neglect the possible formation of $Cu^{2+}S^{2-}$ ion pairs.

6.22. In water at 298 K the solubility product of beryllium hydroxide, Be(OH)$_2$, is equal to 7×10^{-22} mol^3 dm^{-9} and the overall dissociation constant of BeOH$^+$ is equal to 3.2×10^{-10} mol dm^{-3}. What is the solubility of Be(OH)$_2$ in water at 298 K?

CHAPTER 7

Complexation Reactions

7.1. Introduction

The preceding chapter was devoted to Lowry–Brønsted acid-base chemistry; this one is devoted to Lewis acid-base chemistry. The two have a great deal in common. As was suggested in Section 6.2, there is really no fundamental difference between the reaction of pyridine, behaving as a Lewis base, with a metal ion to form a complex

$$Cu(OH_2)_4^{2+} + : N\bigcirc = \left[(H_2O)_3\, Cu : N\bigcirc\right]^{2+} + H_2O$$

and the reaction between pyridine, behaving as a Lowry–Brønsted base, with hydronium ion

$$H_3O^+ + : N\bigcirc = \left[H : N\bigcirc\right]^+ + H_2O$$

The distinction between "complexation reactions" and "acid-base reactions" is purely arbitrary and is made here only as a matter of convenience. The fact that they are essentially identical led to the appearance in Chapter 6 of many of the ideas and techniques that will be needed in this chapter.

7.2. Fundamental ideas

PREVIEW: This section defines a ligand and a complex, and classifies both ligands and complexes in several different ways.

A *complexation reaction* may be defined as a reaction that involves a metal ion, which behaves as a Lewis acid, and a *ligand*, which behaves as a Lewis base. The ligand donates a pair (or, sometimes, two or more pairs) of electrons to the metal ion, and a *complex* is formed. As was said in Section 6.2, one kind of complexation reaction is that in which a "bare" or uncoordinated metal ion accepts electron pairs from successive molecules of a polar solvent to yield the solvated ion:

$$6H_2O : +Ni^{2+} = 5H_2O : +Ni(OH_2)^{2+} = 4H_2O : +Ni(OH_2)_2^{2+}$$
$$= \ldots = Ni(OH_2)_6^{2+}.$$

The product of such a series of reactions may be further solvated by a reaction like

$$Ni(OH_2)_6^{2+} + (n-6)\,H_2O = Ni(OH_2)_n^{2+} \tag{7.1}$$

whose product has the structure

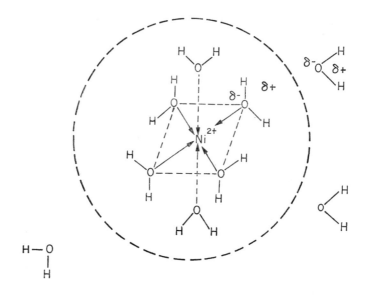

The arrows represent pairs of electrons donated by molecules of water. There is an *inner sphere*, whose surface is represented by the dashed circle, in which the molecules of water (or another solvent) are held to the metal ion by chemical bonds. The additional molecules in the *outer sphere* may be held merely by electrostatic forces resulting from their dipolar nature and the orientation of the dipolar molecules in the inner sphere. The nature of this electrostatic attraction is indicated in the upper right-hand corner of the sketch. It is weaker than the forces that act on molecules in the inner sphere, and a molecule in the outer sphere is much more easily detached when a free molecule of the solvent or some other particle collides with it. As a result, the total number of molecules of water attached to any particular nickel ion fluctuates from one instant to the next. Equation (7.1) suggests that the composition of $Ni(OH_2)_n^{2+}$ is more constant and well defined than it really is.

The *hydration number* (or, in a polar non-aqueous solvent, the *solvation number*) n is the average number of molecules of water (or solvent) attached to a metal ion. The average may be for a single ion over a long period of time or for a great many ions at the same instant. The hydration number of a metal ion depends on both its charge and its radius, because each of these affects the electrostatic potential at the surface of the inner sphere, and consequently affects the force that is exerted on a molecule in the outer sphere. As the radius of the metal ion increases, the electrostatic potential decreases, and therefore large ions have fewer molecules of water in their outer spheres, and hold them less firmly, than small ions do. A very large ion may not be able to retain a molecule of solvent even in its inner sphere. The radius of a lithium ion, as determined from X-ray crystallographic data, is only 0.68 Å (1 Å = 10^{-8} cm), and the hydration number of lithium ion in dilute solutions is about 6. For rubidium and cesium ions, the largest of the alkali-metal ions, the radii are about 1.48 Å and the hydration numbers of these ions are well below 1. This has the rather unexpected consequence that a (hydrated) lithium ion in an aqueous solution is actually larger than a rubidium or cesium ion. Consequently a lithium ion moves through an aqueous solution much more slowly under the influence of a concentration gradient or an electric field than a rubidium or cesium ion does. Increasing the charge on a metal ion has the same effect as decreasing its radius. The radii of the sodium and lanthanum (La^{3+}) ions are almost the same, 0.97 and 1.02 Å respectively, but because the lanthanum ion is more highly charged its hydration number is about 13 while that of sodium ion is only about 4.

There is no appreciable tendency for metal ions to become solvated in non-polar solvents such as benzene and carbon tetrachloride, and the chemist who studies their reactions with ligands in such solvents can therefore define complexation as a process in which a Lewis acid reacts with a Lewis base. To the chemist interested in reactions that occur in polar solvents, however, solvation is a matter of past history: it has already reached equilibrium before those reactions begin. Complexation reactions in such solvents therefore involve, not the simple addition of an ion or molecule of a ligand to a bare metal ion, but the replacement of one coordinated ion or molecule with another. For the sake of brevity these reactions are usually represented by equations like $Cu^{2+} + Cl^- = CuCl^+$, but the brevity is achieved at the cost of concealing the details of the process.

Complexes are classified in many different ways. One, emphasizing the rates of their reactions, divides them into labile and inert complexes. A *labile complex* is one in which ligands in the inner sphere can be rapidly replaced by others; in an *inert complex* such replacements are very slow. The first inert complexes to be discovered were those of chromium(III) and cobalt(III) almost a century ago, when Werner discovered that the reaction $Cr(NH_3)_6^{3+} + 6\,H_3O^+ = Cr(OH_2)_6^{3+} + 6\,NH_4^+$ could not be made to occur at a detectable rate even by boiling with concentrated sulfuric acid, even though the similar reaction $Cu(NH_3)_4^{2+} + 4\,H_3O^+ = Cu(OH_2)_4^{2+} + 4\,NH_4^+$ appeared to be instantaneous. Techniques based on the use of radioisotopes and nuclear magnetic resonance have been developed since then, and make it possible to study the rates of reactions like

$$H_2O^*(aq) + M(OH_2)_6^{n+} = M(O^*H_2)(OH_2)_5^{n+} + H_2O(l)$$

where the asterisk denotes an atom of O^{17} or O^{18}. If M is chromium(III) the complex $Cr(OH_2)_6^{3+}$ is said to be inert. This exchange of one molecule of solvent for another has a half-time of about 2 months at 25°, but the manganese(II) complex $Mn(OH_2)_6^{2+}$

is said to be labile because the half-time is only a few hundredths of a microsecond: the rates differ by a factor of about 10^{14}. Of course this means that the energies of activation are different: they are about 115 kJ mol^{-1} for chromium(III) and 35 kJ mol^{-1} for manganese(II). The majority of complexation reactions lie between these two, and the difference between inert and labile complexes is a highly arbitrary one.

Another classification draws a distinction between *inner-sphere complexation* and *outer-sphere complexation*. Inner-sphere complexation involves the replacement of one ligand that is bonded directly to the metal ion with another. All of the equations given in the last paragraph were for inner-sphere complexations. Outer-sphere complexation involves the substitution of one ligand for another in the outer sphere, and looks like this:

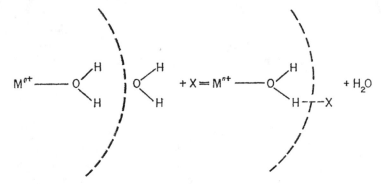

As this representation suggests, the product may be stabilized by hydrogen bonding if X is a strong Lowry–Brønsted base. If the complex is labile, ions or molecules of X can undergo rapid exchange with molecules of water in the inner sphere, and then the nature of the final product depends on the strengths of the bonds that are involved. When sulfate ion is added to an aqueous solution of copper(II) ion, the product is $Cu(OH_2)_4^{2+} SO_4^{2-}$, an outer-sphere complex or ion pair, because the formation of a copper–sulfate bond in the inner sphere does not liberate enough energy to rupture a copper–water bond. If chloride ion is added instead of sulfate ion, so much more energy is liberated on forming a copper–chloride bond that a molecule of water can be displaced from the inner sphere.

A third classification divides complexes into those that are mononuclear and those that are polynuclear. A *mononuclear complex* consists of a single metal atom with the ligands attached to it. All of the complexes mentioned so far have been mononuclear ones. In a *polynuclear complex* there are two or more "central" atoms that are joined in some way, usually by bonding through one or two atoms that their inner spheres have in common. Polynuclear complexes in turn are of two kinds: *isopolynuclear* and *heteropolynuclear*, usually called *isopoly complexes* and *heteropoly complexes* for short. In an isopolynuclear complex all of the "central" atoms are the same. A typical one, which is represented by the formula $Fe_2(OH)_2^{4+}$, exists in weakly acidic solutions of iron(III) and has the structure

$$
\left[\quad (H_2O)_4Fe \begin{array}{c} H \\ | \\ O \\ \diagdown \quad \diagup \\ \diagup \quad \diagdown \\ O \\ | \\ H \end{array} Fe(OH_2)_4 \quad \right]^{4+}
$$

Each of the ferric ions is octahedral and the two oxygen atoms in the hydroxide ions lie on an edge that is shared between the two octahedra, of which the other corners are occupied by molecules of water. Dichromate ($Cr_2O_7^{2-}$), pyrophosphate ($P_2O_7^{4-}$), tetraborate ($B_4O_7^{2-}$), and many other isopolyanions have similar structures. In the isopolynuclear complex Ag_2Cl^+ only the single chlorine atom is common to the coordination fields of the two silver atoms. In a heteropolynuclear complex the "central" atoms are different. One heteropolynuclear complex is formed at the transition state in the reaction

$$
Co^{III}Cl^{2+} + Cr^{2+} = [Co^{III} - Cl - Cr^{II}]^{4+} = Co^{2+} + Cr^{III}Cl^{2+}
$$

from which molecules of water are omitted. The chlorine atom must serve as a bridge between the cobalt and chromium atoms because all of the chloride originally contained in the inert cobalt(III) complex is found in the inert chromium(III) complex when the reaction is over. This particular heteropolynuclear complex is unstable because the tendencies for cobalt(III) to accept an electron and for chromium(II) to donate one are so great that reduction and oxidation occur by the transfer of an electron through the bridge, but others are much more stable. Again there are hetero-polyanions containing only oxygen in addition to the central atoms and having generally similar structures.

A fourth classification is based on the nature and structure of the ligand. A ligand that can form only a single bond with a metal ion is said to be *monodentate*. Ammonia and pyridine are typical monodentate ligands: each can donate a pair of electrons to a metal ion in forming a bond with it, but has only one pair of electrons to share. Other monodentate ligands, including chloride ion, the other halide ions, and cyanide and thiocyanate ions, may have two or more pairs of electrons available. A chloride ion has four unshared pairs of electrons but can donate only one of them to a single metal

ion, because to form a structure like $M \begin{array}{c} \diagup \diagdown \\ \diagdown \diagup \end{array} Cl$ would place an impossibly large stress on

the angle between the vacant bonding orbitals of the metal ion, which are represented by the solid lines, and the filled ones of the chloride ion, which are represented by the dashed lines. However, as was said in the preceding paragraph, a chloride ion can link two different metal ions, and the distinction between monodentate and polydentate ligands is meaningful only for mononuclear complexes. *Polydentate ligands* or *chelating agents* are ligands that can donate two or more pairs of electrons to a single metal ion, and are usually further subdivided into *bidentate ligands* (such as oxalate ion and ethylenediamine, $H_2\ddot{N}CH_2CH_2\ddot{N}H_2$), *tridentate ligands* (such as diethylenetriamine, $H_2\ddot{N}CH_2CH_2\ddot{N}HCH_2CH_2\ddot{N}H_2$), and so on. Chelating agents and their abilities to form chelates by reacting with metal ions were discussed very briefly in Section 4.5.

A ligand that can donate enough pairs of electrons to a metal ion to occupy all of its coordination positions is called a *chelon,* and the complex that is formed when it reacts with a metal ion is called a *chelonate.* Whereas a monodentate or bidentate ligand L must react with a metal ion in a series of steps ($M+L = ML$, $ML+L = ML_2$, etc.), a chelon Y tends to react in a single overall step ($M+Y = MY$). The chelon triethylenetetramine, which is called "trien" for short, reacts with the tetraaqua-copper(II) ion as shown by the equation

The commonest chelon is ethylenediaminetetraacetate ion, usually abbreviated $EDTA^{4-}$:

On reacting with a metal ion M^{n+} that can form strong bonds with both oxygen and nitrogen atoms it can behave as a hexadentate ligand and yield a chelonate such as

Most transition-metal ions form extremely stable chelonates with $EDTA^{4-}$, as is indicated by the equilibrium constants of the reactions

$$Pb^{2+} + EDTA^{4-} = Pb(EDTA)^{2-}; \quad K = 1.2 \times 10^{17} \text{ mol}^{-1} \text{ dm}^3,$$

$$Fe^{3+} + EDTA^{4-} = Fe(EDTA)^{-}; \quad K = 1.2 \times 10^{25} \text{ mol}^{-1} \text{ dm}^3.$$

Iron is needed in plant nutrition, but administering a salt such as iron(II) or (III) chloride is inefficient because the oxidation of iron(II) by atmospheric oxygen and the reaction of iron(III) with natural soils lead to the precipitation of compounds of iron(III) that are too slightly soluble to be of much use to plants. The iron(III)–$EDTA^{4-}$ chelonate is so stable that these compounds do not precipitate, and its administration therefore supplies the necessary iron in a soluble and readily available form. The calcium–$EDTA^{4-}$ chelonate is administered orally to patients suffering from poisoning by lead or other heavy metals; inside the body it gives rise to reactions like $Ca(EDTA)^{2-} + Pb^{2+} = Pb(EDTA)^{2-} + Ca^{2+}$, because the heavy-metal ions give chelonates much more stable than that of calcium ion, and these heavy-metal chelonates are readily excreted from the body.

A final classification is into simple and mixed complexes. A *simple complex* is one in which all of the ligands are the same, ignoring any molecules of solvent. The complex ion usually represented by the formula $Cu(NH_3)_4^{2+}$ is a simple complex even though it may have one or two molecules of water in the inner coordination sphere in addition to the four molecules of ammonia. A *mixed complex* is one that contains two or more different ligands, such as $Fe^{III}BrCl^{+}$ or $Ag(NH_3)Br_3^{2-}$.

These five classifications raise more possibilities than this chapter can consider. The following sections will be confined to inner-sphere coordination equilibria, and thus by implication to labile complexes because the rates of reactions of inert complexes are more interesting than their equilibria. They will stress the behavior that arises from stepwise complex formation and the differences between this and the single-step reaction that occurs with chelons. Many other interesting and important topics are suggested by these classifications but must be left for more advanced courses.

> SUMMARY: A complex is the product of a reaction in which a ligand donates at least one pair of electrons to a metal ion, forming a bond with it. Water and other polar solvents act as ligands, forming hydrated or solvated metal ions. Ligands may be either monodentate or polydentate, depending on the number of bonds that they can form with a single metal ion; a chelon is a polydentate ligand that can occupy all of the coordination positions of a metal ion. Complexes may be classified in different ways: as labile or inert; inner-sphere or outer-sphere; mononuclear or polynuclear; complexes with monodentate ligands, chelates, or chelonates; and simple or mixed.

7.3. Stepwise complex formation

> PREVIEW: This section outlines the algebra of stepwise complex formation and gives distribution curves showing how the proportions of different complexes depend on the concentration of the free ligand.

The conventions that are used in expressing the equilibrium constants for the formations of metal complexes were outlined in Section 2.9. The reactions between a monodentate ligand L and a metal ion M (omitting charges for simplicity) are characterized by the *stepwise* (or *successive*) *formation constants* or *stability constants* of the complexes:

$$M + L = ML; \qquad K_1 = [ML]/[M][L],$$
$$ML + L = ML_2; \qquad K_2 = [ML_2]/[ML][L] \tag{7.2}$$

and so on until the *maximum coordination number j* of the metal ion is fully satisfied by the addition of the *j*th ligand:

$$ML_{j-1} + L = ML_j; \qquad K_j = [ML_j]/[ML_{j-1}][L].$$

The maximum coordination number of a metal ion is the largest number of ions or molecules of any monodentate ligand that can be added to it. Since the majority of metal ions are octahedral, most metal ions have maximum coordination numbers of 6. However, some metal ions (such as silver ion) have maximum coordination numbers of 4, and the highest coordination number that is actually observed in a particular system may be smaller than the maximum coordination number of the metal ion if the ligand is very bulky or very weakly bound. Bidentate ligands behave in the same general way, except that only $j/2$ ions or molecules of a bidentate ligand are needed to occupy j positions in the inner coordination sphere.

For each of the successive complexes there is also an *overall formation constant*, which for any particular complex ML_i is defined by

$$M + iL = ML_i; \qquad \beta_i = [ML_i]/[M][L]^i = K_1 K_2 \ldots K_i. \qquad (7.3)$$

Overall formation constants are useful in some calculations but are less often found in tables because values of the stepwise constants provide more detailed information and are always given when they are known. There is a short table of stepwise formation constants in Appendix V.

For the silver–chloride complexes the values of the stepwise formation constants are given by

$$Ag^+ + Cl^- = AgCl(aq); \qquad \log K_1 = 2.7;$$
$$K_1 = [AgCl(aq)]/[Ag^+][Cl^-] = 5 \times 10^2 \ mol^{-1} \ dm^3. \qquad (7.4a)$$
$$AgCl(aq) + Cl^- = AgCl_2^-; \qquad \log K_2 = 2.1;$$
$$K_2 = [AgCl_2^-]/[AgCl(aq)][Cl^-] = 1.2 \times 10^2 \ mol^{-1} \ dm^3. \qquad (7.4b)$$
$$AgCl_2^- + Cl^- = AgCl_3^{2-}; \qquad \log K_3 = 0.7;$$
$$K_3 = [AgCl_3^{2-}]/[AgCl_2^-][Cl^-] = 5 \ mol^{-1} \ dm^3. \qquad (7.4c)$$
$$AgCl_3^{2-} + Cl^- = AgCl_4^{3-}; \qquad \log K_4 = 0.5;$$
$$K_4 = [AgCl_4^{3-}]/[AgCl_3^{2-}][Cl] = 3 \ mol^{-1} \ dm^3. \qquad (7.4d)$$

As in many other systems, each of the constants is smaller than the preceding one: the first ion (or molecule) of the ligand is added more readily than the second, the second is added more readily than the third, and so on. Exceptions to this generalization are numerous. For the mercury(II)–chloride system the successive values of $\log K$ are 5.3, 7.5, 1.1, and 1.0. These numbers show that the addition of the second chloride ion, in the reaction $HgCl^+ + Cl^- = HgCl_2(aq)$, is accompanied by a larger decrease of free energy than the addition of the first. Some consequences of that fact will appear on page 195.

The overall dissociation constant K_c of $AgCl(aq)$ is equal to the reciprocal of its formation constant K_1:

$$K_c = [Ag^+][Cl^-]/[AgCl(aq)] = 1/K_1 = 2 \times 10^{-3} \ mol \ dm^{-3}$$

so that $\log K_1 = -\log K_c = pK_c$. The value of K_c is so much smaller than the values given in Table 5.1 for ion pairs containing two univalent ions that fairly strong chemical bonding must be involved in the formation of $AgCl(aq)$.

Even though four equilibrium constants are needed to describe the silver–chloride system (a fifth one, the solubility product of silver chloride, is needed as well if precipitation occurs), some important facets of its behavior are easy to discern. Imagine that chloride ion is added to a solution containing silver ion at a concentration so low that silver chloride will not precipitate. As long as the concentration of chloride ion is very low, almost all of the dissolved silver(I) will be present as Ag^+. On increasing the concentration of chloride ion, the concentration of Ag^+ will begin to decrease while the concentration of $AgCl(aq)$ increases, and these two concentrations will become equal when the concentration of free chloride ion is given by $[Cl^-] = 1/K_1 = 2 \times 10^{-3}$ M. If the concentration of free chloride ion is increased above 2×10^{-3} M, the concentration of AgCl(aq) will exceed that of Ag^+ because the equilibrium of eq. (7.4a) will shift still farther toward the right. However, the values of K_1 and K_2 are so little different that there will be an appreciable concentration of $AgCl_2^-$ even in a solution containing 2×10^{-3} M free chloride ion, and increasing the concentration of free chloride ion above this value will increase the concentration of $AgCl_2^-$ as well as that of AgCl(aq). These two concentrations become equal when $[Cl^-] = 1/K_2 = 8 \times 10^{-3}$ M. In other words, the solution contains more Ag^+ than AgCl(aq) if the concentration of free chloride ion is below 2×10^{-3} M, and it contains more $AgCl_2^-$ than AgCl(aq) if the concentration of free chloride ion is above 8×10^{-3} M. Only within this narrow range of chloride-ion concentrations can AgCl(aq) be the predominating one of the dissolved complexes. For $AgCl_2^-$ the situation is a little different because the values of K_2 and K_3 are quite far apart: there is more $AgCl_2^-$ than AgCl(aq) at any concentration of free chloride ion above 8×10^{-3} M, and there is more $AgCl_2^-$ than $AgCl_3^{2-}$ at any concentration of free chloride below $[Cl^-] = 1/K_3 = 0.2$ M. Throughout this range the solution will contain more $AgCl_2^-$ than any other single species. Similarly, it can be deduced that $AgCl_3^{2-}$ can be the chief constituent only if the concentration of free chloride ion is between about 0.2 and 0.3_3 M, and that $AgCl_4^{3-}$ will predominate at concentrations above about 0.3_3 M.

These and other facts about the system are even more easily seen from graphical representations of its behavior. Figure 7.1 shows how the concentrations of the differ-

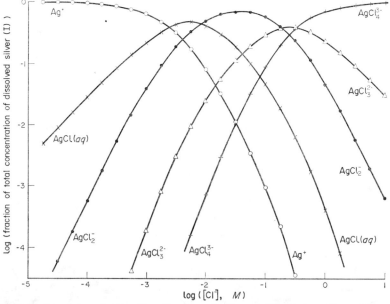

FIG. 7.1. Distribution curves showing how the concentration of chloride ion affects the fractions of dissolved silver(I) present as Ag^+, AgCl(aq), $AgCl_2^-$, $AgCl_3^{2-}$, and $AgCl_4^{3-}$.

ent species depend on the logarithm of the concentration of chloride ion. It confirms that AgCl(aq) and AgCl$_3^{2-}$ are important only over narrow ranges of the chloride-ion concentration, and it also shows that neither of them can ever be overwhelmingly predominant. In a solution containing 4×10^{-3} M free chloride ion, where the proportion of AgCl(aq) is at its highest, there is half as much Ag$^+$ as AgCl(aq), and there is also half as much AgCl$_2^-$ as AgCl(aq), so that half of the dissolved silver(I) is present as AgCl(aq). It is therefore impossible to deduce the properties of AgCl(aq) from data on the behavior of any solution unless corrections are made for the contrir butions of Ag$^+$ and AgCl$_2^-$ to that behavior, and the necessity of making such corrections decreases the reliability of the information that can be obtained abou- AgCl(aq). Even AgCl$_2^-$, with its much wider range of stability, never accounts fot more than about 70% of the dissolved silver(I).

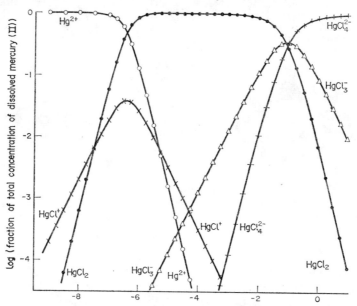

FIG. 7.2. Distribution curves showing how the concentration of chloride ion affects the fractions of mercury(II) present as Hg^{2+}, HgCl$^+$, HgCl$_2$(aq), HgCl$_3^-$, and HgCl$_4^{2-}$.

The behavior of the mercury(II)–chloride system is portrayed by Fig. 7.2. There are two ways in which it differs from the behavior of the silver–chloride system. One is that log K_1 (= 5.3) here is smaller than log K_2 (= 7.5). Because this is so, HgCl$^+$ does not predominate at any concentration of chloride ion and is never more than a very minor constituent. The other is that log K_2 is very much larger than log K_3 (= 1.1), which causes HgCl$_2$(aq) to predominate over a very wide range of chloride-ion concentrations. It is relatively easy to study the properties of HgCl$_2$(aq), but it would be extremely difficult to secure reliable information about those of HgCl$^+$ or HgCl$_3^-$.

14

Curves like the ones in these diagrams are calculated in the same way as those in Section 6.9. For the silver–chloride system the total concentration c_{Ag} of dissolved silver(I) is given by

$$c_{Ag} = [Ag^+] + [AgCl(aq)] + [AgCl_2^-] + [AgCl_3^{2-}] + [AgCl_4^{3-}]. \tag{7.5}$$

To obtain an expression for the formation function for any of these species it is necessary to express the concentrations of all the other species in terms of the concentration of the one that is of interest and the master variable, which would naturally be taken as the concentration of free chloride ion. This may be done by using eqs. (7.4). If it is Ag^+ that is of interest, these equations would be written

$$[AgCl(aq)] = K_1[Ag^+][Cl^-]; \qquad\qquad [AgCl_2^-] = K_1K_2[Ag^+][Cl^-]^2;$$
$$[AgCl_3^{2-}] = K_1K_2K_3[Ag^+][Cl^-]^3; \qquad [AgCl_4^{3-}] = K_1K_2K_3K_4[Ag^+][Cl^-]^4$$

and then combined with eq. (7.5) to obtain

$$c_{Ag} = [Ag^+] + K_1[Ag^+][Cl^-] + K_1K_2[Ag^+][Cl^-]^2 + K_1K_2K_3[Ag^+][Cl^-]^3$$
$$+ K_1K_2K_3K_4[Ag^+][Cl^-]^4. \tag{7.6}$$

A more compact version of eq. (7.6) can be obtained by using overall formation constants instead of successive ones, and rearranged to give

$$\alpha_{Ag^+} = \frac{[Ag^+]}{c_{Ag}} = \frac{1}{1 + \beta_1[Cl^-] + \beta_2[Cl^-]^2 + \beta_3[Cl^-]^3 + \beta_4[Cl^-]^4}. \tag{7.7a}$$

To obtain a similar expression for $AgCl_2^-$, eqs. (7.4) could be transformed into

$$[Ag^+] = [AgCl_2^-]/K_1K_2[Cl^-]^2; \qquad [AgCl(aq)] = [AgCl_2^-]/K_2[Cl^-];$$
$$[AgCl_3^{2-}] = K_3[AgCl_2^-][Cl^-]; \qquad\qquad [AgCl_4^{3-}] = K_3K_4[AgCl_2^-][Cl^-]^2$$

which can be combined with eq. (7.5) to yield

$$c_{Ag} = [AgCl_2^-]\left(\frac{1}{K_1K_2[Cl^-]^2} + \frac{1}{K_2[Cl^-]} + 1 + K_3[Cl^-] + K_3K_4[Cl^-]^2\right)$$
$$= [AgCl_2^-]\left(\frac{1 + \beta_1[Cl^-] + \beta_2[Cl^-]^2 + \beta_3[Cl^-]^3 + \beta_4[Cl^-]^4}{\beta_2[Cl^-]^2}\right),$$

or

$$\alpha_{AgCl_2^-} = \frac{[AgCl_2^-]}{c_{Ag}} = \frac{\beta_2[Cl^-]^2}{1 + \beta_1[Cl^-] + \beta_2[Cl^-]^2 + \beta_3[Cl^-]^3 + \beta_4[Cl^-]^4}. \tag{7.7b}$$

Equations (7.7) differ slightly from eqs. (6.50), but only because the former employ formation constants while the latter employ dissociation constants. The overall similarities between these two sets of equations should be obvious.

SUMMARY: If a metal ion M reacts with a ligand L to form a series of successive complexes ML, ML_2, ..., ML_j, the concentrations of free M and all of the complexes vary with the concentration of free L. Increasing the concentration of L decreases the concentration of M and increases the concentration of ML_j, but causes the concentration of each intermediate complex to pass through a maximum value.

7.4. Solubility and complex formation

PREVIEW: Complex formation influences the solubility of a precipitate in a solution containing one of its own ions: the precipitate is more soluble than it would be if complexes did not form, and its solubility may be quite large if the concentration of the common ion is very high.

Section 5.3 described the common-ion effect on the solubility of a precipitate but did not take the possibility of complex formation into account. According to what was said there, the solubility of silver chloride would decrease continuously toward zero as the concentration of chloride ion increased. Sections 5.5 to 5.7 added the idea that a complex or ion pair having the formula $AgCl(aq)$ exists in this system, and showed that its existence would give rise to a lower limit below which the solubility of silver chloride would not be decreased no matter how high the concentration of chloride ion might be. This lower limit would be equal to the equilibrium constant of the reaction

$$AgCl(s) = AgCl(aq); \qquad K_0 = [AgCl(aq)].$$

Equation (5.11) showed how this equilibrium constant is related to K_{AgCl}, the solubility product of silver chloride, and K_c, the overall dissociation constant of $AgCl(aq)$. Another formulation, employing the formation constant of $AgCl(aq)$ instead of its dissociation constant, is

$$K_0 = K_1 K_{AgCl}. \tag{7.8}$$

With $K_1 = 5 \times 10^2$ mol^{-1} dm^3 (eq. (7.4a)) and $K_{AgCl} = 1.8 \times 10^{-10}$ mol^2 dm^{-6}, this gives a value of 9×10^{-8} M for the solubility that would be approached asymptotically, if $AgCl_2^-$ and the higher complexes did not form, as the concentration of chloride ion was increased.

Since anionic complexes do form, the solubility of silver chloride must be represented by the equation

$$S = [Ag^+] + [AgCl(aq)] + [AgCl_2^-] + [AgCl_3^{2-}] + [AgCl_4^{3-}]. \tag{7.9}$$

This differs from eq. (7.5) only in that the symbol S is used to denote the solubility of silver chloride and emphasize the fact that we are now considering solutions that are in equilibrium with solid silver chloride. Expressing the concentrations of all the complexes in terms of the concentrations of silver and chloride ions, as was done in obtaining eq. (7.6), and replacing the product $[Ag^+][Cl^-]$ whenever it occurs by K_{AgCl}

14*

because the solutions are saturated with silver chloride, we can obtain

$$S = K_{AgCl} \left(\frac{1}{[Cl^-]} + \beta_1 + \beta_2[Cl^-] + \beta_3[Cl^-]^2 + \beta_4[Cl^-]^3 \right). \qquad (7.10)$$

The first two of the terms inside the parentheses appeared in Section 5.5; the last three arise from the anionic complexes. As the concentration of chloride ion increases, the first term decreases, the second remains constant, and the last three increase. Figure 7.3 shows the result. If the concentration of chloride ion is very low, increasing it slightly causes the solubility of silver chloride to decrease. This is because the complexes $AgCl_2^-$, $AgCl_3^{2-}$, and $AgCl_4^{3-}$ account for only a very small fraction of the total concentration of dissolved silver(I) if the chloride-ion concentration is small, as was shown by Fig. 7.1. Their concentrations increase as the concentration of chloride ion rises, but these increases are small and are overwhelmed by the decrease that occurs

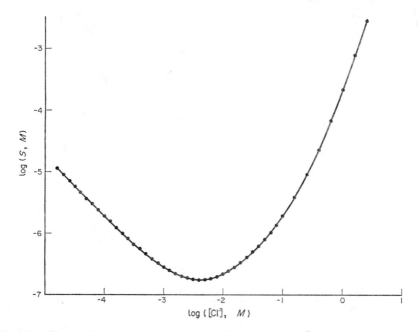

FIG. 7.3. The effect of the total concentration of chloride ion on the solubility of silver chloride.

in the concentration of silver ion, which is the major constituent of the solution. At higher concentrations of chloride ion exactly the opposite is true, and increasing the concentration of chloride ion increases the concentrations of the anionic complexes much more than it decreases the tiny concentration of silver ion.

Consequently the solubility of silver chloride goes through a minimum as the concentration of chloride ion increases. The concentration of free chloride ion at this minimum may be found by differentiating eq. (7.10) and setting $dS/d[Cl^-]$ equal to zero.

$$\frac{dS}{d[Cl^-]} = K_{AgCl}\left(-\frac{1}{[Cl^-]^2} + \beta_2 + 2\beta_3[Cl^-] + 3\beta_4[Cl^-]^2\right) = 0. \qquad (7.11)$$

This equation is satisfied by $[Cl^-] = 4\times10^{-3}$ M. The solubility of silver chloride at the minimum may be found by substituting $[Cl^-] = 4\times10^{-3}$ M into eq. (7.10); the result is $S = 1.8\times10^{-7}$ M, which is exactly twice the limiting value that was predicted by ignoring the existence of the anionic complexes. The concentrations of the individual species at the minimum may be found by evaluating the successive terms between the parentheses in eq. (7.10); they are

Species	Ag^+	$AgCl(aq)$	$AgCl_2^-$	$AgCl_3^{2-}$	$AgCl_4^{3-}$
Concentration, M	4.5×10^{-8}	9.0×10^{-8}	4.4×10^{-8}	9×10^{-10}	1×10^{-11}

The uncharged complex $AgCl(aq)$ accounts for half of the solubility at the minimum, while Ag^+ and $AgCl_2^-$ each account for a quarter of it. The concentration of chloride ion is so low that only traces of $AgCl_3^{2-}$ and $AgCl_4^{3-}$ are present.

Adding chloride ion to a saturated solution of silver chloride decreases the solubility of the silver chloride only if the concentration of free chloride ion is below 4×10^{-3} M. At higher concentrations of chloride ion the solubility of silver chloride rises again, and becomes very large if the concentration of chloride ion is very high. The following values may be obtained from eq. (7.10):

$[Cl^-]$, M	0.5	1	2	3	5	10
S_{AgCl}, M	4×10^{-5}	1.7×10^{-4}	1.5×10^{-3}	4.9×10^{-3}	2.2×10^{-2}	0.17

Activity effects (Chapter 13) were ignored in calculating these values, and somewhat different ones would have been obtained if those effects had been taken into account. Nevertheless, the experimentally observed fact is that the solubility of silver chloride is enormously increased by the presence of a large excess of chloride ion. Even in a solution containing 0.5 M chloride ion the solubility is several times as high as it is in pure water; with 2 M chloride ion it is about 100 times as large as in water; with 10 M chloride ion it is about 10^4 times as large as in water.

All these numerical values are different for different precipitates, but there are few precipitates that do not redissolve, to some extent at least, if too large an excess of reagent is added. The common-ion effect is a very incomplete and unreliable description of the behavior that is actually observed.

SUMMARY: Because of complex formation, most precipitates are appreciably soluble in solutions containing high concentrations of one of their ions. If you are trying to precipitate a compound ML by adding L ions to a solution containing M ions, a small excess of L is usually beneficial because it represses the solubility of ML, but a large excess is likely to redissolve some of the ML.

7.5. Will a precipitate form?

PREVIEW: There are two somewhat different ways in which you can decide whether a compound ML will precipitate on mixing M and L ions at known total concentrations if they react to form complexes.

It is often necessary to decide whether, and at what point, a precipitate will appear in a chemical system, and this can be done with the aid of the ideas and equations developed in the preceding sections.

A chemist adds 1×10^{-6} mole of silver ion to 1 dm^3 of a solution containing 0.1 M bromide. Will silver bromide precipitate? The solubility product of silver bromide is $K_{AgBr} = 5.2 \times 10^{-13}$ mol^2 dm^{-6} and the stepwise formation constants of the complexes AgBr(aq), $AgBr_2^-$, $AgBr_3^{2-}$, and $AgBr_4^{3-}$ are $K_1 = 1.6 \times 10^4$ mol^{-1} dm^3, $K_2 = 1 \times 10^3$ mol^{-1} dm^3, $K_3 = 6.3 \times 10^1$ mol^{-1} dm^3, and $K_4 = 2$ mol^{-1} dm^3.

One approach is to say that the concentration of silver ion would be 10^{-6} M, and that the product $[Ag^+][Br^-]$ would be equal to 10^{-7} mol^2 dm^{-6}, if a precipitate of silver bromide did not form. Since this product would be much larger than K_{AgBr}, it would be concluded that silver bromide must precipitate. The approach and the conclusion are wrong because they ignore the complexes that are formed.

The complexes could be taken into account by using an equation like eq. (7.10) to calculate the solubility of silver bromide:

$$S = K_{AgBr} \left(\frac{1}{[Br^-]} + \beta_1 + \beta_2[Br^-] + \beta_3[Br^-]^2 + \beta_4[Br^-]^3 \right).$$

With $[Br^-] = 0.1$ M, $\beta_1 = K_1 = 1.6 \times 10^4$ mol^{-1} dm^3, $\beta_2 = K_1K_2 = 1.6 \times 10^7$ mol^{-2} dm^6, $\beta_3 = K_1K_2K_3 = 1 \times 10^9$ mol^{-3} dm^9, and $\beta_4 = K_1K_2K_3K_4 = 2 \times 10^9$ mol^{-4} dm^{12}, this gives $S = 7.1 \times 10^{-6}$ M. Thus 7.1×10^{-6} mole of silver bromide would dissolve in 1 dm^3 of the solution, and the total concentration of silver(I) in a saturated solution would be 7.1×10^{-6} M. Consequently the solution will not be saturated if it contains silver(I) at a total concentration of only 1×10^{-6} M, and a precipitate of silver bromide cannot be present when equilibrium is reached.

An equivalent approach would use an equation like eq. (7.7a) to calculate the fraction of the dissolved silver(I) that will be present as Ag^+. The result is

$$\alpha_{Ag^+} = \frac{1}{1 + \beta_1[Br^-] + \beta_2[Br^-]^2 + \beta_3[Br^-]^3 + \beta_4[Br^-]^4} = \frac{1}{1.36 \times 10^6} = 7.3 \times 10^{-7}.$$

Assuming that silver bromide does not precipitate, the total concentration of dissolved silver(I), c_{Ag^+}, will be equal to 1×10^{-6} M; the concentration of free silver ion will be equal to $(7.3 \times 10^{-7})(1 \times 10^{-6}) = 7.3 \times 10^{-13}$ M; and the product $[Ag^+][Br^-]$ will be equal to $(7.3 \times 10^{-13})(0.1) = 7.3 \times 10^{-14}$ mol^2 dm^{-6}. As this is smaller than K_{AgBr} ($= 5.2 \times 10^{-13}$ mol^2 dm^{-6}), silver bromide will not precipitate. This conclusion is consistent with the assumption that was made to obtain it.

Either of these approaches would enable us to find the total concentration of silver(I) that would have to be added in order to yield a precipitate. The first one gives the answer directly: since as much as 7.1×10^{-6} mole of silver bromide can be dissolved in 1 dm³ of the solution, 7.1×10^{-6} mole of silver(I) can be added to 1 dm³ of it without causing precipitation. At that point the solution is saturated with silver bromide, and consequently a precipitate will form if any larger amount of silver(I) is added. If we choose the second approach, we must begin by calculating the concentration of free silver ion at which the solution just becomes saturated. This is given by $[Ag^+] = K_{AgBr}/[Br^-] = 5.2 \times 10^{-13}/0.1 = 5.2 \times 10^{-12} M$. Since $[Ag^+] = \alpha_{Ag+}c_{Ag+}$, the corresponding total concentration of silver(I) is given by $c_{Ag+} = [Ag^+]/\alpha_{Ag+} = 5.2 \times 10^{-12}/7.3 \times 10^{-7} = 7.1 \times 10^{-6} M$. Again this is the highest concentration of silver(I) that could be added without causing precipitation.

> SUMMARY: If M and L ions react to form complexes, the equilibrium concentration of M in a solution containing L will be smaller than the total concentration of M, and the stabilities of the complexes as well as the solubility product of ML must be taken into account in deciding whether ML will precipitate.

7.6. Formation curves and the evaluation of formation constants

> PREVIEW: This section describes several ways in which data on the equilibrium compositions of solutions containing a metal ion M and a ligand L can be used to identify the complexes that are formed and to evaluate their formation constants.

Another kind of graphical representation of the behaviors of systems like these is provided by the *formation curve*. We define a quantity $\bar{n}$ as the average number of ions or molecules of ligand that are bound to a single metal ion. A formation curve is a plot of $\bar{n}$ against the logarithm of the concentration of free ligand. Formation curves are of interest because the formation constants are often easy to obtain from them, and also because they are closely related to the titration curves that will be discussed in Chapter 11.

The simplest kind of formation curve is the one obtained for a system that involves only a single complex. Let us consider the general reaction

$$M + L = ML(aq); \qquad K_1 = [ML(aq)]/[M][L] \qquad (7.12)$$

in which M represents a metal ion and L a ligand, which might be a chelon. As in some of the preceding equations, charges are omitted for simplicity. The symbols c_M and c_L will be used to represent the *total* concentrations of M or L, respectively. For any particular mixture the values of c_M and c_L are easy to calculate from the known amounts of M and L that were used in preparing it. There are a number of different ways in which it might be possible to evaluate the concentration of M, L, or ML that is present at equilibrium. Often one of these will absorb visible or ultraviolet radiation of a wavelength to which the other two are transparent, and measuring the extent to which radiation of that wavelength is absorbed in passing through the mixture will then make it possible to deduce the concentration of the species that is responsible for the absorption (Chapter 15). In other cases it is possible to find an indicator electrode whose electrical potential depends on the concentration of M or L (Chapter 14). There are many other possibilities, and once any of the three equilibrium concentrations is known the other two can be calculated from the equations

$$c_M = [M] + [ML], \tag{7.13a}$$
$$c_L = [L] + [ML]. \tag{7.13b}$$

The quantity $\bar{n}$, which is called the *ligand number*, is given for this reaction by

$$\bar{n} = [ML]/c_M \; (= \alpha_{ML}). \tag{7.14}$$

If $\bar{n} = 0$, then $[ML] = 0$ and there is no L bound to any of the M ions; if $\bar{n} = 1$, then $[ML] = c_M$ and every M ion has one ion or molecule of L bound to it. Obtaining the values of $\bar{n}$ for many different mixtures and plotting them against log [L] gives the formation curve, which will have the shape shown in Fig. 7.4.

The curves in Fig. 7.4 differ because they are drawn for different values of K_1, but their similarities are more important than their differences. For each one the value of $\bar{n}$

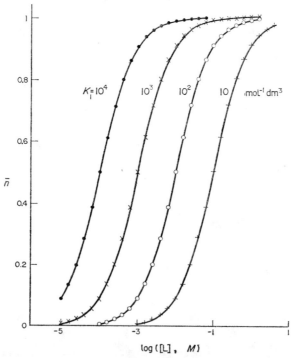

FIG. 7.4. Formation curves showing how $\bar{n}$, the average number of ligands bound to each metal ion, varies with the concentration of free ligand if only a single complex ML is formed. Different curves represent different values of the formation constant K_1.

approaches 0 at very small concentrations of L and approaches 1 at very large ones. For each one the width of this range is the same as for any other, which is to say that all these curves have the same shape. This is because eqs. (7.14), (7.13a), and (7.12) can be combined to give

$$\bar{n} = \frac{[ML]}{[M]+[ML]} = \frac{K_1[M][L]}{[M]+K_1[M][L]} = \frac{K_1[L]}{1+K_1[L]} \qquad (7.15)$$

from which it can be seen that changing the value of K_1 merely shifts the curve along the log [L] axis but does not affect the form of the dependence on [L] or log [L].

Setting $\bar{n}$ equal to 1/2 in eq. (7.15) and solving for $K_1[L]$ gives $K_1[L] = 1$, which means that $K_1 = 1/[L]$, or log $K_1 = $ pcL, at the point where $\bar{n} = 1/2$. This is a very easy way to obtain the value of K_1 from experimental data that show how the concentration of M or ML varies with the concentration of free L. Another way, which is more complicated but which uses the data much more efficiently and provides a much more reliable value, is suggested by the general solution of eq. (7.15) for $K_1[L]$:

$$K_1[L] = \frac{\bar{n}}{1-\bar{n}}$$

which may be rewritten as

$$\log \frac{\bar{n}}{1-\bar{n}} = \log K_1 + \log [L]. \qquad (7.16)$$

A plot of log $[\bar{n}/(1-\bar{n})]$ against log [L] will be a straight line. Its slope will be equal to 1 and the point at which log $[\bar{n}/(1-\bar{n})] = 0$ will be the point at which log $K_1 = -$ log [L]. Figure 7.5 shows such a plot for one of the curves of Fig. 7.4.

Very different behavior is observed when stepwise complex formation occurs. If there are two complexes, ML and ML_2, the system is described by the equations

$$M + L = ML; \qquad K_1 = [ML(aq)]/[M][L], \qquad (7.17a)$$
$$ML + L = ML_2; \qquad K_2 = [ML_2]/[ML(aq)][L], \qquad (7.17b)$$

and the conservation equations become

$$c_M = [M]+[ML]+[ML_2], \qquad (7.18a)$$
$$c_L = [L]+[ML]+2[ML_2] \qquad (7.18b)$$

while $\bar{n}$ is given by

$$\bar{n} = \frac{[ML]+2[ML_2]}{c_M} \qquad (= \alpha_{ML}+2\alpha_{ML_2}). \qquad (7.19)$$

The factors of 2 that appear in eqs. (7.18b) and (7.19) reflect the fact that ML_2 contains two ions or molecules of the ligand bound to each ion of M. Comparing eq. (7.19) with eq. (7.7b) enables us to write

$$\bar{n} = \frac{K_1[L]+2K_1K_2[L]^2}{1+K_1[L]+K_1K_2[L]^2}. \qquad (7.20)$$

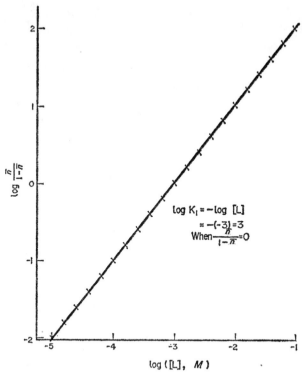

$$\log K_1 = -\log [L]$$
$$= -(-3) = 3$$
$$\text{When} \frac{\bar{n}}{1-\bar{n}} = 0$$

FIG. 7.5. Plot of $\log \bar{n}/(1-\bar{n})$ against the concentration of free ligand, constructed from the data shown in the curve for $K_1 = 10^3 \text{ mol}^{-1} \text{ dm}^3$ in Fig. 7.4.

The corresponding equation (eq. (7.15)) for a system containing only a single complex gave rise to curves that had the same shape and differed only in their positions along the log [L] axis for different values of K_1. Equation (7.18) yields more complicated curves because the terms containing $[L]^2$ do not have the same coefficients as those containing [L]. It can be recast in a more revealing form by dividing both the numerator and the denominator of its right-hand side by $\sqrt{K_1 K_2}$, which yields

$$\bar{n} = \frac{\sqrt{K_1/K_2}[L] + 2\sqrt{K_1 K_2}[L]^2}{1/\sqrt{K_1 K_2} + \sqrt{K_1/K_2}[L] + \sqrt{K_1 K_2}[L]^2} . \qquad (7.21)$$

This shows that the shape of the curve depends on both the ratio K_1/K_2 and the product $K_1 K_2$ of the two formation constants. Figure 7.6 shows the effect of changes in $K_1 K_2$ while K_1/K_2 remains constant: as in Fig. 7.4, the curves have the same shape and are merely shifted along the log [L] axis. The curves in Fig. 7.6 are less steep than those in Fig. 7.4: the change of $\bar{n}$ from zero to its limiting value always occurs over a wider range of values of log [L] for stepwise complex formation than it does when only one complex is formed. Figure 7.7 shows the effect of changes in K_1/K_2 while K_1/K_2 remains constant. Here the curve changes shape while the value of log [L] at

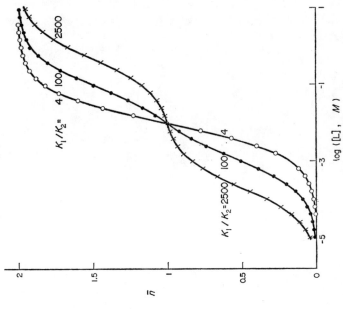

Fig. 7.7. Formation curves showing how $\bar{n}$ varies with the concentration of free ligand if two complexes, ML and ML_2, are formed. Different curves represent different values of the ratio K_1/K_2 of the stepwise formation constants of the complexes, but their product K_1K_2 is the same for all the curves and is equal to 1×10^4 mol^{-2} dm^6. The curve labeled "$K_1/K = 100$" in this figure is identical with the one labeled "$K_1K_2 = 10^{4}$" in Fig. 7.6.

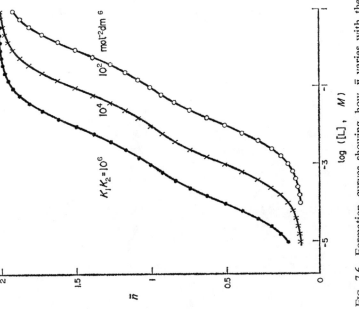

Fig. 7.6. Formation curves showing how $\bar{n}$ varies with the concentration of free ligand if two complexes, ML and ML_2, are formed. Different curves represent different values of the product $K_1K_2 (= \beta_2)$ of the stepwise formation constants of the complexes, but their ratio K_1/K_2 is the same for all the curves and is equal to 100.

its midpoint (where $\bar{n} = 1$) remains unaffected. If K_1/K_2 is very large there is a wide range of concentrations of L over which n is very close to 1; throughout this range the formation of ML is virtually complete whereas that of ML_2 is inappreciable. As K_1/K_2 decreases this range becomes narrower, until finally the two steps overlap so much that they appear to merge into a single overall reaction.

Multiplying both sides of eq. (7.20) by $(1+K_1[L]+K_1K_2[L]^2)$ and collecting the terms containing identical powers of [L] gives

$$\bar{n} = (1-\bar{n})\,K_1[L] + (2-\bar{n})\,K_1K_2[L]^2$$

from which

$$\frac{\bar{n}}{(1-\bar{n})[L]} = K_1 + \frac{2-\bar{n}}{1-\bar{n}}\,K_1K_2[L]. \tag{7.22}$$

This provides a way of evaluating both K_1 and K_2. After [L] (the concentration of *free* L) and $\bar{n}$ have been determined in each of a number of different equilibrium mixtures, $\bar{n}/\{(1-\bar{n})[L]\}$ is plotted against $(2-\bar{n})[L]/(1-\bar{n})$. The plot is a straight line having a slope equal to K_1K_2, and its intercept on the ordinate axis [where $(2-\bar{n})[L]/(1-\bar{n}) = 0$] is equal to K_1.

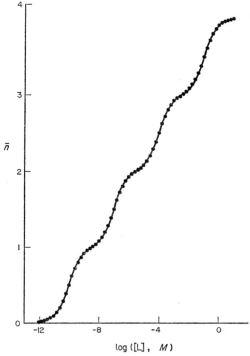

FIG. 7.8. Formation curve showing how $\bar{n}$ varies with the concentration of free ligand for a hypothetical system in which there are four complexes: ML $(K_1 = 1\times10^{10})$, ML_2 $(K_2 = 1\times10^7)$, ML_3 $(K_3 = 1\times10^4)$, and ML_4 $(K_4 = 10)$. Each value of K has the units mol^{-1} dm^3.

If more than two complexes are formed the equations become still more complicated. For the silver–bromide system the formation curve is described by the equation

$$\bar{n} = \frac{[\text{AgBr}(aq)] + 2[\text{AgBr}_2^-] + 3[\text{AgBr}_3^{2-}] + 4[\text{AgBr}_4^{3-}]}{c_{\text{Ag}}}$$

$$= \frac{\beta_1[\text{Br}^-] + 2\beta_2[\text{Br}^-]^2 + 3\beta_3[\text{Br}^-]^3 + 4\beta_4[\text{Br}^-]^4}{1 + \beta_1[\text{Br}^-] + \beta_2[\text{Br}^-]^2 + \beta_3[\text{Br}^-]^3 + \beta_4[\text{Br}^-]^4} \cdot$$

If each of the successive formation constants were very much larger than the next one the curve would have the shape shown in Fig. 7.8: there would be four plateaus (at $\bar{n} = 1$, 2, 3, and 4), and the values of K would be easy to find. For example, K_1 is equal to the reciprocal of the concentration of free bromide ion when $\bar{n}$ is equal to 1/2. For the silver–bromide complexes, however, both K_1/K_2 and K_2/K_3 are equal to only 16, while K_3/K_4 is equal to only 31, and the curve, which is shown in Fig. 7.9, is completely featureless. There is no obvious indication of the formation of any of the intermediate complexes; and because activity effects become important at the highest concentrations of bromide ion and are difficult to take into account, it might even be difficult to decide whether such a curve really has an asymptote at

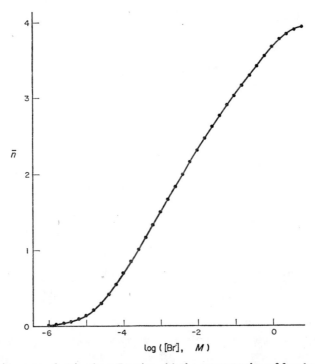

FIG. 7.9. Formation curve showing how $\bar{n}$ varies with the concentration of free bromide ion in the silver(I)–bromide system, in which there are four complexes: AgBr(aq) ($K_1 = 1.6 \times 10^4$ mol^{-1} dm^3), AgBr$_2^-$ ($K_2 = 1.0 \times 10^3$ mol^{-1} dm^3), AgBr$_3^{2-}$ ($K_3 = 6 \times 10^1$ mol^{-1} dm^3), and AgBr$_4^{3-}$ ($K_4 = 2$ mol^{-1}dm^3).

$\bar{n} = 4$ or whether traces of some still higher complexes might be formed as well. An enormous amount of research has been done on such systems, but there are many questions that remain unsettled and much more work is needed.

SUMMARY: A formation curve is a plot of the ligand number $\bar{n}$ against the concentration of free ligand. With the aid of formation curves it is often easy to identify the complexes that are formed and to evaluate their formation constants.

7.7. Acidity and complex formation

PREVIEW: Most ligands are Lowry–Brønsted bases of appreciable strength and can therefore react with hydrogen ions as well as metal ions. This section gives a qualitative description of the resulting behavior.

Ligands owe their coordinating ability to the fact that they are Lewis bases, possessing electron pairs that they can share with metal ions. They can also behave as Lowry–Brønsted bases, sharing these electron pairs with hydrogen ions to form their conjugate acids. Since every aqueous solution must contain some hydrogen ions, there is always a competition between metal ions and hydrogen ions for possession of a ligand. Sometimes the competition is one-sided, as it is with ligands such as the halide ions and thiocyanate, which are such weak Lowry–Brønsted bases that they have no appreciable tendency to react with hydrogen ion. For such ligands the effect of acidity on complexing ability can be ignored, but there are few ligands for which this is possible, and reactions like $ML + H_3O^+ = M(OH_2) + HL$ almost always have to be taken into account.

The situation is further complicated by the fact that metal ions are acidic, so that reactions like $M(OH_2)_6^{n+} + OH^- = M(OH)(OH_2)_5^{(n-1)+} + H_2O(l)$ can also occur. Because they do, there is always a competition between a ligand and hydroxide ion for the possession of a metal ion. All other things being equal, the likelihood that this competition will be won by hydroxide ion increases as the Lowry–Brønsted acidity of the metal ion increases. The acidic strengths of metal ions differ widely, as do the basic strengths of ligands. In general, the acidic strengths of metal ions increase as the charge n increases. Univalent ions such as the alkali-metal ions are weaker acids than divalent ions such as the alkaline earth metal ions, and these in turn are much weaker acids than tri- and higher-valent ions such as aluminum and iron(III) ions. Like the chlorine atom in a molecule of chloroacetic acid (Section 6.6), a metal ion tends to attract electrons. By doing so it weakens the oxygen–hydrogen bonds in the molecules of water in its inner coordination sphere, and thereby increases the ease with which the hydrated ion can lose protons to Lowry–Brønsted bases. This electrostatic effect becomes greater as the charge on the metal ion increases. It also becomes greater as the radius of the anhydrous metal ion decreases, because molecules of water can approach a small metal ion more closely than a large one and because the strength of the electric field to which they are subjected in the inner coordination sphere increases as their distance from the metal ion decreases. For this reason magnesium ion (whose radius r when anhydrous is 0.66 Å) is a much stronger acid than barium ion (for which $r = 1.34$ Å), and the fact that the radius of a metal ion decreases as its charge increases (so that $r_{Fe^{2+}} = 0.74$ Å while $r_{Fe^{3+}} = 0.64$ Å, and $r_{Sn^{2+}} = 0.93$ Å while $r_{Sn^{4+}} = 0.71$ Å) contributes materially to the effect of charge on acidity. Of course there are also effects attributable to the electronic structure of the metal ion, so that the lead(II) ion (for which $r = 1.20$ Å) is a much stronger acid than magnesium ion although it has the same charge and a much larger radius.

Similar though they are in nature, the basicity of the ligand and the acidity of the metal ion have rather different effects. That of the basicity of the ligand is the simpler of the two. When a ligand reacts with hydrogen ion, some of its conjugate acid is produced and the concentration of the free unprotonated ligand decreases. This simply causes the extent of the reaction between the ligand and the metal ion to decrease. When a metal ion reacts with hydroxide ion, the formation of $M(OH)^{(n-1)+}$ similarly decreases the extent of its reaction with the ligand, but it also gives rise to the possibility that $M(OH)^{(n-1)+}$ may react with the ligand to yield a mixed complex such as $M(OH)L$.

This discussion will be confined to the effects that arise from the basicity of the ligand. It will illustrate them by describing the behavior of the copper(II)–triethanolamine system. Triethanolamine has the formula $N(CH_2CH_2OH)_3$. Like ammonia, it is a weak base. Its basicity can be described by means of the acidic dissociation constant of its conjugate acid, triethanolammonium ion:

$$^+HN(CH_2CH_2OH)_3 + H_2O(l) = H_3O^+ + N(CH_2CH_2OH)_3(aq);$$
$$K_a = 1.73 \times 10^{-8} \text{ mol dm}^{-3}. \qquad (7.23a)$$

The reaction between unprotonated triethanolamine and copper(II) ion can be described by the equation

$$Cu^{2+} + N(CH_2CH_2OH)_3(aq) = Cu[N(CH_2CH_2OH)_3]^{2+};$$
$$K_1 = 1.6 \times 10^4 \text{ mol}^{-1} \text{ dm}^3. \qquad (7.23b)$$

Other processes may occur as well but will not be considered here.

For brevity we may represent triethanolamine, triethanolammonium ion, and the copper(II)–triethanolamine complex by L, HL^+, and CuL^{2+}, respectively. For convenience we may define a quantity $[L]'$ by the equation

$$[L]' = c_L - [CuL^{2+}] \qquad (7.24)$$

which should be compared with eq. (7.13b). The symbol c_L denotes the total concentration of the ligand, as does c_L in eq. (7.13b); and $[CuL^{2+}]$ denotes the concentration of the complex, as does $[ML]$ in eq. (7.13b). The quantity $[L]'$ represents the total concentration of triethanolamine that is not bound to copper(II) ion, just as $[L]$ in eq. (7.13b) represented the concentration of the ligand that was not bound to the metal ion M. The difference between the two situations is simply that $[L]'$ in eq. (7.24) represents the sum of the concentrations of free triethanolamine and its conjugate acid

$$[L]' = [L] + [HL^+] \qquad (7.25)$$

because any triethanolamine that is not bound to copper(II) ion may be present in either of these forms, whereas $[L]$ in eq. (7.13b) represented the concentration of free ligand because protonation was ignored there.

Combining eq. (7.25) with eq. (6.48) gives

$$[L] = \alpha_L [L]' \qquad \text{where} \qquad \alpha_L = \frac{K_a}{K_a + [H_3O^+]} \qquad (7.26)$$

which in turn may be combined with the expression for K_1, the formation constant of the complex, to yield

$$\frac{[CuL^{2+}]}{[Cu^{2+}][L]'} = \alpha_L K_1 = \frac{\bar{n}}{(1-\bar{n})[L]'} . \qquad (7.27)$$

This is exactly analogous to eq. (7.16), except that it contains the product $\alpha_L K_1$ in place of K_1 alone. The value of α_L depends on the experimental conditions—and in particular on the pcH—and therefore the value of $\alpha_L K_1$ also depends on these conditions. Often the product $\alpha_L K_1$ is called a *conditional formation constant*. Equation (7.27) may be solved for $\bar{n}$ to obtain

$$\bar{n} = \frac{\alpha_L K_1 [L]'}{1 + \alpha_L K_1 [L]'} \qquad (7.28)$$

which is just like eq. (7.15) except that $\alpha_L K_1$ has again replaced K_1.

Figure 7.4 showed how variations of the formation constant affect the formation curve when protonation of the ligand does not occur. Variations of the conditional constant $\alpha_L K_1$ have exactly parallel effects when protonation does occur. It is the effect of pcH that is most important. If the pcH is much smaller than pK_a, α_L is very small (eq. 7.26), and then the value of the conditional formation constant is also small. Almost all of the triethanolamine is bound to hydrogen ion, and very little of it is available to react with copper(II) ion. Hardly any complexation can occur unless an enormous excess of triethanolamine is present. As the pcH increases the value of the conditional formation constant increases and the complex forms more readily. However, the value of α_L cannot exceed 1, and therefore the conditional formation constant cannot exceed K_1. If $pcH = pK_a$, the conditional constant is equal to $0.5 K_1$; if $pcH = pK_a + 1$, the conditional constant is equal to $0.91 K_1$. As the pcH is increased above the latter value, α_L continues to increase toward 1, but its increase is too slight to have much effect on the formation curve. This means that the formation curve is affected by pcH if $pcH < pK_a + 1$, but is virtually independent of pcH in more strongly basic solutions. Since $pK_a = 7.8$ for the triethanolammonium ion, the reaction

$$Cu[N(CH_2CH_2OH)_3]^{2+} + H_3O^+ = Cu^{2+} + {}^+HN(CH_2CH_2OH)_3 + H_2O(l);$$
$$K = 1/K_a K_1 = 3.6 \times 10^3$$

begins to become important at about $pcH = 9.0$, and the complex becomes less and less stable as the pcH decreases below this value. However, increasing the pcH above 9.0 has little effect on the stability of the complex because nearly all of the triethanolamine is already in the unprotonated form at this pcH.

SUMMARY: When a basic ligand L reacts with a metal ion M to form a complex ML, the formation curve is affected by changes of pcH. There is no appreciable effect as long as the pcH is larger than about $(pK_a + 1)$, where K_a is the overall acidic dissociation constant of the protonated form HL of the ligand. In more acidic solutions hydrogen ions can compete with M ions for possession of the ligand, and the formation curve shifts to lower pcH-values. The acidity of metal ions in briefly discussed, but the further complications that result from it are not described.

Problems

Answers to some of these problems are given on page 520.

7.1. The formation constant of $CuBr^+$ is given by $\log K_1 = 0.3$. Exactly 0.1 mole of potassium bromide is added to 1 dm^3 of a solution containing 1.00×10^{-3} M copper(II) perchlorate. What fraction of the copper(II) will be present as $CuBr^+$ when equilibrium is reached?

7.2. The stepwise formation constants of the copper(II)–ammonia complexes are given by $\log K_1 = 4.3$, $\log K_2 = 3.7$, $\log K_3 = 3.0$, $\log K_4 = 2.3$, and $\log K_5 = -0.5$.

(a) Over what range of concentrations of ammonia would the concentration of $Cu(NH_3)_3^{2+}$ be larger than the concentration of any other copper–ammonia complex?

(b) What would be the largest fraction of the copper(II) in a solution that could be converted into $Cu(NH_3)_3^{2+}$?

7.3. The values of the stepwise formation constants of the copper(II)–oxalate complexes are $K_1 = 1.6 \times 10^6$ mol^{-1} dm^3 and $K_2 = 60$ mol^{-1} dm^3. The solubility product of copper(II) oxalate is equal to 2.3×10^{-8} mol^2 dm^{-6}.

(a) What is the solubility of copper(II) oxalate in pure water?

(b) What would be the concentration of oxalate ion that would cause the solubility of copper(II) oxalate to attain its minimum value?

(c) Compare the solubility of copper(II) oxalate in a solution containing 0.10 M free oxalate ion with its solubility in pure water.

7.4. The solubility product of magnesium oxalate is equal to 1.0×10^{-8} mol^2 dm^{-6} and the formation constants of the magnesium–oxalate complexes are given by $\log K_1 = 3.4$ and $\log K_2 = 1.0$. A chemist adds 1.0×10^{-4} mole of magnesium ion to 1 dm^3 of a solution containing 0.10 M oxalate ion. Will magnesium oxalate precipitate?

7.5. The ion pair $TlCl(aq)$ is so unstable in aqueous solutions at 298 K that its formation constant is too small to evaluate, but the following values have been obtained in solutions saturated with thallium(I) chloride:

$$TlCl(s) = Tl^+ + Cl^-; \qquad K(= K_{TlCl}) = 1.7 \times 10^{-4} \ mol^2 \ dm^{-6}$$
$$TlCl(s) + Cl^- = TlCl_2^-; \qquad K = 1.8 \times 10^{-4}$$
$$TlCl(s) + 2 \ Cl^- = TlCl_3^{2-}; \qquad K = 2.0 \times 10^{-4} \ mol^{-1} \ dm^3.$$

(a) What concentration of thallium(I) must be added to a solution containing hydrochloric acid at a total concentration of 0.100 M to yield a precipitate of thallium(I) chloride?

(b) Plot the solubility of thallium(I) chloride against the concentration of free chloride ion present in the saturated solution.

(c) Plot the logarithm of the solubility of thallium(I) chloride against the logarithm of the concentration of free chloride ion present in the saturated solution.

15

(d) If these plots had been obtained from experimental data, how could they be used to show that $TlCl_3^{2-}$ had been formed?

7.6. If 0.10 mole of sodium thiocyanate is added to 1 dm^3 of a solution containing 1.00×10^{-4} M zinc perchlorate, $Zn(ClO_4)_2$, the concentration of free zinc ion decreases from 1.00×10^{-4} to 2.00×10^{-5} M. If $ZnSCN^+$ is the only complex formed, what are the values of

(a) $\bar{n}$?   (b) The formation constant of $ZnSCN^+$?

7.7. A chemist prepares a number of different solutions containing tin(II) and fluoride ions. Each solution contains 1.00×10^{-3} mol dm^{-3} of tin(II), but different solutions contain different amounts of fluoride. For each solution the first line below gives the number of moles of fluoride used in preparing 1 dm^3 of the solution, while the second line gives the concentration of free fluoride ion that is present at equilibrium:

Total concentration of F^-, M	1.00×10^{-4}	2.00×10^{-4}	4.00×10^{-4}	1.00×10^{-3}	2.00×10^{-3}
Equilibrium concentration of F^-, M	1.0×10^{-5}	2.8×10^{-5}	4.8×10^{-5}	2.6×10^{-4}	1.1×10^{-3}

(a) Plot log $[\bar{n}/(1-\bar{n})]$ against log $[F^-]$, and plot $\bar{n}/\{(1-\bar{n})[F^-]\}$ against $(2-\bar{n})[F^-]/(1-\bar{n})$.
(b) On the basis of these plots decide whether SnF^+ is the only complex formed or whether $SnF_2(aq)$ appears to be formed as well.
(c) Evaluate the formation constant(s) of the complex(es) with the aid of the appropriate plot.
(d) Why is tin(II) fluoride used in toothpaste in preference to, say, potassium fluoride?

7.8. The formation constant of the zinc-glycolate complex, $Zn(OOCCH_2OH)^+$, is equal to 80 mol^{-1} dm^3 and the overall acidic dissociation constant of glycolic acid, $HOCH_2COOH$, is given by $pK_a = 3.83$. What is the concentration of each species present at equilibrium in a solution prepared by dissolving 1.00×10^{-3} mole of zinc(II) perchlorate and 1.00×10^{-2} mole of glycolic acid in enough water to yield 1 dm^3 of solution? Neglect the formation of zinc-perchlorate ion pairs.

CHAPTER 8

Redox Processes

8.1. Introduction

This group of chapters concludes with an examination of redox processes, which are widely important in both nature and human activity. The photosynthetic reactions that occur in plants consist of the reduction of carbon dioxide and the oxidation of water to yield oxygen and various organic compounds and transform the energy of sunlight into forms that living creatures can use. Those reactions are non-spontaneous; the reverse reactions, in which organic matter is oxidized and oxygen reduced, are spontaneous under ordinary conditions and are the basis of human life and civilization alike. They occur in human and other animal metabolism; in furnaces and fireplaces; in steam, gasoline, and diesel engines; and in most electrical generating stations. All the metals except the noble ones are obtained from natural raw materials by redox processes, and they return to nature by corrosion, which is also a redox process. Dry cells, storage batteries, and fuel cells are all based on redox reactions, and localized redox reactions provide the energy that the brain and muscles consume in thought and activity.

8.2. The nature of a redox reaction

PREVIEW: This section defines a number of terms that are important in redox chemistry and stresses the similarity between redox reactions and Lowry–Brønsted acid–base reactions.

A redox reaction is a reaction in which one substance, called a *reducing agent*, donates electrons to another substance, called an *oxidizing agent*. The oxidizing agent gains electrons and is said to be *reduced* by the reducing agent, while the reducing agent loses electrons and is said to be *oxidized* by the oxidizing agent. A typical redox reaction is the one described by the equation $Ag(s) + Fe^{3+} = Ag^+ + Fe^{2+}$. Each atom of silver, the reducing agent, donates an electron to an iron(III) ion, the oxidizing agent, and the result is that silver is oxidized and iron(III) ion is reduced. If this reaction is allowed to come to equilibrium its rate will become equal to that of the reverse reaction, in which silver ion behaves as an oxidizing agent and accepts an electron from an iron(II) ion, which behaves as a reducing agent. In the forward reaction one oxidizing agent (Fe^{3+}) and one reducing agent $(Ag(s))$ are consumed, and another oxidizing agent (Ag^+) and another reducing agent (Fe^{2+}) are formed. In any redox reaction one oxidizing agent reacts with one reducing agent to form a different oxidizing agent and a different reducing agent. The oxidizing agent that is formed is called the *oxidized form* of the reducing agent that is consumed, so that silver ion is the oxidized form of elemental silver, and the reducing agent that is formed is called the *reduced form* of the oxidizing agent that is consumed, so that iron(II) ion is the reduced form of iron(III) ion. An oxidizing agent and its reduced form together are called a *redox couple*.

Every one of these statements is exactly parallel to one that was made in Section 6.2 about Lowry–Brønsted acids and bases. Redox reactions and Lowry–Brønsted acid-base reactions differ merely in that redox reactions involve the transfer of electrons whereas Lowry–Brønsted acid-base reactions involve the transfer of protons. The similarity between these two kinds of reactions is one of the main themes of this chapter.

To pursue that similarity a little farther, a reaction between an acid and a base can be described by dividing the overall equation, say $NH_3(aq) + CH_3COOH(aq) = NH_4^+ + CH_3COO^-$, into two parts, one for each of the two conjugate acid-base pairs involved:

$$NH_3(aq) + H_3O^+ = NH_4^+ + H_2O(l),$$
$$CH_3COOH(aq) + H_2O(l) = H_3O^+ + CH_3COO^-.$$

The first of these depicts the ability of the base to accept protons, and the second depicts the ability of the acid to donate them. Similarly, a redox reaction can be described by dividing its overall equation, say $Ag(s) + Fe^{3+} = Ag^+ + Fe^{2+}$, into two half-reactions, one for each of the redox couples involved:

$$Fe^{3+} + e = Fe^{2+},$$
$$Ag(s) = Ag^+ + e,$$

in which the symbol "e" denotes the electron. Some chemists prefer to write "e^-" to make sure that the negative charge is not overlooked. The first of these half-reactions depicts the ability of the oxidizing agent to accept electrons, and the second depicts the ability of the reducing agent to donate them. Different acids and bases have different strengths, and the strengths of an acid and its conjugate base can be described quantitatively by giving the value of a quantity, either K_a or pK_a, that is a constant for any particular conjugate acid-base pair at any particular temperature and pressure. Different oxidizing and reducing agents have different strengths, and the strengths of an oxidizing agent and its reduced form can be described quantitatively by giving the value of a quantity that is a constant for any particular redox couple at any particular temperature and pressure.

If that constant were the equilibrium constant of a half-reaction the parallel would be exact. Indeed, there is a strong and influential body of chemical opinion in favor of that choice as this is written, and it may eventually become the prevailing one because it makes equilibrium calculations very simple. Only a short section (Section 8.10) is devoted to it here, however, because it is not the general choice now, and many chemists are not yet aware of its advantages. Although the free electron is reasonably stable in a few solvents such as liquid ammonia, it is much too reactive to survive for long in most others. Not only can it react with dissolved oxidizing agents, but it can also decompose by reactions such as $H_2O(l) + e = H^{\cdot\dagger} + OH^-$ or $(C_2H_5)_2O + e = C_2H_5O^- + {}^{\cdot}CH_2CH_3$, which are followed by others like $2H^{\cdot} = H_2(g)$ or, in solutions

containing other solutes, like $CH_3C\!\!\begin{array}{c}\diagup O \\ \diagdown H\end{array} + {}^{\cdot}H = CH_3\dot{C}\!\!\begin{array}{c}\diagup OH \\ \diagdown H\end{array}$. Consequently an

expression like $K = [Fe^{2+}]/[Fe^{3+}][e]$, for the reduction of iron(III) ion, involves a species (the dissolved electron) that is known to be unstable, and this puts it on a different footing from one like $K_a = [H^+][OAc^-]/[HOAc]$ for the "dissociation" of an acid, where the proton is solvated but does not decompose chemically.

The constant that is generally used to describe the strengths of the oxidizing agent and reducing agent in any particular couple is instead the *standard potential E^0* of that couple. The standard potential will be defined in Section 8.4.

Redox reactions also have much in common with Lewis acid-base reactions. Sections 3.10 and 3.11 discussed the mechanism of the redox reaction

$$I_3^- + H_2O(l) + H_3AsO_3(aq) = H_3AsO_4(aq) + 2H^+ + 3I^-.$$

There is a fast prior equilibrium in which the triiodide ion reacts with water to give HOI, and this is followed by the reaction

$$H\!-\!\underset{\underset{I}{|}}{O}\!: + As(OH)_3(aq) = H\!-\!\underset{\underset{I}{|}}{O}\!:As(OH)_3$$

that yields the activated complex. This is a Lewis acid-base reaction, and of course there is an equilibrium between the activated complex and the reactants from which it is formed. What makes the redox reaction possible is the fact that the activated complex can undergo an internal electronic rearrangement, in which the electron density around the iodine atom increases while that around the arsenic atom decreases. The oxygen–iodine bond is thereby weakened and the oxygen–arsenic bond strengthened, and a different Lewis acid-base equilibrium

$$H\!-\!\underset{\underset{I}{|}}{O}\!:As(OH)_3 = I^- + {}^+H\!-\!O\!:As(OH)_3$$

becomes possible. The overall redox reaction consists of two Lewis acid-base reactions having a common adduct that can have two different electronic structures enabling it to decompose in two different ways.

---

† The dot represents a single unpaired electron, and the symbol $H^{\cdot}$ represents an atom of hydrogen.

At the other extreme, where a Lewis acid-base reaction occurs but a redox reaction does not, the adduct is so unlikely to have any electronic structure except the original one that it can decompose in only one way, regenerating the substances from which it was formed. This is the case with an iron(III)–fluoride complex such as FeF^{2+}; fluoride ion has so little tendency to lose electrons that a structure like Fe: F, which might decompose into iron(II) ions and atoms of fluorine, is wildly improbable. A similar structure is much more likely for the corresponding bromide or thiocyanate complex, $FeBr^{2+}$ or $FeSCN^{2+}$,[†] and such a complex, in which the transfer of one or more electrons from one atom or group of atoms to another is almost complete enough to make reduction and oxidation possible, is called a *charge-transfer complex*. Charge-transfer complexes are in general intensely colored, and are much used in spectrophotometric determinations of metal ions (Chapter 15) for that reason. Going one step farther, the affinity of an iron(III) atom for electrons so far exceeds that of an iodide ion that FeI^{2+} has only a transient existence, and the behavior of the iron(III)–iodide system must therefore be represented by the redox equation $2\,Fe^{3+} + 3\,I^- = 2\,Fe^{2+} + I_3^-$ rather than by the Lewis acid-base equation $Fe^{3+} + I^- = FeI^{2+}$ that would be appropriate if electrons were not transferred between the atoms in FeI^{2+}.

SUMMARY: A redox reaction is a reaction in which one oxidizing agent accepts electrons from one reducing agent, and in which another oxidizing agent and another reducing agent are formed. Two couples are involved: one consists of the original oxidizing agent and its reduced form, and the other consists of the original reducing agent and its oxidized form. Redox reactions differ from Lowry–Brønsted acid-base reactions in that they involve the transfer of electrons rather than protons, and from Lewis acid-base reactions in that they involve internal electron transfer in the adduct.

8.3. Balancing redox equations

PREVIEW: A procedure is described for balancing the equation for a redox reaction of any degree of complexity.

To continue the discussion it will be necessary to write balanced equations for redox reactions and half-reactions of varying degrees of complexity, and this section is inserted here to make sure that you will be able to see how those equations are obtained and able to write others as you need them. No doubt you have already learned a technique for balancing redox equations, and you may feel that you are the master of it. If so, try your hand at the equation for the reaction between copper(I) thiocyanate and iodate ion in a solution containing hydrochloric acid:

$$CuSCN(s) + IO_3^- = Cu^{2+} + SO_4^{2-} + HCN(aq) + ICl_2^-.$$

† It is now believed that this complex is more correctly represented by the formula $FeNCS^{2+}$ because the iron(II) is bonded to nitrogen rather than to sulfur.

Measure the time it takes you. If you obtain the result given at the end of this section within 3 minutes you should go on to Section 8.4; otherwise you will find the technique described here to be better and simpler than the one you are trying to use.

The equation for any redox reaction can be balanced by carrying out the following steps. They are illustrated for the reaction between oxalic acid and permanganate ion, which occurs in strongly acidic solutions and yields carbon dioxide and manganese(II) ion.

1. Divide the overall equation into the equations for two half-reactions. It will almost always be obvious how to do this. For the reaction $MnO_4^- + H_2C_2O_4(aq) = Mn^{2+} + CO_2(aq)$, one of the two half-reactions must certainly involve the permanganate and manganese(II) ions while the other involves the oxalic acid and carbon dioxide, and therefore you would write

$$MnO_4^- = Mn^{2+} \qquad \text{and} \qquad H_2C_2O_4(aq) = CO_2(aq).$$

You would surely not write $MnO_4^- = CO_2(aq)$ or $H_2C_2O_4(aq) = Mn^{2+}$.

The only situation that can be puzzling is the one in which some reactant or product must appear in both half-reactions. For the reaction whose skeleton is $Pb(s) + PbO_2(s) = PbSO_4(s)$, which occurs during the discharge of a lead storage battery (where solid lead sulfate is formed on each of the electrodes), you might begin by writing $Pb(s) = PbSO_4(s)$ and $PbO_2(s) = ?$, and wonder what to put on the right-hand side of the second equation. It would have to become $PbO_2(s) = PbSO_4(s)$ because there is no other product that would account for the atom of lead on the left-hand side. Similarly, if you were trying to write an equation for the decomposition of manganate(VI) ion, MnO_4^{2-}, into permanganate ion and manganese(IV) dioxide you would have to begin with $MnO_4^{2-} = MnO_4^-$ and $MnO_4^{2-} = MnO_2(s)$.

Before going any farther make sure that the formulas of all the reactants and products are correct. You cannot obtain the right result if you write MnO_4^{2-} or MnO_5^- for permanganate ion, and you will obtain a misleading one if you write $C_2O_4^{2-}$ instead of $H_2C_2O_4(aq)$ for a strongly acidic solution.

One of the two equations resulting from this step will have the oxidizing agent, and the other will have the reducing agent, on its left-hand side. You will usually know which is which, but don't worry if you don't, for it will become clear in step 2(d).

2. Balance the equation for each of the half-reactions. This is done in the following way.

(a) Make the numbers of atoms of each element, *except oxygen and hydrogen,* equal on the two sides of the equation. Do this by introducing whatever numerical coefficients are required. In the equation $MnO_4^- = Mn^{2+}$ the numbers of manganese atoms on the two sides are already equal, but the equation $H_2C_2O_4(aq) = CO_2(aq)$ must be changed to $H_2C_2O_4(aq) = 2\,CO_2(aq)$ to equalize the numbers of carbon atoms.

You may find that some element is missing on one side of one of the equations, and if so you should supply it by adding some substance containing that element and known to be present in the solution. In a hydrochloric acid solution the reaction $Fe^{2+} + ClO_3^- = FeCl^{2+} + ClO_2(g)$ can occur. It would be divided into $Fe^{2+} = FeCl^{2+}$ and $ClO_3^- = ClO_2(g)$ in the first step, and the first of these should now be changed to $Fe^{2+} + Cl^- = FeCl^{2+}$, which can be done because the solution contains chloride ion. Do not add a substance to either equation if it already appears in the other; if you try to supply the chlorine atom by writing $Fe^{2+} + ClO_3^- = FeCl^{2+}$ you will not get out alive.

(b) Make the numbers of atoms of oxygen equal on the two sides of the equation. Do this by adding molecules of water to the side that contains the smaller number of atoms of oxygen. In the equation $MnO_4^- = Mn^{2+}$ four molecules of water must be added to the right-hand side, giving $MnO_4^- = Mn^{2+} + 4 H_2O(l)$, but the two sides of the equation $H_2C_2O_4(aq) = 2 CO_2(aq)$ already contain equal numbers of oxygen atoms.

(c) Make the numbers of atoms of hydrogen equal on the two sides of the equation. Do this by adding protons to the side that contains the smaller number of hydrogen atoms. When this is done the equations for our two half-reactions will be

$$MnO_4^- + 8H^+ = Mn^{2+} + 4H_2O(l) \qquad \text{and} \qquad H_2C_2O_4(aq) = 2H^+ + 2CO_2(aq)$$

Most chemists use protons, rather than hydronium ions, in balancing redox equations because the basicity of water and the hydration of the proton, which are emphasized by writing H_3O^+ instead of H^+, are not of prime concern in redox processes. If you prefer to write H_3O^+, you should find how many atoms of hydrogen have to be supplied on one side of an equation and add that number of hydronium ions to that side and the same number of water molecules to the other. For the equation $MnO_4^- = Mn^{2+} + 4 H_2O(l)$, where eight atoms of hydrogen must be added to the left-hand side, this would give $MnO_4^- + 8 H_3O^+ = Mn^{2+} + 12 H_2O(l)$.

(d) Make the electrical charges equal on the two sides. Do this by adding electrons to the side on which the sum of the ionic charges is more positive (or less negative). For the permanganate–manganese(II) ion half-reaction the result of the preceding step has seven positive charges on its left-hand side but only two on the right; five electrons must be added to the left-hand side, and the result is

$$MnO_4^- + 8H^+ + 5e = Mn^{2+} + 4H_2O(l).$$

For the oxalic acid–carbon dioxide half-reaction two electrons must be added on the right-hand side, giving

$$H_2C_2O_4(aq) = 2H^+ + 2CO_2(aq) + 2e.$$

The equations for the half-reactions are now balanced, and it can be seen that permanganate ion is the oxidizing agent because it accepts electrons, while oxalic acid is the reducing agent because it loses electrons.

(e) If the reaction occurs in a neutral or basic solution remove any hydrogen (or hydronium) ions from the equation. Do this by adding to each side of the equation a number of hydroxide ions that is equal to the number of hydrogen ions appearing in the equation. Combine the hydrogen ions with the hydroxide ions to form molecules of water, and drop equal numbers of molecules of water from the two sides if possible.

In neutral or moderately basic solutions permanganate ion is reduced to manganese(IV) dioxide by many reducing agents. The preceding steps would give $MnO_4^- + 4H^+ + 3e = MnO_2(s) + 2H_2O(l)$ as the equation for this half-reaction. In the present one that equation would become successively

$$MnO_4^- + 4H^+ + 4OH^- + 3e = MnO_2(s) + 2H_2O(l) + 4OH^-,$$
$$MnO_4^- + 4H_2O(l) + 3e = MnO_2(s) + 2H_2O(l) + 4OH^-,$$
$$MnO_4^- + 2H_2O(l) + 3e = MnO_2(s) + 4OH^-.$$

This step makes it possible for a chemist to recognize that the equation $MnO_4^- + 8H^+ + 5e = Mn^{2+} + 4H_2O(l)$ refers to a reaction that occurs in an acidic solution, whereas the one $MnO_4^- + 2H_2O(l) + 3e = MnO_2(s) + 4OH^-$ refers to one that occurs in a neutral or basic one. Omitting this step would make it impossible to deduce this important information.

3. Combine the balanced equations for the two half-reactions. There will be electrons on the left-hand side of one equation and on the right-hand side of the other. Unless the numbers of electrons appearing in the equations are already equal, make them equal by multiplying the equations by suitable factors. Finally add the two equations, dropping the electrons and anything else that can be dropped from both sides. For the reaction between permanganate ion and oxalic acid we would first obtain

$$2[MnO_4^- + 8H^+ + 5e = Mn^{2+} + 4H_2O(l)]$$
$$\underline{5[H_2C_2O_4(aq) = 2H^+ + 2CO_2(aq) + 2e]}$$
$$2MnO_4^- + 16H^+ + 10e + 5H_2C_2O_4(aq) = 2Mn^{2+} + 8H_2O(l) + 10H^+ + 10CO_2(aq) + 10e.$$

Ten hydrogen ions can be dropped as well as the electrons, and the final result is

$$2MnO_4^- + 6H^+ + 5H_2C_2O_4(aq) = 2Mn^{2+} + 8H_2O(l) + 10CO_2(aq).$$

Reading about this process takes much longer than doing it. Here, shorn of all the discussion, is another example. Its starting point is the equation $CuSCN(s) + IO_3^- = Cu^{2+} + SO_4^{2-} + HCN(aq) + ICl_2^-$, which appeared in the first paragraph of this section.

Step 1: $CuSCN(s) = Cu^{2+} + SO_4^{2-} + HCN(aq)$ $\Big|$ $IO_3^- = ICl_2^-.$

Step 2(a): $CuSCN(s) = Cu^{2+} + SO_4^{2-} + HCN(aq)$ $\Big|$ $IO_3^- + 2Cl^- = ICl_2^-$

Step 2(b): $CuSCN(s) + 4H_2O(l) = Cu^{2+} + SO_4^{2-} + HCN(aq)$ $\Big|$ $IO_3^- + 2Cl^- = ICl_2^- + 3H_2O(l)$

Step 2(c): $CuSCN(s) + 4H_2O(l) =$ $\qquad$ $Cu^{2+} + SO_4^{2-} + HCN(aq) + 7H^+$ $\Big|$ $IO_3^- + 2Cl^- + 6H^+ = ICl_2^- + 3H_2O(l)$

Step 2(d): $CuSCN(s) + 4H_2O(l) =$ | $IO_3^- + 2Cl^- + 6H^+$
$\qquad = Cu^{2+} + SO_4^{2-} + HCN(aq) + 7H^+ + 7e$ | $+4e = ICl_2^-$
     | $+3H_2O(l)$

Step 3: $4[CuSCN(s) + 4H_2O(l) = Cu^{2+} + SO_4^{2-} + HCN(aq) + 7H^+ + 7e]$
$\qquad 7[IO_3^- + 2Cl^- + 6H^+ + 4e = ICl_2^- + 3H_2O(l)]$

$$4CuSCN(s) + 16H_2O(l) + 7IO_3^- + 14Cl^- + 42H^+ + 28e =$$
$$4Cu^{2+} + 4SO_4^{2-} + 4HCN(aq) + 28H^+ + 28e + 7ICl_2^- + 21H_2O(l)$$
$$4CuSCN(s) + 7IO_3^- + 14Cl^- + 14H^+ =$$
$$4Cu^{2+} + 4SO_4^{2-} + 4HCN(aq) + 7ICl_2^- + 5H_2O(l)$$

SUMMARY: You can balance the equation for any redox reaction by
1. dividing its reactants and products between two half-reactions;
2. balancing the equations for the half-reactions separately, using molecules of water to balance oxygen atoms, hydrogen ions to balance hydrogen atoms, and electrons to balance electrical charges, then adding hydroxide ions to consume any hydrogen ions if the reaction occurs in a neutral or basic solution and
3. combining the equations for the two half-reactions so that one produces the same number of electrons as the other consumes.

8.4. Electrode potentials and the Nernst equation

PREVIEW: This section explains how the potential of an electrode arises, presents and discusses the Nernst equation for the potential of an electrode, and describes the condition for chemical equilibrium in an electrochemical cell.

If a piece of platinum or some other substance that is both chemically inert and a good conductor of electricity is brought into contact with a solution, it will acquire an electrical potential reflecting the tendency for the solution to accept or donate electrons. If the solution contains iron(III) and iron(II) ions, there will be a tendency for iron(III) ions to accept electrons from the electrode and a tendency for iron(II) ions to donate electrons to the electrode. If the electrode and solution form part of a complete circuit through which an electric current can flow, the tendency for iron(III) ions to accept electrons will create a tendency for electrons to flow toward the electrode in the external conductor by which it is attached to the rest of the circuit. On reaching the electrode, these electrons would be consumed in reducing iron(III) ions at the interface between the electrode and the solution. Conversely, the tendency for iron(II) ions to donate electrons will create a tendency for iron(II) ions to be oxidized at this interface, liberating electrons that would flow away from the electrode in the external conductor. The potential acquired by the electrode is a measure of the net tendency for electrons to flow toward or away from it in the external circuit, and it depends on the strengths of the tendencies for iron(III) ions to accept electrons from the electrode and for iron(II) ions to donate electrons to the electrode. All this is summarized diagrammatically in Fig. 8.1.

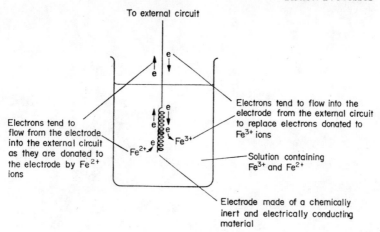

To external circuit

Electrons tend to flow into the electrode from the external circuit to replace electrons donated to Fe^{3+} ions

Electrons tend to flow from the electrode into the external circuit as they are donated to the electrode by Fe^{2+} ions

Solution containing Fe^{3+} and Fe^{2+}

Electrode made of a chemically inert and electrically conducting material

FIG. 8.1. The processes that can occur when an electrode is brought into contact with a solution containing the oxidized and reduced forms of a redox couple.

Both the natures and the activities of the oxidizing and reducing agents affect the potential of the electrode. Chromium(III) ion is a much weaker oxidizing agent than iron(III) ion, and chromium(II) ion is a much stronger reducing agent than iron(II) ion. Because chromium(III) ion has a smaller tendency to accept electrons than iron(III) ion does, there will be a smaller tendency for electrons to flow toward an electrode in a solution containing chromium(III) ion than there is for them to flow toward one in a solution containing iron(III) ion at the same activity. Because chromium(II) ion has a greater tendency to donate electrons than iron(II) ion does, there will be a greater tendency for electrons to flow away from an electrode in a solution of chromi-um(II) ion than there is for them to flow away from an electrode in a solution containing iron(II) ion at the same activity. These statements are summarized in Fig. 8.2.

The net result is that electrons have a much higher tendency to flow away from an electrode in a solution containing chromium(III) and chromium(II) ions at any partic-ular activities than to flow away from one in a solution containing iron(III) and iron(II) ions at the same activities. Since an increasing tendency for electrons to flow away from any point in an electrical circuit corresponds to an increasingly negative potential at that point, the potential of an electrode in a solution of chromi-um(III) and chromium(II) ions is more negative than the potential of an electrode in a solution of iron(III) and iron(II) ions at the same activities. The potential of an elec-trode becomes more negative as the tendency for a solution to donate electrons to it increases, and becomes more positive as the tendency for a solution to accept elec-trons from it increases.

The effects of the activities of the dissolved substances on the potential obey the same general principle. If an electrode is placed in a solution containing iron(III) and iron(II) ions it will acquire a certain potential. Increasing the activity of iron(III) ion in the solution will increase the tendency for iron(III) ions to accept electrons

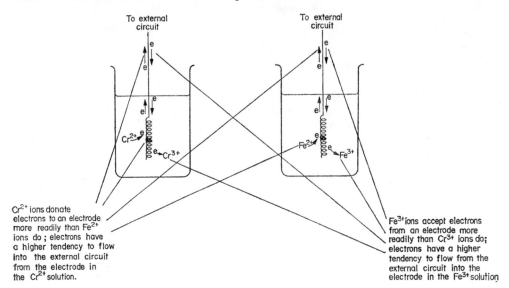

To external circuit

To external circuit

Cr^{2+} ions donate electrons to an electrode more readily than Fe^{2+} ions do ; electrons have a higher tendency to flow into the external circuit from the electrode in the Cr^{2+} solution.

Fe^{3+}ions accept electrons from an electrode more readily than Cr^{3+} ions do; electrons have a higher tendency to flow from the external circuit into the electrode in the Fe^{3+} solution

FIG. 8.2. Two electrochemical half-cells. Because Cr^{2+} ions donate electrons to an electrode more readily than Fe^{2+} ions do, and because Fe^{3+} ions accept electrons from an electrode more readily than Cr^{3+} ions do, electrons would flow spontaneously from the electrode in the Cr^{3+}–Cr^{2+} solution to the one in the Fe^{3+}–Fe^{2+} solution if the electrical circuit were completed.

from the electrode, in accordance with le Chatelier's principle. This in turn increases the tendency for electrons to flow toward the electrode in the external conductor; the potential of the electrode becomes more positive. Increasing the activity of iron(II) ion causes the potential to become more negative.

Both of these effects are described by the Nernst equation, which is

$$E = E^0(Fe^{3+}, Fe^{2+}) - \frac{RT}{nF} \ln \frac{a_{Fe^{2+}}}{a_{Fe^{3+}}} \tag{8.1a}$$

for the iron(III)–iron(II) couple. The symbol E represents the potential, and $E^0(Fe^{3+}, Fe^{2+})$ is a constant, called the *standard potential*, that characterizes the strengths of the oxidizing and reducing agents involved in the couple. Both E and the standard potential are expressed in volts (V). The gas constant R is expressed in joules K^{-1} mol^{-1}, as it was in the equations in Chapter 2; T is the temperature (in kelvins); n is the number of electrons appearing in the balanced equation for the half-reaction; and F is the number of coulombs per faraday. A faraday is a mole (Avogadro's number) of electrons, and it contains 96,485 coulombs. Since 1 joule = 1 watt second = 1 volt ampere second = 1 volt coulomb, the factor RT/nF has the units of (volt coulomb K^{-1} mol^{-1}) (K)/(coulomb mol^{-1}) = volts. Converting from natural to decadic logarithms, inserting the numerical values of $R(= 8.3143$ J K^{-1} $mol^{-1})$ and F, and letting $T = 25°$ C $= 298.15$ K, eq. (8.1a) becomes

$$E = E^0(Fe^{3+}, Fe^{2+}) - \frac{0.059\ 16}{n} \log_{10} \frac{a_{Fe^{2+}}}{a_{Fe^{3+}}} \tag{8.1b}$$

Of course $n = 1$ for the half-reaction $Fe^{3+}+e = Fe^{2+}$. More generally, for the half-reaction $Ox+ne = Red$ at 298 K

$$E = E^0(Ox, Red) - \frac{0.059\,16}{n} \log_{10} \frac{a_{Red}}{a_{Ox}}. \tag{8.2}$$

If the equation for the half-reaction has been written with the oxidized form of the couple and the electrons on the left-hand side and the reduced form on the right-hand side, the argument of the logarithmic term, which is abbreviated here as a_{Red}/a_{Ox}, will look just like an equilibrium constant for the half-reaction—except that it will not contain the electrons. For the half-reaction $MnO_4^- + 2\,H_2O(l) + 3\,e = MnO_2(s) + 4\,OH^-$ the Nernst equation would be

$$E = E^0(MnO_4^-, MnO_2(s)) - \frac{0.059\,16}{3} \log \frac{a_{OH^-}^4}{a_{MnO_4^-}}$$

since the activity of solid manganese(IV) dioxide is equal to 1 if the solid is pure and since the activity of liquid water is indistinguishable from 1 if the solution is not too concentrated. The symbol $E^0(MnO_4^-, MnO_2(s))$ looks clumsy, but it identifies the oxidized and reduced forms of the couple and reminds the writer and reader of the direction in which the half-reaction has been written. Identifying the oxidized and reduced forms of the couple is important because one or both may be missing from the logarithmic term. No one could tell whether an equation like

$$E = E^0 - 0.059\,16 \log a_{Cl^-}$$

pertained to the couple $AgCl(s)+e = Ag(s)+Cl^-$ or to the one $PbCl_2(s)+2\,e = Pb(s)+2\,Cl^-$, and the standard potentials of these couples are very different.

The standard potential of a couple is equal to the potential E if the logarithmic term in the Nernst equation is equal to zero. For the iron(III)–iron(II) couple the logarithmic term will be equal to zero if the ratio $a_{Fe^{2+}}/a_{Fe^{3+}}$ is equal to 1; if activities and concentrations were exactly identical, the standard potential would be equal to the potential obtained with any solution in which the concentrations of iron(II) and iron(III) ions were the same. For other couples the matter is complicated by exponents that appear in the argument of the logarithmic term: to say that $E^0(MnO_4^-, MnO_2(s)) = E$ if $a_{OH^-}^4 = a_{MnO_4^-}$ is true but not so easy to interpret. We can always regard the standard potential as being equal to the potential that would be obtained if every substance appearing in the equation for the half-reaction were present at unit activity.

Different couples have different standard potentials. The interpretation of the standard potential will be discussed in Sections 8.5 and 8.6. Meanwhile we need only note that the standard potential of the iron(III)–iron(II) couple is 1.18 V more positive than that of the chromium(III)–chromium(II) couple. This is a large difference. It means that a solution containing iron(III) and iron(II) ions at unit activity is a much more powerful oxidizing agent—and a much less powerful reducing agent—than one containing chromium(III) and chromium(II) ions at unit activity.

If all of these ideas were put together on a laboratory bench, the result would be the electrochemical cell shown in Fig. 8.3. This comprises two half-cells. One half-cell consists of an inert electrode immersed in a solution containing chromium(III) and chromium(II) ions; the other consists of another inert electrode immersed in a solution containing iron(III) and iron(II) ions. There is an external electrical circuit through which electrons can flow, and the circuit is completed by a "salt bridge", which contains a solution of potassium chloride or some other non-reactive electrolyte whose ions conduct current between the solutions in the two half-cells. If the external circuit includes a potentiometer or a voltmeter with a very high resistance, the difference between the potentials of the two electrodes can be measured. If the cell is short-circuited by simply connecting the electrodes together, a current will flow and a spontaneous chemical reaction will occur.

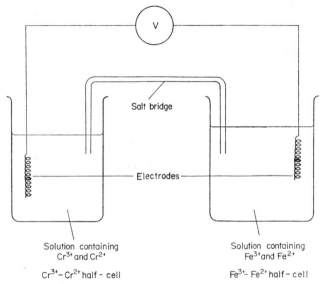

Fig. 8.3. An electrochemical cell. The two half-cells are identical with those shown in Fig. 8.2. They have been combined by connecting the electrodes through a potentiometer (Fig. 14.9) or a high-resistance voltmeter V, and also by connecting the two solutions through a salt bridge, which is an ionic conductor that makes it possible for electricity to flow between the solutions.

Since the electrode in the chromium(III)–chromium(II) solution has the more negative potential, electrons will flow from it to the other electrode if the cell is short-circuited. These electrons will be produced by the oxidation of chromium(II) ions; on reaching the electrode in the iron(III)–iron(II) solution they will be consumed by the reduction of iron(III) ions. The chemical reaction $Cr^{2+} + Fe^{3+} = Cr^{3+} + Fe^{2+}$ will occur spontaneously, just as it would if the two solutions were mixed. As it proceeds, the ratio $a_{Cr^{2+}}/a_{Cr^{3+}}$ will decrease because chromium(II) ions are being consumed while chromium(III) ions are being produced, and therefore the potential of the electrode in the chromium(III)–chromium(II) solution will become less negative.

At the same time, the ratio $a_{Fe^{2+}}/a_{Fe^{3+}}$ will increase, because iron(II) ions are produced and iron(III) ions are consumed in the iron(III)–iron(II) solution, and the potential of the electrode in this solution will become more negative. The two potentials approach each other as the reaction proceeds, and become equal when equilibrium is reached. At equilibrium there can be no net tendency for electrons to flow in either direction through the external conductor.

Equality of the two potentials is the fundamental criterion of equilibrium in a redox process. Section 8.6 will show how it enables us to correlate standard potentials with the equilibrium constants of redox and other half-reactions.

SUMMARY: The potential of an electrode that is immersed in a solution containing a redox couple reflects the tendency for the oxidized form of the couple to gain electrons and the tendency for the reduced form to lose them, and is described by the Nernst equation. The difference between the potentials of the two electrodes of an electrochemical cell is equal to zero if the corresponding redox reaction is at equilibrium.

8.5. The meanings of standard potentials

PREVIEW: The standard potential of a redox couple reflects the strengths of the oxidized and reduced forms of the couple.

The potential of an electrochemical cell, like the one in Fig. 8.3, is the *difference* between the potentials of its electrodes. Section 8.6 will show that the equilibrium constant of a redox reaction is related to the *difference* between the standard potentials of the redox couples involved in it. There is no way in which these differences can be separated to obtain the value of any one potential or standard potential. Similarly, a surveyor can measure the difference between the elevations at two points, but cannot measure (or even define) the "absolute" altitude at any single point. An arbitrary reference point is needed so that a numerical value can be assigned to the height of a mountain or to the standard potential of a redox couple. The surveyor chooses mean sea level; for use in all solvents except inert ones the chemist chooses the standard potential of the half-reaction $2 H^+ + 2e = H_2(g)$, and defines $E^0(H^+, H_2(g))$ as being equal to ± 0.0000 V at any temperature. Since other reference points could be chosen, and are more convenient in experimental work (Section 14.6), the numerical value of a standard potential is usually accompanied by an indication of the reference point. Taking $E^0(H^+, H_2(g))$ as ± 0.0000 V, the standard potentials of the iron(III)–iron(II) and chromium(III)–chromium(II) couples at 298 K are $E^0(Fe^{3+}, Fe^{2+}) = 0.771$ V and $E^0(Cr^{3+}, Cr^{2+}) = -0.41$ V. This information is conveyed by writing $E^0(Fe^{3+}, Fe^{2+}) = +0.771$ V vs. N. H. E. and $E^0(Cr^{3+}, Cr^{2+}) = -0.41$ V vs. N. H. E. The abbreviation "N. H. E." stands for "normal hydrogen electrode", a hypothetical hydrogen electrode having a potential identical with the standard potential of the hydrogen ion–hydrogen gas couple. Appendix VI lists the numerical values of the standard potentials of a number of common couples, all referred to the N. H. E.

Saying that $E^0(Fe^{3+}, Fe^{2+}) = +0.771$ V vs. N. H. E. means that the standard potential of the iron(III)–iron(II) couple is 0.771 V more positive than that of the hydrogen ion–hydrogen gas couple. In the preceding section it was shown that the reaction $Fe^{3+} + Cr^{2+} = Fe^{2+} + Cr^{3+}$ is spontaneous when all of the reactants and products are present at unit activity because $E^0(Fe^{3+}, Fe^{2+})$ is more positive than $E^0(Cr^{3+}, Cr^{2+})$. Similarly, the fact that $E^0(Fe^{3+}, Fe^{2+})$ is more positive than $E^0(H^+, H_2(g))$ means that the reaction $2 Fe^{3+} + H_2(g) = 2 Fe^{2+} + 2 H^+$ is spontaneous when all of the substances appearing in it are present at unit activity. The same reasoning shows that, because $E^0(Cr^{3+}, Cr^{2+})$ is more negative than $E^0(H^+, H_2(g))$, the chromium(III)–chromium(II) couple is a more powerful reducing agent than the hydrogen ion–hydrogen gas couple if all of the activities are equal to 1, and therefore that the reaction $2 Cr^{2+} + 2 H^+ = 2 Cr^{3+} + H_2(g)$ is spontaneous under these conditions.

Since the standard potential of a couple reflects the strengths of the oxidizing and reducing agents contained in that couple, we can estimate the magnitude of the equilibrium constant for any particular redox reaction by comparing the standard potentials of the two half-reactions involved in it. Section 8.6 will show how the numerical value of K can be calculated, and will also show that the numerical calculation is greatly facilitated by making the estimate before doing the arithmetic.

Before dealing with redox couples it is advantageous to recall the behaviors of Lowry–Brønsted acids. Arranging a few acids in the order of their overall dissociation constants gives a list like

	Acid	Base	K_a (in H_2O at 298 K), mol dm^{-3}
increasing strength ↑	$HCl\,(aq) + H_2O(l) = H_3O^+ + Cl^-$ $HOAc(aq) + H_2O(l) = H_3O^+ + OA^-$ $NH_4^+ + H_2O(l) = H_3O^+ + NH_3(aq)$ $H_2O(l) + H_2O(l) = H_3O^+ + OH^-$	increasing strength ↓	very large 1.8×10^{-5} mol dm^{-3} 5.5×10^{-10} mol dm^{-3} 1.0×10^{-14} mol^2 dm^{-6}

Each acid is stronger than all the ones below it, and each base is stronger than all the ones above it. In every acid-base reaction an acid and a base are consumed, and another acid and another base are produced. In the reaction $HOAc(aq) + NH_3(aq) = OAc^- + NH_4^+$, the acid that is consumed is a stronger acid than the one that is produced, because K_a is larger for acetic acid than it is for ammonium ion. In addition, the base that is consumed is a stronger base than the one that is produced, because K_a is smaller for the conjugate acid of ammonia than it is for the conjugate acid of acetate ion. The fact that K_a is larger for acetic acid than for ammonium ion means that this reaction consumes an acid that is stronger than the one it produces, and also that it consumes a base that is stronger than the one it produces.

In short, a relatively strong acid and a relatively strong base are consumed in this reaction, and a weaker acid and a weaker base are produced. If all four of these substances are present at unit activity, the reaction will have a greater tendency to proceed in the forward than in the reverse direction. Its equilibrium constant must be larger than 1. Conversely, in the reaction $NH_4^+ + Cl^- = HCl(aq) + NH_3(aq)$, the acid that is consumed is weaker than the one that is produced because K_a is smaller for ammonium ion than for hydrochloric acid, and the base that is consumed is also weaker than the one that is produced, again because K_a is smaller for ammonium ion than for hydrochloric acid. In this reaction a relatively weak acid and a relatively weak base are consumed, and a stronger acid and a stronger base are produced. This reaction has a greater tendency to proceed in the reverse than in the forward direction, and therefore its equilibrium constant is smaller than 1.

Exactly similar reasoning can be applied to redox reactions. Arranging a few half-reactions in the order of their standard potentials gives

	Oxidized form	Reduced form	E^0, V vs. N.H.E.
increasing strength	$MNO_4^- + 8\,H^+ + 5e = Mn^{2+} + 4\,H_2O$		$+1.51$
	$Fe^{3+} + e = Fe^{2+}$	increasing strength	$+0.771$
	$2\,H^+ + 2e = H_2(g)$		±0.0000
	$Cr^{3+} + e = Cr^{2+}$		-0.41

In the reaction $Fe^{3+} + Cr^{2+} = Fe^{2+} + Cr^{3+}$, the oxidizing agent that is consumed is stronger than the one produced, and the reducing agent consumed is also stronger than the one produced. Both these things are true because $E^0(Fe^{3+}, Fe^{2+})$ is more positive than $E^0(Cr^{3+}, Cr^{2+})$. The reaction must therefore have a higher tendency to occur in the forward direction than to occur in the backward direction, and its equilibrium constant must be greater than 1. The reaction $5\,Cr^{3+} + Mn^{2+} + 4\,H_2O(l) = 5\,Cr^{2+} + MnO_4^- + 8\,H^+$, however, consumes a weak oxidizing agent and a weak reducing agent and produces a stronger oxidizing agent and a stronger reducing agent, and its equilibrium constant must be smaller than 1.

SUMMARY: The standard potentials of redox couples are referred to the normal hydrogen electrode, which is arbitrarily said to have a standard potential equal to ±0.0000 V. Of two different couples, the one having the more positive standard potential includes the stronger oxidizing agent and the weaker reducing agent. Knowing the standard potentials of the couples involved in a redox reaction therefore enables the chemist to decide whether the equilibrium constant of the reaction must be larger or smaller than 1.

8.6. Standard potentials and equilibrium constants

PREVIEW: The equilibrium constant for any chemical reaction can be evaluated from the difference between the standard potentials of the redox couples that the reaction involves.

Suppose that the cell shown in Fig. 8.3 is short-circuited, so that current flows through it and the reaction $Fe^{3+}+Cr^{2+} = Fe^{2+}+Cr^{3+}$ occurs spontaneously. Equilibrium will have been reached when the two potentials become equal and the current becomes zero. Then

$$E^0(Cr^{3+}, Cr^{2+})-0.059\ 16 \log \frac{a_{Cr^{2+}}}{a_{Cr^{3+}}} = E^0(Fe^{3+}, Fe^{2+})-0.059\ 16 \log \frac{a_{Fe^{2+}}}{a_{Fe^{3+}}}.$$

Collecting the standard potentials on one side and the logarithmic terms on the other, and separating the factor 0.059 16 gives

$$0.059\ 16 \left(\log \frac{a_{Fe^{2+}}}{a_{Fe^{3+}}} - \log \frac{a_{Cr^{2+}}}{a_{Cr^{3+}}} \right) = E^0(Fe^{3+}, Fe^{2+})-E^0(Cr^{3+}, Cr^{2+}).$$

Since $\log a - \log b = \log (a/b)$,

$$0.059\ 16 \log \frac{a_{Fe^{2+}}+a_{Cr^{3+}}}{a_{Fe^{3+}}+a_{Cr^{2+}}} = E^0(Fe^{3+}, Fe^{2+})-E^0(Cr^{3+}, Cr^{2+}).$$

The argument of the logarithmic term is equal to the equilibrium constant K of the reaction, and therefore

$$\log K = \frac{E^0(Fe^{3+}, Fe^{2+})-E^0(Cr^{3+}, Cr^{2+})}{0.059\ 16} = \frac{+0.771-(-0.41)}{0.059\ 16} = 19.9_6$$

so that $K = 9 \times 10^{19}$. This is larger than 1, as was concluded in Section 8.5.

The equilibrium constant of any redox reaction can be calculated in this way if the standard potentials of the two half-reactions are known, but there is a small algebraic complication if the values of n in the two half-reactions are not the same. If the overall reaction is $2\ MnO_4^-+5\ Zn(s)+16\ H^+ = 2\ Mn^{2+}+5\ Zn^{2+}+8\ H_2O(l)$, $n = 5$ for the permanganate–manganese(II) ion half-reaction while $n = 2$ for the zinc ion–zinc half-reaction. Equating the Nernst expressions for the potentials of these two couples because they must be identical at equilibrium,

$$E^0(Zn^{2+}, Zn(s))-\frac{0.059\ 16}{2} \log \frac{1}{a_{Zn^{2+}}} = E^0(MnO_4^-, Mn^{2+})-\frac{0.059\ 16}{5} \log \frac{a_{Mn^{2+}}}{a_{MnO_4^-}a_{H^+}^8}$$

or

$$\frac{0.059\ 16}{5} \log \frac{a_{Mn^{2+}}}{a_{MnO_4^-}a_{H^+}^8} + \frac{0.059\ 16}{2} \log a_{Zn^{2+}} = E^0(MnO_4^-, Mn^{2+})-E^0(Zn^{2+}, Zn(s)).$$

To make the two numerical factors the same, we may multiply both sides of the equation by 10 and employ the relationship $a \log b = \log b^a$:

$$0.059\ 16 \times 2 \log \frac{a_{Mn^{2+}}}{a_{MnO_4^-}a_{H^+}^8} +0.059\ 16 \times 5 \log a_{Zn^{2+}}$$
$$= 10[E^0(MnO_4^-, Mn^{2+})-E^0(Zn^{2+}, Zn(s))]$$
$$= 0.059\ 16 \times \log \frac{a_{Mn^{2+}}^2}{a_{MnO_4^-}^2 a_{H^+}^{16}} +0.059\ 16 \log a_{Zn^{2+}}^5$$
$$= 0.059\ 16 \log \frac{a_{Mn^{2+}}^2+a_{Zn^{2+}}^5}{a_{MnO_4^-}^2 a_{H^+}^{16}}.$$

Consequently

$$\log K = \frac{10[E^0(MnO_4^-, Mn^{2+})-E^0(Zn^{2+}, Zn(s))]}{0.059\ 16}$$

$$= \frac{10[+1.51-(-0.763)]}{0.059\ 16}$$

$$= 384.2$$

so that $K = 1.6 \times 10^{384}$ mol^{-11} dm^{33}.

From these two results we can induce a general relationship:

$$\log K = nF\Delta E^0/2.303RT(= n\Delta E^0/0.059\ 16 \text{ at } 298 \text{ K}) \qquad (8.3)$$

in which **n** is the number of electrons that would appear on each side of the balanced equation for the half-reaction if they had not been dropped in writing it, and ΔE^0 is the difference between the standard potentials of the two half-reactions. Usually **n** is equal to the least common multiple of the numbers of electrons that appear in the two half-reactions, but it is safer to rely on the definition in the preceding sentence. This is because the equation for the reaction between sodium and water, for example, might be written as either $Na(s)+H_2O(l) = Na^++OH^-+\frac{1}{2} H_2(g)$ or $2 Na(s)+2 H_2O(l) = 2 Na^++2 OH^-+H_2(g)$. Since the equilibrium constant for the second of these is the square of that for the first, **n** must be taken to be 2 for the second but only 1 for the first.

Suppose that you wish to calculate the equilibrium constant at 298 K of the reaction $Ag(s)+Cl^-+Fe^{3+} = AgCl(s)+Fe^{2+}$, which occurs when metallic silver is added to a solution containing chloride ion and iron(III) ion. The standard potentials of the two half-reactions are given by

$$AgCl(s)+e = Ag(s)+Cl^-; \qquad E^0(AgCl(s), Ag(s)) = +0.222 \text{ V.}$$
$$Fe^{3+}+e = Fe^{2+}; \qquad E^0(Fe^{3+}, Fe^{2+}) = +0.771 \text{ V.}$$

Evidently **n** = 1, but is ΔE^0 equal to $0.222-0.771 = -0.549$ V or to $0.771-0.222 = +0.549$ V? The answer is provided by the considerations outlined in Section 8.5. Iron(III) ion is a stronger oxidizing agent than silver chloride, and silver is a stronger reducing agent than iron(II) ion. The equilibrium constant of the reaction must be larger than 1, and log K must be positive. You must therefore select the value of ΔE^0 that will yield a positive value of log K, and on doing so you obtain log $K = +0.549/0.059\ 16 = 9.28$, so that $K = 1.9 \times 10^9$ mol^{-1} dm^3. If you had wanted the value of K for the reverse reaction, $Fe^{2+}+AgCl(s) = Fe^{3+}+Ag(s)+Cl^-$, you would have reasoned that it must be smaller than 1 because the oxidizing and reducing agents that are consumed in this reaction are weaker than the ones that are produced. You would therefore have sought a negative value of log K, and obtained it by writing log $K = -0.549/0.059\ 16 = -9.28$, so that $K = 5.2 \times 10^{-10}$ mol dm^{-3}.

All of this can be expressed concisely by the relationships

$$\Delta G^0 = -nF\Delta E^0 = -RT \ln K \qquad (8.4)$$

16*

which include eq. (2.12) as well as eqs. (8.1) and (8.3). These equalities make it possible to pack a very large amount of information about chemical equilibria into a very short table of standard potentials, and to include equilibria that appear to have nothing whatever to do with reduction and oxidation.

To obtain the solubility product of silver chloride, for example, one has only to divide the equation $AgCl(s) = Ag^+ + Cl^-$ into the equations for two half-reactions that appear in such a table. This can be done by writing

$$AgCl(s) + e = Ag(s) + Cl^-,$$
$$Ag(s) = Ag^+.$$

The two standard potentials are $E^0(AgCl(s), Ag(s)) = +0.2224$ V and $E^0(Ag^+, Ag(s)) = +0.7994$ V, respectively. Adding the equations for the two half-reactions gives

$$AgCl(s) + Ag(s) = Ag(s) + Cl^- + Ag^+$$

which is identical with the desired equation except for the atom of silver on each side. The arrows will help us to distinguish the two different half-reactions. The oxidizing agent (silver chloride) on the left-hand side is weaker than the oxidizing agent (silver ion) on the right-hand side. In addition, the reducing agent (silver) on the left-hand side is weaker than the reducing agent (silver) on the right-hand side, which sounds odd but means that it is more difficult to oxidize an atom of silver to yield a silver ion than it is to oxidize an atom of silver to yield a "molecule" of silver chloride. Hence $\log K$ must be negative, and its numerical value is given by

$$\log K = \frac{+0.2224 - 0.7994}{0.059\,16} = -\frac{0.5770}{0.059\,16} = -9.75_3$$

whence $K = 1.77 \times 10^{-10}$ mol^2 dm^{-6}.

Conversely, an equilibrium constant can be combined with a standard potential to obtain the value of a new standard potential. A chemist attempting to improve or modify a process for the electroplating of silver, or to recover the silver(I) dissolved in the effluent from a silver-plating shop, would probably wish to know the standard potential of the half-reaction $Ag(CN)_2^- + e = Ag(s) + 2\,CN^-$ but might be unable to find it in a table. It could be evaluated by combining the Nernst equation for the silver ion–silver couple

$$E = E^0(Ag^+, Ag(s)) - 0.059\,16 \log \frac{1}{a_{Ag^+}}$$

with the expression for the overall formation constant of $Ag(CN)_2^-$

$$Ag^+ + 2CN^- = Ag(CN)_2^-; \qquad \beta_2 = a_{Ag(CN)_2^-}/a_{Ag^+} a_{CN^-}^2$$

which is written here with activities, instead of with concentrations as in Chapter 7, to avoid scrambling activities and concentrations in the following steps. Solving the second of these equations for a_{Ag^+} and substituting the result into the Nernst equation gives

$$E = E^0(Ag^+, Ag(s)) - 0.059\ 16 \log \frac{\beta_2 a_{CN^-}^2}{a_{Ag(CN)_2^-}}.$$

The value of β_2 is known: $1.2 \times 10^{21}\ mol^{-2}\ dm^6$. Separating it from the variable activities that appear in the logarithmic term,

$$E = E^0(Ag^+, Ag(s)) - 0.059\ 16 \log \beta_2 - 0.059\ 16 \log \frac{a_{CN^-}^2}{a_{Ag(CN)_2^-}}.$$

Comparing this with the Nernst equation for the potential of the half-reaction $Ag(CN)_2^- + e = Ag(s) + 2\ CN^-$

$$E = E^0(Ag(CN)_2^-, Ag(s)) - 0.059\ 16 \log \frac{a_{CN^-}^2}{a_{Ag(CN)_2^-}}$$

shows that

$$E^0(Ag(CN)_2^-, Ag(s)) = E^0(Ag^+, Ag(s)) - 0.059\ 16 \log \beta_2$$

or, inserting the numerical values of the quantities on the right-hand side,

$$E^0(Ag(CN)_2^-, Ag(s)) = +0.7994 - 0.059\ 16 \log (1.2 \times 10^{21}) = -0.448\ V.$$

Finally, you can sometimes combine two standard potentials to obtain the value of a third. Tables of standard potentials contain the following information about indium:

$$In^{3+} + 3e = In(s); \qquad E^0(In^{3+}, In(s)) = -0.33\ V,$$
$$In^{3+} + 2e = In^+; \qquad E^0(In^{3+}, In^+) = -0.40\ V,$$

but do not give the standard potential of the half-reaction $In^+ + e = In(s)$. It could be evaluated in the following way. The Nernst equation for this half-reaction

$$E = E^0(In^+, In(s)) - 0.059\ 16 \log \frac{1}{a_{In^+}} \qquad (8.5a)$$

contains the activity of In^+, and so does the Nernst equation for the In^{3+}–In^+ couple,

$$E = E^0(In^{3+}, In^+) - \frac{0.059\ 16}{2} \log \frac{a_{In^+}}{a_{In^{3+}}}. \qquad (8.5b)$$

However, the latter also contains the activity of In^{3+}, which does not appear in eq. (8.5a). We may use the Nernst equation for the In^{3+}–$In(s)$ couple

$$E = E^0(In^{3+}, In(s)) - \frac{0.059\ 16}{3} \log \frac{1}{a_{In^{3+}}} \qquad (8.5c)$$

to eliminate $a_{In^{3+}}$ from eq. (8.5b). Solving eq. (8.5c) for $0.059\ 16 \log a_{In^{3+}}$ gives

$$0.059\ 16 \log a_{In^{3+}} = 3E - 3E^0(In^{3+}, In(s)).$$

Substituting this into eq. (8.5b), we obtain

$$E \left(= E^0(In^{3+}, In^+) - \frac{0.059\,16}{2} \log a_{In^+} + \frac{0.059\,16}{2} \log a_{In^{3+}} \right)$$

$$= E^0(In^{3+}, In^+) - \frac{0.059\,16}{2} \log a_{In^+} + \left[\frac{3}{2} E - \frac{3}{2} E^0(In^{3+}, In(s)) \right].$$

Rearranging, multiplying each term by -2, and replacing $\log a_{In^+}$ by $-\log (1/a_{In^+})$,

$$E = 3E^0(In^{3+}, In(s)) - 2E^0(In^{3+}, In^+) - 0.059\,16 \log \frac{1}{a_{In^+}}. \qquad (8.5d)$$

Comparing this with eq. (8.5a) shows that

$$E^0(In^+, In(s)) = 3E^0(In^{3+}, In(s)) - 2E^0(In^{3+}, In^+)$$
$$= 3(-0.33) - 2(-0.40) = -0.99 + 0.80 = -0.19 \text{ V}.$$

In this procedure we have assumed that the potential E is the same for all three of these half-reactions. Assuming that it is the same for two of them, say $In^{3+} + 3\,e = In(s)$ and $In^{3+} + 2\,e = In^+$, is equivalent to assuming that equilibrium has been attained in the reaction $3\,In^+ = In^{3+} + 2\,In(s)$, which is obtained by adding the equations for these two half-reactions. This assumption is necessary because we cannot make any equilibrium calculation without assuming that equilibrium has been attained. The same overall reaction is obtained by combining either of these two half-reactions with the third one:

$$
\begin{array}{c}
In(s) = In^{3+} + 3e \\
3[In^+ + e = In(s)] \\
\hline
3In^+ = In^{3+} + 2In(s)
\end{array}
\qquad \text{or} \qquad
\begin{array}{c}
In^+ = In^{3+} + 2e \\
2[In^+ + e = In(s)] \\
\hline
3In^+ = In^{3+} + 2In(s)
\end{array}
$$

If equilibrium has been attained, the half-reaction $In^+ + e = In(s)$ must therefore have the same potential as each of the others. If two of the potentials are equal, the third cannot differ from them.

The result can be summarized by saying that the standard potentials of the three half-reactions

$$
\begin{array}{ll}
C + n_{C,\,B}\,e = B; & E^0 = E^0(C, B) \\
B + n_{B,\,A}\,e = A; & E^0 = E^0(B, A) \\
\hline
C + n_{C,\,A}\,e = A; & E^0 = E^0(C, A)
\end{array}
$$

are related by the equation

$$n_{C,\,A}E^0(C, A) = n_{C,\,B}E^0(C, B) + n_{B,\,A}E^0(B, A). \qquad (8.6)$$

Exactly the same result can be obtained by means of the first equality in eq. (8 4):

$$
\begin{array}{ll}
C + n_{C,\,B}\,e = B; & \Delta G^0_{C,\,B} = -n_{C,\,B}FE^0(C, B) \\
B + n_{B,\,A}\,e = A; & \Delta G^0_{B,\,A} = -n_{B,\,A}FE^0(B, A) \\
\hline
C + n_{C,\,A}\,e = A; & \Delta G^0_{C,\,A} = \Delta G^0_{C,\,B} + \Delta G^0_{B,\,A} = -n_{C,\,A}FE^0(C, A)
\end{array}
$$

where, for example, $\Delta G^0_{C,\,B} = G^0_B - G^0_C$.

SUMMARY: Values of the standard potentials of two redox couples can be combined to calculate the equilibrium constant of the redox reaction involving those couples, or the standard potential of another couple involving some of the same substances.

8.7. Formal potentials

PREVIEW: The formal potential of a redox couple has the same significance as the standard potential but is easier to evaluate and to use in practical calculations. This section defines the formal potential and discusses the factors that enter into it.

In much practical work it is difficult to use the Nernst equation in the form that has appeared in the prior sections of this chapter. This is partly because the logarithmic term contains activities rather than concentrations, partly because those activities pertain to individual species that may not be the predominating ones in real solutions, and partly because the standard potential is impossible to measure directly.

All these difficulties have the same cause. The activity (in mol dm^{-3}) of the ith species dissolved in a solution is given by

$$a_i = y_i c_i \tag{8.7}$$

where c_i is the concentration of that species, in mol dm^{-3}, and y_i is a dimensionless number called its *molarity activity coefficient*. Since y_i is defined in such a way that it approaches 1 as the total ionic concentration of a solution approaches zero, the activity and the concentration are identical in an "infinitely dilute" solution but differ in any real one.

The potential E of a metallic silver electrode in a solution containing silver ion at 25° C is given by

$$E = E^0(Ag^+, Ag(s)) + 0.059\,16 \log a_{Ag^+}$$
$$= E^0(Ag^+, Ag(s) + 0.059\,16 \log y_{Ag^+} + 0.059\,16 \log [Ag^+]. \tag{8.8}$$

If we knew the activity coefficient and concentration of silver ion in a solution of, say, silver nitrate, we could measure the potential of a silver electrode in that solution and calculate the value of $E^0(Ag^+, Ag(s))$ from the result. In any real solution of silver nitrate the activity coefficient of silver ion will differ from 1, and there is no way of evaluating it.

In addition, the concentration of silver ion will differ from the total concentration of silver nitrate because silver-nitrate ion pairs are formed. Consequently the standard potential of the half-reaction can be evaluated only by a much more circuitous procedure, which will be described in Section 14.3. It amounts to measuring E with each of a number of solutions of silver nitrate having different concentrations, calculating the value of the quantity $(E - 0.05916 \log c_{AgNO_3})$ for each solution, and extrapolating these values to $c_{AgNO_3} = 0$. As the concentration of silver nitrate decreases, $\log y_{Ag^+}$ approaches zero, and in addition $[Ag^+]$ approaches c_{AgNO_3} because the ion pairs become more and more completely dissociated. The extrapolated value is the standard potential.

Similar problems arise in attempting to use the standard potential once it has been evaluated. If we could measure the potential of a silver electrode in a solution containing an unknown concentration of silver nitrate, we could calculate the activity of silver ion from eq. (8.8), but we could not calculate the equilibrium concentration of silver ion without knowing the value of y_{Ag^+} in that particular solution—and even then we could not calculate the total concentration of silver nitrate without knowing both the concentration of nitrate ion and the formation constant of the silver–nitrate ion pair. The difficulty is compounded by the fact that the unknown solution may contain other solutes in addition to silver nitrate. These will certainly affect the activity coefficients of both silver ion and nitrate ion, and if another nitrate is present the extent of ion-pair formation may be even more drastically affected. Pure solutions are rarely if ever encountered outside the laboratory, and even inside the laboratory mixtures of different solutes are often needed for chemical reasons. In work with solutions containing iron(III) it is usually desirable to add either an acid or a complexing agent to suppress the acidic dissociation of iron(III) ion and prevent the precipitation of hydrous iron(III) oxide, and it would clearly be impossible to study complex formation without adding a ligand to a solution containing a metal ion.

For reasons like these it is often advantageous to write the Nernst equation in a different form. We assume that a relatively large concentration of some acid, base, complexing agent, or buffer system is present in addition to the redox couple being studied. Experiments in which the concentrations of the oxidized and reduced forms of the couple must exceed about 0.01 M are rare, and "relatively large" may therefore mean anything above about 0.1 M. In studying the iron(III)–iron(II) couple we might select 1 M sulfuric acid. The sulfuric acid is called a *supporting electrolyte*. Among other purposes, the supporting electrolyte serves to keep the total ionic concentration of the solution virtually constant even though the much smaller concentrations of iron(III) and iron(II) may vary over wide ranges, and thus it keeps the activity coefficients of the iron(III) and iron(II) ions fixed. Various kinds of complexes may exist in such a solution: for iron(III) these include $FeOH^{2+}$, $FeSO_4^+$, and $Fe(SO_4)_2^-$. The total concentration of iron(III) in all its forms is given by

$$[Fe(III)] = [Fe^{3+}]+[FeOH^{2+}]+[FeSO_4^+]+[Fe(SO_4)_2^-]+\cdots$$
$$= [Fe^{3+}](1+\beta_{1,\,OH}-[OH^-]+\beta_{1,SO_4^{2-}}[SO_4^{2-}]+\beta_{2,SO_4^{2-}}[SO_4^{2-}]^2+\cdots)$$

The dots mean that other complexes may be present as well, and $\beta_{n,\,L}$ $(= [FeL_n]/[Fe^{3+}][L]^n)$ is the overall formation constant of the nth complex with the ligand L.

Since the concentration of sulfuric acid is large and constant, the concentrations of hydroxide and sulfate ions are constant, and we can therefore write

$$[Fe^{3+}] = [Fe(III)] \times \frac{1}{1+\beta_{1,\,OH}-[OH^-]+\beta_{1,\,SO_4^{2-}}[SO_4^{2-}]+\beta_{2,\,SO_4^{2-}}[SO_4^{2-}]^2+\cdots}$$
$$= \alpha_{Fe^{3+}}[Fe(III)].$$

Whatever the values of the formation constants may be, and even if the solution contains complexes that we know nothing about, the formation function $\alpha_{Fe^{3+}}$ will have some fixed value. Combining a similar equation for iron(II) with the Nernst equation,

$$E = E^0(Fe^{3+}, Fe^{2+}) - 0.059\ 16 \log \frac{a_{Fe^{2+}}}{a_{Fe^{3+}}}$$

$$= E^0(Fe^{3+}, Fe^{2+}) - 0.059\ 16 \log \frac{y_{Fe^2+}[Fe^{2+}]}{y_{Fe^3+}[Fe^{3+}]}$$

$$= E^0(Fe^{3+}, Fe^{2+}) - 0.059\ 16 \log \frac{y_{Fe^2+}\alpha_{Fe^2+}[Fe(II)]}{y_{Fe^3+}\alpha_{Fe^3+}[Fe(III)]}$$

$$= E^0(Fe^{3+}, Fe^{2+}) - 0.059\ 16 \log \frac{y_{Fe^2+}\alpha_{Fe^2+}}{y_{Fe^3+}\alpha_{Fe^3+}} - 0.059\ 16 \log \frac{[Fe(II)]}{[Fe(III)]}. \qquad (8.9)$$

We can now define a new quantity, the *formal potential $E^{0\prime}$* of the iron(III)–iron(II) couple in 1 M sulfuric acid, by the equation

$$E^{0\prime}(Fe(III), Fe(II);\ 1\ M\ H_2SO_4) = E^0(Fe^{3+}, Fe^{2+}) - 0.059\ 16 \log \frac{y_{Fe^2+}\alpha_{Fe^2+}}{y_{Fe^3+}\alpha_{Fe^3+}}. \qquad (8.10)$$

The formal potential of any particular couple in any particular supporting electrolyte is a constant because all of the quantities appearing on the right-hand side of eq. (8.10) are constants. Changing the nature or concentration of the supporting electrolyte will, however, affect the values of the ys and αs, and therefore the symbol of the formal potential includes not only a prime, to distinguish it from a standard potential, but also an identification of the supporting electrolyte to which it refers.

Combining eqs. (8.9) and (8.10) gives

$$E = E^{0\prime}(Fe(III), Fe(II);\ 1\ M\ H_2SO_4) - 0.059\ 16 \log \frac{[Fe(II)]}{[Fe(III)]} \qquad (8.11)$$

or, in general, for the half-reaction $Ox + ne = Red$,

$$E = E^{0\prime}(Ox, Red;\ supporting\ electrolyte) - \frac{0.059\ 16}{n} \log \frac{[Red]}{[Ox]} \qquad (8.12)$$

where [Red] and [Ox] denote the *total* concentrations of the reduced and oxidized forms of the couple.

All this has two advantages and one disadvantage. One of the advantages is that a formal potential can be evaluated simply by measuring the potential obtained with a single solution containing the supporting electrolyte and known total concentrations of iron(III) and iron(II). Of course a chemist would actually make several different measurements with different solutions, both to guard against accidental errors and to obtain some idea of the precision of the result, but far less time and effort have to be expended on evaluating a formal potential than are needed to evaluate a standard potential. The other advantage is that the concentration, or ratio of concentrations, appearing in the logarithmic term can be calculated directly from the formal potential and the value of E obtained with an unknown solution. This is in marked contrast to the difficulty of finding any concentration from eq. (8.2) and the standard potential.

The disadvantage is that the formal potential of a couple differs from one supporting electrolyte to another. The formal potentials of the iron(III)–iron(II) couple in some supporting electrolytes are

Supporting electrolyte	$E^{\circ\prime}$ (Fe(III), Fe(II)), V vs. N.H.E.
0.5 M HCl	$+0.71$
10 M HCl	$+0.53$
1 M HClO$_4$	$+0.735$
1 M H$_2$SO$_4$	$+0.68$
1 M K$_2$C$_2$O$_4$, pH 5	$+0.01$
10 M NaOH	-0.68

whereas the standard potential of the iron(III)–iron(II) couple is $+0.771$ V vs. N. H. E. The differences among the formal potentials are due partly to activity effects but chiefly to complex formation. Iron(III) is a weaker oxidizing agent in 1 M perchloric acid than at infinite dilution because the activity coefficient of the Fe^{3+} ion in 1 M perchloric acid is smaller than that of the Fe^{2+} ion. (Neither iron(III) nor iron(II) is complexed to any significant extent by hydroxide or perchlorate ion in this supporting electrolyte.) It is a still weaker oxidizing agent in 1 M sulfuric acid because iron (III) is more strongly complexed by sulfate ion than iron(II) is: the formation constants of the iron(III)–sulfate complexes are $K_1 = 1 \times 10^3$ mol^{-1} dm^3 and $K_2 = 10$ mol^{-1} dm^3 while the formation constant of FeSO$_4(aq)$ is only 2.5 mol^{-1} dm^{-3}, which is unusually small for an ion pair of this charge type. The values in chloride, oxalate, and hydroxide media reflect similar differences between the extents of complexation of the two oxidation states. The necessity of measuring the formal potential in each new supporting electrolyte is not an unmixed disadvantage because useful information about complex formation can often be obtained from the result.

The supporting electrolyte often serves to maintain a constant concentration of hydronium ion or some other substance involved in a hal-freaction. When it does, the Nernst equation can be further simplified by separating the constant concentration from the logarithmic term and incorporating it into the formal potential. For the arsenic(V)–arsenic (III) couple in 1 M sulfuric acid, for example, the half-reaction is H$_3$AsO$_4(aq)+2$ H$^+ +2$ e $=$ H$_3$AsO$_3(aq)+$H$_2$O(l), and the concentration (or activity) of hydronium ion is fixed because the concentration of sulfuric acid is fixed. We would write

$$E = E^0(\text{H}_3\text{AsO}_4(aq),\ \text{H}_3\text{AsO}_3(aq)) - \frac{0.059\ 16}{2} \log \frac{a_{\text{H}_3\text{AsO}_4(aq)}}{a_{\text{H}_3\text{AsO}_4(aq)}a_{\text{H}^+}^2}$$

$$= E^0(\text{H}_3\text{AsO}_4(aq),\ \text{H}_3\text{AsO}_3(aq)) + 0.059\ 16 \log a_{\text{H}^+}$$

$$- \frac{0.059\ 16}{2} \log \frac{y_{\text{H}_3\text{AsO}_3}\alpha_{\text{H}_3\text{AsO}_3}}{y_{\text{H}_3\text{AsO}_4}\alpha_{\text{H}_3\text{AsO}_4}} - \frac{0.059\ 16}{2} \log \frac{[\text{As(III)}]}{[\text{As(V)}]}$$

$$= E^{0\prime}(\text{As(V)},\ \text{As(III)};\ 1M\ \text{H}_2\text{SO}_4) - \frac{0.059\ 16}{2} \log \frac{[\text{As(III)}]}{[\text{As(V)}]}$$

where the αs reflect the existence of species such as $H_2AsO_3^-$, $H_2AsO_4^-$, and $H_4AsO_4^+$ in addition to $H_3AsO_3(aq)$ and $H_3AsO_4(aq)$. If the concentration of sulfuric acid is changed, or if another acid is used in its place, the values of the activity coefficients and formation functions may change, and so may the formal potential. This is another reason why the formal potential of a couple may depend on the supporting electrolyte. Similarly, the Nernst equation for the tin(IV)–tin(II) couple, which is usually represented by the equation $SnCl_6^{2-}+2\,e = Sn^{2+}+6\,Cl^-$, would be written as

$$E = E^{0\prime}(Sn(IV), Sn(II); 1\,M\,HCl) - \frac{0.059\,16}{2}\log\frac{[Sn(II)]}{[Sn(IV)]}$$

if 1 M hydrochloric acid were used as the supporting electrolyte. The activity of chloride ion would be constant and its value would be included in the formal potential.

Formal potentials can be interpreted and combined in the same ways as standard potentials (Sections 8.5 and 8.6). The results differ slightly because of the factors that are included in the formal potential but not in the standard potential. The equilibrium constant of the reaction between metallic silver and iron(III) can be calculated by combining either the standard potentials or the formal potentials of the couples involved. Combining the standard potentials, which are $E^0(Ag^+, Ag(s)) = +0.7994$ V and $E^0(Fe^{3+}, Fe^{2+}) = +0.771$ V, gives $K = a_{Ag^+}a_{Fe^{2+}}/a_{Fe^{3+}} = 3.3\times10^{-1}$ mol dm^{-3}. In 1 M sulfuric acid the formal potentials are $E^0(Ag(I), Ag(s); 1\,M\,H_2SO_4) = +0.77$ V and $E^{0\prime}(Fe(III), Fe(II); 1M\,H_2SO_4) = +0.68$ V, and combining these gives $K = [Ag(I)][Fe(II)]/[Fe(III)] = 3.0\times10^{-2}$ mol dm^{-3}. The second result takes into account the values of the activity coefficients and formation functions in 1 M sulfuric acid and is valid only for that particular supporting electrolyte, but it provides much the more convenient description of the extent to which the reaction actually takes place. Since the reaction yields one millimole of iron(II) along with each millimole of silver(I), it must be true that $[Ag(I)] = [Fe(II)]$, but there is no reason to suppose that the activities of Ag^+ and Fe^{2+} are anywhere near equal.

If the formal potential of one couple in the reaction medium is unknown, the standard potential can often be used as an approximation to it. The value of K obtained in this way will also be only an approximation, but may be useful none the less. You should never try to combine two formal potentials that pertain to different supporting electrolytes.

SUMMARY: The formal potential of a redox couple in a particular kind of ionic solution depends on the standard potential of the couple, and also on the activity coefficients and formation functions of the oxidized and reduced forms. It therefore provides a truer picture of the oxidizing and reducing abilities of the couple than the standard potential alone.

8.8. Some consequences of the Nernst equation

PREVIEW: This section describes some additional similarities between redox chemistry and Lowry–Brønsted acid-base chemistry.

Section 8.2 discussed the similarity between redox reactions, in which electrons are transferred from reducing agents to oxidizing agents, and Lowry–Brønsted acid-base reactions, in which protons are transferred from acids to bases. Nowhere is the similarity more pronounced than it is in their algebraic descriptions. The strengths of an acid and its conjugate base, and the effects of their concentrations on the pcH, appear in the Henderson–Hasselbalch equation

$$\text{pcH} = \text{p}K_a - \log \frac{[\text{acid}]}{[\text{base}]} \qquad (8.13)$$

while the strengths of a reducing agent and its oxidized form, and the effects of their concentrations on the potential, appear in the Nernst equation

$$E = E^{0\prime}(\text{Ox}, \text{Red}; \text{supporting electrolyte}) - \frac{2.303RT}{nF} \log \frac{[\text{Red}]}{[\text{Ox}]} \qquad (8.14)$$

These two equations have exactly the same form. The only difference between them arises from the fact that a pcH-value is a dimensionless number while a potential has the units of volts, so that the dimensionless logarithmic term in the Nernst equation must be multiplied by a factor having the same units. All these similarities have several important consequences.

Every solution containing an acid (or a base) also contains some of its conjugate base (or acid), and every solution containing a reducing (or oxidizing) agent also contains some of its oxidized (or reduced) form. Even if an acid HB is an extremely weak acid, and even if an enormous excess of a strong acid is added to it, the reaction $B^- + H_3O^+ = HB(aq) + H_2O(l)$ in an aqueous solution, or the reaction $B^- + H_2So^+ = H_2So + B^-$ in a non-aqueous one, can never be driven all the way to completion. A little B^- must always be present, though there will be very little of it indeed if the pcH is more than three or four units smaller than $\text{p}K_a$. This is the result of the logarithmic dependence in eq. (8.13). Only by making the pcH infinitely negative could the concentration of B^- be decreased to zero, and this is impossible. Similarly, a solution that contained the reduced form of a redox couple but none at all of the oxidized form would have an infinitely negative potential according to eq. (8.14), and therefore it would be an infinitely powerful reducing agent. Even if atmospheric oxygen were kept away from it, hydrogen ions or molecules of the solvent would be reduced by reactions like $\text{Red} + H^+ = \text{Ox} + \frac{1}{2}H_2(g)$ or $\text{Red} + H_2O(l) = \text{Ox} + OH^- + \frac{1}{2}H_2(g)$. Conversely, a solution that contained an oxidizing agent but none at all of its reduced form would have an infinitely positive potential and would attack the solvent by a reaction like $2\text{Ox} + H_2O(l) = 2\,\text{Red} + 2\,H^+ + \frac{1}{2}O_2(g)$.

Every amphoteric solvent exerts levelling effects on acids and bases that are dissolved in it. An aqueous solution cannot contain an appreciable concentration of an acid that is stronger than hydronium ion, or of a base that is stronger than hydroxide ion. There are similar limits to the strengths of oxidizing and reducing agents that are stable in any solvent that, like water, can be both oxidized and reduced. Redox chemistry in non-aqueous solvents is being studied intensively by electrochemists among others, but is still much less well understood than redox chemistry in water, and this discussion will therefore be limited to aqueous solutions. At 298 K the standard potentials of the couples $H^+ + e = \frac{1}{2}H_2(g)$ and $O_2(g) + 4 H^+ + 4 e = 2 H_2O(l)$ in aqueous solutions are ± 0.0000 V and $+1.229$ V, respectively. In the presence of 1 M hydrogen ion the reaction $Red + H^+ = Ox + \frac{1}{2}H_2(g)$ will have an equilibrium constant equal to 1 if the standard potential of the couple $Ox + e = Red$ is equal to ± 0 V vs. N. H. E. This equilibrium constant is so large that a solution of the reduced form of the couple would tend to react extensively with hydrogen ion—especially if the hydrogen gas is allowed to escape into the atmosphere. If the standard potential of the couple is more negative than ± 0 V vs. N. H. E. the equilibrium constant will be even larger and the reaction between Red and hydrogen ion will be still more extensive. The reaction may be slow, and solutions of some quite strong reducing agents, such as vanadium(II) and chromium(II), can be kept for some time in the presence of 1 M hydrogen ion although the standard potentials of the vanadium(III)–vanadium(II) and chromium(III)–chromium(II) couples are -0.255 V and -0.41 V vs. N. H. E., respectively. Nevertheless these solutions are inherently unstable. The same thing is true of oxidizing agents having standard potentials more positive than about $+1.2$ V, for these tend to be reduced by water in the presence of 1 M hydrogen ion. Again, such reactions are often slow. Indeed, solutions of cerium(IV) in 1 M sulfuric acid have been kept for decades without observable decomposition although $E^{0\prime}(Ce(IV)$ Ce(III); 1 M $H_2SO_4) = +1.44$ V, which is so much more positive than the potential of the oxygen–water couple that almost 99.99% of the cerium(IV) would be reduced at equilibrium. The redox analog of the acid-base levelling effect is much more elastic because redox processes often reach equilibrium far less rapidly than acid-base reactions do.

Buffers were discussed in Section 6.10. If a small amount of a base or an acid is added to a solution containing an acid HB and its conjugate base B^-, it will be consumed by reacting with one or the other of these. The ratio $[HB]/[B^-]$ will change, and so will the pcH, but the change of pcH will be smaller than it would have been in the absence of the HB and B^-. The buffer capacity of the solution increases as the sum of the concentrations of HB and B^- increases; for any particular value of the sum, the buffer capacity is highest if the two concentrations are equal. The redox equivalent of buffering is called *poising*. A solution containing the oxidized and reduced forms Ox and Red of a redox couple can react with a reducing or oxidizing agent that is added to it. The reaction alters the concentrations of Ox and Red, and consequently the potential changes, but its change is smaller than it would have been if the Ox and Red had not been present. The change decreases if the concentrations of Ox and Red are increased, or if they are made more nearly equal. Poising is exactly analogous to buffering, and is equally important in biochemistry. Hemoglobin and its oxidized form constitute the chief overall poising system in the human body, and other couples, including the adenosinetriphosphate–adenosinediphosphate couple, are intimately involved in local muscular and neural actions.

SUMMARY: The similarity between the Nernst equation and the Henderson equation; the fact that a solution containing one substance appearing in a redox couple must also contain at least a trace of the other, just as a solution containing an acid (or a base) must also contain some of its conjugate base (or acid); and the existence of redox processes resembling the levelling effect and buffering reinforce the general resemblance between redox chemistry and Lowry–Brønsted acid-base chemistry.

8.9. The rates and mechanisms of redox reactions

PREVIEW: This section stresses the differences between the mechanisms of redox reactions and those of Lowry–Brønsted acid-base reactions.

The oxidation of iodide ion by hypochlorite ion, and the oxidation of arsenic(III) acid by triiodide ion, were discussed in some detail in Chapter 3 but were not identified as redox reactions. This section will deal with the mechanisms of redox reactions in a more general way, emphasizing the differences between these and the mechanisms of Lowry–Brønsted acid-base reactions to counteract the emphasis in Section 8.8 on the similarities between these two kinds of reactions.

A Lowry–Brønsted acid-base reaction can be represented by the general equation $B + HA = BH^+ + A^-$. It involves the transfer of one proton, and it involves the intermediate BHA and its ionized form BH^+A^-. Reactions like $H_2C_2O_4(aq) + H_2NCH_2CH_2NH_2 = C_2O_4^{2-} + {}^+H_3NCH_2CH_2NH_3^+$, which involve the transfer of two protons, can be written. However, no such reaction is known in which both protons are transferred in a single step, and the formation of an intermediate like

$$
\begin{array}{c}
\overset{\displaystyle O}{\underset{\displaystyle \text{C}}{\diagup}}\text{—OH}:NH_2\diagdown \\
| \qquad\qquad CH_2 \\
| \qquad\qquad | \\
\underset{\displaystyle O}{\text{C}}\text{—OH}:NH_2\diagup CH_2
\end{array}
$$

(which contains a ten-membered ring) is extremely unlikely. Changes of solvation may accompany the transfer of protons. The products BH^+ and A^- may be solvated even if the reactants B and HA are not, or vice versa, and the hydration of either reactant in an aqueous solution may give rise to complications like

$$
B: + \overset{H}{\underset{H}{\diagup}}O:HA \; = \; \overset{B:H}{\underset{H}{\diagup}}O:HA \left\langle
\begin{array}{l}
\longrightarrow BH^+ + \overset{H}{\diagup}O\text{—}HA^- \\
\\
\longrightarrow BH\diagdown O\text{—}H^+ + A^- \\
\qquad\quad H\diagup
\end{array}
\right.
$$

These may be ignored for our purposes: the crux of the matter is that one proton is transferred, and the energy of activation involved in its transfer is almost always negligibly small. There are many proton–transfer reactions that are slow, but with only a few kinds of exceptions they are slow simply because the free energies of the reactants are smaller than those of the products. This is true, for example, of a reaction such as $C_6H_5OH(aq)+H_2O(l) = C_6H_5O^-+H_3O^+$, for which the forward rate constant is only about 0.1 s^{-1}; for an even weaker acid the rate constant would be even smaller. However, the reverse reaction is extremely fast: its rate constant is roughly $10^{10} \text{ dm}^3 \text{ mol}^{-1} \text{ s}^{-1}$, which is in the range expected for diffusion-controlled reactions. That could not be true if the reverse reaction involved an appreciable energy of activation, and therefore the energy of the transition state must be indistinguishable from the sum of the energies of $C_6H_5O^-$ and H_3O^+. No additional expenditure of energy is needed to reach the transition state in the forward reaction. Reactions involving so-called "C-acids", in which the proton that is transferred is originally attached to a carbon atom, are among the Lowry–Brønsted acid-base reactions that may involve appreciable energies of activation, for they may be accompanied by internal electronic rearrangements. Typical examples are the reactions of nitromethane and acetone with hydroxide ion, for which the overall equations are

$$CH_3NO_2(aq)+OH^- = CH_2{=}NO_2^- +H_2O(l)$$

$$\text{and } CH_3{-}\overset{\overset{\textstyle O}{\|}}{C}{-}CH_3(aq)+OH^- = CH_3{-}\overset{\overset{\textstyle O^-}{|}}{C}{=}CH_2+H_2O(l).$$

Redox reactions behave differently in three ways. One is that more or less extensive structural changes are very common instead of being exceptional. Both of the redox reactions discussed in Chapter 3 involved such changes. The simplest redox reactions are those, such as $Cr(OH_2)_6^{2+}+Fe(CN)_6^{3-} = Cr(OH_2)_6^{3+}+Fe(CN)_6^{4-}$ in which the inner coordination spheres remain intact. Changes of solvation may still occur in the outer ones, but because such *charge-transfer processes* involve only the transfer of electrons they may be extremely fast. There is even some evidence to indicate that the transfer of electrons may occur without actual collision between the reactants. *Atom-transfer reactions*, like $OCl^-+I^- = OI^-+Cl^-$—which seems to involve the transfer of an atom of oxygen, although the atom of oxygen finally attached to the iodine atom may not be the same one that was originally attached to the chlorine atom—are often far slower, as is the reaction $2 \text{ MnO}_4^-+5 \text{ H}_2C_2O_4(aq)+6 \text{ H}^+ = 2 \text{ Mn}^{2+}+10 \text{ CO}_2(aq)+8 \text{ H}_2O(l)$, where the structural havoc is evident. To prevent hasty generalization it must be added that the reaction $Fe^1(H_2O)_6^{2+}+Fe^2(H_2O)_6^{3+} = Fe^1(H_2O)_6^{3+}+Fe^2(H_2O)_6^{2+}$, which may look like a charge-transfer reaction involving nothing more than the transfer of an electron from one atom of iron to the other, is fairly slow, so that some sort of intermediate disruption of the coordination shells must be involved.

The second difference is a consequence of the first: the majority of redox reactions involve substantial energies of activation, and very slow redox reactions are therefore very common. The extreme stabilities of acidic solutions of cerium(IV), which were mentioned in Section 8.8, reflect the fact that the reaction $4 Ce(IV) + 2 H_2O(l) = 4 Ce(III) + 4 H^+ + O_2(g)$ is too slow to be detected, even in boiling solutions. There are a host of other redox reactions that have large equilibrium constants but that occur only very slowly, if at all, in the absence of substances that catalyze them.

A third difference is that the transfer of two electrons at once is quite common. Chapter 3 gave two examples in which two electrons were transferred simultaneously from one atom to another in the activated complex. Some curious consequences result. The reaction between mercury(II) and tin(II) ions, $Hg_2^{2+} + Sn^{2+} = 2 Hg(l) + Sn(IV)$ is fairly rapid; the reaction between iron(III) and tin(II) ions, $2 Fe^{3+} + Sn^{2+} = 2 Fe^{2+} + Sn(IV)$, is much slower. The frequency of collision is higher with mercury(II) ion, because this is less highly charged, and therefore less strongly repelled, than iron(III) ion, but this is responsible for only a fraction of the difference of rate. In the activated complex Hg_2Sn^{4+} a two-electron transfer can occur, and yields mercury(0) and tin(IV) as stable products. In the reaction with iron(III) ion the activated complex is probably $FeSn^{5+}$, which can decompose in three ways: one regenerates the Fe^{3+} and Sn^{2+} ions from which it was formed, the second involves the transfer of one electron and gives Fe^{2+} and Sn^{3+}, and the third involves the transfer of two electrons and gives Fe^+ and Sn^{4+}. Either Fe^{2+} or Sn^{4+} would be a stable product, but Fe^+ and Sn^{3+} are so unstable that they cannot be detected in solutions at equilibrium. Both Fe^+ and Sn^{3+} must have very high free energies of formation, and neither is a likely product. Actually it appears that Sn^{3+} does form, and that the reaction occurs in two successive one-electron steps, but its formation is slow because few of the $FeSn^{5+}$ ions decompose in this way.

Similarly, the reaction between cerium(IV) and chromium(III) is slow because an ion of cerium(IV) can accept only a single electron, and because intermediates containing two or three cerium atoms are improbable, as was Fe_2Sn^{8+} in the preceding example. Hence an atom of chromium can lose only one electron at a time, and the reaction must yield Cr(IV) first and then Cr(V) before the stable product chromium (VI) is produced. Since both chromium(IV) and chromium(V) are extremely unstable, the reactions that produce them are slow, and of course the overall reaction $3 Ce(IV) + Cr(III) = 3 Ce(III) + Cr(VI)$ is slow as a result.

One of the reactions in such a chain may yield a product that decomposes or reacts in some other way before it can take part in the next step. If a carbonyl compound, such as acetone, is treated with a reducing agent that can donate two electrons simultaneously, the result can be represented by the equation

$$CH_3-\overset{\overset{\displaystyle O}{\|}}{C}-CH_3 \underset{-2e}{\overset{+2e}{\rightleftharpoons}} CH_3-\overset{\overset{\displaystyle \cdot\cdot}{O}}{\underset{\cdot\cdot}{C}}-CH_3 \underset{-2H^+}{\overset{+2H^+}{\rightleftharpoons}} CH_3-\overset{\overset{\displaystyle OH}{|}}{\underset{|}{C}}-CH_3$$
$$H$$

If an ion or molecule of the reducing agent can provide only one electron, the intermediate in this reaction cannot be obtained in a single step, and another intermediate must be formed first:

$$CH_3-\overset{\overset{\textstyle O}{\|}}{C}-CH_3 \underset{-e}{\overset{+e}{\rightleftharpoons}} CH_3-\overset{\overset{\textstyle \ddot{O}}{|}}{\underset{|}{C}}-CH_3 \underset{-e}{\overset{+e}{\rightleftharpoons}} CH_3-\overset{\overset{\textstyle \ddot{O}}{|}}{C}-CH_3 \underset{-2H^+}{\overset{+2H^+}{\rightleftharpoons}} CH_3-\overset{\overset{\textstyle OH}{|}}{\underset{\underset{\textstyle H}{|}}{C}}-CH_3$$

$$\frac{1}{2}\begin{array}{c}CH_3-\overset{\overset{\textstyle \ddot{O}}{|}}{C}-CH_3 \\ | \\ CH_3-\overset{}{\underset{\underset{\textstyle \ddot{O}}{\cdot}}{C}}-CH_3\end{array} \underset{-H^+}{\overset{+H^+}{\rightleftharpoons}} \frac{1}{2}\begin{array}{c}CH_3-\overset{\overset{\textstyle OH}{|}}{C}-CH_3 \\ | \\ CH_3-\overset{}{\underset{\underset{\textstyle OH}{|}}{C}}-CH_3\end{array}$$

In this scheme the upper line depicts one possible reaction path. If the radical anion

$$CH_3-\overset{\overset{\textstyle \ddot{O}}{|}}{\underset{\cdot}{C}}-CH_3$$

dimerizes before it accepts a second electron, an entirely different product will be obtained.

> SUMMARY: Redox reactions differ from Lowry–Brønsted acid-base reactions in several ways. Extensive structural rearrangements and substantial energies of activation are more common in redox reactions. Two electrons can be transferred simultaneously (whereas two protons cannot), and side reactions of intermediates are also possible.

8.10. Equilibrium constants of half-reactions

> PREVIEW: This section describes an alternative way of making equilibrium calculations for redox reactions.

A half-reaction, such as $Fe^{3+}+e = Fe^{2+}$, may be regarded as a process in which an electron, acting as a "ligand", is added to the oxidized form of a couple. A sort of equilibrium constant can be written for that process:

$$K(Fe^{3+}, Fe^{2+}) = \frac{a_{Fe^{2+}}}{a_{Fe^{3+}}a_e}. \tag{8.15}$$

To assign a numerical value to such a constant, it is necessary to define the standard state of the electron, in which its activity will be considered to be equal to 1. Consistency with the traditional assumption that the hydrogen ion–hydrogen gas couple has a standard potential of ± 0.0000 V is attained by saying that the electron is in its standard state when it is in equilibrium with hydrogen ion and hydrogen gas, both at unit activity. This gives

$$K(H^+, H_2(g)) = \frac{a_{H_2(g)}}{a_{H^+}^2 a_e^2} = 1 \text{ atm mol}^{-2} \text{ dm}^6 \tag{8.16}$$

in which the activity of the electron is arbitrarily taken to be dimensionless (it is certainly not equal to 1 M!).

On this basis eq. (8.4) can be used to assign a numerical value to the equilibrium constant of the iron(III)–iron(II) half-reaction, or any other half-reaction for which the standard potential is known. At 25°, since $E^0(Fe^{3+}, Fe^{2+}) = +0.771$ V vs. N. H. E.,

$$\log K(Fe^{3+}, Fe^{2+}) = E^0(Fe^{3+}, Fe^{2+})/0.059\ 16 = 0.771/0.059\ 16 = 13.0_3$$

whence $K(Fe^{3+}, Fe^{2+}) = 1.1 \times 10^{13}$. In the same way, since $E^0(Cr^{3+}, Cr^{2+}) = -0.41$ V vs. N. H. E., $K(Cr^{3+}, Cr^{2+}) = 1.2 \times 10^{-7}$. In general,

$$\log K(Ox, Red) = nFE^0(Ox, Red)/2.303RT$$
$$(= nE^0(Ox, Red)/0.059\ 16 \text{ at } 298 \text{ K}) \qquad (8.17a)$$

and a similar expression

$$\log K'(Ox, Red;\ \text{supporting electrolyte})$$
$$= nFE^{0\prime}(Ox, Red;\ \text{supporting electrolyte})/2.303\ RT$$
$$(= nE^{0\prime}/0.059\ 16 \text{ at } 298 \text{ K}) \qquad (8.17b)$$

relates the conditional equilibrium constant of a half-reaction in any particular supporting electrolyte to the corresponding formal potential. The formulations in parentheses may be compared with eq. (8.4). Values of K and K' for a number of redox couples are given along with their standard and formal potentials in Appendix VI.

If the equation for one chemical reaction can be obtained by subtracting the equation for another from the equation for a third, the equilibrium constants of the three reactions are related by

$$\begin{array}{lll} A = B; & K = K_1 & = a_B/a_A, \\ C = B; & K = K_2 & = a_B/a_C, \\ \hline A = C; & K = K_1/K_2 = a_C/a_A. \end{array}$$

The equilibrium constant for the reaction between iron (III) and chromium (II) ions may be evaluated by writing

$$\begin{array}{ll} Fe^{3+} + e = Fe^{2+}; & K = K(Fe^{3+}, Fe^{2+}) = 1.1 \times 10^{13} \\ Cr^{3+} + e = Cr^{2+}; & K = K(Cr^{3+}, Cr^{2+}) = 1.2 \times 10^{-7} \\ \hline Fe^{3+} + Cr^{2+} = Fe^{2+} + Cr^{3+}; & K = K(Fe^{3+}, Fe^{2+})/K(Cr^{3+}, Cr^{2+}) \\ & = 1.1 \times 10^{13}/1.2 \times 10^{-7} \\ & = 9 \times 10^{19} = a_{Fe^2} + a_{Cr^3} + /a_{Fe^3} + a_{Cr^2} +. \end{array}$$

which is much easier and shorter than the procedure described in the first paragraph of Section 8.6. Combining the values of K' for these two couples in 5 M hydrochloric acid would give a value of the conditional equilibrium constant, $K' = [Fe(II)][Cr(III)]/[Fe(III)][Cr(II)]$, in 5 M hydrochloric acid. From the values of K' given in Appendix VI, the conditional equilibrium constant is found to be equal to 3.8×10^{17}. This is smaller than the value of K just given because complexation with chloride ion decreases the strength of iron(III) as an oxidizing agent.

For the reaction between permanganate ion and zinc, whose equilibrium constant was calculated in the second paragraph of Section 8.6, simply subtracting the equations for the two half-reactions does not give the desired result because the numbers of electrons are not the same. Here one must write

$$2(MnO_4^- + 8H^+ + 5e = Mn^{2+} + 4H_2O(l)); \qquad K = K_1 = a_{Mn^2+}^2/a_{MnO_4^-}a_{H^+}^{16}a_e^{10},$$
$$\underline{5(Zn^{2+} + 2e = Zn(s)); \qquad\qquad\qquad\qquad K = K_2 = 1/a_{Zn^2+}^5 a_e^{10},}$$
$$2MnO_4^- + 5Zn(s) + 16H^+ = 2Mn^{2+} + 5Zn^{2+} + 8H_2O(l); \quad K = K_1/K_2.$$

Since $K(MnO_4^-, Mn^{2+}) = a_{Mn^2+}/a_{MnO_4^-}a_{H^+}^8 a_e^5$, $K_1 = [K(MnO_4^-, Mn^{2+})]^2$; similarly, $K_2 = [K(Zn^{2+}, Zn(s))]^5$. The final result,

$$\begin{aligned} K &= [K(MnO_4^-, Mn^{2+})]^2/[K(Zn^{2+}, Zn(s))]^5 \\ &= (4.2 \times 10^{127})^2/(1.6 \times 10^{-26})^5 \\ &= 1.6 \times 10^{384} \, mol^{-11} \, dm^{33}, \end{aligned}$$

is again identical with the one obtained in Section 8.6. In general, if the two half-reactions are $Ox_1 + n_1 e = Red_1$ and $Ox_2 + n_2 e = Red_2$, the equilibrium constant for the reaction between Ox_1 and Red_2 is given by

$$K = [K(Ox_1, Red_1)]^{n/n_1}/[K(Ox_2, Red_2)]^{n/n_2} \qquad (8.18)$$

which is analogous to eq. (8.3), and where **n** is again the number of electrons that would have appeared on each side of the balanced equation for the reaction if they had not been dropped in writing it.

Other kinds of calculations are facilitated as well. The equilibrium constant of the half-reaction $Ag(CN)_2^- + e = Ag(s) + 2\,CN^-$ can be obtained very easily by writing

$$Ag^+ + e = Ag(s); \qquad\qquad K = K(Ag^+, Ag(s)),$$
$$\underline{-(Ag^+ + 2CN^- = Ag(CN)_2^-); \qquad K = \beta_2,}$$
$$Ag(CN)_2^- + e = Ag(s) + 2CN^-; \qquad K = K(Ag(CN)_2^-, Ag(s))$$
$$= K(Ag^+, Ag(s))/\beta_2.$$

Very much more space had to be expended to obtain an equivalent result in Section 8.6.

Nothing can be done in this way that cannot also be done with standard potentials, but most equilibrium calculations are easier with the equilibrium constants of half-reactions than they are with standard potentials. There is another advantage, and it is a more important one. Tables of standard potentials now contain a great deal of equilibrium information that has been obtained in a great many different ways. They contain the standard potentials of half-reactions, such as $Pb^{2+} + 2\,e = Pb(s)$, which were originally deduced by measuring the potentials of electrodes and can be used to predict the potentials of electrodes. They also contain the standard potentials of other half-reactions, such as $Ni^{2+} + 2\,e = Ni(s)$, which were obtained by evaluating the equilibrium constants of chemical reactions and combining them with other standard potentials. The potential of a nickel electrode in an aqueous solution containing nickel(II) ion cannot be predicted by the Nernst equation, because the rate at which electrons are transferred between solid nickel and dissolved nickel(II) ions is so low that other very slow processes affect that potential. Some standard potentials are relevant to electrical measurements while others are not. It would be convenient to separate them, reserving the standard potential for couples that do obey the Nernst equation in electrochemical experiments. If this is done, it will be done slowly, and you will probably find standard potentials and equilibrium constants of half-reactions being used interchangeably for many years.

> SUMMARY: Many equilibrium calculations are facilitated by assigning values to the equilibrium constants of redox half-reactions. The values are derived by saying that the electron is in its standard state when it is in equilibrium with hydrogen gas and hydrogen ion, both at unit activity.

Problems

Use the table of standard and formal potentials in Appendix VI (page 534) in solving these problems. Answers to some of these problems are given on page 520.

8.1. Write balanced equations for the following redox reactions under the conditions specified. Assume that all of the solutions are aqueous.

(a) $Cr_2O_7^{2-} + Fe^{2+} = Cr^{3+} + Fe^{3+}$ acidic solution.
(b) $OsO_4(aq) + H_3AsO_3(aq) = OsO_2(s) + H_3AsO_4(aq)$ acidic solution
(c) $MnO_4^- + C_2O_4^{2-} + Ba^{2+} = BaMnO_4(s) + CO_3^{2-}$ alkaline solution.
(d) $CuI(s) + Ce(SO_4)_3^{2-} = Cu^{2+} + ICl_2^- + Ce^{3+}$ hydrochloric acid solution.
(e) $NaBiO_3(s) + Mn^{2+} = Bi^{3+} + MnO_4^-$ acidic solution.

8.2. Calculate the value of the equilibrium constant for each of the following reactions, using the standard potentials of the couples involved:

(a) $Ce(IV) + Ag(s) = Ce(III) + Ag^+$.
(b) $BrO_3^- + 5 Br^- + 6 H^+ = 3 Br_2(l) + 3 H_2O(l)$.
(c) $2 H_2(g) + O_2(g) = 2 H_2O(l)$.
(d) $PbO_2(s) + Pb(s) + 2 H^+ + 2 HSO_4^- = 2 PbSO_4(s) + 2 H_2O(l)$.
(e) $Cr_2O_7^{2-} + 6 Cr^{2+} + 14 H^+ = 8 Cr^{3+} + 7 H_2O(l)$.
(f) $Hg_2Cl_2(s) = Hg_2^{2+} + 2 Cl^-$.

8.3. The formation constant of the copper(II)-triethanolamine complex, $Cu[N(CH_2CH_2OH)_3]^{2+}$, is equal to 1.6×10^4 mol^{-1} dm^3. What is the value of the standard potential of the half-reaction $Cu[N(CH_2CH_2OH)_3]^{2+} + 2 e = Cu(s) + N(CH_2CH_2OH)_3(aq)$?

8.4. The standard potential of the half-reaction $CuY^{2-} + 2 e = Cu(s) + Y^{4-}$, where Y^{4-} represents the ethylenediaminetetraacetate ion (page 190) is equal to $+0.13$ V vs. N.H.E. at pH 4.5. What is the value of the conditional formation constant of the copper(II)–ethylenediaminetetraacetate chelonate at this pH?

8.5. Evaluate the standard potentials of the following half-reactions:

(a) $Fe^{3+} + 3 e = Fe(s)$.   (b) $Cu^+ + e = Cu(s)$.

8.6. The standard potential of the half-reaction $Fe^{3+} + e = Fe^{2+}$ is equal to $+0.771$ V vs. N.H.E. The formal potential of the iron(III) – iron (II) couple in 0.5 M hydrochloric acid is equal to $+0.71$ V vs. N.H.E. If the difference between these figures is entirely due to the formation of $FeCl^{2+}$, calculate the value of the formation constant of this complex.

8.7. The reaction $Fe^{3+} + Ag(s) = Fe^{2+} + Ag^+$ occurs when an 0.0500 M solution of iron(III) perchlorate is shaken with an excess of metallic silver. When equilibrium is reached it is found that half of the iron(III) has been reduced. What is the value of the equilibrium constant of the reaction?

8.8. An excess of metallic silver is added to a solution that contains 1 M hydrochloric acid and 1.00×10^{-3} M iron(III). Estimate the fraction of the iron(III) that will remain unreduced when equilibrium is reached.

8.9. A solution contains 1.00×10^{-4} M potassium thiocyanate and is saturated with silver thiocyanate. The potential of a silver electrode immersed in the solution is found to be $+0.326$ V vs. N.H.E. Calculate the value of the solubility product of silver thiocyanate.

8.10. If 0.030 mole of zinc is added to 1 dm^3 of the solution described in Problem 8.9, the potential of the silver electrode changes to $+0.346$ V vs. N.H.E. If $ZnSCN^+$ is the only complex that is formed, what is the value of its formation constant?

CHAPTER 9

Chemical Analysis and Stoichiometry

9.1. Introduction

The last five chapters described some of the fundamental ideas that are associated with a number of different kinds of chemical equilibria. This chapter is the first of several that will describe some of the ways in which those ideas can be put to use. It will provide an introduction to chemical analysis, an outline of two important and widely used analytical techniques, and a description of the stoichiometric calculations that these—and many other kinds of chemical processes—involve.

9.2. Chemical analysis and analytical chemistry

PREVIEW: This section defines chemical analysis and the broader field of analytical chemistry, and discusses some of the problems with they are concerned.

Chemical analysis is the branch of human knowledge and ability that enables us to discover what constituents a substance contains, and in what proportions. The substance may be of any nature whatever. It may be a solution in which a chemical reaction has taken place, and its analysis may be aimed at identifying a product of that reaction or at obtaining data from which the equilibrium constant of the reaction can be evaluated. It may be a solution in which a reaction is proceeding while the analysis is being performed, and its analysis may be aimed at elucidating the mechanism or evaluating the rate constant of the reaction. It may be a sample of a rock or some other naturally occurring material, and may be analyzed to obtain information about the conditions under which it must have been formed or to discover whether it contains enough of some useful constituent to repay the cost and trouble of separating that constituent from the others that accompany it. It may be a small volume of someone's blood or spinal fluid, and may be analyzed to reveal whether there is some upset in the body chemistry or whether a desired level of medication is being maintained. It may be a portion of the water or smoke that is discharged from an industrial plant, and may be analyzed to discover whether any toxic or otherwise undesirable materials are present and, if they are, to provide the information that is needed in designing a process for removing them. Chemical analysis is needed in answering most of the fundamental questions that were raised in Chapter 1, and also in coping with almost any problem of applied chemistry.

Chemical analysis is in turn the applied branch of a broader field called analytical chemistry. Whether or not a child is suffering from lead poisoning is easy to decide after the concentration of lead in the child's blood has been found by chemical analysis. Skill, experience, and care are likely to be needed in performing the analysis. To make analysis and interpretation possible, there must be a procedure that will give accurate and reliable results. Analytical chemistry is concerned with devising such procedures, with obtaining and understanding the fundamental information on which they are based, and with devising and improving the instruments and apparatus employed in their execution. The properties of lead must be known and so must the natures and properties of the other constituents of blood. Some of these constituents might act as masking agents and cause the concentration of lead to seem to be smaller than it really is; others might behave in so nearly the same way that lead does that they cause the result to be erroneously high. If such problems arise and cannot be circumvented in some way, it becomes necessary to devise a procedure for separating the lead from the interfering constituents, and therefore techniques of separation must be devised, and studied until their principles and effects are thoroughly understood. In determining the amount or concentration of a substance, one must measure some quantity that is related in some way to that amount or concentration; the nature of the relation must be known so that the result of the measurement can be interpreted correctly, and the theory underlying that relation must be appreciated so that errors can be eliminated or minimized. The measurement may be made with a device as simple as a buret or with one that contains a great deal of sophisticated electronics, and it may involve the recording of some sort of curve and may require computer processing of the data obtained. Developing new kinds of instruments and new ways of handling data, and improving the sensitivity, reliability, and convenience of existing ones, is an important part of progress in analytical chemistry.

Chemical analysis is traditionally divided into several different areas. *Qualitative analysis* is aimed at revealing what constituents are present in a sample; *quantitative analysis* is aimed at revealing their amounts or concentrations. Qualitative analysis provides answers to questions like "Does this compound contain a carbonyl $\left(\diagup C{=}O \right)$ group?" and "Does this water contain fluoride?" Questions like the first one are vitally important to synthetic chemists. Questions like the second one are less important unless they are very carefully phrased, for it does us very little good to know that there is some fluoride in a sample of water: depending on whether the concentration of fluoride is $1\ M$, $10^{-5}\ M$, or $10^{-10}\ M$, one might want to recover the fluoride from the water, drink it, or fluoridate it. *Semiquantitative analysis*, which also provides some rough indication of the proportion of each constituent that is found to be present, is much more useful. A semiquantitative analysis might indicate that the concentration of fluoride was between 3×10^{-6} M and 3×10^{-5} M and consequently that the water was fit to drink; a quantitative one might show that the concentration of fluoride was 1.22×10^{-5} M and consequently that the fluoridation machinery was operating properly. Nowadays most semiquantitative analyses are made by techniques such as emission spectroscopy, X-ray fluorimetry, and flame photometry for elements and mass spectrometry and infrared, ultraviolet, and Raman spectroscopy for compounds and functional groups. The majority of these are outside the scope of this book, but the analytical chemist who designs a procedure for the quantitative analysis of a sample must generally take into account, and the analyst who selects and executes such a procedure must generally have, the kind of prior knowledge about the composition of the sample that a semiquantitative analysis would provide. The concentration of lead in a saturated aqueous solution of lead chloride can be determined in ways that would be quite useless for determining the traces of lead that are found in the blood of a person suffering from acute lead poisoning, and a procedure that worked well for either of these purposes would need drastic modification before it could be successfully applied to the determination of lead in a copper ore.

Chemical analyses can also be divided into *classical analysis* and *instrumental analysis*. As its name implies, classical analysis is the more fundamental, and it is easier to explain and understand. Its two main quantitative branches are *gravimetric analysis* and *volumetric analysis*, to which the next two chapters will be devoted. In gravimetric analysis the substance that is being determined is isolated from the other constituents of the sample by precipitating it with a suitable reagent, and is eventually obtained as a pure compound of known composition. The mass of this compound is measured by means of a balance, and is used to calculate the amount of the substance that was originally present. Since the result is unlikely to be correct if any of that substance is lost during the process, or if there are any contaminants in the compound whose mass is measured, gravimetric analysis has been described as the quantitative synthesis of pure substances. (Sections 10.6 through 10.9 will show the description to be rather optimistic.) Volumetric analysis is a branch of titrimetric analysis. In titrimetric analysis a solution of the substance being determined is treated with a solution of a reagent with which it reacts in some known and reproducible fashion. The reaction may involve neutralization, precipitation, complexation, reduction and oxidation, or any other sort of chemical reaction. The process is called *titration*. In volumetric analysis, which is sometimes called visual titrimetry, the addition of reagent is continued until a change in the color of the solution occurs. The change of color may be due to the appearance of the first small excess of an intensely colored reagent, such as permanganate ion, in a solution that is otherwise colorless or nearly so. More often, it is due to the action of an *indicator*, which is a substance that undergoes some chemical or physical change, and therefore changes color, at or very near the point where the substance being titrated and the reagent are present in chemically equivalent amounts. This point is the *equivalence point* of the titration; the point at which the change of color is perceived and the addition of reagent is stopped is the *end point*. It is desirable that the end point and the equivalence point should coincide as closely as possible. If they do differ, it may be possible to correct for the difference between them, but such corrections are undesirable because they always entail some uncertainty. Usually a buret is employed to measure the volume of the reagent solution that is needed to reach the end point, and the amount of the substance titrated is calculated from this volume, the known concentration of the reagent solution, and the balanced chemical equation for the reaction that occurred.

Gravimetric and volumetric analysis differ in that a small excess of the reagent is always used in gravimetric analysis to ensure that precipitation will be virtually complete, whereas in volumetric analysis the aim is to add no more than the exact amount of reagent that is required to react with the substance being determined. Despite this difference, each of these techniques is based on the occurrence of some chemical reaction. Instrumental analyses, on the other hand, are generally based on the measurement of some physical property. In some instrumental analyses a chemical reaction may be carried out in order to convert the substance being determined into a species having the property that the analyst wishes to measure, but in others this is unnecessary because the substance being determined already possesses the desired property.

Instrumental analyses may take any of a great many different forms. Two kinds of instrumental analyses, absorption spectrophotometry and potentiometry, will be described briefly in later chapters. The first is based on the abilities of some substances to absorb radiation of various wavelengths, and the second involves the measurement of the potential of an electrochemical cell in which one of the two electrodes is immersed in the solution being analyzed. Each of these kinds of measurement can be made to provide information about the composition of a solution, and so can many others. Titrimetric analysis includes not only volumetric analysis but also titrations in which the weight of the reagent solution is measured instead of its volume, and procedures in which some instrumental technique is used to follow the progress of a titration and locate the equivalence point; such instrumental titrimetry eliminates the difficulty and judgment that are involved in deciding whether an indicator has changed color or not.

Instrumental analyses are nowadays far more common than classical ones. They are usually faster and they are much more amenable to automation. Many of them enable the chemist to determine a number of different constituents simultaneously, whereas by classical analysis each individual constituent has to be determined separately. These things mean that an instrumental analysis costs less to perform than a classical one does, an advantage that may be partially counterbalanced by the higher price of the apparatus that is needed in an instrumental analysis. Other considerations, which cannot be counterbalanced, are that instrumental analyses are generally more sensitive than classical analyses and are also useful for determining many substances that are nearly or completely immune to classical analyses. Gravimetric and volumetric analysis are at their best with amounts of substance between about 0.5 and 5 millimoles, and become more difficult and less accurate and precise with smaller amounts. Their lowest practical limits, even with special equipment such as microbalances and microburets, are of the order of 0.001 millimole. At the other extreme, there are instrumental techniques that can be used to determine substances present at concentrations as low as 10^{-10} or 10^{-11} M in as little as one or two drops of solution: these figures correspond to amounts between about 10^{-11} and 5×10^{-13} millimole. There is no reasonable way in which a compound like cyclohexane could be determined by gravimetric or volumetric analysis, especially in the presence of such closely related compounds as cyclopentane and methylcyclohexane, but there are a number of instrumental techniques, notably gas chromatography, by which such analyses can be made very easily, cheaply, and reliably.

As against these disadvantages, the classical techniques are capable of much better precision than the great majority of instrumental ones. The *precision* of a set of results is a measure of their agreement among themselves; their *accuracy* is a measure of their agreement with the truth, or with the value that is believed to be correct.

Their precision reflects the stability and reproducibility of the behavior that is observed and of the apparatus with which the measurements are made, and it also reflects the care that is taken to make each of those measurements in exactly the same way as the others. Their accuracy reflects the care that has been taken to eliminate errors that would be the same from one measurement to the next. The percentage of silica that is present in a rock can be determined by dissolving a weighed amount of the rock with a suitable combination of reagents, then evaporating the solution repeatedly with concentrated hydrochloric acid. Dissolved silicate ions are converted to anhydrous SiO_2 by this treatment, and the SiO_2 can be filtered off, dried, and weighed. A sloppy analyst might take different samples from different portions of the rock that actually had different compositions because of segregation while the rock was being formed, might dissolve or dehydrate some samples more thoroughly than others, might spill portions of some of the mixtures and allow dirt to fall into others, might dry some of the precipitates of SiO_2 more thoroughly than others, or might make the weighings carelessly. All of these errors would lower the precision that was attained, but because some of them would make the percentage of silica in the rock seem to be too high while others made it seem too low, the average of a large number of determinations might be reasonably accurate—that is, reasonably close to the accepted value. If, however, the rock contained some tungsten, WO_3 would be dehydrated and precipitated along with the SiO_2 if nothing were done to prevent this. Then the results would be inaccurate, even though they might be very precise if all of the samples were treated in exactly the same way. No matter what question we are trying to answer, we need results that are both accurate enough and precise enough to settle it. The precision of scientific measurements is so important that it will be discussed at greater length in Chapter 12.

Gravimetric and titrimetric analyses, when properly designed and executed, can ususally be made to yield relative precisions of the order of 0.1 to 0.2%. This means that the difference between any single result and the average of a number of results may be as small as 0.001 to 0.002 times the average. There are a few instrumental techniques that have comparable precisions, and a very few that have even better ones, but for most the comparable figure is of the order of 2 to 5%. The purpose of an analysis governs the precision that is needed. The circulation of water in the oceans can be followed by observing tiny differences between the concentrations of chloride ion at different points. If a relative precision of 5% were the best that could be obtained, the oceans would appear to be perfectly homogeneous everywhere except near the mouths of rivers. Much more precise analyses are needed to trace a current in mid-ocean. However, a relative precision of 5% would surely suffice if samples of ocean water were being analyzed to obtain the data needed in designing a desalination plant.

All these things—speed, cost, sensitivity, applicability to the determination of the particular substance that is sought in the presence of the other constituents that accompany it, and the precision and accuracy that are needed to settle the question being asked—must be considered in designing, selecting, or using an analytical procedure. The continued improvement of instruments and instrumental techniques will cause instrumental analyses to become even more predominant than they already are. However, classical analyses will continue to serve as the standard by which the accuracies of instrumental ones are gauged, and the ideas and manipulations that they involve will continue to be fundamental and essential.

> SUMMARY: Chemical analyses can be classified as qualitative, semiquantitative, or quantitative-
> or as classical and instrumental. Here attention is concentrated on classical quantitac
> tive analysis, of which the main branches are gravimetric analysis and volumetri;
> analysis.

9.3. Preliminary steps in a gravimetric or volumetric analysis

> PREVIEW: This section discusses the steps that must be taken in preparing a sample for a gravi-
> metric or volumetric determination of one of its constituents.

Very few materials have compositions so simple that they can be analyzed without prior treatment. Many are inhomogeneous, and obtaining representative samples of them may be more difficult and time-consuming than the actual analysis. Many contain other constituents whose presence would interfere with the determination of the one that is of interest, and such constituents must be eliminated or rendered innocuous before the analysis is begun. The discussions and illustrations in this section are taken from gravimetric and volumetric procedures, but the same steps are needed in instrumental procedures as well.

1. *Sampling.* Obtaining a sample that is truly representative of the material being analyzed is much easier when that material is homogeneous—as many gases, liquids, and solutions are—than when it is heterogeneous, as are suspensions and almost all solids. In either case sampling is very much easier in the laboratory than outside it. Given a solution contained in a glass flask, the analyst need only shake it to make sure that it really is homogeneous, and then withdraw a portion of it with a pipet. But the effluent from a factory is likely to vary in composition from one day, or even from one hour, to the next. To obtain a single representative sample it may be necessary to combine portions of a large number of individual samples taken over a long period of time, in proportion to the rate at which the effluent was being discharged when each individual sample was taken. In taking a sample of a metal casting, in which oxides, silicates, and other impurities are likely to be concentrated on the outside, it is necessary to obtain material from the interior as well as from the surface, and to obtain it without oxidizing it or contaminating it with fragments of the tools that are used. Care and common sense are needed in dealing with all such problems.

2. *Measurement of the amount of sample.* Solid samples are always weighed; liquid ones may be either weighed or measured with a pipet. If this measurement is not made with a relative precision that is considerably better than the one desired from the overall analysis, the fight may be lost before it is fairly begun. The precision of a single weighing on a typical analytical balance is about 0.1 to 0.2 mg, but most samples are weighed by difference: a weighing bottle or other container is weighed together with the sample, then the sample is removed and the container is reweighed, and the weight of the sample is found from the difference between these two weights. Consequently two individual weighings are necessary. The uncertainty in their difference will exceed the uncertainty in either one alone, and might be as large as 0.4 mg. If an overall precision of 0.1% is wanted, it is dangerous to begin with a sample weighing much less than about 1 g. If this contains an inconveniently large amount of the substance being determined, the solution prepared from it may be diluted to an exactly known volume in a volumetric flask and an *aliquot*, or known fraction, of this may be taken with a pipet for the actual analysis.

Similarly, since a precision of about 0.003 cm³ can be obtained in measuring out a volume of liquid with a volumetric pipet, a volume no smaller than 10 cm³ of a liquid or solution should be taken. Some liquid always remains in a volumetric pipet after it is emptied, partly in a droplet at the tip and partly in a thin film adhering to the walls. Since the size of the droplet and the thickness of the film depend on the surface tension, density, and viscosity of the liquid, the same pipet will deliver slightly different volumes of different liquids. To take full advantage of the precision that can be obtained, a pipet should therefore be calibrated with the liquid for which it is to be used. Calibration is done by measuring the density of the liquid and the weight of it that is delivered by the pipet.

It is always best to take four or five samples at once and carry them through the procedure together, for this not only allows for an accident that might ruin the analysis of a single sample but also enables the analyst to evaluate the reliability of the average result (Chapter 12). The chemist who has learned how to work efficiently in the laboratory needs very little more time to handle four or five samples together than to handle a single one alone.

3. *Dissolution.* If the sample is a solid it must be completely dissolved, for any undissolved matter may enclose an appreciable amount of the substance that is being determined and prevent it from reacting with the reagent. Some samples can be dissolved in water or another pure solvent or a mixture of solvents (such as water and ethanol), and there are many substances that will not dissolve in pure water but that can be brought into solution if a strong acid or base is added. Heating is often necessary but must be done with care to avoid loss of material, either by volatilization or by bumping or spattering. Mixtures of acids are useful for some special purposes: a mixture of nitric and hydrochloric acids will dissolve gold and the platinum metals, and a mixture of hot concentrated nitric and sulfuric acids will char and oxidize organic matter while bringing most inorganic constituents into solution. More refractory samples may have to be fused with acidic or basic substances such as potassium pyrosulfate or sodium carbonate. Such fusions involve reactions like

$$3K_2S_2O_7(l) + Al_2O_3(s) = 3K_2SO_4 + Al_2(SO_4)_3,$$
$$Na_2CO_3(l) + As_2O_5(s) = 2NaAsO_3 + CO_2(g)$$

by which insoluble materials are converted into substances that will dissolve in water after the fused mass has cooled. Other materials may have to be burned in an atmosphere of oxygen or fused with sodium peroxide in a sealed "bomb" to obtain products that are readily dissolved.

The product of any such treatment is cooled and dissolved in a suitable solvent. The entire solution may be used in the next step, or an accurately known fraction of it may be taken as described in step 2.

4. *Preliminary separations and masking.* Most samples contain many different constituents, and the one that is being determined is often not the only one that can react with the reagent to be employed. Suppose that a sample of natural water is to be analyzed for magnesium. There are a number of reactions on which the analysis might be based. Three of them are

1. $Mg^{2+} + NH_3(aq) + HPO_4^{2-} + 6H_2O(l)$

$$= MgNH_4PO_4 \cdot 6H_2O(s) \xrightarrow{1350-1400K} \tfrac{1}{2}Mg_2P_2O_7(s) + NH_3(g) + 6.5H_2O(g)$$

2. $Mg^{2+} + 2$ 8 - quinolinol $= $ $(s) + 2H^+$

3. $Mg^{2+} + EDTA^{4-} = Mg(EDTA)^{2-}$

The first involves the precipitation of magnesium ammonium phosphate hexahydrate and its ignition to give magnesium pyrophosphate, which is finally weighed. Conversion to the pyrophosphate is necessitated by the fact that the precipitate may be contaminated by other products, such as $MgHPO_4$, $Mg_3(PO_4)_2$, and anhydrous $MgNH_4PO_4$, all of which yield $Mg_2P_2O_7$ on ignition. The second involves the precipitation of magnesium 8-quinolinolate, which is filtered off and washed until it is free from excess 8-quinolinol and the other constituents of the solution. It can then be dried at 380 K and weighed, or it can be dissolved in acid, treated with excess bromide ion, and titrated with a solution of potassium bromate. In this titration the 8-quinolinol that was contained in the precipitate undergoes bromination:

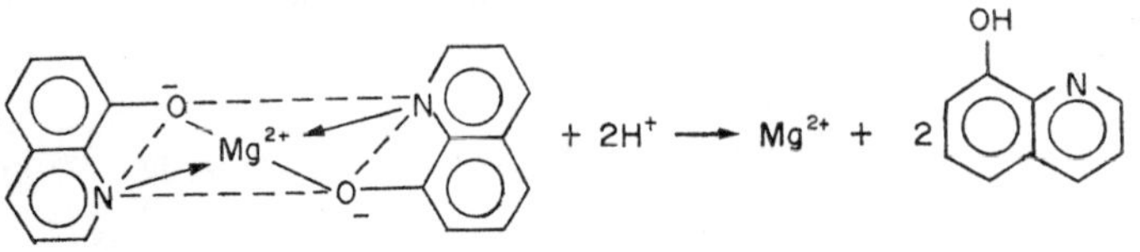

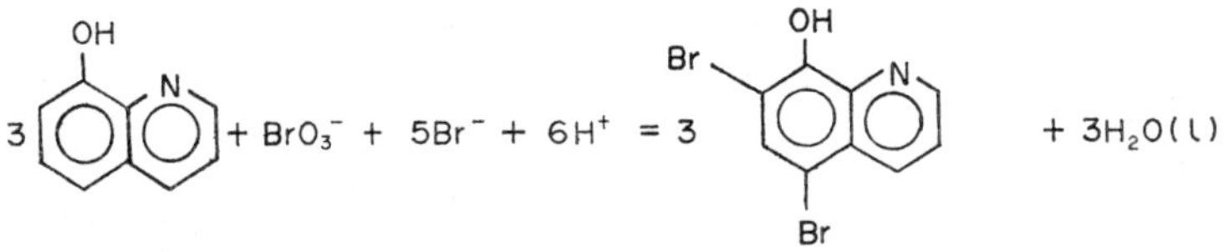

For each millimole of bromate ion consumed in the second reaction there must have been 3 millimoles of 8-quinolinol and 1.5 millimole of magnesium contained in the precipitate, and therefore the number of millimoles of magnesium in the sample must be 1.5 times as large as the number of millimoles of bromate ion used in the titration. The third reaction involves the titration of magnesium ion with ethylenediamine-tetraacetate, which must be done in a moderately basic solution because the formation constant of the chelonate is not very large and because the ethylenediaminetetra-acetate ion is a fairly strong Lowry–Brønsted base (Section 7.7).

Each of these procedures gives excellent results when it is applied to a solution of a magnesium salt alone, but none of them could be applied to the analysis of a natural water without taking some preliminary steps. These are necessary because other metal ions that might be present would react in various ways. In the first procedure iron(III), aluminum, and calcium would all precipitate along with the magnesium, giving products like $Fe_2O_3 \cdot xH_2O$, $Al_2O_3 \cdot xH_2O$, $AlPO_4$, $CaHPO_4$, and $Ca_3(PO_4)_2$. The ignited product would contain many substances in addition to $Mg_2P_2O_7$, and its weight would provide no useful information about the amount of magnesium in the sample. Similar interferences would arise in the other two procedures.

Usually it is necessary to separate the substance being determined from the interfering constituents of the sample before the actual analysis can be begun. Aluminum and iron(III) might be precipitated together as the hydrous oxides by adding ammonia to neutralize the originally acidic sample to a pH around 7 or 8; after they had been filtered off, calcium might be precipitated as $CaC_2O_4 \cdot H_2O$ by adding oxalic acid and ammonia. The filtrate from the precipitation of calcium oxalate would be ready for a determination of magnesium. The necessary separations could be effected in other ways, such as liquid-liquid extraction or ion-exchange chromatography.

The preliminary separations may consume as much time and effort as the entire remainder of an analytical procedure. Sometimes separations can be rendered unnecessary by the use of a masking agent or a combination of masking agents. Aluminum and iron(III) will not interfere in a titration of magnesium ion with ethylenediaminetetraacetate if the solution has a pH around 12 and contains an excess of triethanolamine. Probably the triethanolamine acts as a bidentate ligand at this pH:

$$Fe(III) + p\, N(CH_2CH_2OH)_3 + p\, OH^- = Fe\left[\begin{array}{c} (CH_2CH_2OH)_2 \\ N \\ \diagdown \\ CH_2 \\ | \\ \bar{O} - CH_2 \end{array} \right]_p^{3-p} + p\, H_2O(\ell)$$

Whatever the structure of the complex may be, it is too stable to react with ethylenediaminetetraacetate. Calcium and magnesium will be titrated together at this pH, so that only the sum of their concentrations could be determined. It might be possible to titrate magnesium alone by adding oxalate, to mask calcium by precipitating it as $CaC_2O_4 \cdot H_2O$, as well as triethanolamine. Another approach, if the concentration of calcium is not too much higher than that of magnesium, would be to determine the sum of the two concentrations by one titration, then perform a second titration in a still more strongly alkaline solution, containing about 1 M hydroxide ion as well as excess triethanolamine. In this second titration magnesium would be masked by the precipitation of magnesium hydroxide, and only calcium would be titrated. The concentration of magnesium would be obtained by subtracting the concentration of calcium found in the second titration from the sum of the concentrations of calcium and magnesium found in the first.

The success of this scheme will depend on the ratio of the two concentrations. If each titration has a relative precision of 0.2% and if the sample contains much more magnesium than calcium, the result might be something like

Sum of Ca^{2+} and Mg^{2+} concentrations (from the first titration)	$= 0.02000 \pm 4 \times 10^{-5}\ M$
Ca^{2+} concentration (from the second titration)	$= 0.002000 \pm 4 \times 10^{-6}\ M$
Difference $= Mg^{2+}$ concentration	$= 0.01800 \pm 4.4 \times 10^{-5}\ M$

In the first two lines each uncertainty is equal to 0.2% of the corresponding concentration; in the third line the uncertainty is the sum of the two preceding uncertainties.

The relative precision of the difference is $(4.4 \times 10^{-5}/0.018) \times 100 = 0.24\%$, which is not appreciably worse than the relative precision of 0.2% that might have been obtained by titrating the magnesium alone. However, if the sample contains much more calcium than magnesium, the result might be

Sum of Ca^{2+} and Mg^{2+} concentrations	$= 0.02000 \pm 4 \times 10^{-5} M$
Ca^{2+} concentration	$= 0.01800 \pm 3.6 \times 10^{-5} M$
Difference = Mg^{2+} concentration	$= 0.00200 \pm 7.6 \times 10^{-5} M$

Now the relative precision of the difference is $(7.6 \times 10^{-5}/0.002) \times 100 = 3.8\%$. In the second of these two cases a prior separation of the calcium and magnesium might be impossible to avoid.

5. *Adjustment of the experimental conditions.* After interfering substances have been removed or masked, further treatment of the sample may be necessary so that the desired reaction can take place. In the gravimetric determination of magnesium by the first of the three methods outlined above, after aluminum, iron(III), and calcium had been removed it would be necessary only to acidify the solution and add an excess of ammonium dihydrogen phosphate to it. In the next step precipitation would be effected by adding ammonia slowly:

$$Mg^{2+} + 2NH_3(aq) + H_2PO_4^- + 6H_2O(l) = MgNH_4PO_4 \cdot 6H_2O(s) + NH_4^+.$$

The reason for adding the reagents in this order will be explained in Chapter 10.

More elaborate manipulations are sometimes necessary. Suppose that iron had to be determined in an alloy that also contained chromium, nickel, and cobalt. After reviewing the methods that have been devised for the determination of iron in various materials and the ways in which these other elements would behave in those methods, we might select one in which iron is brought into solution as iron(II) and titrated with a strong oxidizing agent such as cerium(IV):

$$Fe(II) + Ce(IV) = Fe(III) + Ce(III).$$

We might begin by dissolving a sample of the alloy in hydrochloric acid, adding a little nitric acid to make sure that dissolution was complete. Dissolved oxides of nitrogen resulting from this treatment would interfere, either by oxidizing iron(II) or by reducing cerium(IV), and would have to be removed, for example by adding either concentrated hydrochloric acid or perchloric acid and boiling until all of the nitric acid had been decomposed and all of the nitrogen oxides driven off. At this point the iron would be more or less completely oxidized to iron(III), which would of course have to be reduced to iron(II) before the titration could be begun. To do this an excess of some reducing agent would have to be added, and then, after the reduction was complete, the excess of reducing agent would have to be removed so that cerium-(IV) would not be consumed by reacting with it. It would be essential to choose a reducing agent that is strong enough to reduce the iron quantitatively but too weak to reduce a measurable fraction of the chromium(III) that would be present, for any chromium(II) that was formed would react with the cerium(IV). Too much cerium(IV) would then be consumed, and the calculated percentage of iron in the alloy would be too high. Fortunately there are a number of reducing agents that meet these requirements. Sulfur dioxide might be bubbled through the solution until the reduction was complete, and the excess might be removed by boiling; or the solution might be passed through a column of finely divided metallic copper supported on a glass-wool plug, or shaken with a liquid amalgam containing an excess of metallic bismuth.

All of the preceding steps are needed in both gravimetric and volumetric analyses; only at this point do the two begin to differ. Section 9.4 will describe the subsequent steps in a gravimetric analysis, and Section 9.5 will describe those in a volumetric analysis.

SUMMARY: Preparing a sample for analysis entails five steps:
1. Obtaining a representative portion of the material.
2. Measuring the amount of it that will be used in the analysis.
3. Dissolving it in a suitable solvent.
4. Separating or masking other constituents that may interfere.
5. Adjusting the experimental conditions to accord with the requirements of the analytical procedure.

9.4. The execution of a gravimetric analysis

PREVIEW: This section describes the steps that are necessary in completing a gravimetric analysis

The following additional steps are needed to complete a gravimetric analysis:

1. *Precipitation.* The actual formation of the precipitate is brought about by adding a dilute solution of the reagent, as slowly as possible and with continuous stirring, and usually to a hot solution containing the substance being determined. Chapter 10 will show why these precautions cause the precipitate to form slowly and why its slow formation is desirable. A small excess of the reagent is always added. A deficiency of reagent must be avoided at all cost, because some of the substance being determined would then escape precipitation, and a small excess is usually helpful because it represses the solubility of the precipitate, but a large excess may be harmful for the reason given in Section 7.4. The usual procedure is to continue adding the reagent as long as its addition seems to lead to the formation of more precipitate. There comes a point at which so much precipitate is already present that the formation of a little more cannot be discerned. Then the stirring is discontinued, the solution is allowed to come to rest, and one or two more drops of the reagent solution are added. If no more precipitate forms at the point where they enter the solution, enough reagent has been added; if more precipitate can be seen to form, the solution is stirred and then tested for complete precipitation again in the same way. If the result of the test is

18

doubtful the solution should be allowed to stand for a minute or two so that most of the precipitate can settle out of it before the test is repeated.

Except with amorphous precipitates or in unusual cases where postprecipitation or delayed simultaneous precipitation (Section 10.6) is feared, the mixture is now put aside for a few hours or days to undergo the process called aging. Aging will be described in Section 10.11. It often leads to increases of both the purity of the precipitate and the sizes of its particles. An increase of particle size is desirable because it lessens the risk that tiny particles will run through the filter and be lost during filtration and washing, and also because it may enable the analyst to use a coarser, and consequently faster, filter.

2. *Filtration.* The precipitate is next separated from the supernatant liquid by filtration, most conveniently by means of a filtering crucible that has a disc made of finely divided particles of glass or porcelain sintered together into a rigid but porous mass. Sintered (or "fritted") glass can be used for precipitates that need not be heated to temperatures much above 525 K (about 250° C), but porcelain must be used if higher temperatures are required. Before the crucible is used it must have been thoroughly cleaned, dried by heating it to the same temperature for the same length of time and in the same way as the precipitate will be heated, allowed to cool to room temperature in a desiccator (step 4 below), and weighed: then the heating, cooling, and weighing are repeated until successive weights agree within the precision of weighing. This process, called *bringing to constant weight*, is necessitated by the fact that the crucible and precipitate will later be weighed together, and the weight of the empty crucible must be accurately known so that it can be subtracted from the total to find the weight of precipitate.

The filtration is begun by transferring some of the clear solution above the precipitate into the filtering crucible. If solution is simply poured into the crucible, some of it will get where it is not wanted, and the transfer should therefore be made by allowing liquid to flow along a stirring rod that touches the lip of the beaker and extends almost to the sintered disc. The crucible should never be filled more than halfway with liquid to avoid later "creeping" of precipitate up its walls. Suction is usually needed to cause liquid to pass through the sintered disc at a reasonable rate. When almost all of the clear supernatant solution has passed through the filter, the remainder is mixed with the precipitate and used to transfer it to the filtering crucible. Especially with gelatinous precipitates such as the hydrous oxides of iron(III) and aluminum, the rate at which liquid can be sucked through the disc decreases dramatically as the disc becomes covered with precipitate, and therefore it is advantageous to delay adding any of the precipitate to the filtering crucible until the last possible moment.

3. *Washing.* After the beaker in which the precipitation was conducted has been emptied into the filtering crucible, a wash liquid is needed to help in getting any last

traces of precipitate out of the beaker and onto the filter, and also to remove any non-volatile impurities contained in the liquid with which the precipitate and filtering crucible are wet. Pure water is rarely useful because many precipitates dissolve in it to an appreciable extent or, what is worse, revert to the colloidal state and run through the fritted disc unless some electrolyte is present: such reversion is called *peptization*. The electrolyte that is chosen must be one that will volatilize at the temperature to which the precipitate is heated, and if possible it should include a small concentration of one of the ions of the precipitate so as to repress the solubility. For example, if silver ion were being determined by precipitating silver chloride, a very dilute—perhaps 0.004 to 0.01 M (see Section 7.4)—solution of hydrochloric acid might be used as the wash liquid. On the other hand, if it were chloride ion that was being determined, a solution of silver nitrate could not be used because any silver nitrate that was left at the end of the washing process would give a solid residue on heating and thereby increase the weight of the product finally obtained. Neither could a solution of hydrochloric acid be used, because it would react with any excess reagent in the crucible to yield more of the precipitate, and so a very dilute solution of nitric acid might be used to prevent peptization. A precipitate of hydrous iron(III) oxide, which must be heated to at least 775 K to render it thoroughly anhydrous, might be washed with a solution containing ammonia and ammonium chloride: the latter to prevent peptization and the former to make the solution alkaline so that the precipitate would not redissolve. Any ammonium chloride remaining in the crucible when washing is complete will decompose into gaseous ammonia and hydrogen chloride on heating to 775 K, and will therefore have no effect on the final weight.

When the last of the original mixture has been poured into the filtering crucible, the stirring rod that was used to transfer it from the beaker is washed down into the beaker with a few drops of wash liquid, scrubbed with a "rubber policeman" (a short length of gum rubber tubing that covers the tip of a stirring rod and is sealed at the bottom) to remove any trace of precipitate adhering to it, and set aside. The level of the liquid in the filtering crucible is allowed to fall to just above the top of the layer of precipitate, and this portion of wash liquid is then added to the crucible by letting it flow down the rod to which the rubber policeman is attached. Several more small portions of wash liquid are used to remove any precipitate clinging to the walls of the beaker or to the rubber policeman, which is used to scrub down those walls, and each is transferred to the filtering crucible when most of the preceding one has passed through it. When the beaker and rubber policeman have been thoroughly cleaned, wash liquid is added to the crucible until it is about a third full, most of this is allowed to run through, another portion of wash liquid is added, and so on until washing is believed to be complete. At least five washings are usually needed, and more should be made if the sum of the concentrations of the non-volatile solutes in the original precipitation mixture was above about 0.1 M. Finally suction is continued until the crucible and precipitate are as dry as they can be gotten.

18*

4. *Drying.* From some precipitates the water that still remains can be removed by washing with several small portions of ethanol followed by several more of diethyl ether. The crucible is then disconnected from the suction apparatus and allowed to stand in the open air, or in a desiccator, for a few minutes while the ether evaporates. This procedure is simple and convenient, but of course it cannot be used if the laboratory air is so humid that water vapor will condense onto the crucible as it is cooled by the evaporation of the ether.

Drying is usually done by allowing the crucible and precipitate to stand for 20–30 minutes in an electric oven at 380–395 K (about 105–120° C). This removes surface moisture, and it also removes water of hydration from some (though not all) hydrates. Much higher temperatures are needed to remove water from hydrous metal oxides, to bring about deep-seated chemical changes such as the ones in which $MgNH_4PO_4$ and $MgHPO_4$ are converted into $Mg_2P_2O_7$, and to decompose and volatilize any ammonium salts that were present in the wash liquid and would remain as solid contaminants after drying at 380–395 K. These higher temperatures are most conveniently obtained with special electric ovens or muffle furnaces.

The crucible and precipitate then have to be cooled. By tradition a desiccator is used for this purpose. A desiccator is a container that has a tightly fitting cover and a chamber into which a drying agent can be put to absorb[†] moisture from the air and prevent sorption[†] of water by the crucible and precipitate. A few grams of anhydrous calcium sulfate ("Drierite") may be used. If this came to equilibrium with the atmosphere inside a desiccator—which it never does in practice because weeks would be required—it would maintain the partial pressure of water vapor at a value given by the equilibrium constant of the reaction

$$2CaSO_4 \cdot \tfrac{1}{2}H_2O(s) = 2CaSO_4(s) + H_2O(g); \qquad K = p_{H_2O}.$$

This corresponds to a concentration of water vapor of only about 5 μg dm^{-3}.

To make an airtight seal, a film of grease is placed on the ground surfaces at which the cover and body of a desiccator join. Before a crucible is placed in a desiccator it should be allowed to cool to, or below, about 475 K (200° C) in the open air. The cover

---

† *Absorption* is a process in which one substance (such as water) is taken up by another (such as a precipitate) and becomes more or less equally distributed throughout the latter. A blotter absorbs water by capillary action; anhydrous magnesium oxide absorbs water by forming magnesium hydroxide. *Adsorption* is a process in which one substance is taken up by, and remains on the surface of, another. Glass and porcelain adsorb water from the atmosphere, and the water molecules may be found on their surfaces. The surface area of a finely divided powder can be measured by determining how much nitrogen it can adsorb. *Chemisorption* is a kind of adsorption in which there is chemical bonding between the adsorbed substance and the surface; the adsorption of water by glass is largely due to the occurrence of the reaction

$$\text{\Large$>$}Si = O(s) + H_2O(g) = \text{\Large$>$}Si\!\!\big\langle{}^{OH}_{OH}\ (s).$$

Sorption is the general term; it includes both absorption and adsorption, and is used when the location of the material that is taken up is unknown or unimportant.

of the desiccator is then removed, the crucible is placed inside it, and the cover is replaced and tightly sealed without delay. At least 30 minutes should be allowed for the crucible to cool to room temperature, for if it is weighed while it is still even slightly warm, convection currents will arise in the balance and the weighing will be erroneous and erratic. When it has cooled, the desiccator should be opened *slowly* by sliding its cover along the ground surface of its body. The pressure inside the desiccator drops as the initially warm air sealed inside it cools, and if the cover is removed rapidly there will be a rush of air that may blow some precipitate out of the crucible.

5. *Weighing*. Once the desiccator has been opened successfully the crucible is transferred to the pan of a balance. The desiccator should be resealed at once, and the crucible should be weighed as quickly as possible because a thoroughly dried crucible and precipitate always sorb some water on exposure to the atmosphere of the laboratory.

Heating, cooling, and weighing must then be repeated, just as they were with the empty crucible, until the crucible and precipitate together have been brought to constant weight. The constant weight of the empty crucible is subtracted from this to obtain the weight of the final product.

Finally the crucible should be cleaned, either with a suitable solvent (such as concentrated hydrochloric acid for silver chloride, or a dilute solution of strong acid for magnesium 8-quinolinolate) or by scrubbing with detergent and water, rinsed thoroughly by sucking a great deal of distilled water through the disc, and allowed to dry in the open air.

> SUMMARY: A gravimetric analysis consists of five steps:
> 1. Precipitating a compound containing the substance being determined.
> 2. Filtering the resulting mixture to separate the precipitate from the supernatant solution.
> 3. Washing the precipitate to remove supernatant liquid from its surface.
> 4. Drying, either at a low temperature to remove water or at a higher one to convert the precipitate into a form more suitable for weighing.
> 5. Weighing the precipitate to find the amount of it that was obtained.

9.5. The execution of a volumetric analysis

> PREVIEW: This section describes the steps that are necessary in completing a volumetric analysis.

We return to the point that had been reached at the end of Section 9.3, where the sample had just been prepared for the final determination. For a volumetric analysis the solution should usually be contained in an erlenmeyer flask at this point. Often a beaker can be used, but since it facilitates the absorption of carbon dioxide from the laboratory air and the analyst's breath it makes good results difficult to obtain in many acid-base titrations. Stirring is essential and is most conveniently provided by a magnetic stirrer, but if this is not available a stirring rod can be used in a beaker, while the contents of an erlenmeyer flask can be mixed by swirling gently if the flask is no more than about a third full.

All the glassware must be scrupulously clean. Grease and dirt inside the calibrated portion of the buret that is used to measure out the reagent solution are especially harmful, for they prevent the solution from draining evenly down the walls as it is withdrawn, and make accurate measurements of volume impossible. Sometimes a buret can be cleaned by scrubbing it with detergent and tap water, followed by thorough rinsing with distilled water. The test is to fill it with distilled water and observe what happens when it is emptied. If the film of water drains evenly down its walls without leaving any droplets behind, a piece of glassware is clean enough to use; otherwise more thorough cleaning is required.

The best way to get a piece of glassware thoroughly clean is to let it stand in contact with a solution containing 1 to 2 M sodium hydroxide together with enough potassium permanganate to impart a moderately deep purple color. The hydroxide ion dissolves organic grease and the permanganate ion oxidizes it. Depending on the properties and thickness of the grease, a few minutes may suffice, or a day or more may be needed. The glassware is rinsed, first with distilled water and then with concentrated hydrochloric acid, which removes insoluble manganese compounds (such as MnO_2) adhering to its surface. If the acid does not drain evenly down its walls, the glassware is not yet clean, and should be rinsed with distilled water and returned to the alkaline permanganate solution for further treatment. The older and better known "cleaning solution" containing concentrated sulfuric acid and potassium dichromate, though viciously corrosive, has relatively little effect on grease, especially after it has become contaminated with a little water, and it contaminates glass surfaces with chromium compounds that no known treatment will remove.

The clean empty buret is mounted in a buret clamp and a few cm^3 of reagent is poured into it and drained out through the stopcock into a beaker reserved for waste. This is repeated twice to remove water adhering to the walls, and then the stopcock is closed and the buret is filled to a point some distance above the zero mark. Air must now be removed from the delivery tip and the hole in the plug. Allowing solution to run out rapidly through the wide-open stopcock into the waste beaker for a moment may suffice; sometimes the tip must be rapped sharply at the same time to dislodge a stubborn air bubble. If the level of the reagent is appreciably below the top of the graduated portion of the buret, more may have to be added. Any fraction of a drop that may be clinging to the tip of the buret is removed by touching it with a clean stirring rod. When at least 30 seconds has elapsed since the last manipulation of the stopcock or addition of reagent, the position of the bottom of the meniscus should be estimated to the nearest tenth of the smallest graduated division. Holding your index finger behind the buret just below the meniscus makes the estimation easier.

The titration vessel is placed underneath the buret and the buret is lowered so that its tip is as close to the surface of the solution as is convenient. The indicator is usually added at this point if one is needed, and a "titration thief" may be added as well to increase the speed and confidence with which the titration can be made. A titration thief is a piece of glass tubing open at both ends, having an internal diameter of 6–10 mm, and just long enough to protrude above the top of the vessel. It serves to protect a small fraction of the solution from the reagent until the end point has almost been reached, and it is left undisturbed until very nearly the end of the titration.

Addition of the reagent and stirring are now begun. The rate at which the reagent can be added depends on the rate of the reaction that occurs, and also on the speed with which the reagent is mixed with the solution being titrated. If these are very high, there may be no visible change at the point where the stream of reagent enters the solution, and the addition may then be as rapid as the buret will permit. If the reaction is slower, a change of color will be visible around the point of entry because of the local excess of reagent and its effect on the indicator, and then the addition should be slow enough to keep this change confined to a small fraction of the entire volume. Even if the reaction is very fast originally, it will become slower as the substance being titrated is consumed, and the change of color around the point of entry will become apparent after the addition of some fraction of the total volume of reagent that will be needed to complete the titration. As more reagent is added, the rate of addition should be decreased whenever the change of color seems to be spreading throughout a larger fraction of the solution. Near the end of the titration the rate of addition may have to be as small as one drop (about 0.05 cm^3) in 5–10 seconds, so that the effect of each drop can be seen before the next one falls in. There will come a point at which the change of color spreads throughout the entire solution and persists for a few seconds, and the stopcock must be closed as soon as this point is reached.

The titration thief is now lifted out of the mixture, releasing the original solution that was trapped inside it, and is rinsed down into the mixture with a stream of water from a wash bottle and set aside. The buret tip, the walls of the vessel, and the stirring rod (if one is used) are rinsed down as well. The indicator should now have returned to its original color; if it has not, the titration is spoiled, either because the reagent was added with insufficient care in approaching the preliminary end point, because the titration thief was too small, or because its contents were allowed to mix with the main volume of the solution prematurely. If the original color has returned, the titration should be continued very cautiously, adding just a drop or a fraction of a drop at a time until the final change of color just occurs. As little as 0.005 cm^3 of reagent can be added at a time by opening the stopcock to the smallest possible extent and closing it as soon as a tiny droplet begins to protrude from the tip of the buret, then either touching the stirring rod to the tip and using it to convey the droplet into the solution being titrated, or using a wash bottle to place a drop or two of water on the outside of the delivery tip, from which it will run down into the titration mixture, carrying the droplet of reagent with it. Occasionally the reaction is so slow near the end point that the reagent reacts more rapidly with the indicator than with the substance being titrated. When this happens, the original color returns within a few seconds after having changed throughout the solution, and another fraction of a drop of reagent should be added. The end point, where the addition of reagent is finally stopped, is generally taken as the first point at which the change of color persists for at least 15 seconds. Unless the titration reaction is known to be extremely slow, failure of the color change to persist for a longer time reflects instability of the final form of the indicator, absorption of carbon dioxide or some other substance from the air, or some other side reaction.

When the end point has been reached the buret reading must be noted. It should again be estimated to the nearest tenth of the smallest division on the buret: for example, to 0.01 cm³ with an ordinary 50-cm³ buret that is graduated in 0.1-cm³ intervals. The buret may be refilled with the reagent if another titration is to be performed at once. Otherwise its contents should be discarded by drainage through the stopcock. All of the glassware should be rinsed thoroughly with distilled water and allowed to drain before being put away. The top of a buret should be covered during storage to keep dirt and grease out of it.

The procedure described here is that for a *direct titration*, which consists simply of the addition of the reagent to the substance being determined. Sometimes a direct titration is impractical, either because the reaction becomes inconveniently slow near the end point or for some other reason. Then a *back titration* may be made instead. In a back titration the solution being analyzed is first treated with a known amount of one reagent that exceeds the amount that would be equivalent to the substance being determined. After the resulting reaction has reached equilibrium, the amount of the reagent that remains unreacted is found by titrating it with another solution. Hydrogen carbonate ion may be determined by a direct titration with a strong acid, such as hydrochloric acid; in such a titration the reaction that occurs is $HCO_3^- + H_3O^+ = = CO_2(aq) + 2H_2O(l)$. It gives rise to a supersaturated solution of carbon dioxide, which comes to equilibrium with the air only very slowly, and which has a fairly high buffer capacity. Consequently a small excess of either hydrogen carbonate ion or hydronium ion has little effect on the pH, and the color change of an acid-base indicator occurs so gradually that acceptably precise results are nearly impossible to obtain. It is much better to add a small excess of acid, boil the solution for a few minutes to expel the dissolved carbon dioxide, and back titrate with a solution of sodium hydroxide. At the equivalence point of the back titration the solution contains only dissolved sodium chloride, and its buffer capacity is so small (eq. (6.55)) that the color change of an indicator will be very sharp. Section 9.7 will show how the amount of hydrogen carbonate ion in the original solution could be calculated from the data obtained.

To make that calculation, or any other, the concentration of the reagent solution must be known. It is usually found by a process called *standardization*, which is a measurement of the volume that is consumed in titrating a known amount of some substance, such as a weighed amount of a pure chemical called a *primary standard* or a measured volume of some other solution, called a *secondary standard*, whose concentration has already been found by using it to titrate a primary standard. Secondary standards are more convenient because they eliminate the necessity of obtaining, drying, weighing, and dissolving primary standards, but since any error or uncertainty in the concentration of a secondary standard will also afflict the concentration of another solution standardized against it there is some sacrifice of accuracy and precision involved in their use.

Of course it would be still simpler and more convenient if a reagent of accurately known concentration could be prepared by dissolving a weighed amount of the solute in enough water or other solvent to give a known total volume, but there are surprisingly few substances with which this is possible. Silver nitrate, sodium chloride, potassium dichromate, and disodium dihydrogen ethylenediaminetetraacetate dihydrate are among those for which it is. Most other useful reagents are either too impure in the ordinary grade and too expensive in a specially purified one, too difficult to dry or bring to some reproducible and known state of hydration, too likely to react in some way (such as by gaining or losing water, carbon dioxide, or some other substance, or by undergoing oxidation) on exposure to the air during weighing, or too unstable in solution to be handled in this way. Solutions of many reagents can be bought in ampules of which each is guaranteed to contain an exactly specified amount of solute, so that a solution of known concentration can be prepared simply by transferring the contents of an ampule to a volumetric flask of the appropriate size and diluting to the mark with pure water. However, even a solution thus prepared should be standardized to guard against accidental errors, and in any event a solution must be restandardized occasionally unless it is known to be completely stable. Solutions of bases and strong oxidizing and reducing agents are especially susceptible to deterioration on long standing, though this can sometimes be prevented by careful preparation and storage.

To be useful as a primary standard, a substance must react with some common reagent in a known and highly reproducible way, and some indicator must be available for locating the equivalence point of its titration with that reagent. It should be either commercially available in a state of high purity or readily amenable to purification by some simple technique such as recrystallization. It must not decompose during storage, exposure to ordinary laboratory air, or dissolution; substances that are hygroscopic or readily oxidized by atmospheric oxygen are useless. It should preferably be anhydrous and non-volatile so that it can be dried by heating, but a hydrate may also be useful if it can be brought to a definite hydration state by long standing in an atmosphere of constant relative humidity and then retains this hydration state while it is being weighed. The weight of it that is required to react with a millimole of reagent should be large so that a large amount of it can be used for each standardization to minimize the effects of unavoidable errors in weighing. Finally, it should be reasonably inexpensive. Some of the substances commonly used as primary standards are borax ($Na_2B_4O_7 \cdot 10\,H_2O$) and tris(hydroxymethyl)aminomethane [$H_2NC(CH_2OH)_3$] for standardizing solutions of acids, potassium hydrogen iodate

[KH(IO₃)₂] and potassium hydrogen phthalate

$$\left[\underset{\displaystyle \bigcirc}{\bigcirc} \begin{matrix} \text{COOH} \\[2ex] \text{COOK} \end{matrix} \right]$$

for standardizing solutions of bases, arsenic trioxide for standardizing solutions of oxidizing agents, and potassium dichromate and potassium hydrogen iodate for standardizing solutions of reducing agents.

> SUMMARY: This section described and discussed the procedures used in carrying out direct titrations and back titrations, and explained how reagents used in volumetric analysis are standardized.

9.6. The calculations of gravimetric analysis

> PREVIEW: This section presents some of the fundamental ideas of stoichiometry and shows how they are applied in gravimetric analysis.

In a gravimetric analysis the chemist measures a weight or volume of sample that contains some unknown amount of a substance A, performs a series of manipulations that convert the A into some other substance Z without loss or contamination, and measures the weight of Z that is obtained. Finally the desired information must be calculated from the measured weight of Z. Such a calculation is said to be a *stoichiometric* one; *stoichiometry* is the branch of arithmetic that deals with relationships among the amounts of different chemical substances.

Usually the desired information is the percentage or concentration of A in the sample, but there are other possibilities as well. Imagine that a rock consisting chiefly of the mineral chromite, $FeCr_2O_4$, has been analyzed. A geologist might be content to know the percentage of $FeCr_2O_4$ in the rock, but a miner might prefer to know the percentage of chromium, and a manufacturer might want to know how much potassium dichromate could be obtained from a ton of rock, how much $NH_4Fe(SO_4)_2 \cdot 12H_2O$ could be obtained as a by-product, or how much ammonium sulfate would be needed to obtain the $NH_4Fe(SO_4)_2 \cdot 12H_2O$. The concentration of a solute in a solution may have to be expressed in mol dm⁻³, mg cm⁻³, or lb gal⁻¹. Stoichiometric calculations may be very diverse, and only the most important possibilities can be considered here.

A basic idea is that of the *formula weight*, which is the sum of the atomic masses represented by a chemical formula. The formula weights of Na and Cl are identical with their atomic weights, 22.9898 and 35.453, respectively; the formula weight of NaCl is 58.443; and the formula weight of Cl_2 is 70.906. Although the number 58.443 is sometimes called the "molecular weight" of sodium chloride, it is not easy to define a "molecule" of sodium chloride. Confusion about the meaning of "molecular weight" can also arise from the fact that this term is widely used to denote the result of an experimental measurement, which often depends on the conditions under which the measurement is made. The "molecular weight" of chlorine can be found by measuring the pressure that is exerted by a known volume of gaseous chlorine at a known temperature. It is equal to 70.906 at low temperatures, but it decreases as the temperature rises and the endothermic reaction $Cl_2(g) = 2Cl(g)$ proceeds to a larger extent, and it would approach 35.453 as this reaction approached completion. For these reasons the meaning of "molecular weight" is often unclear, but a formula weight is always unambiguous once the corresponding formula has been specified. It is essential to give the formula, for although "the formula weight of Cl" and "the formula weight of Cl_2" are both unambiguous, "the formula weight of chlorine" is not.

A formula weight is the weight of one particle (atom, ion, molecule, group of atoms, etc.) in atomic mass units; a *gram-formula weight* is the weight of $6.0221_7 \times 10^{23}$ particles. That number (Avogadro's number) of particles is called a *mole*. A formula weight has the units of atomic mass units (amu) per particle; a gram-formula weight has the units of g mol^{-1}. The gram-formula weight of a substance is always numerically equal to its formula weight: since the formula weight of NaCl is 58.443 amu particle^{-1}, its gram-formula weight is 58.443 g mol^{-1}. The numerical value remains unaffected if these units are changed in the same proportions: the gram-formula weight of NaCl can also be expressed as 58.443 mg $mmol^{-1}$, 58.443 μg μmol^{-1}, or 58.443 kg $kmol^{-1}$.

There are two other fundamental facts about the gram-formula weight. If n is a number of millimoles and W is a gram-formula weight, their product nW is a weight in milligrams:

$$n \text{ (mmol)} \times W \text{ (mg mmol}^{-1}) = nW \text{ (mg)}. \tag{9.1}$$

Conversely, if w is a weight in milligrams and W is a gram-formula weight, their ratio w/W is a number of millimoles:

$$w \text{ (mg)}/W \text{ (mg mmol}^{-1}) \quad \left[= w \text{ (mg)} \times \frac{1}{W} \text{ (mmol mg}^{-1}) \right] = w/W \text{ (mmol)}. \tag{9.2}$$

As these equations may suggest, it is essential to keep track of the units throughout every stoichiometric calculation.

The stoichiometric calculation that follows any gravimetric analysis can be made by carrying out four steps.

1. Combine the weight of the final product Z with its gram-formula weight to find the number of millimoles of Z obtained in the analysis. Use eq. (9.2), being careful to express the weight of Z in mg. If 0.6574 g ($= 657.4$ mg) of $PbCrO_4$ (formula weight $= 323.18$) was weighed at the end of an analysis, there was 657.4/323.18 millimole of $PbCrO_4$. There is no advantage of performing the division at this stage.

2. Write a balanced equation for the overall chemical reaction in which Z appears on the right-hand side and the substance sought appears on the left-hand side. The

substance sought is the substance whose weight (or percentage or concentration, etc.) you wish to calculate. It may be a substance that was present in the sample analyzed, or one that can react with or be made from such a substance. The desired equation, or at least a part of it, could be obtained by adding the balanced equations for all of the individual steps involved, including those that occurred during the analysis, but time would be wasted in obtaining those balanced equations because most of the atoms in them are irrelevant. The following procedure is much simpler. It will be illustrated by supposing that the lead chromate mentioned in step 1 had been obtained by fusing a sample of $FeCr_2O_4$ with sodium peroxide (thereby oxidizing the iron to Fe_2O_3 and the chromium to Na_2CrO_4), bringing the resulting chromate ion into an aqueous solution, and precipitating all of it by adding an excess of lead nitrate under suitable conditions.

(a) Write a skeleton equation with Z on the right-hand side and the substance sought on the left-hand side. If the analysis was done to find the percentage of $FeCr_2O_4$ in the sample, write $FeCr_2O_4 = PbCrO_4$; if it was done to find the percentage of chromium, write $Cr = PbCrO_4$; if it was done to find the percentage of iron (assuming that all of the iron in the sample is contained in the $FeCr_2O_4$), write $Fe = PbCrO_4$.

(b) If the formulas on the two sides have any one element in common, go on to step 2(c). Do not count hydrogen or oxygen unless one of the two formulas contains no other element. If there is no common element, look for a substance that was involved in the process and that contains one element in common with the substance sought on the left-hand side and another element in common with the substance weighed on the right-hand side, and insert its formula in the middle of the skeleton equation. If you wrote $FeCr_2O_4 = PbCrO_4$ or $Cr = PbCrO_4$ in step 2(a), nothing need be done here because chromium appears on both sides of each of these, but if step 2(a) gave $Fe = PbCrO_4$ this one will give $Fe = FeCr_2O_4 = PbCrO_4$. Two or even more intermediate substances may be needed: if you were trying to find how much ammonium sulfate would have to be used to make $NH_4Fe(SO_4)_2 \cdot 12H_2O$ from the iron in some amount of the sample, step 2(a) would have given $(NH_4)_2SO_4 = PbCrO_4$, and since you must ignore oxygen you would have to write $2(NH_4)_2SO_4 = NH_4Fe(SO_4)_2 \cdot 12H_2O = FeCr_2O_4 = PbCrO_4$.

Sometimes you need to look for a substance that reacted with the one on the left-hand side of an equality. If you are trying to calculate how much elemental zinc would be needed to reduce the iron(III) contained in the $NH_4Fe(SO_4)_2 \cdot 12H_2O$ that could be prepared from a given amount of sample, you would write $Zn = PbCrO_4$ in step 2(a) and should change this to $Zn + Fe(III) = FeCr_2O_4 = PbCrO_4$ here. Of course the first of the equalities in this result does not mean that iron(III) reacts with zinc to give $FeCr_2O_4$; it means that the $FeCr_2O_4$ is the source of the iron(III) that reacts with the zinc. You might prefer to write

$$Zn + Fe(III) =$$
$$\uparrow$$
$$FeCr_2O_4 \rightarrow PbCrO_4$$

to avoid getting confused in such a situation. The important thing is to obtain a chain of reactions that connect the substance sought with the substance weighed and that contain any intermediates needed to represent the chemistry of the process.

(c) If, in every one of the resulting equalities, the substance on the right-hand side contains all of the atoms of any one element contained in the substance on the left-hand side, go on to step 2(d). Again ignore hydrogen or oxygen unless one of the two formulas contains no other element. If some of the element that the two substances have in common does disappear in the process, look for the last substance that contained all of it, and insert the formula of that substance in the middle of the equality. In this step you must look at the chemistry of the process. If a sample of $FeCr_2O_4$ is analyzed in the fashion described above, the equations $FeCr_2O_4 = PbCrO_4$, $Cr = PbCrO_4$, and $Fe = FeCr_2O_4 = PbCrO_4$ given in step 2(a) will all survive this step without change: the first two because there is no loss of chromium in the process, and the third because of the assumption that all of the iron in the sample is contained in the $FeCr_2O_4$. But if the analysis were made by a procedure that involved the precipitation of $Cr^{III}(OH)CrO_4$ at some intermediate stage, followed by its dissolution in a non-oxidizing solution and precipitation of lead chromate, you would need to change them all to $FeCr_2O_4 = Cr^{III}(OH)CrO_4 = PbCrO_4$, $Cr = Cr^{III}(OH)CrO_4 = PbCrO_4$, and so on, because the chromium(III) contained in the $Cr^{III}(OH)CrO_4$ would not appear in the lead chromate finally weighed.

(d) Add numerical coefficients as needed to balance all of the equalities, taking them one at a time and continuing to ignore hydrogen or oxygen unless there is no other element on one side. Most often there will be just one other element on both sides of any one equality, and then that is the only element for which the equality needs to be balanced. The equations $FeCr_2O_4 = 2\,PbCrO_4$ and $Cr = PbCrO_4$ would result; in each the numbers of chromium atoms are equal on the two sides, and no useful purpose would be served by balancing them with respect to oxygen, iron, or lead. Sometimes it is an ion or group that should be considered: in $Fe = FeCr_2O_4 = Cr^{III}(OH)CrO_4 = PbCrO_4$ it is the chromate ions that should be balanced in the last step and it would be wrong to write $Cr^{III}(OH)CrO_4 = 2\,PbCrO_4$ because only one mole of lead chromate would be obtained from a mole of $Cr^{III}(OH)CrO_4$ in the procedure described above. In $2\,(NH_4)_2SO_4 = NH_4Fe(SO_4)_2 \cdot 12\,H_2O = FeCr_2O_4 = 2\,PbCrO_4$, the first 2 is needed to supply the sulfate ions contained in the ammonium iron(III) sulfate; it would be wrong to write $(NH_4)_2SO_4 = 2\,NH_4Fe(SO_4)_2 \cdot 12\,H_2O$ (balancing the ammonium ions) unless you knew that there was some other source of sulfate ions.

If there are two substances on the same side of any equality you should begin by balancing the equation for the reaction between them, then adjust the other coefficients correspondingly. The scheme

$$Zn + Fe(III) =$$
$$\uparrow$$
$$FeCr_2O_4 \rightarrow PbCrO_4,$$

which arose in step 2b, involves the reaction $Zn + 2\,Fe(III) = Zn^{2+} + 2\,Fe^{2+}$, and the result would be

$$Zn + 2Fe(III) =$$
$$\uparrow$$
$$2FeCr_2O_4 \rightarrow 4PbCrO_4.$$

Two millimoles of iron(III) are involved in the reaction with 1 millimole of zinc, 2 millimoles of $FeCr_2O_4$ are needed to supply 2 millimoles of iron(III), and 4 millimoles of lead chromate can be obtained from 2 millimoles of $FeCr_2O_4$.

3. Combine the number of millimoles of the substance weighed, which was calculated in step 1, with the equation obtained in step 2 to find the corresponding number of millimoles of the substance sought. To do this,

(a) Ignore all of the intermediate equalities in the result of step 2. The very first term gives a number of millimoles of the substance sought, and the very last one gives a number of millimoles of the substance weighed. Write a fraction having that number of millimoles of the substance sought in the numerator and that number of millimoles of the substance weighed in the denominator. If step 2 gave $FeCr_2O_4 = 2\,PbCrO_4$ the fraction should be 1 mmol of $FeCr_2O_4$/2 mmol of $PbCrO_4$, and it would mean that 1 mmol of $FeCr_2O_4$ is needed to obtain 2 mmol of $PbCrO_4$. For $Fe = FeCr_2O_4 = 2\,PbCrO_4$ it would be 1 mmol of Fe/2 mmol of $PbCrO_4$, and would mean that 1 mmol of iron can be obtained from the same amount of $FeCr_2O_4$ that is needed to yield 2 mmol of $PbCrO_4$.

(b) Multiply the number of millimoles calculated in step 1 by the fraction obtained in step 3(a). The result will be

$$n \text{ millimoles of the substance weighed} \times \frac{s \text{ millimoles of the substance sought}}{w \text{ millimoles of the substance weighed}}$$
$$= \frac{ns}{w} \text{ millimoles of the substance sought.}$$

You can often abbreviate steps 2 and 3, and you certainly should abbreviate them whenever you can. You might analyze a mixture of sodium nitrate and sodium chloride by dissolving a weighed sample of it in water, adding silver nitrate, and finally weighing the silver chloride that precipitated. Then you might wish to calculate the percentage of sodium chloride in the sample. Of course the number of millimoles of sodium chloride in your sample would be equal to the number of millimoles of silver chloride, which you would have calculated in step 1. To go painstakingly through steps 2 and 3 would be like shooting a fly with a cannon. However, even in this very simple problem you would really be doing some of the steps in your head [the equation $NaCl = AgCl$ (step 2a) contains a chlorine atom on each side (step 2(b)), it is balanced with respect to chlorine atoms (step 2(d)), and it involves equal numbers of millimoles of sodium and silver chlorides (step 3)], and you would be dismissing the other steps as being irrelevant or unnecessary. You need to think about the individual steps only if you are attacked by a problem too big for a fly swatter.

4. Combine the result of step 3 with any other available information to calculate the desired quantity. Some of the commonest possibilities are

(a) Number of millimoles of the substance sought $\div$ volume of a solution containing the substance sought (cm^3) = concentration of the substance sought (mmol cm^{-3}, or M),

(b) Number of millimoles of the substance sought $\times$ gram-formula weight of the substance sought (mg $mmol^{-1}$) = weight of the substance sought (mg). This is copied from eq. (9.1), and might be followed by

(i) weight of the substance sought (mg) $\div$ weight of the sample (mg) = fraction of the substance sought, or $100 \times$ weight of the substance sought (mg) $\div$ weight of the sample (mg) = percentage of the substance sought, or

(ii) weight of the substance sought (mg) $\div$ volume of a solution containing the substance sought (cm^3) = concentration of the substance sought (mg cm^{-3}, or g dm^{-3}). This is called a weight-volume concentration to distinguish it from a molar or molal concentration.

Any of these results might be converted into other units. A weight-volume concentration is most easily calculated in mg cm^{-3}, but a physician might prefer to have the result of a clinical analysis expressed in mg per 100 cm^3, and others might insist on ounces per gallon or scruples per firkin. All such oddities can be handled by multiplying the original result of step 4 by one or more conversion factors, with due attention to keeping the units straight.

SUMMARY: The formula weight of a chemical substance is the sum of the atomic masses represented by its formula; its gram-formula weight is the weight of a mole in grams. A four-step procedure is described for calculating any desired quantity from the weight of precipitate obtained in a gravimetric analysis. The essential step is the one in which a relation is derived between the number of millimoles of the precipitate and the number of millimoles of the other substance that is of interest.

9.7. The calculations of volumetric analysis

PREVIEW: This section describes the stoichiometric calculations associated with direct titrations, back titrations, and standardizations.

Most of the calculations of volumetric analysis can be made in exactly the same general fashion as those of gravimetric analysis. There are a few minor differences, which arise from the fact that the amount of a reactant is measured in volumetric analysis whereas the amount of a product is measured in gravimetric analysis. These differences will be outlined first, with frequent reference to the steps listed in Section 9.6. The calculations involved in standardizations and back titrations will then be described separately because they do not fit the pattern of most other stoichiometric calculations.

In an ordinary volumetric analysis one titration is made, and the weight (or percentage, etc.) of the substance sought is calculated from the measured volume and known concentration of the reagent. Four steps are necessary, and the following outline of them is exactly parallel to the one in Section 9.6. To provide a specific example, it will be supposed that a weighed amount of a sample containing iron(III) oxide (but no other iron compound) has been analyzed by dissolving it in acid, reducing all of the iron to iron(II) ion, removing any excess of reducing agent, and titrating with an 0.01976 M solution of potassium permanganate, of which 35.79 cm^3 is needed to reach the end point under conditions such that manganese(II) and iron(III) ions are formed.

1. Combine the volume of the reagent with its concentration to find the number of millimoles that were used. The equation

$$V \text{ (cm}^3) \times c \text{ (mmol cm}^{-3}, \text{ or } M) = Vc \text{ (millimoles)} \qquad (9.3)$$

has already been stated in words (on p. 89). In this titration (35.79) (0.01976) millimoles of potassium permanganate were used.

2. Write a balanced equation for the overall chemical reaction. Begin with a balanced equation for the reaction that occurred during the titration. If the substance sought appears in that equation, go on to step 3. Otherwise write the formula of the substance sought, followed by an equals sign, to the left of the equation and then go through steps 2(a) through 2(d) of Section 9.6. For the titration described above you would first write $5 \text{ Fe}^{2+} + \text{MnO}_4^- + 8 \text{ H}^+ = 5 \text{ Fe}^{3+} + \text{Mn}^{2+} + 4 \text{ H}_2\text{O}(l)$. If the analysis had been made to determine the percentage of iron in the sample, you would go on to step 3 because iron appears in this equation. If it had been made to determine the percentage of iron(III) oxide, you would write $\text{Fe}_2\text{O}_3 = 5 \text{ Fe}^{2+} + \text{MnO}_4^- + \ldots$, which includes all of the useful part of the equation, and steps 2(a)–(d) would give $\frac{5}{2}\text{Fe}_2\text{O}_3 = 5 \text{ Fe}^{2+} + \text{MnO}_4^- + \ldots$. It is safer to write a fractional coefficient like 5/2 than to tamper with the coefficients in the original balanced equation.

3. Combine the number of millimoles of reagent, which was calculated in step 1, with the equation obtained in step 2 to find the corresponding number of millimoles of the substance sought. This is done in the same way as step 3 in a gravimetric calculation, except that here you ignore everything in the equation except the substance sought and the reagent, and write a fraction having the number of millimoles of the substance sought in the numerator and the number of millimoles of the reagent in the denominator. If you seek the percentage of iron in the sample you will have obtained $5 \text{ Fe}^{2+} + \text{MnO}_4^- + \ldots$ from step 2, and should multiply the number of millimoles of permanganate by the ratio 5 mmol of iron/1 mmol of permanganate. If it is the percentage of ferric oxide that you want, you should multiply the number of millimoles of permanganate by the ratio (5/2) mmol of ferric oxide/1 mmol of permanganate.

4. Combine the result of step 3 with other information to calculate the desired

quantity. Once you have calculated the number of millimoles of the substance sought, it no longer matters whether the analysis was gravimetric or volumetric, and this step is exactly the same as the corresponding one in Section 9.6.

In an ordinary volumetric analysis you know the concentration of the reagent and calculate the amount (or concentration, etc.) of the substance sought. In a standardization you know how much primary or secondary standard was used and and calculate the concentration of the reagent; in effect, the reagent itself is the substance sought. The procedure will be illustrated with a specific example. Suppose that 0.2361 g of pure arsenic trioxide (As_2O_3, formula weight 197.84) is dissolved in a solution of sodium hydroxide, and that this solution is acidified and titrated with a solution of potassium permanganate, of which 43.88 cm³ is used. The reaction that occurs during the titration is

$$2MnO_4^- + 5H_3AsO_3(aq) + H_3O^+ = 2Mn^{2+} + 5H_2AsO_4^- + 4H_2O(l).$$

The number of millimoles of arsenic trioxide used is given by eq. (9.2):

$$236.1 \text{ mg of } As_2O_3 \div 197.84 \text{ mg of } As_2O_3 \text{ (mmol of } As_2O_3)^{-1}$$

$$= \frac{236.1}{197.84} \text{ mmol of } As_2O_3.$$

We need a relationship between this and the number of millimoles of permanganate needed to react with it. To obtain one we must connect the arsenic trioxide with the titration reaction, and we can do this by writing

$$As_2O_3$$
$$\downarrow$$
$$2MnO_4^- + 5H_3AsO_3(aq) + H_3O^+ = 2Mn^{2+} + 5H_2AsO_4^- + 4H_2O(l).$$

Since 5/2 millimole of As_2O_3 will be needed to obtain 5 millimoles of H_3AsO_3,

$$\frac{236.1}{197.84} \text{ mmol of } As_2O_3 \times \frac{2 \text{ mmol of } MnO_4^-}{5/2 \text{ mmol of } As_2O_3} = \frac{(236.1)(2)}{(197.84)(2.5)} \text{ mmol of } MnO_4^-.$$

This amount of permanganate is contained in 43.88 cm³ of the permanganate solution, and therefore

$$\frac{(236.1)(2)}{(197.84)(2.5)} \text{ mmol of } MnO_4^- \div 43.88 \text{ cm}^3 \text{ of } MnO_4^-$$

$$= \frac{(236.1)(2)}{(197.84)(2.5)(43.88)}$$

$$= 0.021\ 76 \text{ mmol of } MnO_4^-(\text{cm}^3 \text{ of } MnO_4^-)^{-1}$$

$$= 0.021\ 76 \ M.$$

In a direct titration all of the reagent is consumed in reacting with the substance that is titrated, but in a back titration some of the reagent is consumed in another reaction. A known amount of the reagent is added to the solution that is being ana-

19

lyzed; some of the reagent reacts with the substance being determined, and the rest of it remains unreacted. The titration serves as a way of determining the amount of unre-acted (or "excess") reagent. That amount is subtracted from the total to find the amount consumed by the substance being determined. In the procedure described at the end of Section 9.5, a sample of sodium hydrogen carbonate weighing 0.9391 g is dissolved in water, 50.00 cm³ of 0.2012 M hydrochloric acid is added, and the solution is boiled to expel dissolved carbon dioxide and then back titrated with 0.05031 M sodium hydroxide, of which 1.97 cm³ is needed to reach the end point. What is the percentage of sodium hydrogen carbonate in the sample?

The number of millimoles of hydroxide ion used in the back titration is given by eq. (9.3):

$$1.97 \text{ cm}^3 \text{ of } OH^- \times 0.050\ 31 \text{ mmol of } OH^- (\text{cm}^3 \text{ of } OH^-)^{-1}$$
$$= (1.97)(0.050\ 31) \text{ mmol of } OH^-.$$

The back titration involved the reaction $H_3O^+ + OH^- = 2\ H_2O(l)$; 1 millimole of hydronium ion reacts with 1 millimole of hydroxide ion, and therefore $(1.97)(0.05031)$ millimole of hydronium ion took part in the back titration. The total number of millimoles of hydronium ion used was

$$50.00 \text{ cm}^3 \text{ of } HCl \times 0.2012 \text{ mmol of } H_3O^+ (\text{cm}^3 \text{ of } HCl)^{-1}$$
$$= (50.00)(0.2012) \text{ mmol of } H_3O^+.$$

All of this reacted with hydrogen carbonate ion except the $(1.97)(0.05031)$ millimole that was consumed in the back titration, which is to say that $(50.00)(0.2012) - (1.97)(0.05031)$ millimoles of hydronium ion reacted with hydrogen carbonate ion. Since the equation for that reaction is $H_3O^+ + HCO_3^- = CO_2(aq) + 2\ H_2O(l)$, 1 millimole of hydrogen carbonate ion reacted with 1 millimole of hydronium ion, and 1 millimole of sodium hydrogen carbonate is needed to provide 1 millimole of hydrogen carbonate ion. The final result, obtained by including the factors needed to convert the number of millimoles of sodium hydrogen carbonate into the weight and thence into the percentage of sodium hydrogen carbonate, is

$$[(50.00)(0.2012) - (1.97)(0.05031)] \text{ mmol of } H_3O^+ \times \frac{1 \text{ mmol of NaHCO}_3}{1 \text{ mmol of H}_3O^+}$$

$$\times \frac{84.00 \text{ mg of NaHCO}_3}{1 \text{ mmol of NaHCO}_3} \div 939.1 \text{ mg of sample} \times 100$$

$$= \frac{[(50.00)(0.2012) - (1.97)(0.05031)](84.00)(100)}{939.1} = 89.1\% \text{ NaHCO}_3.$$

Stoichiometric calculations are not really difficult, but are often troublesome for those who have not yet acquired the experience that is needed to perform them with ease and confidence. This chapter is followed by a great many typical problems that will help you to become thoroughly familiar with the practical applications of the ideas and techniques that these two sections have described.

SUMMARY: In the calculations of volumetric analysis the essential step is the one in which a relation is derived between the number of millimoles of the reagent and the number of milli-moles of the substance titrated (or of some other substance chemically related to it).

Problems

Answers to some of these problems are given on page 520.

9.1. An excess of silver nitrate is added to 40.00 cm³ of a solution of hydrochloric acid, and the silver chloride that precipitates is filtered off, washed, dried, and weighed. It weighs 0.5000 g.

(a) How many millimoles of silver chloride were obtained?
(b) How many millimoles of hydrochloric acid did the original solution contain?
(c) What was the concentration of hydrochloric acid in that solution?

9.2. Exactly 30 cm³ of a solution of sodium sulfate is treated with an excess of barium nitrate. After filtration, washing, and drying, the precipitated barium sulfate is found to weigh 0.2700 g.

(a) How many millimoles of barium sulfate were precipitated?
(b) How many millimoles of sodium sulfate did the sample contain?
(c) What was the concentration of sodium sulfate in the sample?

9.3. A solution of potassium chromate, K_2CrO_4, is buffered at pH 5.0 to suppress the reaction $PbCrO_4(s)+H_3O^+ = Pb^{2+}+HCrO_4^-+H_2O(l)$, and is then treated with an excess of lead nitrate. The lead chromate that precipitates is filtered off, washed, dried, and found to weigh 0.6000 g,

(a) How many millimoles of lead chromate were obtained?
(b) How many millimoles of potassium chromate did the sample contain?
(c) What weight of potassium chromate was present in the sample?

9.4. A solution of oxalic acid, $H_2C_2O_4$, is treated with an excess of calcium chloride, and an excess of ammonia is then added to the mixture. The resulting precipitate of $CaC_2O_4 \cdot H_2O$ is filtered off. washed, and dried at 380 K, a temperature at which it retains its water of hydration. It is then found to weigh 0.3456 g.

(a) How many millimoles of calcium oxalate were obtained?
(b) How many millimoles of ammonium ion were formed during the process?
(c) How many millimoles of ammonium chloroplatinate(IV), $(NH_4)_2PtCl_6$, could have been obtained from the filtrate?
(d) What weight of ammonium chloroplatinate(IV) could have been obtained?
(e) If the precipitate had been heated to 775 K it would have decomposed in accordance with the equation $CaC_2O_4 \cdot H_2O(s) = CaCO_3(s)+CO(g)+H_2O(g)$. What weight of calcium carbonate would have been obtained?

9.5. A sample of iron ore weighing 0.8901 g is dissolved in hydrochloric acid. An excess of nitric acid is then added and the solution is boiled to ensure that all of the iron is present as iron(III). Hydrous iron(III) oxide, $Fe_2O_3.xH_2O$, is then precipitated by adding excess ammonia. The precipitate is filtered off, washed, and converted to anhydrous Fe_2O_3 by ignition at 975 K. The product weighs 0.1089 g.

(a) How many millimoles of Fe_2O_3 were obtained?
(b) How many millimoles of iron did the sample contain?
(c) What weight of iron did the sample contain?
(d) What was the percentage of iron in the sample?

9.6. A fertilizer contains some potassium monohydrogen phosphate, K_2HPO_4. A sample of the fertilizer, weighing 1.2653 g, is dissolved in water and the solution is treated with excess magnesium chloride, ammonium chloride, and ammonia. The reaction $Mg^{2+}+2NH_3(aq)+H_2PO_4^-+6H_2O(l) =MgNH_4PO_4 \cdot 6H_2O(s)+NH_4^+$ occurs. The precipitate is filtered off, washed, and converted to $Mg_2P_2O_7$ (page 255) by igniting it to constant weight at 1400 K. The magnesium pyrophosphate weighs 0.2653 g.

19*

(a) How many millimoles of magnesium pyrophosphate were obtained?
(b) How many millimoles of phosphorus did the sample contain?
(c) What weight of phosphorus did the sample contain?
(d) What was the percentage of phosphorus in the sample?
(e) How many millimoles of P_2O_5 might have been obtained from the sample?
(f) What weight of P_2O_5 might have been obtained from the sample?
(g) The sample was taken from a bag of fertilizer whose label reads "Phosphorus, minimum 13.00% as P_2O_5". Is the label accurate?

9.7. A sample of a mixture of solid sodium and potassium chlorides weighs 0.1260 g. The sample is dissolved in water, and the solution is treated with an excess of silver nitrate. After filtration, washing, and drying, the silver chloride is found to weigh 0.2867 g.

(a) How many millimoles of silver chloride were obtained?
(b) How many millimoles of chloride did the sample contain?
(c) What was the sum of the number of millimoles of sodium ion and the number of millimoles of potassium ion in the sample?
(d) What weight of chloride did the sample contain?
(e) What was the sum of the weights of sodium and potassium ions in the sample?
(f) How many millimoles of sodium ion did the sample contain?
(g) How many millimoles of sodium chloride did it contain?
(h) What weight of sodium chloride did it contain?
(i) What were the percentages of sodium and potassium chlorides in the sample?

9.8. 22.00 cm³ of 0.1010 M sodium hydroxide is needed to reach the equivalence point in the titration of a certain solution of hydrochloric acid.

(a) How many millimoles of hydroxide ion were used in the titration?
(b) How many millimoles of hydrochloric acid were titrated?
(c) What weight of hydrogen chloride did the hydrochloric acid solution contain?

9.9. A 5.000-cm³ sample of vinegar is titrated with 0.2105 M sodium hydroxide, of which 21.05 cm³ is needed to reach the equivalence point.

(a) How many millimoles of hydroxide ion were used?
(b) How many millimoles of acetic acid did the sample contain?
(c) What was the total concentration of acetic acid in the sample?

9.10. An 0.2609-g sample of impure oxalic acid, $H_2C_2O_4 \cdot 2 H_2O$, is dissolved in water, and the solution is acidified with sulfuric acid and titrated with 0.01850 M potassium permanganate. In acidic solutions permanganate ion reacts with oxalic acid to form manganese(II) and carbon dioxide. 39.00 cm³ of the permanganate solution was needed to reach the equivalence point.

(a) How many millimoles of permanganate ion were used?
(b) How many millimoles of oxalic acid did the sample contain?
(c) What weight of $H_2C_2O_4 \cdot 2 H_2O$ did the sample contain?
(d) What was the percentage of $H_2C_2O_4 \cdot 2 H_2O$ in the sample?

9.11. A 1.2000-g sample of an iron ore is dissolved in hydrochloric acid and all of the iron contained in the solution is reduced to Fe^{2+}. 21.26 cm³ of 0.01925 M potassium permanganate is consumed in titrating this, and the products of the reaction that occurs during the titration are Mn^{2+} and Fe^{3+}.

(a) How many millimoles of permanganate ion were used in the titration?
(b) How many millimoles of iron(II) ion were oxidized in the titration?
(c) What weight of iron did the sample contain?
(d) What was the percentage of iron in the sample?

9.12. When manganese(II) reacts with permanganate ion in a weakly alkaline solution, the manganese(II) is oxidized to MnO_2 and the permanganate is reduced to MnO_2.

(a) A 1.993-g sample of a manganese ore is dissolved in acid, the dissolved manganese is converted to Mn^{2+}, and the solution is diluted to a volume of 250.0 cm³. A 50.00-cm³ aliquot of the solution is neutralized and titrated with 0.01122 M potassium permanganate, of which 22.33 cm³ is needed to reach the end point. What is the percentage of manganese in the ore?

(b) What weight of MnO_2 was formed in the titration described in part (a)?

9.13. A chemist employed by a mining company is assigned to supervise a technician in analyzing several thousand samples of iron ore by the procedure described in problem 9.11. What weight of ore should be taken for each analysis so that the percentage of iron in it will exactly equal to the number of cm³ of 0.01925 M potassium permanganate used in the titration?

9.14. An 0.2510-g portion of pure arsenic(III) oxide, As_2O_3, is dissolved in a solution of sodium hydroxide, and an excess of hydrochloric acid is added to convert the arsenic(III) to $H_3AsO_3(aq)$. Then a trace of potassium iodide is added and the solution is titrated with a solution of calcium hypochlorite. The reaction of hypochlorite ion with arsenic(III) acid is catalyzed by iodide ion, and its final products are chloride ion and arsenic(V) acid. The end point of the titration is reached when 32.20 cm³ of the hypochlorite solution has been added.

(a) How many millimoles of arsenic(III) oxide were used in this standardization?
(b) How many millimoles of hypochlorite ion, OCl^-, were used?
(c) What was the concentration of hypochlorite ion in the solution being standardized?

9.15. Chromium(III) reacts with ethylenediaminetetraacetate ion, Y^{4-}, to form the chelonate CrY^-, but the reaction is very slow. A solution containing an unknown concentration of chromium(III) is analyzed by back titration in the following way. A 50.00-cm³ aliquot of the unknown solution is treated with 50.00 cm³ of 0.0942 M Na_2H_2Y, and the mixture is boiled until the formation of CrY^- has reached completion. It is then cooled, and the excess of reagent is determined by titrating it with 0.1067 M zinc chloride, of which 22.22 cm³ is required. The reaction that occurs in the back titration is $Zn^{2+} + Y^{4-} = ZnY^{2-}$; the extent of the reaction $Zn^{2+} + CrY^- = ZnY^{2-} + Cr(III)$ is too small to detect.

(a) How many millimoles of zinc ion were used in the back titration?
(b) How many millimoles of ethylenediaminetetraacetate survived the reaction with chromium (III)?
(c) How many millimoles of chromium(III) did the original solution contain?
(d) What was the concentration of chromium(III) in the original solution?

CHAPTER 10

Precipitation and Coprecipitation

10.1. Introduction

Chapter 5 described some of the chemical equilibria that are associated with the formation and dissolution of precipitates. This chapter deals with the mechanisms of those processes: with the ways in which particles of precipitate form and grow, with the factors that affect the rates of formation and growth, and with the effects of their rates on the physical forms and properties of precipitates. It is especially concerned with impurities in precipitates and the processes that are responsible for their presence. All of these matters are important in gravimetric analysis, and they are also important in the synthesis and purification of solids.

10.2. Nucleation, supersolubility, and particle size

PREVIEW: Nucleation, the first step in the formation of a precipitate, occurs when the supersolubility of the precipitate is exceeded by the addition of reagent, and is followed by particle growth.

Homogeneous solutions of two substances are mixed, a solid phase appears, and a number of particles of precipitate are eventually obtained. Two different processes have occurred. *Nucleation* is the step in which the first tiny solid particles, or nuclei, are formed. It is followed by *particle growth*, in which ions are added to the nuclei, and which continues until equilibrium is reached. Nucleation is discussed in this section and the following one; particle growth is discussed in Section 10.4.

Clear supersaturated solutions of many substances can be prepared and kept for long periods of time without change, but their supersaturation may be destroyed by adding a small particle of the solute. A particle of dust or some other foreign material often serves equally well. Precipitation occurs at the surface of the added nucleus, and the concentration of solute decreases toward the equilibrium value. For each solute, solvent, and temperature there is a limiting concentration, called the *supersolubility*, that cannot be exceeded for long even if foreign nuclei are scrupulously excluded. If the supersolubility is exceeded, nuclei of the solute appear of themselves, and then grow at the expense of dissolved material until the concentration has decreased to the equilibrium solubility.

The phenomenon is represented by Fig. 10.1, in which curve a represents the solubility and curve b represents the supersolubility. A solution whose concentration and temperature correspond to a point below curve a is unsaturated; one for which they correspond to a point on curve a is saturated. Either of these solutions is stable. A solution whose concentration lies between the equilibrium solubility on curve a and the supersolubility on curve b is metastable. A metastable system is one that undergoes no change even though it is not at equilibrium. Precipitation from a supersaturated solution is spontaneous (that is, it is accompanied by a decrease of free energy) but does not occur in the absence of nuclei. Supercooled liquids are metastable, and so are solutions of cerium(IV) in dilute sulfuric acid (Sections 8.8 and 8.9). In a metastable system the spontaneous process has a very large energy of activation. Solutions whose concentrations exceed the supersolubility are unstable because they cannot be kept for long without change.

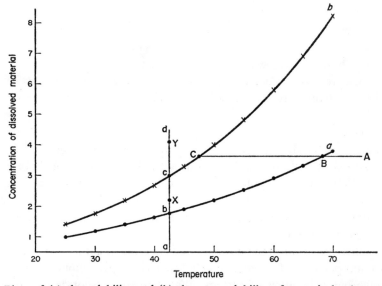

FIG. 10.1. Plots of (a) the solubility and (b) the supersolubility of a typical substance against temperature. Line ABC represents the cooling of a hot unsaturated solution free from foreign nuclei; line abcd represents precipitation by addition of a reagent.

Supersaturated solutions can be obtained in either of two ways. If the solubility increases as the temperature rises, as in Fig. 10.1, one is to begin with a hot unsaturated solution, such as the one represented by point A. It may be passed through a very fine filter, or ultracentrifuged, to remove all undissolved solid matter. As it cools, its behavior is represented by line ABC. Between point B on the solubility curve and point C on the supersolubility curve it is supersaturated but remains clear because it is metastable. When the temperature reaches point C, however, the solution becomes unstable, and nuclei will be formed and a precipitate will appear.

The other way of obtaining a supersaturated solution is more relevant to gravimetric analysis. It consists of adding a solution of a reagent to a solution of a substance with which it can react to form a precipitate. Suppose that silver ion is added in very small portions to a solution containing chloride ion. Initially there is no dissolved silver chloride; this corresponds to point a. Line abcd is followed on adding silver ion. If foreign nuclei are absent (which they never are in practical analysis because they are very difficult to remove), a precipitate will not appear until point c on the supersolubility curve is reached. Then nuclei will form, and silver chloride will precipitate until point b on the solubility curve is reached.

What will happen when the next portion of silver ion is added? If it is very small, and if the supersolubility considerably exceeds the solubility, point X will be reached. The solubility will be exceeded but the supersolubility will not. No new nuclei will be formed. Precipitation must occur, and can occur only by the growth of the particles already present. If the portion is larger, or if the solubility and supersolubility curves are close together, the supersolubility will be exceeded and point Y will be reached. More nuclei will be formed, and they will grow along with the original ones. In either case the total amount of precipitate will be the same, but it will be divided among different numbers of particles. If the supersolubility is exceeded only when the first portion of the reagent is added, only a few particles will be formed, and their average size will be relatively large. If the supersolubility is exceeded with each successive addition of reagent, there will be many more particles and their average size will be much smaller.

There are three factors that determine whether the supersolubility will be exceeded by the addition of any portion of reagent after the first, and therefore govern the average particle size of the precipitate that is obtained. They are the magnitude of the difference between the supersolubility and the equilibrium solubility, the rate of addition of the reagent, and the efficiency with which the solution is stirred while the precipitation is carried out.

As the difference between the supersolubility and the solubility increases, it becomes easier to add portions of reagent so small that the supersolubility is not exceeded. The equilibrium solubilities of barium sulfate and silver chloride are not much different: their solubility products are equal to 1.3×10^{-10} $mol^2\ dm^{-6}$ and 1.8×10^{-10} $mol^2\ dm^{-6}$, respectively. For barium sulfate the ratio of the supersolubility to the equilibrium solubility is approximately 30 in the presence of foreign nuclei such as particles of dust; for silver chloride it is very little larger than 1. It is much easier to precipitate barium sulfate without exceeding its supersolubility over and over again than to do the same thing for silver chloride. Adding a solution of barium chloride slowly to a solution of sulfuric acid gives a product consisting of relatively large crystals; precipitating silver chloride in a similar way gives a "curdy" precipitate in which very tiny individual particles are loosely stuck together. If a substance is very sparingly soluble, the difference between its supersolubility and its solubility will be very small even if their ratio is well above 1, and then it will be very difficult to add portions of reagent so tiny that the supersolubility will not be exceeded on each addition. Very sparingly soluble precipitates are usually obtained in extremely small particles; fairly soluble ones are usually obtained as much larger crystals. Hence the particle size can generally be increased by increasing the solubility. Lead chromate is more soluble in acidic solutions than in neutral ones, and its crystals are much larger when they are precipitated from acidic solutions.

The rate of addition of reagent is important for similar reasons. Adding very tiny portions of reagent, and waiting until equilibrium is attained after each addition, yields relatively few nuclei and causes the average particle to grow until it is fairly large. The same result is attained by adding the reagent continuously but so slowly that particle growth removes it as fast as it is added. Adding large portions, or a rapid continuous stream, of reagent yields more and smaller particles.

When a solution of a reagent is added to a mixture in which a reaction is occurring, the concentration of the reagent is highest around the point where it is added. If a drop of a solution containing silver ion is added to a much larger volume of a solution containing chloride ion, there will be a small volume of the mixture that contains most of the silver ions until these are dispersed throughout the whole volume by stirring. In volumetric analysis this is an advantage because it helps the analyst to judge when the end point is being approached; in gravimetric analysis it is a disadvantage because it increases the probability that the supersolubility will be exceeded at the point where the reagent enters the mixture. Many additional nuclei will be formed if the rate of nucleation is high and the stirring inefficient.

There are a number of reasons why the particle size of a precipitate is important. Dissolved impurities are much more extensively adsorbed by small particles than by large ones. Large particles can be separated from their *mother liquor*, the solution in which a precipitate is formed and with which it is in equilibrium when precipitation is complete, by means of a coarse filter, through which liquid passes rapidly. Small particles not only necessitate the use of a finer, and therefore slower, filter but also tend to make it still slower by clogging its pores. Large particles also settle more rapidly, facilitating the filtration of most of the clear mother liquor before any of the precipitate is transferred to the filter. So little mother liquor adheres to the surfaces of large particles when filtration is complete that washing is easy and rapid; with small particles it is harder and slower. With curdy precipitates, such as silver chloride, special care must be taken in washing to avoid peptization, in which the loosely bound aggregates of colloidal particles are broken up into fragments that can run through the filter. Large particles can be dried more quickly than small ones, and when dry have less tendency to adsorb water vapor from the atmosphere because they have smaller exposed areas. For all these reasons it is far easier to work with large particles than with small ones.

SUMMARY: The average size of the particles of a precipitate depends on the difference between the supersolubility and the equilibrium solubility, on the rate of addition of the reagent, and on the efficiency of stirring.

10.3. The mechanism of nucleation

PREVIEW: This section discusses the formation of nuclei, defines the critical nucleus, and outlines two theories about its size.

Nucleation is clearly a vital step in the formation of any precipitate, and yet surprisingly little is known about its details. This is partly because the complete removal of foreign nuclei is far easier to imagine than to accomplish. Crystallization from a supersaturated solution can often be induced merely by scratching the walls of the container. It is often difficult to assess the relative importances of homogeneous nucleation and deposition onto bits of dust or imperfections in the walls of the vessel. Moreover, although a single nucleus is too small to detect, particle growth begins as soon as the first nucleus appears, and thereafter particle growth proceeds side by side with the formation of additional nuclei. This makes it difficult to separate the rate of nucleation from the rate of particle growth. The rate and mechanism of particle growth are easier to study because it is often easier to achieve conditions under which it occurs while nucleation does not.

If nucleation is regarded from the thermodynamic viewpoint, the most important fact is that the free energy of an ion depends on its location in a crystal. An ion embedded in the center of a crystal is subjected to forces exerted by the ions surrounding it, and these forces are symmetrical: those exerted by the ion below it are counterbalanced by those exerted by the one above it, and so on. For an ion at the face of a crystal the forces are asymmetrical: those exerted by the liquid are not the same as those exerted by its neighbors in the interior of the crystal. The free energy of an ion on the face of a crystal is greater than that of an ion in the interior. The asymmetry and free energy are even higher for an ion at an edge, and are highest of all for an ion at a corner.

A nucleus is so small that it has little or no interior: most or all of its ions are at faces, edges, and corners. Its free energy is larger than either that of an ion pair at one extreme or that of a large crystal at the other. At some intermediate size the free energy has a maximum value. The *critical nucleus* is a nucleus that has this size. A nucleus smaller than the critical nucleus tends to redissolve and disappear; one larger than the critical nucleus tends to grow. The critical nucleus corresponds to the activated complex in an ordinary reaction. By correlating the surface tension of the solid, which is a measure of the average excess free energy of the exposed ions, with particle size it can be concluded that the critical nucleus of a typical precipitate contains something like 100 ions, but substantial difficulties are involved in making the correlation and its result is very uncertain.

There is another interpretation that leads to a very different result. It is based on the commonly observed fact that a precipitate cannot be detected until some time has elapsed after solutions of the reacting ions are mixed. If the concentrations of the two ions in the mixture are both equal to c, the duration t of this induction period obeys an empirical equation of the form $t = k/c^n$, where k and n are constants for any particular precipitate under any particular conditions. Typical values of the order n range from about 9 for calcium fluoride to 2.5 for potassium perchlorate. On the assumption that the length of the induction period is governed entirely by the rate of nucleation and not at all by the rate of growth of the critical nuclei into particles large enough to detect, these figures suggest that the critical nucleus is $(CaF_2)_3$ for calcium fluoride and either $KClO_4$, $K_2ClO_4^+$, or $K(ClO_4)_2^-$ for potassium perchlorate. Since the values of n for most other precipitates lie between these limits, the critical nucleus of a typical precipitate is judged to contain between two and nine ions.

It is not yet possible to say that one of these conflicting interpretations is right and the other is wrong. The first depends on assumptions about the surface tensions of very small particles that are not easy to prove; the second assumes that the size of the critical nucleus is independent of c. For the reasons given in the first paragraph of this section, it would be very difficult to devise an experiment that would make it possible to decide between these theories, and the question must be left in this obscure and unsatisfactory state.

SUMMARY: The critical nucleus is the smallest particle that can grow spontaneously. There are two different ways of estimating its size: one suggests that it may contain about 100 ions, and the other suggests that it contains only two to nine.

10.4. The mechanism of particle growth

PREVIEW: The rate of particle growth can be governed by either diffusion or crystallization, and these can affect the shape and form of the particles obtained.

More is known about particle growth than about nucleation, but even particle growth is a complex process. It involves two steps. A dissolved ion must approach the surface of the growing particle, and then it must be incorporated into the crystal lattice. Either of these can be rate-determining.

There are two extreme situations. One is illustrated by the very slow evaporation of a saturated solution of sodium chloride, in which a single small crystal of sodium chloride is suspended to serve as a nucleus, and from which foreign nuclei are carefully excluded. A large and nearly perfect cubical crystal is obtained. On increasing the rate of crystallization by evaporating the solution more rapidly, the supersolubility may be exceeded and more nuclei may form. More and smaller crystals will result, but these too will be cubical in form. Under these conditions it is the incorporation of ions into the lattice that is rate-determining. If the rate of crystallization is increased very greatly—for example, by adding concentrated hydrochloric acid to the saturated solution of sodium chloride—the crystals are no longer cubical. The lattice structure is not changed, but the shape and form of the particles are.

The rate of crystallization depends on the concentrations of the reacting ions. If these are very high, the rate of crystallization is also high. The ions in a very thin layer of solution adjacent to the surface of a growing particle are consumed quickly. The rate of further growth is limited by the rate at which ions can diffuse to the surface from the undepleted solution some distance away. The rate of diffusion depends on many factors; what is most important here is that it differs from one part of the surface to another. The ions that diffuse to the face of a crystal are drawn from a limited volume of solution nearby. Edges and corners are more exposed, and ions can diffuse to them from larger volumes of solution. This is illustrated by Fig. 10.2. Growth at edges and corners is therefore promoted by increasing the concentrations of the reacting ions. Needles, whiskers, dendrites, and other features having large surface areas result.

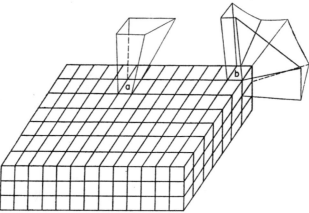

Fig. 10.2. A comparison of the volume from which an ion in solution can diffuse to a site a on the face of a crystal with the volume from which it can diffuse to a site b at a corner.

When precipitation is slow, there is very little depletion of the layer of solution next to the surface. The concentrations of the ions will be uniform over the entire surface, and growth can occur as rapidly at faces as at edges and corners. Different faces may grow at different rates, and if a foreign substance is more strongly adsorbed onto one face than onto another it may alter the crystal habit by changing the relative rates of growth of different faces. In any event, decreasing the rate of precipitation, so that crystallization rather than diffusion becomes rate-determining, generally yields more compact particles and decreases their specific surface area. The specific area is the ratio of the total area to the amount of precipitate. As was explained in Section 10.2, decreasing the rate of precipitation also decreases the specific area by decreasing the number of nuclei, but this is a quite different effect even though it has a similar result. The rate of precipitation affects the specific area in both of these ways, but in one case it does so by affecting the number of ions contained in the average particle, while in the other case it does so by affecting the manner in which the ions are arranged in the particles.

Growing a nearly perfect crystal is a difficult task, and one that consumes much time and requires very close control of the experimental conditions. In gravimetric analysis crystals never approach perfection. They contain impurities, holes, and defects of many other kinds. One defect that affects the rate of particle growth is the screw dislocation. Imagine a perfect crystal at an instant when a face has just been completed; how can the next ion be added? In Section 10.3 it was argued that the free energy of an ion at the corner of a crystal is much higher than that of an ion near the center, which has a larger number of near neighbors. An ion that was deposited anywhere on a perfectly flat surface would have only a single nearest neighbor, and its free energy would be so high that crystal growth in this way would be far too slow to detect.

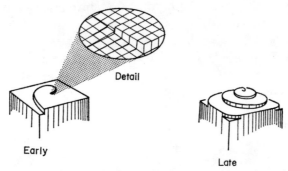

FIG. 10.3. Successive stages in the growth of a crystal at a screw dislocation. The emerging ramp of the dislocation provides a step at which newly arriving ions deposit in preference to other sites on the surface. The rate of deposition is approximately equal over the whole length of the step, and the new ions therefore form a spiral on which the ramp sweeps around the face of the crystal. From H. A. Laitinen and W. E. Harris, *Chemical Analysis*, McGraw-Hill Book Co., Inc., New York, 2nd ed., 1975. By permission of McGraw-Hill Book Company, Inc.

A screw dislocation is shown in Fig. 10.3. It arises from the presence of a few ions slightly displaced from the positions they would have in a perfect crystal. A discontinuity in the surface results, and persists as the crystal grows. Ions from the solution are deposited at the discontinuity, which consequently sweeps around the face of the crystal in an advancing spiral path. Spiral growth patterns are discernible in many crystals, and screw dislocations and other imperfections play a predominant role in particle growth.

SUMMARY: Particle growth involves two successive steps: diffusion of ions to the surface of a particle, then crystallization. In a very slow precipitation crystallization is likely to be the rate-determining step. It tends to give compact particles having small specific surface areas. The rates at which crystals grow are strongly influenced by the imperfections they contain.

10.5. The colloidal state

PREVIEW: Very small particles have very high specific areas, and their behaviors are greatly affected by processes that occur at their surfaces. This section is a brief introduction to some of the fundamental ideas of colloid science.

A particle is said to be colloidal if its diameter is between roughly 1 nm and 1 μm (10^{-7}–10^{-4} cm). Many precipitations yield particles no larger than this. Even when larger particles are finally obtained, they must pass through the colloidal state as they grow. Silver chloride has a gram-formula weight of 143.3 g mol^{-1} and a density of 5.56 g cm^{-3}. If a gram of silver chloride, which occupies a volume of 0.18 cm^3, were obtained as a single cube, its surface area would be 1.9 cm^2. The same amount of silver chloride might be divided into 1.8×10^{20} cubes 1 nm on each edge. Each cube would have a volume of 10^{-21} cm^3 and a surface area of 6×10^{-14} cm^2. The total surface area would be 1.1×10^7 cm^2, which is not much different from the area of a football field or a hockey rink. Adsorption and other processes occurring at their surfaces exert large effects on the properties of such finely divided substances.

A colloidal particle acquires an electrical charge when it adsorbs ions, and the charge stabilizes the particle by causing it to repel other particles having charges of the same sign. Suppose that silver chloride is being precipitated by adding silver nitrate to a solution of hydrochloric acid. Precipitates tend to adsorb their own ions. A precipitate of silver chloride can adsorb either silver ion or chloride ion, but it cannot adsorb much hydronium or nitrate ion because neither of these is strongly bound to either of the ions in the precipitate. Until the equivalence point is reached the concentration of chloride ion in the solution will be some orders of magnitude larger than that of silver ion. The particles of precipitate will adsorb chloride ions in preference to silver ions because there are so many more chloride ions available. As they adsorb chloride ions they become negatively charged, and a stable colloidal dispersion results because they repel each other. As the equivalence point is approached, the concentrations of dissolved silver and chloride ions become more nearly equal, and the excess of chloride ion adsorbed on each particle decreases. Somewhere near the equivalence point the electrostatic forces among the particles, which tend to keep them apart because they all have charges of the same sign, become smaller than the short-range forces of attraction that tend to bring them together. When the short-range forces predominate, the particles *flocculate*, giving particle clusters or aggregates that are large enough to settle out rapidly, leaving a clear supernatant liquid. The occurrence of flocculation during a titration can be used as a warning that the equivalence point has been very closely approached. Similar behavior would be observed if the precipitation were carried out in the opposite direction by adding chloride ion to a solution of silver ion. Here, however, there would be a large excess of silver ion until most of it had been precipitated. The particles would be positively charged, because the adsorption of silver ion would predominate over that of chloride ion, until the concentration of excess silver ion became very small. In either case the precipitate that settled out near the equivalence point could be peptized—returned to the colloidal state—by stirring vigorously to break up the weakly bound particle clusters and adding an excess of the reagent, thereby providing ions whose adsorption would stabilize the colloidal fragments.

Although a precipitate can adsorb either of its ions, it is unlikely to adsorb them to equal extents even from a solution in which their concentrations are equal. Silver chloride adsorbs chloride ion more strongly than silver ion. The concentrations of silver and chloride ions in a saturated solution of silver chloride would be exactly equal if adsorption did not occur and if traces of species like $AgCl_2^-$ were not present as well. If the saturated solution is in contact with solid silver chloride, chloride ions will be preferentially removed from it and adsorbed onto the solid. With a large particle of the solid the effect is negligibly small because a large particle has a surface area so small that adsorption cannot be extensive. With colloidal silver chloride the effect is far from negligible. Many chloride ions are adsorbed onto the large area that is exposed, and the concentration of chloride ion decreases. Consequently, silver ions must dissolve so that the solubility equilibrium will continue to be satisfied. When equilibrium is reached the concentration of dissolved silver ion will exceed that of chloride ion. If the area of the solid is sufficiently large and the volume of solution sufficiently small, the solution will contain approximately 1×10^{-4} M silver ion and 2×10^{-6} M chloride ion at equilibrium. Silver and chloride ions are equally strongly adsorbed from a solution of this composition. Because they are both univalent, a precipitate that adsorbs them in equal numbers will be electrically neutral. The *point of zero charge* is defined as the point at which a solid adsorbs cations and anions in such proportions that it acquires no net electrical charge. For silver chloride the point of zero charge lies at pcAg $= 4.0$ (and pcCl $= 5.7$, since the solubility product must be satisfied and $pK_{AgCl} = 9.7$). If the pcAg is smaller than 4.0, silver ion is adsorbed in excess and the particles acquire a positive charge; if it is larger than 4.0, chloride ion is adsorbed in excess and the particles become negatively charged.

Once a colloidal particle has acquired an electrical charge by adsorbing one of its own ions, it tends to repel dissolved ions of the same charge and attract oppositely charged ions. If silver chloride is precipitated by adding silver nitrate to a solution of hydrochloric acid, the precipitate will adsorb chloride ions in excess, and will therefore have a negative charge, until the point of zero charge is reached. The negatively charged surface attracts hydronium ions and repels nitrate ions. There will be a layer of solution adjacent to the surface that will contain a higher concentration of hydronium ion, and a lower concentration of nitrate ion, than the bulk of the solution. For brevity the situation is usually represented as $AgCl \cdot Cl^-$ ⦙ H_3O^+. The vertical dashed line represents the surface of the particle. It separates the adsorbed ion, which is strongly bound to the crystal lattice, from the *counter ion*, which may be attached to the adsorbed ion by nothing more than electrostatic attraction.

These facts have curious consequences. If a large amount of perfectly pure colloidal silver chloride is added to a sufficiently small volume of pure water, equilibrium will be reached when the solution contains 1×10^{-4} M silver ion and 2×10^{-6} M chloride ion. At this point the silver chloride is negatively charged because it adsorbed excess chloride ions while equilibrium was being attained. An equal number of hydronium ions will be attracted to the surface as counter ions. If the mixture is centrifuged or passed through a very fine filter, the hydronium ions will accompany the precipitate, and the solution that is left behind will contain an excess of hydroxide ion. Its pcH will have been changed from 7.0 to 10.0 by equilibration with the silver chloride, even though both the acidic strength of silver ion and the basic strength of chloride ion are negligible. If the silver chloride is put back into the solution and recrystallized into one large particle, almost all of the adsorbed and counter ions will be released as the area decreases, and the pcH will return to 7.0. The behavior of colloidal particles may be very different from that of large ones.

The charged surface of a particle and the layer of counter ions together constitute what is called the *electrical double layer*. The electrical double layer resembles an electrical capacitor, and its charge and thickness influence the behavior of the particle in important ways. The quantity of electrical charge stored in the double layer naturally reflects the extent of adsorption, increasing as more ions of the same charge are adsorbed on a given area of the surface. When a precipitate is put into a solution at the point of zero charge, there is no excess of charge on the surface and the double layer does not form. When a double layer does form, its thickness depends on the concentration and charge of the counter ion. Increasing either the concentration or the charge decreases the thickness of the double layer.

All of the double layers surrounding the particles of any precipitate in any particular solution have charges of the same sign. Hence—except at the point of zero charge—there is always an electrostatic repulsion between two particles as they approach each other. The force of attraction between the particles decreases more rapidly with increasing distance than the electrostatic force of repulsion does. If the particles can approach each other very closely, the attractive force may predominate, and if it does it will lead to actual contact and aggregation. Decreasing the thickness of the double layer decreases the distance between the particles at the point where their mutual repulsion becomes appreciable, gives the short-range force of attraction more chance to become effective, and tends to favor flocculation.

It is therefore possible to flocculate a colloidal precipitate by increasing either the concentration of the counter ion or the force between the counter ion and the adsorbed ion. A suspension of silver chloride, negatively charged as the result of adsorption of chloride ions, would be stable indefinitely if the solution surrounding it were sufficiently dilute. Adding enough hydronium ion, sodium ion, potassium ion, or any other univalent cation compresses the double layer and results in flocculation after a little while. The concentration that is required to cause flocculation is called the *coagulation value*, and is almost the same for all univalent ions that are not chemically bonded to the adsorbed ion. The effect of chemical bonding is discussed in the next paragraph. Different univalent ions do have different sizes and therefore yield double layers of slightly different thicknesses, but this is a minor effect. The coagulation value is almost independent of the nature and charge of the anion that is added to the solution along with the counter ion, but it does increase as the extent of adsorption increases and the particles become more highly charged. With silver chloride in a solution containing only a moderate concentration of chloride ion—say, 10^{-3} M—

the coagulation value is of the order of 0.1 M for a univalent cation. For divalent cations the coagulation value is much smaller, because the electrostatic attraction between an adsorbed chloride ion and a cationic counter ion is much larger if the counter ion is divalent than if it is univalent, and therefore the double layer is thinner when the counter ion is divalent. For tri- and tetravalent counter ions the coagulation values are still smaller, for the same reason. The ratio of the coagulation value for a z-valent counter ion to that for a univalent one is close to $(1/z)^6$, a fact that provides coordination chemists and others with a way of estimating the charge that is carried by an ion whose composition and structure are not yet known. Entirely similar considerations would apply to the flocculation of silver chloride that was positively charged because of adsorption of excess silver ion, but here it would be the charge of the anionic counter ion that governed the coagulation value.

The coagulation value is lowered if there is chemical bonding as well as electrostatic attraction between the counter ion and the adsorbed ion. If the concentration of chloride ion in the solution were very low, the coagulation value observed with a negatively charged precipitate of silver chloride would be smaller for lead nitrate than for magnesium nitrate. This reflects the fact that the formation constant of $PbCl^+$ is much higher than that of $MgCl^+$. Lead ions are therefore much more strongly attracted to the adsorbed chloride ions than magnesium ions would be. At a higher concentration of chloride ion, most of the lead ion in the bulk of the solution would be converted into $PbCl^+$, and the coagulation value for lead(II) would rise toward that characteristic of a univalent counter ion.

SUMMARY: A colloidal particle of a precipitate will adsorb one of its own ions from a solution containing an excess of that ion. By doing so it acquires an electrical charge, which is counterbalanced by a layer of oppositely charged counter ions. The adsorbed ions and counter ions constitute the electrical double layer, which stabilizes the particle. There is a point of zero charge at which the concentrations of the cation and anion of the precipitate are such that the ions are adsorbed in equal numbers, causing the double layer to disappear and the particles to flocculate. Flocculation can also be caused by increasing the concentration of the counter ion above the coagulation value.

10.6. Impurities in precipitates: coprecipitation

PREVIEW: No precipitate is completely pure. This section describes the processes that may be responsible for the presence of impurities. Coprecipitation is the most important because it is the most difficult to prevent.

One aim of gravimetric analysis is to obtain a pure product. It is an aim that is never reached; every precipitate is contaminated to some extent. It may contain several different impurities, and may contain each of them for a different reason. This section describes the commonest and most important of the ways in which precipitates become contaminated.

Contamination can result from any of three different kinds of processes: *post-precipitation, simultaneous precipitation,* and *coprecipitation.* In simultaneous precipitation and coprecipitation the impurity comes down at the same time as the precipitate; in postprecipitation the precipitate comes down first and then the impurity forms on its surface. In simultaneous precipitation the impurity comes down because its solubility is exceeded; in coprecipitation it comes down even though its solubility is not exceeded.

Postprecipitation is the least common of the three and can be dismissed with a single paragraph. In most cases it happens because the solution remains supersaturated with the impurity while the precipitate is being formed. Then the particles of precipitate act as nuclei for the impurity, which consequently deposits onto their surfaces. Usually the precipitate and the postprecipitated impurity have an ion of the reagent in common, and it is the adsorption of an excess of this ion onto the particles of precipitate that is responsible for their ability to act as nuclei for the impurity. An example arises in the precipitation of calcium oxalate by adding oxalate ion to a solution containing calcium and magnesium ions. If the original concentrations of the two ions are comparable, calcium oxalate precipitates first because it is less soluble than magnesium oxalate ($K_{CaC_2O_4} = 4 \times 10^{-9}$ mol^2 dm^{-6}, $K_{MgC_2O_4} = 1 \times 10^{-8}$ mol^2 dm^{-6}), but calcium oxalate cannot be completely precipitated without adding so much oxalate that the solubility product of magnesium oxalate is exceeded as well, However, the nucleation of magnesium oxalate is so slow that the precipitation of calcium oxalate can be completed before the precipitation of magnesium oxalate begins. At the end of the precipitation the calcium oxalate adsorbs excess oxalate ions. The concentration of oxalate ion at the surface of the precipitate consequently becomes much higher than it is in the bulk of the solution, and therefore the precipitation of magnesium oxalate begins at the surfaces of the particles of calcium oxalate. Postprecipitation occurring in this way is sometimes called delayed simultaneous precipitation to emphasize the fact that the solubility of the impurity is exceeded in the bulk of the solution, as it always is in simultaneous precipitation. There is a rarer variety of postprecipitation in which the solubility of the impurity is not exceeded in the bulk of the solution, but only in the double layers around the particles of precipitate after an excess of reagent has been added, and this is sometimes called delayed coprecipitation.

To illustrate the difference between simultaneous precipitation and coprecipitation, let us suppose that silver nitrate is added to a solution containing 0.05 *M* chloride ion and 0.05 *M* acetate ion with the intent of precipitating silver chloride so that the amount of chloride ion can be calculated. To simplify matters we may imagine that the addition of silver nitrate does not change the total volume of the solution, so that the concentration of acetate ion will remain equal to 0.05 *M* if silver acetate does not precipitate. Since the solubility product of silver acetate is equal to 4.4×10^{-3} mol^2 dm^{-6}, silver acetate can precipitate if the concentration of silver ion ever exceeds the value given by $[Ag^+] = K_{AgOAc}/[OAc^-] = 4.4 \times 10^{-3}/0.05 = 0.09$ *M*. Simultaneous precipitation of silver acetate and silver chloride will occur if a larger concentration of silver ion is ever present. The addition of a very large excess of silver ion would have to be avoided. It would also be dangerous to add the silver nitrate in large portions and without stirring efficiently, for then silver acetate might precipitate around the point where the reagent was being added and become coated over with silver chloride that would prevent its going back into solution.

Simultaneous precipitation would not be hard to prevent in this case. It certainly could not occur if an 0.09 M solution of silver nitrate, or any more dilute solution, were used as the reagent. Nevertheless, the precipitate of silver chloride will be contaminated with some silver acetate even if this is done, and the presence of the silver acetate will be the result of coprecipitation.

Simultaneous precipitation may be very much more extensive than coprecipitation. The addition of reagent is always continued until no more precipitate results from adding a little more reagent. If the solubility product of silver acetate were exceeded, the addition of silver nitrate would be continued until all of the acetate had been precipitated along with the chloride. A very large error in the determination of chloride would result. The error is smaller in coprecipitation because only a fraction of the impurity is carried down with the precipitate. Sometimes this fraction is so small that the amount of impurity is negligible, but sometimes it is so large that the impurity weighs more than the desired precipitate. The following pages describe the ways in which coprecipitation can occur and the extents of the contaminations they can produce.

Many different processes leading to coprecipitation have been identified, given various names, and defined in various ways. The following classification and definitions will be adopted here.

1. *Inclusion* is the mechanical entrapment of a portion of the solution surrounding a growing particle. It can result from irregularities in crystal growth or from the coalescence of several colloidal particles in such a way that a little solution is enclosed in a pocket formed by their surfaces.

2. *Adsorption* is the carrying down of an impurity on the surface of a particle of precipitate. Usually this occurs in the way described in Section 10.5: one of the ions of the precipitate is adsorbed and attracts an oppositely charged counter ion. The addition of excess sodium chloride to a solution of silver nitrate would yield a precipitate that could be represented as $AgCl \cdot Cl^- \mid Na^+$, and that would contain sodium chloride as an impurity.

It is possible for a precipitate to adsorb an ion that is different from either of its own ions: this is called *specific adsorption*. If barium sulfate is suspended in an acidic solution of iron(III) nitrate, it adsorbs iron(III) ions, which not only outnumber the dissolved barium and sulfate ions but also can form fairly strong bonds with the sulfate ions in the crystals of the precipitate, as is shown by the fact that the formation constant of the $FeSO_4^+$ ion pair is fairly large, approximately 10^3 mol^{-1} dm^3. An exact-

20*

ly similar situation would arise if only a very small excess of barium ion were added in precipitating barium sulfate from an acidic solution of iron(III) sulfate. Some of the excess barium ions would be adsorbed, but iron(III) ions would also be adsorbed and will predominate if their concentration is sufficiently high. A precipitate can always adsorb its own ions, and one of these must be in excess during the precipitation while the other must be in excess after the precipitation is complete. However, other ions may be adsorbed as well, and may be adsorbed to even larger extents, depending on the composition of the solution surrounding the precipitate. In stressing adsorption of the ions of precipitates in gravimetric analysis this chapter underestimates the importance of specific adsorption in other circumstances.

Precipitates can also adsorb uncharged molecules (such as those of some dyes and detergents) and ion pairs, but such adsorption cannot be discussed here.

3. *Occlusion* is the carrying down of an impurity in the interior of a particle of precipitate. It occurs by the adsorption of the impurity onto the surface of the growing particle, followed by further growth of the particle to enclose the adsorbed impurity. It differs from inclusion in that the latter does not involve adsorption of the impurity. During the precipitation of barium sulfate by adding sulfuric acid to a solution of barium chloride, there will be excess barium ions adsorbed on the surfaces of the growing particles until enough reagent has been added to reach the point of zero charge. Chloride ions will be attracted to the barium ions, and the precipitate can be represented as $BaSO_4 \cdot Ba^{2+} \mid Cl^-$. Some of the counter ions will be trapped by further growth of the crystal, and the precipitate is said to be contaminated with occluded barium chloride.

Occlusion is discussed further in Section 10.7.

4. *Solid-solution formation* may be regarded as a kind of occlusion in which a precipitate becomes contaminated with another compound that crystallizes in the same system and is formed from ions whose sizes are nearly equal to those of the ions of the precipitate. The foreign compound often has an ion in common with the precipitate, but this is not a necessary requirement. Lead sulfate and barium sulfate form solid solutions: each gives rhombic crystals and the radii of lead and barium ions, 1.20 and 1.34 Å, respectively, differ by only 12%. Solid solutions are known in which the ionic radii differ by as much as 15%. Potassium hydrogen sulfate also forms rhombic crystals, the radius of a potassium ion is 1.33 Å, and a hydrogen sulfate ion must have almost the same size as a sulfate ion. Consequently potassium hydrogen sulfate can form solid solutions with either barium sulfate or lead sulfate. If barium sulfate is precipitated from a solution containing lead ion, it will be contaminated with lead sulfate; if it is precipitated from an acidic solution containing potassium ion, it will be contaminated with potassium hydrogen sulfate; if it is precipitated from an acidic solution containing lead and potassium ions, it will be contaminated with both lead sulfate and potassium hydrogen sulfate.

Solid-solution formation is discussed further in Section 10.8. Section 10.9 will review the mechanisms of coprecipitation, compare their magnitudes and effects, and indicate what can be done about them.

SUMMARY: There are three processes that can lead to the contamination of a precipitate: postprecipitation, simultaneous precipitation, and coprecipitation. In postprecipitation the impurity comes down after the desired precipitate. Simultaneous precipitation occurs when the solubility of the impurity is exceeded during the precipitation of the desired compound. Coprecipitation can occur by inclusion, adsorption, occlusion, and solid-solution formation.

10.7. Occlusion

PREVIEW: This section discusses the mechanism, extent, and effects of occlusion.

In the preceding section it was said that barium chloride would be occluded in a precipitate of barium sulfate obtained by adding sulfuric acid to a solution of barium chloride. Until the precipitation was nearly complete the concentration of barium ion would be higher than it is at the point of zero charge. Barium ions would therefore be adsorbed onto the precipitate. Some negatively charged counter ion must accompany an adsorbed cation, and chloride ion is the only anion of which a sufficient supply is available.

The identity of the occluded substance will often depend on which of the two reacting solutions is added to the other. If barium chloride is added to a solution of sulfuric acid, the growing particles will have sulfate ions adsorbed on their surfaces. Hydronium ion will act as the counter ion, and at the end of the precipitation there will be some sulfuric acid occluded in the barium sulfate.

These are simple examples because the identity of the counter ion is obvious in each. In an actual gravimetric analysis the solution would probably contain several different cations or anions. The following rules are used in predicting which of two ions will predominate as the counter ion in any given situation:

1. If two possible counter ions are present at the same concentration, the one having the higher charge will predominate. This is because the more highly charged ion is more strongly attracted to the adsorbed ion.

2. If two possible counter ions have the same charge, the one present at the higher concentration will predominate. This is mere weight of numbers.

3. If two possible counter ions have the same charge and are present at the same concentration, the one that forms a stronger bond with the adsorbed ion will predominate. The strengths of the bonds can be gauged by comparing the solubilities of the salts, or the stabilities of the complexes or ion pairs, that the possible counter ions form on reacting with the adsorbed ion. This is called the *Paneth–Fajans–Hahn rule*.

To illustrate the application of these rules, let us consider several different examples. In each one it is supposed that barium sulfate is precipitated by adding barium chloride to a solution that contains sulfate ion and that also contains two different cations.

Sulfate ion will be present in excess on the surfaces of the particles until the point of zero charge is reached. One of the cations will act as the counter ion to a much larger extent than the other, and the sulfate of that cation will be the chief occluded impurity.

In one analysis the precipitation is made from a solution that contains 0.05 M lithium ion and 0.05 M magnesium ion. The concentrations are the same, but magnesium ion will predominate because its charge is higher, and magnesium sulfate will be the chief occluded impurity. In a second analysis the solution contains 0.01 M magnesium ion and 0.1 M cadmium ion. The charges are the same, but cadmium ion will predominate because its concentration is higher, and cadmium sulfate will be the chief occluded impurity. In a third analysis the solution contains 0.001 M magnesium ion and 0.001 M calcium ion. The charges and concentrations are the same, but the solubility of calcium sulfate in water is only about 0.002 M while that of magnesium sulfate is not far below 3 M. Calcium sulfate will be the chief occluded impurity. In a fourth analysis the solution contains 0.05 M zinc ion and 0.05 M iron(II) ion. Again the charges and concentrations are the same. The solubility of zinc sulfate is several times as large as that of iron(II) sulfate, but both exceed 1 M. However, the formation constant of the zinc–sulfate ion pair is approximately equal to 200 mol^{-1} dm^3 while that of the iron(II)–sulfate ion pair is only about 2.5 mol^{-1} dm^{-3}. The bonding between zinc and sulfate ions is much stronger than that between iron(II) and sulfate ions, and zinc sulfate will be the chief occluded impurity.

You should be wondering what happens if both the concentrations and the charges are different. Sometimes the answer is obvious; sometimes it is not. If a fifth analysis were made with a solution containing 0.05 M lithium ion and 0.2 M zinc ion, the zinc ion would predominate because its concentration is higher, its charge is higher, and it gives a more stable ion pair with sulfate ion. If a sixth were made with a solution containing 2 M lithium ion and 0.01 M zinc ion, it would be hard to predict the result with much assurance. Probably the precipitate will contain both occluded lithium sulfate and occluded zinc sulfate. To find out which will be present in the larger amount you would have to perform the precipitation and analyze the precipitate for lithium and zinc.

Some chemists find these rules so enjoyable that they use them all the time, but you should do better than that. If you use them to predict what will happen if potassium perchlorate is precipitated by adding perchlorate ion to a solution that contains 0.01 M ammonium ion and 0.2 M calcium ion in addition to some potassium ion, they will lead you to think that calcium perchlorate will be the impurity to fear. However, ammonium ion has almost the same radius as potassium ion (the respective values are 1.43 Å and 1.33 Å), and both potassium perchlorate and ammonium perchlorate form rhombic crystals. The danger is not that a little calcium perchlorate will be occluded, but that a great deal of ammonium perchlorate may be carried down as a solid solution. You should not begin to worry about how much of an impurity will be occluded until you have assured yourself that it will be occluded instead of being carried down by solid-solution formation or simultaneous precipitation.

You might think that a precipitate must be too heavy if it contains an impurity, but a peculiar feature of occlusion is that it often gives precipitates that are too light. If barium sulfate is precipitated by adding sulfuric acid to a pure solution of barium chloride, virtually all of the barium ions in the original solution will appear in the final precipitate. For the reasons given in the first paragraph of this section, some of them will be accompanied by chloride ions rather than by sulfate ions. Since the formula weights of $BaCl_2$ and $BaSO_4$ are 208.2 and 233.4, respectively, the barium chloride weighs less than the barium sulfate that it replaces, and the precipitate is lighter than it would be if the chloride ions were not present. On the other hand, if calcium sulfate is precipitated by adding calcium ion to a solution of zinc sulfate, the precipitate will contain virtually all of the sulfate ions, but it will contain some of them as occluded zinc sulfate rather than as calcium sulfate, and the precipitate will be too heavy because the formula weight of zinc sulfate is higher than that of calcium sulfate. You must compare the weight of the occluded impurity with the weight of the precipitate that should have been obtained.

SUMMARY: Occlusion is the result of entrapment of counter ions that are attracted to the surface of a growing precipitate. When the solution contains several ions that might be occluded, their concentrations, charges, and the stabilities of the compounds or ion pairs that they form with the oppositely charged adsorbed ion of the precipitate all influence the proportions in which they are occluded. A peculiarity of occlusion is that it sometimes causes a precipitate to be lighter than it would be if it were pure.

10.8. Solid-solution formation

PREVIEW: Each constituent of a solid solution has an activity that is smaller than 1. Expressions that take their activities into account are needed to describe the equilibrium between a solid solution and the liquid solution with which it is in contact. Such expressions show that an impurity can be coprecipitated even though its solubility product is not exceeded, and make it possible to estimate how much of it will be present and how its amount will be affected by carrying out the precipitation in different ways.

A solid solution is often defined as a solution of one solid in another, and alloys are cited as familiar examples. This definition assumes that the reader knows what a solution is. Most of our ideas about solutions are based on liquid solutions because these are much more common than solid solutions. Vodka, antifreeze, battery acid, 1 M sodium hydroxide, and other familiar liquid solutions are invariably homogeneous, or can be made so by shaking for a little while. Hence we usually associate solutions with homogeneity, and indeed a solution is often defined as a homogeneous mixture. However, a solid solution—like a liquid one—may not be homogeneous. Some non-homogeneous solid solutions will be described in this section. A solid solution that was not homogeneous would eventually become so by diffusion, recrystallization, or some other process, but it would certainly not be convenient to define a solution as a mixture that will eventually become homogeneous.

A more complicated definition is needed to avoid such difficulties. We might define a solution as a mixture in which there is no boundary between different phases, so that there is no discontinuity of composition. Another possibility would be to define a solution as a mixture of two (or more) constituents in which the activity of each depends on the composition.[†] In any case a solid solution would then be simply a solution that is a solid.

A mixture of calcium sulfate and silver chloride is not a solid solution, and neither of these compounds affects the solubility of the other. The solubility of silver chloride is governed by the expression $K = a_{Ag^+} a_{Cl^-}/a_{AgCl(s)}$, and does not change because the addition of calcium sulfate does not alter the activity of silver chloride in the solid phase. Nor does the solubility of calcium sulfate change, because its activity is not affected by the addition of silver chloride. A mixture of calcium sulfate and silver sulfate is also not a solid solution. Here each compound does alter the solubility of the other, but only through the common-ion effect. The products $a_{Ca^{2+}} a_{SO_4^{2-}}$ and $a_{Ag^+}^2 a_{SO_4^{2-}}$ have the same values for the mixture as they do for the pure salts.

Mixtures of barium sulfate with potassium permanganate or with lead sulfate are solid solutions. The presence of potassium permanganate decreases the solubility of barium sulfate even though these two compounds do not have an ion in common. The value of the product $a_{Ba^{2+}} a_{SO_4^{2-}}$ is decreased by the presence of the potassium permanganate, which means that the activity of barium sulfate is smaller when the solid contains potassium permanganate than it is for pure barium sulfate. In a mixture of barium sulfate and lead sulfate each compound decreases the activity of the other, and depresses its solubility more than the calculations of Section 5.3 would predict.

In an *ideal solid solution* the activity of each substance would be equal to its mole fraction N. Since the *mole fraction* of either substance is the ratio of the number of moles n of that substance to the total number of moles of material present,

$$N_{PbSO_4} = \frac{n_{PbSO_4}}{n_{PbSO_4} + n_{BaSO_4}} \quad \text{and} \quad N_{BaSO_4} = \frac{n_{BaSO_4}}{n_{PbSO_4} + n_{BaSO_4}}. \quad (10.1)$$

Evidently

$$N_{PbSO_4}/N_{BaSO_4} = n_{PbSO_4}/n_{BaSO_4}. \quad (10.2)$$

To describe the behavior of a mixture of barium and lead sulfates, two equilibrium constants are needed:

$$K_{PbSO_4} = \frac{a_{Pb^{2+}} a_{SO_4^{2-}}}{a_{PbSO_4(s)}} = 1.62 \times 10^{-8} \text{ mol}^2 \text{ dm}^{-6} \quad (10.3a)$$

---

† In an aqueous solution of sodium chloride the activity of sodium chloride depends on its concentration. So does the activity of water, as is shown by the fact that the boiling and freezing points and the partial pressure of water vapor above it, among others, all change with concentration. In taking the activity of water as being equal to 1 in a dilute solution we are not denying that it varies with concentration, but are simply neglecting the variation when it is small.

and

$$K_{BaSO_4} = \frac{a_{Ba^{2+}} a_{SO_4^{2-}}}{a_{BaSO_4(s)}} = 1.35 \times 10^{-10}\,\text{mol}^2\,\text{dm}^{-6}. \tag{10.3b}$$

These are the equations for the solubility products, but the activities of the solids cannot be dropped because neither is equal to 1 in a solid solution. Combining eqs. (10.3), and replacing the activities of the dissolved ions by their concentrations,

$$\frac{a_{PbSO_4(s)}}{a_{BaSO_4(s)}} = \frac{K_{BaSO_4}}{K_{PbSO_4}} \frac{[Pb^{2+}]}{[Ba^{2+}]}. \tag{10.4}$$

Assuming that this solid solution is ideal, which experiments show to be very nearly true, $a_{PbSO_4(s)}/a_{BaSO_4(s)} = N_{PbSO_4}/N_{BaSO_4}$, and eqs. (10.4) and (10.2) can therefore be combined to give

$$\frac{N_{PbSO_4(s)}}{N_{BaSO_4(s)}} = \frac{n_{PbSO_4}}{n_{BaSO_4}} = \frac{K_{BaSO_4}}{K_{PbSO_4}} \frac{[Pb^{2+}]}{[Ba^{2+}]} = 8.3 \times 10^{-3} \frac{[Pb^{2+}]}{[Ba^{2+}]}. \tag{10.5}$$

Let us suppose that a solution containing 0.02 M barium ion and 1 M lead ion is treated with a portion of sulfuric acid so small that the solubility of barium sulfate is only just barely exceeded, and that the addition is made so slowly, and with such efficient stirring, that there is no local excess of sulfate ion at the point where the acid enters the solution. The required concentration of sulfate ion will be about 5×10^{-9} M.[†] It is so small that the solubility of lead sulfate will not be exceeded: the product $[Pb^{2+}][SO_4^{2-}]$ will be equal to $(1)(5 \times 10^{-9}) = 5 \times 10^{-9}\,\text{mol}^2\,\text{dm}^{-6}$, which is less than a third as large as K_{PbSO_4}. Nevertheless, there will be lead sulfate in the precipitate, and its mole fraction will be given by eq. (10.5). Since eqs. (10.1) show that $N_{PbSO_4} + N_{BaSO_4} = 1$, eq. (10.5) may be written

$$\frac{N_{PbSO_4}}{1 - N_{PbSO_4}} = 8.3 \times 10^{-3} \times \frac{1}{0.02} = 0.415$$

whence $N_{PbSO_4} = 0.293$ so that $N_{BaSO_4} = 1 - 0.293 = 0.707$. There will be 0.293 mole of lead sulfate for each 0.707 mole of barium sulfate; of the total weight of the first bit of precipitate, 35% will be lead sulfate and only 65% will be barium sulfate. This is an unfavorable case because the concentration of lead ion is much larger than the concentration of barium ion and because K_{BaSO_4}/K_{PbSO_4} is not very small (which is to say that barium sulfate is not very much less soluble than lead sulfate), but it shows that coprecipitation by solid-solution formation may be very extensive indeed.

If a second portion of reagent were added to the same solution, the concentrations of the dissolved ions will be different from their original values because of the precipitation and coprecipitation that occurred on adding the first portion. The second portion of precipitate will have a different composition from the first. There was less

---

[†] It is given by $[SO_4^{2-}] = K_{BaSO_4} a_{BaSO_4(s)}/[Ba^{2+}]$. We shall presently calculate that $a_{BaSO_4(s)}$ will be equal to 0.71 in the precipitate obtained.

barium ion than lead ion in the original solution, and the first portion of precipitate contained more barium ion than lead ion. The ratio $[Pb^{2+}]/[Ba^{2+}]$ will have been increased by the addition of the first portion of the reagent, and the second portion of precipitate will be even less pure than the first.

There are two extreme possibilities. One is that the precipitate dissolves and re-crystallizes very rapidly and thereby comes to equilibrium with the liquid at every stage of the precipitation. Its composition will become the same throughout its entire volume. This is called *homogeneous distribution*. The other is that dissolution and recrystallization do not occur at detectable rates. Then each portion of precipitate will retain the composition that it had at the instant when it came down, and the precipitate will vary in purity from one point to another. This is called *heterogeneous distribution* or, for reasons that will soon become evident, logarithmic distribution, and it results from the fact that equilibrium is not attained during the precipitation.

Homogeneous distribution is described by eq. (10.5) because all of the precipitated material reaches equilibrium at every point. If n_{Pb}^0 and n_{Ba}^0 are the numbers of milli-moles of the two ions that were present when precipitation began, and if V is the volume (in cm^3) of the solution, then

$$[Pb^{2+}] = \frac{n_{Pb}^0 - n_{PbSO_4}}{V} \quad \text{and} \quad [Ba^{2+}] = \frac{n_{Ba}^0 - n_{BaSO_4}}{V}. \tag{10.6}$$

Until an excess of reagent has been added, x millimoles of precipitate will have been formed by the addition of x millimoles of reagent, and $x = n_{PbSO_4} + n_{BaSO_4}$. Combining this with eqs. (10.5) and (10.6) gives

$$\frac{n_{PbSO_4}}{x - n_{PbSO_4}} = \frac{K_{BaSO_4}}{K_{PbSO_4}} \times \frac{n_{Pb}^0 - n_{PbSO_4}}{n_{Ba}^0 + n_{PbSO_4} - x}. \tag{10.7}$$

This is the most useful, though not the most common, form of the homogeneous distribution law.

Example 10.1. What is the equilibrium composition of the solid solution of silver bromide and silver chloride obtained by adding 1 millimole of silver ion to a solution containing 1 millimole of bromide ion and 1 millimole of chloride ion? The solubility products of silver bromide and silver chloride are 5.25×10^{-13} mol^2 dm^{-6} and 1.78×10^{-10} mol^2 dm^{-6}, respectively.

Answer. Silver bromide is the less soluble of the two halides and silver chloride will be coprecipi-tated with it. By analogy with eq. (10.7),

$$\frac{n_{AgCl}}{x - n_{AgCl}} = \frac{K_{AgBr}}{K_{AgCl}} \times \frac{n_{Cl}^0 - n_{AgCl}}{n_{Br}^0 + n_{AgCl} - x}.$$

Since $n_{Br}^0 = n_{Cl}^0 = x = 1$ and $K_{AgBr}/K_{AgCl} = 2.95 \times 10^{-3}$

$$\frac{n_{AgCl}}{1 - n_{AgCl}} = 2.95 \times 10^{-3} \times \frac{1 - n_{AgCl}}{n_{AgCl}}$$

whence $n_{AgCl}/(1 - n_{AgCl}) = (2.95 \times 10^{-3})^{1/2} = 0.054$ and $n_{AgCl} = 0.052$. The precipitate will contain 0.052 mmol (= 7.4 mg) of silver chloride and 0.948 mmol (= 178.1 mg) of silver bromide. It will contain 4.0% of silver chloride by weight, and 5.2% of the original amount of bromide ion will remain in the solution. Experimentally it is found that 4.8% of the bromide ion remains unprecipitated if the precipitate does not flocculate while it is being formed so that it is always very finely divided and can approach equilibrium with the solution at every stage. This should be compared with the result of Example 10.2 below.

The heterogeneous distribution law is obtained by rewriting eq. (10.5) to describe the composition of the infinitesimal increment of precipitate obtained by adding an infinitesimal portion of reagent at any point during the precipitation:

$$\frac{dn_{PbSO_4}}{dn_{BaSO_4}} = \frac{K_{BaSO_4}}{K_{PbSO_4}} \times \frac{[Pb^{2+}]}{[Ba^{2+}]}.$$

Combining this with eqs. (10.6) and separating the variables yields

$$\frac{dn_{PbSO_4}}{n_{Pb}^0 - n_{PbSO_4}} = \frac{K_{BaSO_4}}{K_{PbSO_4}} \times \frac{dn_{BaSO_4}}{n_{Ba}^0 - n_{BaSO_4}}.$$

Integration yields

$$\log (n^0 - n_{PbSO_4}) = \frac{K_{BaSO_4}}{K_{PbSO_4}} \log (n^0 + n_{PbSO_4-x}) \qquad (10.8)$$

as an expression of the heterogeneous distribution law.

Example 10.2. In Example 10.1, what would have been the composition of the precipitate if infinitesimal aliquots of reagent had been added and if heterogeneous distribution had been attained by preventing recrystallization during the precipitation?

Answer. By analogy with eq. (10.8) and with $n_{Br}^0 = n_{Cl}^0 = x = 1$,

$$\log (1 - n_{AgCl}) = \frac{K_{AgBr}}{K_{AgCl}} \log n_{AgCl} = 2.95 \times 10^{-3} \log n_{AgCl}.$$

Solving this transcendental equation yields $n_{AgCl} = 0.0128$. The precipitate will contain 0.0128 mmol (1.83 mg) of silver chloride and 0.987 mmol (185.4 mg) of silver bromide. It will contain 0.98% of silver chloride by weight, and 1.3% of the bromide ion will remain unprecipitated. In an experiment in which aluminum ion was added to flocculate the colloidal particles, which were negatively charged because of adsorption of halide ions, it was found experimentally that 2.2% of the bromide ion remained in solution. The agreement with the predicted value is not as good as in Example 10.1, partly because the reagent must have been added in finite rather than infinitesimal portions, but chiefly because the attainment of equilibrium by recrystallization was not completely prevented by the flocculation of the particles.

Homogeneous distribution yields precipitates that are less pure than heterogeneous distribution does. The ratio of the concentration of the interfering ion to the concentration of the ion being determined increases steadily as the precipitation proceeds. The first portion of the precipitate is purer than any later one. In heterogeneous distribution the first portion retains its initial purity; in homogeneous distribution it becomes as impure as the last portion. In the next section it will be said that the purity of a precipitate is often increased by the recrystallization that occurs on standing in the mother liquor, but exactly the opposite is true if a solid solution has been formed and if homogeneous distribution has not already been attained by the time the last portion of reagent is added.

SUMMARY: Solid-solution formation is the contamination of a precipitate by another compound that crystallizes in the same system and has ionic sizes nearly equal to those in the desired precipitate. Because the activity of each compound in an ideal solution is equal to its mole fraction, each constituent of a solid solution affects the solubility of the other. Homogeneous distribution of the impurity is obtained if the precipitate remains very finely divided and can recrystallize rapidly throughout the precipitation; heterogeneous distribution is obtained if recrystallization is prevented by flocculation. All other things being equal, the percentage of the impurity is smaller if its distribution is heterogeneous.

10.9. A summary of coprecipitation

PREVIEW: This section summarizes the different kinds of coprecipitation and tells when each is likely to predominate and what can be done about it.

Which of the four kinds of coprecipitation—inclusion, adsorption, occlusion, and solid-solution formation—predominates in any particular case depends on the particular circumstances. Inclusion may be the chief source of contamination in large crystals grown from nearly pure solutions, but is rarely important in gravimetric analysis. Most of the included molecules are molecules of the solvent, and not only weigh less than an equal volume of any other substance that is likely to be coprecipitated, but will escape during thermal aging (Section 10.11) if the precipitate is dried at a sufficiently high temperature.

Adsorption is unimportant if the particle size is large when the precipitation is complete, for large particles have surface areas that are very small in proportion to the amounts of precipitate they contain, but it becomes a major source of contamination if the particles are very small. The extent of adsorption depends on the concentration of excess reagent that is added and on the strength with which the ion in excess is adsorbed, as well as on the particle size. Different precipitates adsorb their own ions to different extents. The amount of chloride ion that is adsorbed by a precipitate of silver chloride from a solution containing excess chloride ion is much smaller than the amount of hydroxide ion that is adsorbed by a precipitate of hydrous ferric oxide or aluminum oxide from a basic solution under comparable conditions. A little sodium chloride, or some other chloride, would be coprecipitated by the silver chloride, but much could not be coprecipitated unless the concentration of excess chloride ion were so high that the solubility of the precipitate would be substantially increased by the formation of $AgCl_2^-$ and other ionic complexes. If hydrous iron(III) oxide were precipitated by adding sodium hydroxide to a solution containing iron(III), a great deal of sodium hydroxide would be coprecipitated with it. The difference between the behaviors of silver chloride and hydrous iron(III) oxide is partly due to a difference between the physical forms in which they are usually obtained. Though the individual particles are small in silver chloride and other curdy precipitates, they tend to be larger and more compact than those of gelatinous precipitates like the hydrous oxides, and thus to have much smaller specific areas.

An adsorbed impurity—unlike an occluded one (see page 297)—always increases the weight of a precipitate, for neither the adsorbed ions nor the counter ions that accompany them belong in the precipitate. There are several ways to decrease the error that this causes. One is to decrease the surface area of the precipitate. Occasionally this can be done by allowing it to stand in the mother liquor, preferably at an elevated temperature, for some time after precipitation is complete, whereupon the average particle size increases by a process called Ostwald ripening, of which there is some discussion in Section 10.11. Decreasing the rate of addition of reagent and avoiding a local excess of it during precipitation is more often successful in achieving the same aim, and is carried to extremes in precipitation from homogeneous solution (Section 10.10). A different approach is to keep the concentration of excess reagent to a minimum, for much of it cannot be adsorbed if very little of it is present. You cannot carry this too far, for the concentration of excess reagent must be large enough to render the solubility of the precipitate negligible. Still another approach is to add just enough excess reagent to reach the point of zero charge, at which the precipitate is uncharged and there is no adsorption. This succeeds only when an excess of reagent is present at the point of zero charge. The point of zero charge for silver chloride lies at $pcAg = 4.0$ and $pcCl = 5.7$. If you add silver ion to an 0.01 M solution of chloride ion until you reach this point, almost all of the chloride ions will be in the precipitate; if you add chloride ion to an 0.01 M solution of silver ion until you reach the same point, at least 1% of the silver ions will still be dissolved, and the figure will be even larger if the volume of the mixture has been increased by adding the reagent to it.

All these approaches are aimed at decreasing the extent of adsorption; a final one is aimed at making it innocuous. Suppose that the precipitate $AgCl \cdot Cl^- \mid Na^+$ has been obtained by adding excess sodium chloride to a solution of silver nitrate. The sodium ions are held on the surface of the precipitate by electrostatic forces alone, and can easily be replaced by other cations, such as hydrogen ions. If the precipitate is washed with a dilute solution of a strong acid, such as perchloric acid or nitric acid, the sodium ions will be exchanged for hydrogen ions. Then it will be contaminated with adsorbed hydrochloric acid, which volatilizes on drying the precipitate and leaves pure silver chloride behind it. Sodium ions can be removed from the precipitate $(Fe_2O_3 \cdot xH_2O) \cdot OH^- \mid Na^+$ in the same general way. You would not want to wash this with nitric acid, but you could wash it with a solution of ammonium nitrate containing enough ammonia to raise its pH to a value at which the solubility of the precipitate was negligible. Sodium ions will be exchanged for ammonium ions. On heating the product, $(Fe_2O_3 \cdot xH_2O) \cdot OH^- \mid NH_4^+$, to a temperature sufficiently high to dehydrate it, the adsorbed and counter ions will be driven off as ammonia and water. If the substance being determined is cationic, the adsorbed ions of the reagent will be anionic, and can generally be driven off on heating if either hydrogen ion or ammonium ion is the counter ion. Volatile compounds of adsorbed metal ions are so rare that one of the other approaches has to be adopted in gravimetric determinations of anions.

Occlusion is far more important with large particles than with small ones. If an occluded counter ion differs in size from the lattice ion it replaces, or if it is much less strongly bound to the oppositely charged ions near it, its presence distorts the crystal structure for some distance. Ions dissolve more rapidly from a distorted surface than from a more nearly perfect one. On standing in the mother liquor the distortion can be relieved as ions dissolve from the layers above the ill-fitting occluded ion and deposit onto other parts of the crystal. If the particle is small, the occluded ion cannot be far from its surface, and the dissolution of only a few nearby ions will expose the occluded ion and enable it to escape into the solution. If the particle is large, the strain caused by the presence of an occluded ion is distributed over a much larger fraction of its surface. The effect on the rate of dissolution is smaller, and dissolution must proceed to a greater depth before the occluded ion becomes exposed. The removal of an occluded impurity from large particles of a precipitate is therefore likely to be very slow. The only practical remedy is to carry out the precipitation more slowly so that counter ions present on the surface can be replaced by ions of the reagent instead of being trapped by the rapid growth of the particle around them.

The effect of particle size on solid-solution formation was mentioned in Section 10.8. All of the ions in a very small particle can redissolve and redeposit in far less time than is needed for all of the ions in a large particle of the same material to do so. Small particles approach homogeneous distribution much more rapidly than large ones, and therefore tend to be more badly contaminated by solid-solution formation. In gravimetric analysis the substance that is coprecipitated is always more soluble than the desired precipitate. If the original precipitate is redissolved in some suitable solution and reprecipitated, the concentration of the interfering ion will be smaller in proportion to that of the ion being determined than it was during the first precipitation, and a purer product will result. A precipitate of silver bromide contaminated with silver chloride might be dissolved in a solution containing a high concentration of ammonia or ethylenediamine, and reprecipitated by neutralizing the base. Double or even triple precipitation is often needed to decrease the mole fraction of the impurity to a tolerable level. However, they consume so much time and effort that it is better to search for a masking agent that will react preferentially with the interfering ion and prevent it from contaminating the first precipitate, to perform a preliminary separation of the two ions, or to employ a volumetric or some other procedure instead of a gravimetric one.

SUMMARY: Inclusion is unlikely to be important in gravimetric analysis. Adsorption is most important with very finely divided particles, and can be minimized by increasing the average particle size or rendered harmless by replacing the original counter ion with another that forms a volatile compound with the adsorbed ion. Occlusion can be minimized by allowing the precipitate to stand in the mother liquor or, better, by carrying out the precipitation very slowly. Solid-solution formation can be decreased by increasing the average particle size or by dissolving and reprecipitating the precipitate, but is best handled by modifying the procedure so as to make it impossible.

10.10. Precipitation from homogeneous solutions

PREVIEW: Precipitation from homogeneous solutions is a technique in which the reagent is produced by some slow chemical reaction in the solution. It minimizes the extents of adsorption and occlusion and yields precipitates consisting of large compact particles that are easy to filter and wash.

Many of the things that have been said in this chapter reflect the importance of the rate of addition of reagent. Rapid addition of reagent is conducive to the formation of many nuclei, and thus to a decrease of the average particle size and an increase of the total surface area. In addition it promotes irregularities of form that further increase the area. Hence it favors the formation of small particles having a large total area, which are not only more difficult to filter and wash but also more likely to be seriously contaminated by adsorbed impurities than large particles are.

Precipitation from homogeneous solution is a technique by which a reagent can be added very slowly and in which a local excess of reagent cannot form. It is based on the slow hydrolysis (or some other reaction) of a solute, yielding a product that reacts to form the desired precipitate. The hydrolysis proceeds uniformly throughout the solution, at a rate that depends on its temperature. Precipitation is easily carried out at a rate so low that nucleation and irregular growth of the particles are minimized.

A typical example is the use of urea to precipitate calcium oxalate in the gravimetric determination of calcium. A substantial excess of oxalic acid is first added to the solution of calcium ion. Calcium oxalate is readily soluble in acidic solutions by virtue of the reaction $CaC_2O_4(s) + H_3O^+ = Ca^{2+} + HC_2O_4^- + H_2O(l)$. If some precipitate does form, hydrochloric acid is added to redissolve it, and an excess of urea is then added. Urea is so weak a base that its addition does not significantly affect the pH, but on heating to 358–373 K (85–100° C) it hydrolyzes

$$\underset{NH_2}{\overset{NH_2}{C}}{=}O \ (aq) + 2H_2O(l) = 2NH_4^+ + CO_3^{2-}$$

and the resulting carbonate ion can react with hydronium ion, hydrogen oxalate ion, and other acids. As the concentration of hydronium ion decreases, that of oxalate ion rises in accordance with the equation

$$[C_2O_4^{2-}] = \frac{K_1 K_2}{[H_3O^+]^2 + K_1[H_3O^+] + K_1 K_2} c$$

where c is the total concentration of oxalic acid and K_1 and K_2 are its overall acidic dissociation constants. At some point the supersolubility of calcium oxalate is exceeded. If the hydrolysis is sufficiently slow, nucleation and particle growth will be capable of removing oxalate ion more rapidly than it is being produced, and then nucleation will never occur again. Large compact crystals will be formed and the extents of adsorption and occlusion will be very small. Lead chromate, magnesium ammonium phosphate, and a number of other common precipitates can be made in the same way.

Neutralization of its conjugate acid is not always the best way of obtaining the anion that is wanted. The value of K_2 for sulfuric acid is so high that it would be difficult to make a mixture of calcium and hydrogen sulfate ions so acidic that calcium sulfate would not precipitate before neutralization began. It is better to generate sulfate ion by the hydrolysis of dimethyl sulfate, $(CH_3)_2SO_4$, or sulfamic acid, H_2NSO_3H, and these are used in determinations of calcium, strontium, barium, and lead. Fluoride ion produced by the hydrolysis of fluorosilicic acid, H_2SiF_6, is used in determinations of calcium and separations of some of the transuranium elements. Other slow homogeneous reactions serve as well as hydrolytic ones. Nickel can be determined by adding biacetyl and hydroxylamine, which react at a rate that is pH-dependent

$$\begin{array}{c} CH_3C{=}O \\ | \\ CH_3C{=}O \end{array} (aq) + 2\,NH_2OH(aq) = \begin{array}{c} CH_3C{=}NOH \\ | \\ CH_3C{=}NOH \end{array}(aq) + 2\,H_2O(l).$$

The dimethylglyoxime thus formed reacts with nickel(II) ion to yield a chelate[†]

$$2\begin{array}{c} CH_3C{=}NOH \\ | \\ CH_3C{=}NOH \end{array}(aq) + Ni^{2+} = \begin{array}{c} OH \\ CH_3C{=}N \diagdown\diagup ON{=}CCH_3 \\ | \quad\quad Ni\quad\quad | \\ CH_3C{=}NO \diagdown N{=}CCH_3 \\ HO \end{array}(s) + 2\,H_3O^+$$

that is very sparingly insoluble if the pH is not too low.

Hydrous oxides, such as those of iron(III) and aluminum among many others, can also be precipitated from homogeneous solutions. Excess urea is added to an acidic solution of the metal ion and the mixture is heated. Precipitation occurs as the pH rises. It is customary to add some weakly basic anion, such as formate ($HCOO^-$) or succinate ($^-OOCCH_2CH_2COO^-$), as well. Some of the basic anion appears in the precipitate but is destroyed during ignition. Very compact particles are obtained and are very easy to filter and wash. Precipitation in the ordinary way, by adding ammonia or another base, gives gelatinous precipitates that are more difficult to handle and are also contaminated to much larger extents by other heavy-metal ions that may be present.

> SUMMARY: Very slow generation of the reagent promotes particle growth at the expense of nucleation, minimizes the extents of adsorption and occlusion, and yields dense compact precipitates that are easy to handle.

[†] The actual mechanism is more complex, and therefore more interesting, than this oversimplified description of it. Nickel(II) accelerates the formation of dimethylglyoxime by reacting with intermediates that are formed during the first of these two overall reactions. Catalysis by metal ions is common in the reactions that lead to the formation of reagents used in precipitation from homogeneous solutions.

10.11. Aging

PREVIEW: Recrystallization and, at sufficiently high temperatures, thermal aging occur after precipitation is complete and affect both the physical form and the purity of a precipitate.

When a precipitate is allowed to stand in its mother liquor it undergoes recrystallization in the fashion described in Section 10.9. This has several important consequences, of which two have already been mentioned: the slow removal of occluded impurities, and an approach to homogeneous distribution if the precipitate is a solid solution. In addition, the free energy of a particle is decreased by perfecting its crystal structure and decreasing its exposed area. Needles, whiskers, fronds, and dendrites redissolve, and the ions contained in them redeposit elsewhere in a more orderly way. The surface area decreases, and so does the extent of adsorption. Another consequence is that colloidal particles loosely joined together in a fragile floc become more tightly bound by the deposition of additional ions near the boundary between them. This is called *cementation*, and makes curdy precipitates, such as the silver halides, easier to handle by increasing their resistance to peptization.

Recrystallization is sometimes accompanied by a process called *Ostwald ripening*, in which small particles redissolve and large ones grow at their expense. Of course the surface area and extent of adsorption are decreased by an increase of the average particle size, but for the reason mentioned in the preceding paragraph they decrease during recrystallization even if the average particle size does not change. There are a very few substances, especially hard ones having high surface tensions, such as lead chromate and barium sulfate, of which small particles have actually been shown to be more soluble than larger ones. For these substances the average particle size increases during recrystallization, but for others Ostwald ripening is unimportant if it occurs at all.

The rate of recrystallization is governed by the rate constants for dissolution and precipitation. It therefore depends on the experimental conditions, and differs from one precipitate to another. For any one precipitate it tends to increase as the solubility increases. Since the dissolutions of most precipitates are endothermic, their solubilities increase as the temperature increases. Hence the rate of recrystallization is usually increased by increasing the temperature; it is decreased by adding ethanol or any other organic solvent that decreases the dielectric constant of an aqueous solution and renders ionic precipitates less soluble. The rates of recrystallization of precipitates containing basic anions, such as lead chromate and barium sulfate, may be increased by acidifying their mother liquors sufficiently. In general, moderately soluble precipitates tend to recrystallize more rapidly than less soluble ones: the recrystallization of silver bromide is very fast, while that of hydrous ferric oxide goes on for months (partly because its composition and structure change as it recrystallizes). However, there are many exceptions to this general trend. Although barium sulfate is more soluble than silver bromide, it usually recrystallizes much less rapidly. This is partly because crystals of barium sulfate are usually larger than crystals of silver bromide prepared under comparable conditions, but chiefly because the rate constant for the dissolution of barium sulfate has a smaller value than that for the dissolution of silver bromide.

Aging can continue in either of two ways while a precipitate is being dried. Some precipitates retain thin films of water on their surfaces even at temperatures as high as 573 K (300° C). Recrystallization can continue as long as such a film is present, and becomes very rapid at such high temperatures. *Thermal aging* continues after all of the moisture has been driven from the surface. It arises from the vibrations of the lattice ions. As the temperature increases, the amplitudes of these vibrations increase. Misplaced ions can approach the positions they would occupy in a perfect crystal, and tend to remain there because those are the positions that minimize their free energies. Defects in the lattice are eliminated, and volatile included and occluded impurities can escape. Thermal aging becomes fairly rapid if the temperature (in kelvins) equals or exceeds about $T_f/2$, where T_f is the temperature (in kelvins) at which the crystal melts. For silver bromide $T_f = 705$ K, and there is evidence that the rate of thermal aging of silver bromide is appreciable even at room temperature, about 300 K. Since thermal aging has nothing to do with the solvent, it can occur at this temperature with colloidal particles of silver bromide that are suspended in an aqueous solution and undergoing recrystallization at their surfaces. For barium sulfate $T_f = 1853$ K; thermal aging does not become rapid until the temperature approaches 873–973 K (600–700° C), where water will already have been removed from the surface so that recrystallization is no longer possible. Hence thermal aging and recrystallization are easy to distinguish for barium sulfate but proceed side by side for silver bromide.

> SUMMARY: Recrystallization occurs when a precipitate stands in its mother liquor, and may be accompanied by cementation or Ostwald ripening. Thermal aging occurs at temperatures so high that the amplitudes of vibration of the lattice ions become appreciable. Each yields more compact particles and tends to remove some of the impurities.

Problems

Answers to some of these problems are given on page 520.

10.1. The iodate contained in a solution of potassium iodate is determined by adding excess barium chloride solution and filtering, washing, drying, and weighing the barium iodate that precipitates. What impurity or impurities is the precipitate most likely to contain, and why?

10.2. A solution contains some sodium acetate and some sodium chloride. An analyst wishing to determine the concentration of chloride in the solution treats a portion of it with excess silver nitrate. Great care is taken to avoid exceeding the solubility product of silver acetate (which is equal to 4×10^{-3} mol² dm⁻⁶) anywhere at any time. Nevertheless, after the precipitate is filtered off, washed, dried, and weighed, it is found to contain a good deal of acetate. How did this get into the precipitate and what could have been done about it?

10.3. One precipitate of lead sulfate is prepared by slowly adding a solution of lead nitrate to a solution of sodium sulfate; another is prepared in exactly the same way except that the solution of sodium sulfate is added to the solution of lead nitrate. One of the two precipitates contains much more sodium ion than the other. Which one is this, and why?

10.4. An 0.01 M aqueous solution of iron(III) chloride, $FeCl_3$, is acidified with 0.01 M nitric acid and an excess of silver nitrate is added to it slowly to precipitate silver chloride. The AgCl is filtered off, washed, dried, and weighed. What is the chief impurity in it, where is the impurity, how did it get there, and what could be done to obtain a purer precipitate?

10.6. A mixture of barium and magnesium phosphates is to be analyzed to determine the percentage of barium in it. A weighed portion of the mixture is dissolved in excess hydrochloric acid, and a solution of sodium sulfate is added to precipitate barium sulfate.

What impurities will the barium sulfate contain? What coprecipitation process will be responsible for the presence of each?

10.7. Should precipitations from homogeneous solutions be carried out at higher temperatures or at lower ones? Give the reasons for your choice.

CHAPTER 11

Titration Curves

11.1. Introduction

A titration curve is a graph that shows how some measured quantity, or some concentration or other variable that is of interest, changes during a titration. Calculated titration curves enable the chemist to select indicators for titrations and to estimate the accuracy and precision that may be attained in the titration step. Experimental titration curves are needed in most instrumental titration procedures so that the end points of titrations can be located, and are also useful in evaluating the equilibrium constants for reactions of many different kinds. This chapter will explain how titration curves may be calculated, and will describe their properties.

11.2. Kinds of titration curves

PREVIEW: Titration curves can be classified according to the kind of reaction that occurs, the technique that is used to follow the titration, or the algebraic nature of the quantity that is measured.

Titration curves can be classified in several different ways. One depends on the kind of reaction that occurs. There are acid-base titration curves, precipitation titration curves, redox titration curves, and so on. These may be further divided. The shape of an acid-base titration curve depends on whether the acid and base are strong or weak, and on whether they are monofunctional or polyfunctional. The number of possibilities is very large.

Another classification emphasizes the technique by which the titration is performed. The redox titration of iron(II) with permanganate could be followed in many different ways. One might measure the temperature of the reaction mixture, which changes during the titration because the reaction is exothermic; the potential of an electrochemical cell in which one of the two electrodes is a platinum electrode immersed in the mixture; the ability of the mixture to transmit radiation of some wavelength, which is absorbed by one of the reactants or products but not by others; or the electrical conductance of the mixture. The volume of reagent, or some quantity proportional to the volume of reagent, is always plotted along the abscissa axis, and the classification depends on the quantity that is measured and plotted along the ordinate axis. The possibilities listed in the next to the last sentence give rise to thermometric, potentiometric, spectrophotometric, and conductometric titration curves.

A third classification emphasizes the nature, rather than the specific identity, of the quantity that is measured and plotted on the ordinate axis. It may be proportional to the concentration of a single substance present in the titration mixture, or it may be the sum of two or more contributions, each proportional to the concentration of a different substance. Many spectrophotometric titrations belong to the first of these classes, for it is often possible to find a wavelength at which only one of the reactants or products will contribute to the absorbance that is measured. Most conductometric titrations belong to the second, because every one of the ionic species that are present in a solution makes some contribution to its conductance. Such titration curves are said to be *segmented*. Ideally they consist of two or more linear segments, each of which intersects the next one at an equivalence point. Typical segmented titration curves are shown in Fig. 11.2. Alternatively, the quantity that is measured and plotted may be a linear function of the logarithm of a concentration or ratio of concentrations ($y = k_1 \pm k_2 \log c$ or $y = k_1 \pm k_2 \log c_1/c_2$). Such curves are said to be *sigmoidal*. Ideally a sigmoidal titration curve has a point of maximum slope at or near the equivalence point. Typical sigmoidal titration curves are shown in Figs. 11.6 and 11.7. Potentiometric titration curves are normally sigmoidal, and the considerations that govern the choice of an indicator are most easily explained and understood with the aid of sigmoidal curves. Most other instrumental titration techniques yield segmented curves.

The third of these classifications is stressed in the pages that follow because it is the most general. It enables emphasis to be placed on fundamental considerations that apply to all titrations, no matter what chemical reactions may occur or what technique is used to follow them. However, frequent reference will be made to the chemical and instrumental classifications to help you see how those fundamental considerations are applied in practice.

> SUMMARY: Titration curves can be classified according to
> 1. the kind of reaction that occurs (acid-base, complexation, redox, etc.);
> 2. the technique used to follow the titration (potentiometric, amperometric, spectro-photometric, etc.); or
> 3. the nature of the quantity that is measured (proportional to concentration or linearly dependent on the logarithm of a concentration or ratio of concentrations).
> It is the third that is stressed in this chapter.

11.3. Some definitions and fundamental ideas

> PREVIEW: This section lists and defines some symbols and terms used in the sections that follow.

The reaction that occurs during a titration can be represented by the equation

$$sS + rR + \ldots = pP + qQ + \ldots \qquad (11.1)$$

where S is the substance that is titrated, R is the reagent, and P and Q are products of the reaction; s, r, p, and q are the stoichiometric coefficients needed to balance the equation. The volume of the solution being titrated will always be represented by the symbol V_S^0 and expressed in cm³. The concentration of the substance being titrated in that solution will always be represented by c_S^0 and expressed in mmol cm⁻³ (M). Similarly, V_R will denote the volume (cm³) of the reagent solution that has been added at any point during the titration, and c_R will denote the concentration (mmol cm⁻³) of the reagent in that solution. The subscripts "S" and "R" denote the substance being titrated and the reagent, respectively, and the superscript zeros on V_S^0 and c_S^0 emphasize that these symbols pertain to the start of the titration.

It is assumed that equilibrium has been reached at every point, and the symbol K_t is used to denote the equilibrium constant of the reaction between the substance that is titrated and the reagent. Since this reaction is described by eq. (11.1), the concentrations of S and R will always appear in the denominator of the expression for K_t.

If a solution of silver ion ($= $ S) is titrated with one of chloride ion ($= $ R), the reaction that occurs, and the expression for K_t, will be

$$Ag^+ + Cl^- = AgCl(s); \quad K_t = \frac{1}{[Ag^+][Cl^-]}$$

$$\left(= \frac{1}{K_{AgCl}} = 5.5 \times 10^9 \text{ mol}^{-2} \text{ dm}^6 \right) \quad (11.2a)$$

$$= \frac{1}{[S][R]} .$$

Equation (11.2a) also applies to the titration of chloride ion with silver ion, but in this titration S is chloride ion and R is silver ion. If a solution of a strong acid is titrated with one of a strong base in water as the solvent,

$$H_3O^+ + OH^- = 2H_2O(l); \quad K_t = \frac{1}{[H_3O^+][OH^-]}$$

$$\left(= \frac{1}{K_w} = 9.9 \times 10^{13} \text{ mol}^{-2} \text{ dm}^6 \right)$$

$$= \frac{1}{[S][R]} . \quad (11.2b)$$

If a solution of a weak acid, such as acetic acid, is titrated with a strong base, the expression for K_t takes a different form:

$$HOAc(aq) + OH^- = OAc^- + H_2O(l); \quad K_t = \frac{[OAc^-]}{[HOAc][OH^-]}$$

$$\left(= \frac{K_a}{K_w} = 1.7 \times 10^9 \text{ mol}^{-1} \text{ dm}^3 \right)$$

$$= \frac{[P]}{[S][R]} . \quad (11.2c)$$

If a solution of iron(III) is titrated with chromium(II),

$$Fe(III) + Cr(II) = Fe(II) + Cr(III); \quad K_t = \frac{[Fe(II)][Cr(III)]}{[Fe(III)][Cr(II)]}$$

$$= 3.8 \times 10^{17} \text{ in } 5 M \text{ HCl}$$

$$= \frac{[P][Q]}{[S][R]} \tag{11.2d}$$

where the numerical value is taken from Section 8.10. Of course P or Q does not appear in the expression for K_t if its activity is equal to 1 throughout the titration. If the concentration or activity of a substance involved in the reaction is constant but different from 1, its value is incorporated into that of K_t, which thereupon becomes a conditional constant. The equilibrium constant for the reaction between iron(II) ion and permanganate ion is given by $K = [Mn^{2+}][Fe^{3+}]^5/[MnO_4^-][Fe^{2+}]^5[H^+]^8 = 2.9 \times 10^{62} \text{ mol}^{-8} \text{ dm}^{24}$, but if iron(II) ion were titrated with permanganate in 0.5 M perchloric acid as the supporting electrolyte we would write

$$K_t = \frac{[Mn^{2+}][Fe^{3+}]^5}{[MnO_4^-][Fe^{2+}]^5} (= K[H^+]^8 = 1.1 \times 10^{60}) = \frac{[P][Q]^5}{[S][R]^5} = \frac{[P]}{[S]} \cdot \left(\frac{[Q]}{[R]}\right)^5. \tag{11.2e}$$

Other kinds of expressions for K_t are possible but less common. The overwhelming majority can be divided into three groups:

1. those that contain [S] and [R] alone [as do eqs. (11.2a) and (11.2b)],
2. those that contain [P]/[S] and [R] [as does eq. (11.2c)], and
3. those that contain [P]/[S] and [Q]/[R] [as do eqs. (11.2d) and (11.2e)].

All the values of K_t quoted above are very large. Later sections will show how the feasibility of a titration depends on the value of K_t.

If a titration were performed experimentally, the titration curve would certainly be constructed by plotting the measured quantity on the ordinate axis against V_R, the volume of reagent that has been added at any point, on the abscissa axis. For constructing calculated curves a more convenient and revealing independent variable is the "titration parameter" f defined by the equation

$$f = V_R/V_R^* \tag{11.3}$$

where V_R^* is the volume of reagent that is needed to reach the equivalence point. According to this definition, f is proportional to V_R in any particular titration. It is equal to zero at the start of a titration and to 1 at the equivalence point. If f is smaller than 1, the equivalence point has not yet been reached and some S remains to be titrated. If f exceeds 1, the equivalence point has been passed and the titration mixture contains an excess of R.

The equivalence point is the point at which the number of millimoles of R that has been added is equal to r/s times the number of millimoles of S that were present initially:

$$V_R^* c_R = (r/s)_i V_S^0 c_S^0. \tag{11.4}$$

Combining this with eq. (11.3) yields another description of f:

$$f = \frac{V_R c_R}{V_R^* c_R} = \frac{V_R c_R}{(r/s) V_S^0 c_S^0} \left(= \frac{(s/r) V_R c_R^{\square}}{V_S^0 c_S^0} \right). \tag{11.5}$$

> SUMMARY: The symbol K_t denotes the equilibrium constant of the reaction between the substance that is titrated and the reagent. The "titration parameter" f is the fraction that has been added of the amount of reagent needed to reach the equivalence point.

11.4. Idealized titration curves

> PREVIEW: The fundamental equations of titration-curve theory are derived and explained in this section.

This section will outline some of the basic considerations that are employed in calculating titration curves. In it we shall make the simplifying assumption that K_t is infinitely large. If this is true there can be no R whatever in the titration mixture as long as any S remains to be titrated—that is, up to the equivalence point. Neither can any S remain unreacted after the equivalence point has been passed and an excess of R is present. The assumption can never be quite correct, for although K_t may be very large it must be finite. Section 11.6 will show what effects this has on the curves obtained here.

Two different and unrelated things happen during a titration. One is that the substance being titrated reacts with the reagent. The other is that the volume of the titration mixture increases as the solution of the reagent is added to it. Both must be taken into account in calculating the concentrations of the substances that are present in the mixture.

It is convenient to divide the curve into two portions, one preceding the equivalence point and the other following it. At any point in the first portion the concentration of the reagent R is equal to zero. All of the R that has been added has been consumed in reacting with S. If V_R cm³ of the reagent has been added, $V_R c_R$ millimole of R has been added and has reacted completely. In reacting it has consumed (s/r) times as many millimoles of S, because the equation for the titration reaction is $sS + rR + \ldots = \ldots$. There were $V_S^0 c_S^0$ millimoles of S at the start of the titration, and $V_S^0 c_S^0 - (s/r) V_R c_R$ millimoles are left. They are dissolved in a total volume of $V_S^0 + V_R$ cm³, and the concentration of S is therefore given by

$$[S] = \frac{V_S^0 c_S^0 - (s/r) V_R c_R}{V_S^0 + V_R} \tag{11.6}$$

which may be combined with eq. (11.5) to yield

$$[S] = c_S^0 (1-f) \frac{V_S^0}{V_S^0 + V_R} \qquad (f < 1). \tag{11.7a}$$

The factor $(1-f)$ is equal to the fraction of the substance being titrated that has not yet been consumed by reacting with the reagent, and the factor $V_S^0/(V_S^0+V_R)$ describes the extent to which the solution has been diluted during the titration. Equation (11.7a) is applicable only up to the equivalence point. Beyond that point

$$[S] = 0 \qquad\qquad (f > 1) \qquad\qquad (11.7b)$$

for the reason that was given in the first paragraph of this section.

Nothing has been said so far about the chemical natures of S and R because they do not really matter. In the titration of a strong acid with a base in an aqueous solution eqs. (11.7) would describe the concentration of hydronium ion; in the titration of acetic acid with a base they would describe the concentration of acetic acid; in the titration of iron(II) ion with an oxidizing agent they would describe the concentration of iron(II) ion, and so on. As a concrete example let us suppose that the concentration of hydronium ion is followed during the titration of 40.00 cm^3 of an 0.1000 M aqueous solution of hydrochloric acid with an 0.1000 M aqueous solution of sodium hydroxide. Then S is hydronium ion, R is hydroxide ion, $V_S^0 = 40.00$ cm^3, and $c_S^0 = c_R^0 = 0.1000$ M. The third column of Table 11.1 gives the concentration of hydronium ion at each of a number of points during the titration, and Fig. 11.1(a) shows a plot of the hydronium-ion concentration against f.

The plot consists of two branches. By combining eqs. (11.7a) and (11.7b) it can be shown that these intersect at the equivalence point, where $f = 1$. If both branches

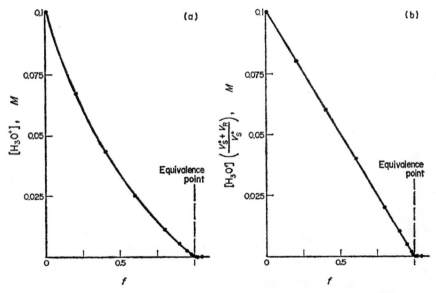

FIG. 11.1. (a) A plot of [S] against f. The curvature of the first branch makes it very difficult to locate the equivalence point. (b) The corresponding plot of $[S](V_S^0+V_R)/V_S^0$ against f. Both branches of this plot are linear, and their point of intersection, which coincides with the equivalence point, is easy to find by extrapolation.

TABLE 11.1. *The coordinates of points on an idealized titration curve*

The text describes the assumptions made in calculating the following values:

f	V_{OH^-} (=V_R) (cm³)	$[H_3O^+]$ (=[S]) (M)	pcH	$[H_3O^+]\times\left(\dfrac{V_S^0+V_R}{V_S^0}\right)$ $\left\{=[S]\times\left(\dfrac{V_S^0+V_R}{V_S^0}\right)\right\}$ (M)	$[OH^-]\times\left(\dfrac{V_S^0+V_R}{V_S^0}\right)$ $\left\{=[R]\times\left(\dfrac{V_S^0+V_R}{V_S^0}\right)\right\}$ (M)	$[P]\times\left(\dfrac{V_S^0+V_R}{V_S^0}\right)$ (M)
0	0	0.1000	1.00	0.1000	0	0
0.2	8.0	0.0667	1.18	0.0800		0.0200
0.4	16.0	0.0429	1.37	0.0600		0.0400
0.6	24.0	0.0250	1.60	0.0400		0.0600
0.8	32.0	0.0111	1.96	0.0200		0.0800
0.9	36.0	5.26×10^{-3}	2.28	1.00×10^{-2}		0.0900
0.95	38.0	2.56×10^{-3}	2.59	5.00×10^{-3}		0.0950
0.98	39 2	1.01×10^{-3}	3.00	2.00×10^{-3}		0.0980
0.99	39.6	5.03×10^{-4}	3.30	1.00×10^{-3}		0.0990
0.995	39.8	2.51×10^{-4}	3.60	5.00×10^{-4}		0.0995
0.999	39.96	5.00×10^{-5}	4.30	1.00×10^{-4}		0.0999
0.9999	39.996	5.00×10^{-6}	5.30	1.00×10^{-5}		0.09999
0.99999	39.9996	5.00×10^{-7}	6.30	1.00×10^{-6}		0.099999
0.999999	39.99996	$5.00\times10^{-8}(?)$	7.30(?)	$1.00\times10^{-7}(?)$		0.0999999
1	40	0 (?)	—	0	(?)	0.100
1.000001	40.00004	0	—		$1.00\times10^{-7}(?)$	0.1000
1.00001	40.0004				1.00×10^{-6}	
1.0001	40.004				1.00×10^{-5}	
1.001	40.04				1.00×10^{-4}	
1.005	40.2				5.00×10^{-4}	
1.01	40.4				1.00×10^{-3}	
1.02	40.8				2.00×10^{-3}	
1.05	42.0				5.00×10^{-3}	

were linear it would be easy to find the equivalence point by extrapolating them to their point of intersection. However, the branch that precedes the equivalence point is not linear; and extrapolation would therefore be difficult or impossible.

To eliminate the curvature and facilitate graphical location of the equivalence point, we need only plot the product $[S](V_S^0+V_R)/V_S^0$, instead of $[S]$ itself, along the ordinate axis. Multiplying $[S]$ by the ratio $(V_S^0+V_R)/V_S^0$ compensates for the dilution that has occurred, and transforms eqs. (11.7) into

$$[S] \frac{V_S^0 + V_R}{V_S^0} = c_S^0(1-f) \qquad (f < 1), \qquad (11.8a)$$

$$[S] \frac{V_S^0 + V_R}{V_S^0} = 0 \qquad (f > 1). \qquad (11.8b)$$

These equations yield the plot shown in Fig. 11.1(b), which is one standard form of the titration curve. It is a segmented curve. Each of its segments is a straight line, and the two straight lines intersect at the equivalence point. In an actual titration one would plot V_R, instead of f, along the abscissa axis, but this does not change the shape of the curve because V_R and f are proportional to each other. Even if you are performing an experimental titration in which you know nothing in advance about the location of the equivalence point, you can always arrange matters so that you know what volume of solution you are titrating and what volume of reagent you have added to it at each point, and therefore you can always calculate the value of $[S](V_S^0 + V_R)/V_S^0$ once you have measured $[S]$. Only a few measurements need be made during the titration to permit plotting such a curve and finding the location of the equivalence point.

Segmented titration curves can also be obtained in other ways. Sometimes it is easier to follow the concentration of the reagent than that of the substance being titrated; sometimes it is easier to follow the concentration of a product than either of these. For the reagent R, the equation

$$[R] = 0 \qquad (f < 1) \qquad (11.9a)$$

follows immediately from the assumption that K_t is infinitely large. Beyond the equivalence point all of the $V_S^0 c_S^0$ millimoles of S that were present at the start of the titration will have been consumed by reacting with R, and $(r/s) V_S^0 c_S^0$ millimoles of R will have been consumed in reacting with it. If $V_R c_R$ millimoles of R have been added, there will be $V_R c_R - (r/s) V_S^0 c_S^0$ millimoles left, and the concentration of R will be given by

$$[R] = \frac{V_R c_R - (r/s) V_S^0 c_S^0}{V_S^0 + V_R} = \frac{(r/s) V_S^0 c_S^0 f - (r/s) V_S^0 c_S^0}{V_S^0 + V_R} = (r/s) c_S^0 (f-1) \frac{V_S^0}{V_S^0 + V_R}$$

$$(f > 1) \qquad (11.9b)$$

in which the second equality is obtained by employing eq. (11.5). Plotting $[R]$, or any measured quantity proportional to it, against f gives a curve in the region where $f > 1$, but a straight line may be obtained by correcting for dilution as was done above for S. The resulting equations

$$[R] \left(\frac{V_S^0 + V_R}{V_S^0} \right) = 0 \qquad (f < 1), \qquad (11.10a)$$

$$[R] \left(\frac{V_S^0 + V_R}{V_S^0} \right) = (r/s) c_S^0 (f-1) \qquad (f > 1) \qquad (11.10b)$$

yield the numerical values shown in the sixth column of Table 11.1 and the titration curve shown in Fig. 11.2(a).

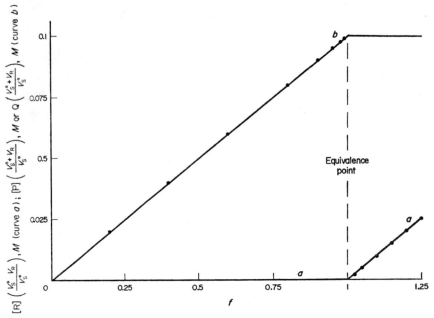

Fig. 11.2. (a) A plot of $[R](V_S^0 + V_R)/V_S^0$ against f; (b) a plot of $[P](V_S^0 + V_R)/V_S^0$ or $[Q](V_S^0 + V_R)/V_S^0$ against f. To make them comparable with Fig. 11.1(b), both curves are plotted for a titration in which $c_S^0 = 0.1000\ M$.

The corresponding equations

$$[P] = (p/s)\,c_S^0 f\,\frac{V_S^0}{V_S^0 + V_R}\ ; \qquad [Q] = (q/s)\,c_S^0 f\,\frac{V_S^0}{V_S^0 + V_R} \qquad (f < 1),\quad (11.11a)$$

$$[P] = (p/s)\,c_S^0\,\frac{V_S^0}{V_S^0 + V_R}\ ; \qquad [Q] = (q/s)\,c_S^0\,\frac{V_S^0}{V_S^0 + V_R} \qquad (f > 1),\quad (11.11b)$$

and

$$[P]\left(\frac{V_S^0 + V_R}{V_S^0}\right) = (p/s)\,c_S^0 f; \qquad [Q]\left(\frac{V_S^0 + V_R}{V_S^0}\right) = (q/s)\,c_S^0 f \qquad (f < 1),\quad (11.12a)$$

$$[P]\left(\frac{V_S^0 + V_R}{V_S^0}\right) = (p/s)\,c_S^0; \qquad [Q]\left(\frac{V_S^0 + V_R}{V_S^0}\right) = (q/s)\,c_S^0 \qquad (f > 1)\quad (11.12b)$$

can be obtained in the same general way. They assume that neither P nor Q is present when the titration begins, and therefore they would not apply to a titration like that of a strong acid with a strong base in an aqueous solution, for the activity or concentration of water, which is the only product of the reaction, would be constant throughout such a titration. If a weak acid HB were titrated with a strong base in an aqueous solution, so that the titration reaction was $HB(aq) + OH^- = B^- + H_2O(l)$, these equations could be used to describe the concentration of B^- but not that of water. Using them to describe the concentration of B^- amounts to assuming that no B^- is present at the start, and in particular that the dissociation of HB is not very extensive. The concentration of water in either of these titrations would be described by eqs. (11.11) and (11.12) if the solutions of the acid and base were anhydrous, and this was assumed to be true in calculating the values in the last column of Table 11.1. A third form of the titration curve, obtained by plotting $[P](V_S^0 + V_R)/V_S^0$ or $[Q](V_S^0 + V_R)/V_S^0$ against f in a titration in which these equations are obeyed, is shown in Fig 11.2(b). Equations (11.11b) and (11.12b) say that the amounts of the products P and Q become constant as soon as the equivalence point is reached, for no more of either can be formed after all the S has been consumed.

Equations (11.7) and (11.8), (11.9) and (11.10), and (11.11) and (11.12) are the heart of titration-curve theory. They can be rewritten and combined in various ways, of which some important ones will appear below. They must be modified to take into account the fact that K_t can never really be infinitely large, the possibility that a titration may involve two or more reactions, and other special circumstances, but these things do not decrease their fundamental importance.

SUMMARY: Two different things happen during a titration: the starting material S reacts with the reagent R to form the products P and Q, and the volume of the titration mixture increases as reagent is added. Taking both of these processes into account, but assuming that K_t is infinitely large (so that $[R] = 0$ before the equivalence point while $[S] = 0$ after it), equations are derived for the concentrations of S, R, P, and Q and used to plot typical segmented titration curves.

11.5. Sigmoidal titration curves

PREVIEW: A sigmoidal titration curve is a plot of the logarithm of a concentration, or a ratio of concentrations, against f. The equations in Section 11.4 enable us to draw only half of a sigmoidal curve, which may be the half that ends at the equivalence point or the half that begins there. This section explains how the other half is obtained, and discusses the errors that arise very close to the equivalence point and that result from approximations made in deriving the equations.

The segmented titration curves shown in Figs. 11.1(b) and 11.2 were obtained by correcting the concentration of S, R, or P for dilution and plotting the corrected value against f. Sometimes it is more natural to plot the logarithm of one of these concentrations, or the logarithm of a ratio of two concentrations, against f. In a titration with or of a strong acid it is natural to be interested in the pcH $(= -\log [H_3O^+] = -\log [R]$ or $-\log [S])$ of the titration mixture. If the titration of chloride ion with a standard solution of silver nitrate is followed by measuring the potential of a silver electrode, the Nernst equation is obeyed and $\log [Ag^+] (= \log [R])$ is interesting. The pcH is also important in the titration of a weak acid, such as acetic acid, with a strong base; the equation $pcH = pK_a + \log[OAc^-]/[HOAc(aq)]$ shows that a plot of pcH against f has the same shape as one of $\log[P]/[S]$ ($= \log[OAc^-]/[HOAc(aq)]$) against f. In the redox titration of iron(II) with an oxidizing agent, the potential varies in a way that is controlled by $\log[Fe(III)]/[Fe(II)]$, which is also equal to $\log[P]/[S]$.

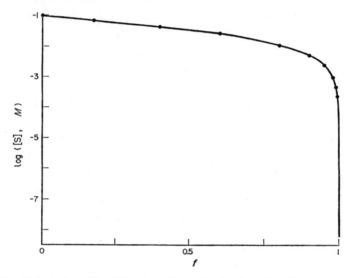

FIG. 11.3. A plot of $\log[S]$ against f for a titration in which $c_S^0 = 0.1000\ M$.

Figure 11.3 shows a plot of the values of $\log[S]$ calculated from eqs. (11.7) with $c_S^0 = c_R = 0.1000\ M$. Values of $-\log[S]$ appeared in the fourth column of Table 11.1. The curve shows how the calculated values of $\log[H_3O^+]$ ($= -pcH$) vary during the titration of $0.1000\ M$ hydrochloric acid with $0.1000\ M$ sodium hydroxide. However, since the nature and properties of the reagent are irrelevant to eqs. (11.7), it also shows how $\log[H_3O^+]$ would vary during the titration of $0.1000\ M$ hydrochloric acid with an $0.1000\ M$ solution of any other base, such as ammonia. Since the nature of the reaction is also irrelevant, it also shows how $\log[Ag^+]$ would vary if an $0.1000\ M$ solution of a completely dissociated silver salt such as silver perchlorate were titrated with $0.1000\ M$ sodium chloride or potassium iodide, or how $\log[Ca^{2+}]$ varies when $0.1000\ M$ calcium chloride is titrated with $0.1000\ M$ ethylenediamine-tetraacetate.

The curve is peculiar because it does not extend as far as the equivalence point. At and beyond the equivalence point the concentration of S is equal to zero according to eqs. (11.7), but of course that is never actually true. This was not important in drawing Fig. 11.1(a). If K_t is very large, the concentration of S will be very small at the equivalence point or beyond it—too small to be distinguished from zero on the scale of Fig. 11.1(a). Finite values of [S] are important in Fig. 11.3 because it is easy to distinguish between the logarithm of a very small but finite quantity and the logarithm of zero.

It is not possible to construct a logarithmic (sigmoidal) titration curve that is a reasonable facsimile of a real one without taking some finite value of K_t into account. This means that the shape of a sigmoidal curve depends on the value of K_t. So does that of a segmented curve, but to a much smaller extent. Once K_t is reasonably large, increasing it further has no visible effect on a segmented curve but does alter a sigmoidal one. Going in the opposite direction, both sigmoidal and segmented curves deteriorate as K_t decreases, but the segmented ones do so more slowly and retain much utility even after the decrease has gone so far that sigmoidal ones are useless.

The simplest way to take the value of K_t into account is to combine the expression for K_t with eq. (11.9b), and with eq. (11.11b) if that is needed as well. By doing this it is possible to calculate the value of [S] at any point beyond the equivalence point. If it were log [R] that was being followed, one would similarly combine the expression for K_t with eq. (11.7a), and if necessary with eq. (11.11a) as well, so that values of [R] could be calculated in the region where f is smaller than 1 and eq. (11.9a) is useless.

To do this for the titration of 0.1000 M aqueous hydrochloric acid with 0.1000 M aqueous sodium hydroxide, for which Fig. 11.3 represents the first part of the titration curve, one would proceed in the following way. The expression for K_t is (eq. (11.2b)) $K_t = 1/[H_3O^+][OH^-] = 1/[S][R]$. Solving this for [S] gives [S] $= 1/K_t[R]$, which in turn can be combined with eq. (11.9b) to obtain

$$[S] = \frac{1}{K_t(r/s)\, c_S^0 (f-1)\dfrac{V_S^0}{V_S^0+V_R}} \qquad (f>1) \qquad (11.13)$$

Equation (11.13) leads to curve b in Fig. 11.4. Curve a in this figure is copied without change from Fig. 11.3, and represents the variation of log [S] up to the equivalence point according to eq. (11.7a).

Figure 11.4 is indistinguishable from the truth except in the immediate vicinity of the equivalence point, where its two branches do not meet. An enormously magnified view of this region is shown in Fig. 11.5. It extends only from $f = 0.9999$ to $f = 1.0001$. For the titration represented by Table 11.1, in which the volume of reagent is 40.00 cm^3 at the equivalence point, the region covered by Fig. 11.5 extends from 39.996 ($= 0.9999 \times 40.00$) to 40.004 ($= 1.0001 \times 40.00$) cm^3 of reagent. In another titration in which the value of K_t was smaller than it is for this one, less magnification would be needed.

Equations more complex than eqs. (11.7a) and (11.13) are needed to obtain accurate values of [S] or [R] in the immediate vicinity of the equivalence point, but values at the equivalence point itself are very easy to obtain. In this titration the solution contains unreacted hydronium ion until the equivalence point has been reached. As long as f is smaller than 1 the concentration of hydronium ion exceeds that of hydroxide ion. After the equivalence point has been passed, there is an excess

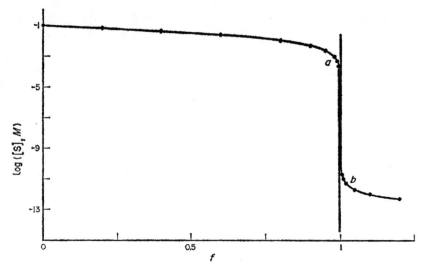

FIG. 11.4. (a) A plot of log [S] against f for a titration in which $c_S^0 = c_R = 0.1000$ M, calculated from eq. (11.7a) and copied from Fig. 11.3. Curve a extends only up to the equivalence point. (b) A plot of log [S] against f beyond the equivalence point of the same titration, assuming $K_t = 1 \times 10^{14}$ mol^{-2} dm^6. The separation between curves a and b is slightly exaggerated for clarity.

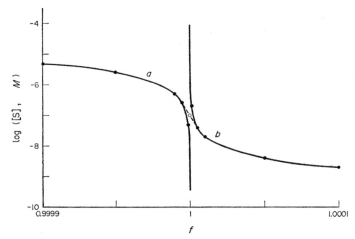

FIG. 11.5. A greatly magnified view of Fig. 11.4 in the immediate vicinity of the equivalence point.

of hydroxide ion, and the concentration of hydronium ion is smaller than that of hydroxide ion. At the equivalence point there is no excess of either, and $[H_3O^+] = [OH^-] = K_w^{1/2} (= 1/K_t^{1/2}) = 1.0 \times 10^{-7}$ mol dm^{-3}, so that log $[H_3O^+] = -7.0$. Exactly the same thing can be said in a different way. The most important single fact about the titration is that 1 millimole of hydroxide reacts with 1 millimole of hydronium ion. Because this is true, $V_S^0 c_S^0 (= V_R^* c_R)$ millimoles of hydroxide ion are needed to reach the equivalence point in the titration of $V_S^0 c_S^0$ millimoles of hydronium ion. For the same reason, the number of millimoles of hydronium ion that have actually reacted at any point during the titration is equal to the number of millimoles of hydroxide ion with which they have reacted. Call each of these numbers of millimoles x. Then the solution at the equivalence point contains $V_S^0 c_S^0 - x$ millimoles of hydronium ion and $V_S^0 c_S^0 - x$ millimoles of hydroxide ion, and the concentrations of the two ions must be the same. The volume and concentration of the solution being titrated do not matter, and neither does the extent of the reaction. The concentrations of hydronium and hydroxide ions must be the same at the equivalence point.

This is because the stoichiometric coefficients for these two ions are equal in the equation for the reaction, $H_3O^+ + OH^- = 2H_2O(l)$. More generally, if the equation is $sS + rR + \ldots = pP + qQ + \ldots$, and if there are $V_S^0 c_S^0$ millimoles of S at the start of the titration, then $(r/s) V_S^0 c_S^0$ millimoles of R will be needed to reach the equivalence point. If x millimoles of S have reacted, $(r/s)x$ millimoles of R have reacted with them. At the equivalence point the solution will contain $V_S^0 c_S^0 - x$ millimoles of S and $(r/s) V_S^0 c_S^0 - (r/s)x$ millimoles of R, and

$$[R] = (r/s)[S] \qquad (f = 1). \tag{11.14}$$

You can always obtain a description of the equivalence point by combining this with the expression for K_t. If that also involves the concentration of P or Q, you can describe these by eqs. (11.11).

Figure 11.5 can now be completed by sketching the dashed portion, which joins curves a and b smoothly and passes through the open circle that represents the equivalence point. The final titration curve, Fig. 11.6(a), is obtained by patching this dashed portion into Fig. 11.4, and Fig. 11.6(b) shows what it would look like if the pcH were chosen as the dependent variable instead of $\log [H_3O^+]$.

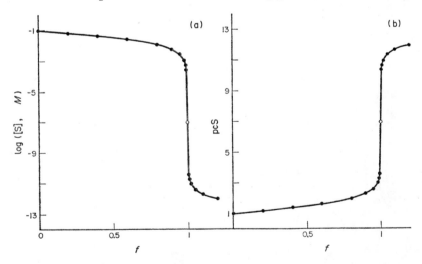

FIG. 11.6. (a) Final plot of $\log [S]$ against f for the titration of S with R with $c_S^0 = c_R = 0.1000\ M$ and $K_t = 1 \times 10^{14}\ mol^{-2}\ dm^6$. (b) The corresponding plot of pcS against f. For the titration of hydrochloric acid with sodium hydroxide described in the text, $\log [S] = \log [H_3O^+]$ and pcS = pcH.

22

TABLE 11.2. *Titration-curve equations*

The first column gives the concentration, or ratio of concentrations, that is followed during a titration. The second, fourth, and sixth columns give expressions for that concentration, or ratio of concentrations, for $f < 1$, $f = 1$, and $f > 1$, respectively. A dash in any of these columns means that the value must be calculated by combining the expression for K_t with expressions for the other concentrations that appear in it. The third, fifth, and seventh columns give the numbers of the corresponding equations in the text.

Quantity followed ($= X$)	$f < 1$ $X =$	eq.	$f = 1$ $X =$	eq.	$f > 1$ $X =$	eq.
[S]	$c_S^0(1-f)\dfrac{V_S^0}{V_S^0+V_R}$	(11.7a)	$[R] = (r/s)[S]$	(11.14)	—	---
[R]	—	---		(11.14)	$(r/s)c_S^0(f-1)\dfrac{V_S^0}{V_S^0+V_R}$	(11.9b)
[P]	$(p/s)c_S^0 f\dfrac{V_S^0}{V_S^0+V_R}$	(11.11a)	$(p/s)c_S^0\dfrac{V_S^0}{V_S^0+V_R}$	(11.11a, b)	$(p/s)c_S^0\dfrac{V_S^0}{V_S^0+V_R}$	(11.11b)
[Q]	$(q/s)c_S^0 f\dfrac{V_S^0}{V_S^0+V_R}\ (=(q/p)[P])$	(11.11a)	$(q/s)c_S^0\dfrac{V_S^0}{V_S^0+V_R}\ (=(q/p)[P])$	(11.11a, b)	$(q/s)c_S^0\dfrac{V_S^0}{V_S^0+V_R}\ (=(q/p)[P])$	(11.11b)
[P]/[S]	$(p/s)\dfrac{f}{1-f}$	(11.11a, 7a)	—	---	—	---
[R]/[Q]	—	---	—	---	$(r/q)(f-1)$	(11.9b,11.11b)

If you were actually constructing a titration curve you would abbreviate this process a great deal. The discussion has been aimed at giving you ideas about what you should and should not do and why. You should begin by inspecting the equation for the reaction that takes place, and identifying the quantity being followed. The possibilities that are most important, because they are the most common, are log [S], log [R], log [P]/[S], and log [R]/[Q]. You should calculate the value of that quantity at the equivalence point. Then, using the equations that are collected in Table 11.2, you should calculate its values at a number of points preceding the equivalence point. You should ignore any value that is less than 0.5 log unit away from that at the equivalence point, so that you will avoid the messy situation represented by Fig. 11.5. Next you should go to the other side of the equivalence point, beginning at $f = 1.5$ or so and working backward toward the equivalence point. Again stop before the value approaches the one at the equivalence point more closely than 0.5 log unit. Now you need only to plot all the values and draw a smooth curve through them.

Sigmoidal titration curves have two important characteristic shapes. One was illustrated by Fig. 11.6. The inversion caused by plotting log $[H_3O^+]$ against f in Fig. 11.6(a) but plotting pcH $(= -\log[H_3O^+])$ against f in Fig. 11.6(b) is trivial. In either case the absolute value of the slope is small at $f = 0$, rises as the titration proceeds, attains a maximum value at or very near the equivalence point, and then decreases again toward zero as excess reagent is added.

The other characteristic shape is shown in Fig. 11.7. This curve was obtained by plotting log [P]/[S] against f. In the titration of acetic acid with a base in water as the solvent, P is acetate ion, S is acetic acid, and

$$\log \frac{[P]}{[S]} = \log \frac{[OAc^-]}{[HOAc(aq)]} = \log \frac{K_a}{[H_3O^+]} = pcH - pK_a. \tag{11.15}$$

In the titration of iron(II) with permanganate ion in, say, 1 M sulfuric acid as the supporting electrolyte, P is iron(III), S is iron(II), and

$$\log \frac{[P]}{[S]} = \log \frac{[Fe(III)]}{[Fe(II)]} = \frac{E - E^{0'}(Fe(III), Fe(II); 1\ M\ H_2SO_4)}{0.059\ 16}. \tag{11.16}$$

In the first of these titrations the variation of log [P]/[S] would parallel that of the pcH; in the second it would parallel that of the potential. In Fig. 11.7 the absolute value of the slope is large at $f = 0$, and it decreases as reagent is added, attaining a minimum value at or near $f = \frac{1}{2}$. Thereafter it increases again, passes through a maximum at or near the equivalence point, and finally decreases again toward zero.

Equation (11.11a) assumes that there is no P whatever in the solution at the start of the titration. Because this is never really possible, the slope at any point near the start of a curve like this one is always finite. For some titrations it is possible to calculate the value of the measured quantity at $f = 0$ and join it to the main portion of the curve in the same way as in Fig. 11.5. In the titration of acetic acid with a

22*

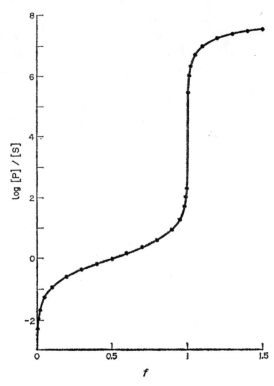

FIG. 11.7. A plot of log [P]/[S] for a titration that involves a reaction of the form S+R = P (as does the titration of acetic acid with a base in an aqueous solution) with $K_t = 10^9$ mol^{-1} dm^3. It is assumed that $c_S^0 = c_R = 0.1000$ M. Values of [P]/[S] up to the equivalence point were obtained from eqs. (11.11a) and (11.7a); beyond the equivalence point they were obtained by calculating [R] from eq. (11.9b) and combining its value with the expression for $K_t([P]/[S]) = K_t[R]$).

base in water as the solvent, there is certainly no difficulty in calculating the pcH of the original solution of acetic acid. In other titrations the calculation is impossible, as it is in the titration of iron(II) with permanganate. There must be some iron(III) in the original solution, but there is no way to guess how much. If there is very little of it, eq. (11.11a) will become very nearly correct as soon as a few drops of reagent have been added, for these will react to form so much more iron(III) than was present initially that the latter can be ignored. For such a titration it is best to begin the calculations at some small finite value of f, such as 0.01.

The point at $f = \frac{1}{2}$, halfway between the start of the titration and the equivalence point, has a special significance on a plot of log [P]/[S], or any equivalent quantity, against f. At this point [P]/[S] = 1. In the titration of acetic acid with a base, eq. (11.15) shows that the pcH is equal to pK_a for acetic acid when [P]/[S] = 1; in the titration of iron(II) with permanganate, eq. (11.16) shows that the potential is equal to the formal potential of the iron(III)–iron(II) couple when [P]/[S] = 1. Similar things are true at $f = 2$, where a 100% excess of the reagent has been added, if the quantity being measured is governed by [R]/[Q] after the equivalence point is passed. In the titration of a strong acid with ammonia, the reaction is $H_3O^+ + NH_3(aq) = NH_4^+ + H_2O(l)$ and the pcH is given by

$$pcH = pK_{a,\,NH_4^+} + \log \frac{[NH_3(aq)]}{[NH_4^+]} = pK_{a,\,NH_4^+} + \log \frac{[R]}{[Q]}. \qquad (11.17)$$

When $f = 2$, $[R]/[Q] = 1$ and the pcH is equal to pK_a for ammonium ion. In the titration of iron(III) with chromium(II), where R is chromium(II) and Q is chromium(III), the potential can be described by the equation

$$E = E^{o\prime}(\text{Cr(III), Cr(II); supporting electrolyte}) - 0.059\,16 \log \frac{[R]}{[Q]}. \qquad (11.18)$$

When $f = 2$ the potential is equal to the formal potential of the chromium(III)–chromium(II) couple. Several different kinds of equilibrium data can be evaluated by very simple experimental procedures based on these consideration.

Most chemists associate curves like the one in Fig. 11.6 with titrations of strong acids and bases and with titrations that involve precipitation or complex formation, and associate curves like the one in Fig. 11.7 with titrations of weak acids and bases and with redox titrations. By very close analysis it is possible to show that none of these associations is invariably correct, but you should make them anyway. They are almost always right, and you can worry about the exceptions to them if you ever see one.

SUMMARY: In constructing a sigmoidal curve, one equation is used to calculate the concentration, ratio of concentrations, or other quantity (such as the pcH or the potential of an electrode) from the start of the titration to the equivalence point, and another equation is used beyond the equivalence point. When it is inconvenient or difficult to calculate the same concentration or ratio of concentrations in the two regions, recourse may be had to an expression for K_t or to the Nernst equations for two different couples. Sigmoidal curves have two different characteristic shapes, depending on whether it is a concentration or a ratio of concentrations that is followed.

11.6. The effects of K_t and concentration

PREVIEW: Changing the concentration of the substance titrated or the value of K_t alters the titration curve and affects the precision that can be attained in measuring the volume of reagent needed in an actual titration.

To show how the shapes of titration curves depend on the value of K_t, it will be assumed that four different portions of an 0.1000 M solution of silver perchlorate are titrated with 0.1000 M solutions of four different reagents: potassium iodide, potassium thiocyanate, potassium iodate, and potassium bromate. The reactions that will occur are

$$Ag^+ + I^- = AgI(s); \quad K_t = 1/K_{AgI} = 1.2 \times 10^{16} \text{ mol}^{-2} \text{ dm}^6$$
$$Ag^+ + SCN^- = AgSCN(s); \quad K_t = 1/K_{AgSCN} = 1.0 \times 10^{12} \text{ mol}^{-2} \text{ dm}^6$$
$$Ag^+ + IO_3^- = AgIO_3(s); \quad K_t = 1/K_{AgIO_3} = 3.3 \times 10^7 \text{ mol}^{-2} \text{ dm}^6$$
$$Ag^+ + BrO_3^- = AgBrO_3(s); \quad K_t = 1/K_{AgBrO_3} = 1.9 \times 10^4 \text{ mol}^{-2} \text{ dm}^6$$

All the values of K_t have the same units, and successive ones differ by factors of approximately 10^4. For the sake of simplicity it will be assumed that silver perchlorate is completely dissociated, and processes such as adsorption and occlusion will be ignored.

Curve a in Fig. 11.8 is the sigmoidal titration curve obtained by plotting the pcAg against f for the titration of silver ion with iodide ion. It has the same general shape as the curve in Fig. 11.6(b). The similarity results from the facts that each reaction is of the form $S + R = P$, that the activity of P is constant throughout each titration (so that each K_t is given by an expression of the form $K_t = 1/[S][R]$), and that $-\log [S]$ is plotted along each ordinate axis.

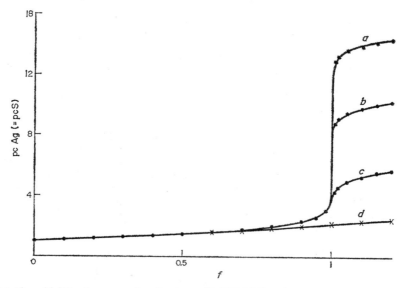

Fig. 11.8. Sigmoidal titration curves for titrations of 0.1000 M silver ion with 0.1000 M (a) iodide ion, (b) thiocyanate ion, (c) iodate ion, and (d) bromate ion. Values of K_t are given in the text. Curves (a)–(c) were constructed by using eq. (11.7a) before the equivalence point and eqs. (11.9b) and (11.2a) after the equivalence point. Those equations are not very useful when K_t is small, and curve (d) was therefore constructed by using eq. (11.47a).

Curves b, c, and d in Fig. 11.8 represent the titrations with thiocyanate, iodate, and bromate, respectively; all are plotted on the same scale as curve a. In every one of these titrations the pcAg is initially equal to 1.00, and does not increase much above that value until a substantial fraction of the silver ion has been precipitated.

Consequently the initial portions of the curves are indistinguishable. However, beyond the equivalence points the values of pcAg are strikingly different. At any particular point beyond the equivalence point, the concentration of R is nearly the same no matter whether R is iodide ion or bromate ion. Because K_t is larger for the titration with iodide ion than for any of the others, the concentration of silver ion is much smaller, and the pcAg is much larger, in a solution containing any given excess of iodide ion than in a solution containing the same excess of any of the other reagents. At the opposite extreme, K_t is so low for the titration with bromate ion that the concentration of silver ion does not become very small, and the pcAg does not become very large, even if a great deal of excess bromate is added.

As the overall change of the pcAg during the titration decreases, so does the slope of the titration curve at the equivalence point or any point near it. The curve for the titration with iodide ion becomes very steep near the equivalence point, steeper even than does the curve in Fig. 11.6(b) because K_t has a larger value in the titration of silver ion with iodide ion than in the titration of hydronium ion with hydroxide ion. For reasons that will be explained in Section 11.7, the accuracy and precision that can be attained in a titration depend on the steepness of the titration curve in the vicinity of the equivalence point. They improve as the curve becomes steeper. Curve a in Fig. 11.8 represents a very favorable situation. On the other hand, the entire change of the pcAg during a titration with bromate (curve d in Fig. 11.8) is so small that the slope cannot become large at the equivalence point. Adding a drop or two of the bromate solution will have only a small effect on the pcAg, on the color of an indicator, on the potential of a silver electrode immersed in the titration mixture, or on any other property that could be observed or measured. The precision will be poor, the accuracy will suffer, and the results will be far from satisfactory.

Changing the concentration of the substance being titrated has the effects shown by Fig.11.9. These curves represent titrations of solutions containing different concentrations of silver ion (ranging from 1.00×10^{-2} M for curve a to 1.00×10^{-5} M for curve d) with thiocyanate ion, so that $K_t = 1.0 \times 10^{12}$ mol^{-2} dm^6 for all of them. Curve b in Fig. 11.8 extends this series to 1.00×10^{-1} M silver ion. Decreasing the initial concentration of silver ion has two effects. One is that it decreases the concentration of silver ion at any point preceding the equivalence point, and thereby causes this portion of the curve to shift toward larger values of pcAg. The other is that it decreases the concentration of excess thiocyanate ion that is present when any particular value of f exceeding 1 has been reached; the corresponding concentration of silver ion increases as a result, and this portion of the curve is shifted toward smaller values of pcAg. The pcAg at the equivalence point is given by $pcAg_{f=1} = (\log K_t)/2$ and is unaffected by changing the initial concentration.

These effects are not the same as those of changing K_t. Decreasing the value of K_t has little effect on the portion of the curve that precedes the equivalence point, but it does decrease the value of pcAg at the equivalence point. It also causes the portion of the curve that follows the equivalence point to shift toward smaller values of pcAg,

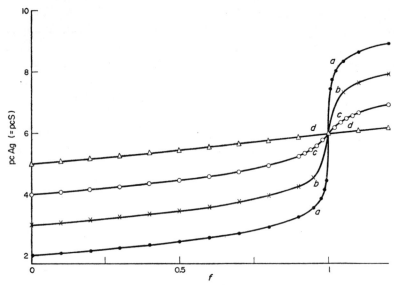

FIG. 11.9. Sigmoidal titration curves for titrations of $c_{Ag^+}^0$ M silver ion with c_{SCN^-} M thiocyanate ion with $c_{Ag^+}^0 = c_{SCN^-} =$ (a) 1.00×10^{-2} M, (b) 1.00×10^{-3} M, (c) 1.00×10^{-4} M, and (d) 1.00×10^{-5} M. Curves (a) and (b) were constructed by using eq. (11.7a) before the equivalence point. Curves (c) and (d) were constructed by using eq. (11.7a) up to $f = 0.5$ and eq. (11.47a) thereafter.

but the shift that results from decreasing K_t by any given factor is smaller than the one that results from decreasing the initial concentration by the same factor. You can confirm all of these statements by inspecting the curves in Figs. 11.8 and 11.9.

Despite the differences of detail between these two effects, their ultimate results are very similar. If either K_t or the initial concentration is decreased, the overall change of pcAg decreases, the curve is compressed along the ordinate axis, the slope at the equivalence point becomes smaller, and the attainable precision becomes poorer.

Figures 11.10 and 11.11 are the segmented titration curves that would be obtained by following the concentration of silver ion in the titrations of Figs. 11.8 and 11.9. Several conclusions may be drawn by comparing these figures. Changing the value of K_t affects segmented curves as well as sigmoidal ones, but affects the segmented ones to a smaller extent. Curves a and c in Fig. 11.10 are almost indistinguishable on this scale even though the values of K_t differ by a factor close to 10^8; the corresponding curves in Fig. 11.8 are conspicuously different. Segmented curves are also much less sensitive to the concentration of the substance being titrated. Both curve d in Fig. 11.10 and curve d in Fig. 11.11 have distinct linear portions that could easily be extrapolated to obtain close estimates of the locations of their equivalence points. Attempts to locate the equivalence points of the corresponding curves in Figs. 11.8 and 11.9 are difficult and unsatisfactory no matter whether they employ indicators or the complete curves plotted from actual experimental data.

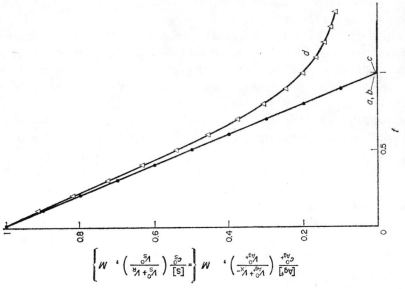

Fig. 11.11. Segmented titration curves for the titrations of Fig. 11.9. On this scale curves (a) and (b) are indistinguishable, and curve (c) deviates from them only in the immediate vicinity of the equivalence point.

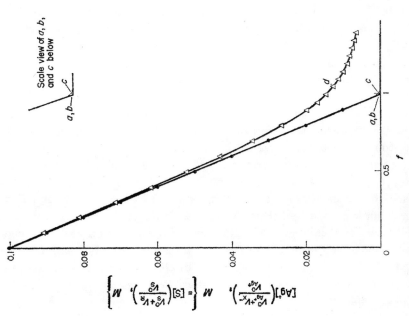

Fig. 11.10. Segmented titration curves for the titrations of Fig. 11.8. On this scale curves (a) and (b) are indistinguishable, and curve (c) deviates from them only in the immediate vicinity of the equivalence point.

If the slope of a sigmoidal curve is small at the equivalence point, the curvature of a segmented one is appreciable. Both of these things reflect the fact that the titration reaction is not quite complete at the equivalence point. You can judge whether a titration curve will give a reasonably well-defined end point from the value of the ratio $c_S^0/[S]^*$, in which the numerator is the initial concentration of S and the denominator is the concentration of S at the equivalence point. As the value of this ratio increases, the reaction becomes more nearly complete at the equivalence point: a sigmoidal titration curve becomes steeper and the curvature of a segmented one decreases. For curve d in each of the four preceding figures, $c_S^0/[S]^*$ is equal to 10. Figures 11.10 and 11.11 show that this value just suffices in a segmented-curve technique, but Figs. 11.8 and 11.9 show that a higher one is needed if the titration is based on a sigmoidal curve. For curve c in Fig. 11.9, $c_S^0/[S]^*$ is equal to 10^2. This curve is also very drawn out, and a reasonably distinct inflection does not appear until curve b is reached. Here $c_S^0/[S]^* = 10^3$, which is about the lower limit of feasibility in potentiometric titrations. An even higher value, perhaps 10^4, is needed to render a relative precision of $\pm 0.1\%$ attainable with an indicator.

In applying this criterion to an actual titration you would proceed in the following way. If, like the titrations represented by these figures, it is based on a reaction of the form $S + R = P$ ($a_P =$ constant), then $[S]^* = [R]^* = 1/K_t^{1/2}$. Hence

$$c_S^0/[S]^* = K_t^{1/2}c_S^0. \tag{11.19}$$

If you propose to use an indicator you have only to see whether $K_t^{1/2}c_S^0$ is larger than about 10^4 or not. If it is, you should be able to obtain satisfactory results provided that you choose the right indicator. Sections 11.7 and 11.8 will explain how to choose it. If $K_t^{1/2}c_S^0$ is much smaller than 10^4, titration with an indicator will be unsatisfactory, agonizing, or impossible, depending on how much smaller it is. If it is between 10^4 and 10^3 you can make the titration potentiometrically (Chapter 14), and locate its end point from a sigmoidal plot of potential against volume of reagent, provided that you can find a satisfactory indicator electrode; if it is between 10^3 and 10 you will have to construct some sort of segmented curve. Thermometric, spectrophotometric, conductometric, amperometric, and various other kinds of titrations yield segmented curves very easily, and Section 14.9 will show how you can secure a segmented curve from the data obtained in a potentiometric titration. If $K_t^{1/2}c_S^0$ is smaller than 10 you should look around for some way of increasing it, either by changing the conditions so as to increase K_t or by separating the S from the solution and getting it into a smaller volume of a more concentrated solution. If these things are difficult or impossible you should look around for another way of determining S.

The algebraic form of eq. (11.19) is governed by that of the expression for K_t. If the titration reaction is $2S + R = P$ ($a_P =$ constant), you would write $[S]^* = 2[R]^* = 1/(4K_t^{1/3})$ and obtain

$$c_S^0/[S]^* = 4K_t^{1/3}c_S^0 \tag{11.20}$$

instead of eq. (11.19).

SUMMARY: Decreasing either the concentration of the substance titrated or the value of K_t decreases the slope of a sigmoidal titration curve in the neighborhood of the equivalence point. The same changes lead to curvature around the equivalence point on a segmented curve. In either case the accuracy and precision become poorer, but segmented curves are still useful under conditions so unfavorable that sigmoidal ones are not.

11.7. Titration errors

PREVIEW: This section introduces the ideas of systematic and random errors, defines the titration error, and describes the consequences of the fact that the color change of an indicator is spread over a color-change interval instead of occurring sharply at a single point.

Both the accuracy and the precision of an analytical method are important. Accuracy and precision were defined in Section 9.2. The chemist who designs or modifies a method seeks to optimize both, and can hardly do so without having ways of estimating them. The chemist who uses a method attempts to obtain the best accuracy and precision that it can yield, and must be able to judge what these are in order to gauge whether the attempt has been successful.

Every measurement involves two kinds of errors: systematic errors and random errors. A *systematic error* is one that recurs in one measurement after another, and has the same sign and the same magnitude in each of a series of successive identical measurements. A *random error* may be almost absent in one measurement but quite large in the next, and both its sign and its magnitude are different in different measurements. Systematic errors affect the accuracy; random errors affect the precision. Systematic errors arise from doing the same wrong thing over and over again; random errors arise from doing something differently in different measurements.

Systematic and random errors will be considered in some detail in Chapter 12; this section deals only with the information that titration curves give about them.

Let us first imagine that the titration of an 0.1000 M solution of silver perchlorate with 0.1000 M potassium iodide is performed with a hypothetical indicator that changes color when the pcAg becomes equal to 5.000. The titration curve, curve a of Fig. 11.8, shows that this point precedes the equivalence point. There will be a systematic error: one titration after another will be stopped before enough potassium iodide has been added to reach the equivalence point. The error will be small, since the titration curve shows that f is almost equal to 1 at this end point. When the measured volume of reagent is too small, as it is here, we shall say that the error is negative. If the same indicator were used in titrating another portion of the silver perchlorate solution with 0.1000 M potassium iodate, the systematic error would be different. The titration curve, curve c of Fig. 11.8, now shows that the end point (where pcAg = 5.000) follows the equivalence point (where pcAg = 3.8), and also that the distance between these two points along the f-axis is quite large. The volume of potassium iodate used in the titration will be much larger than it should be, and the systematic error will be large and positive.

A real indicator does not behave in the way we have imagined. Instead of changing color sharply at a single point, it undergoes a gradual change of color over a range of concentrations of the substance to which it responds. The reasons why this is so will be examined in Section 11.8. The range over which the color changes is called the *color-change interval* or *transition interval* of the indicator. Because the color change is gradual rather than sudden, different titrations will be stopped at different points within the transition interval, and this gives rise to random errors superimposed on the systematic one.

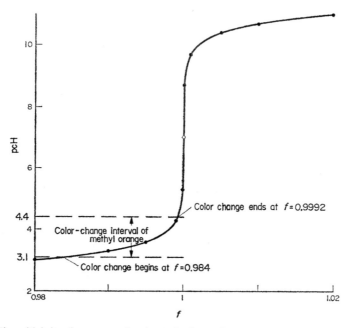

FIG. 11.12. Sigmoidal titration curve, showing only the region very near the equivalence point, for the titration of an 0.1000 M aqueous solution of a strong acid with an 0.1000 M aqueous solution of a strong base. The horizontal dashed lines represent the beginning and the end of the color-change interval of methyl orange, which is red at pcH $\leqslant$ 3.1 and orange at pcH $\geqslant$ 4.4. The points at which the dashed lines intersect the curve give the values of f at the extremes of the color-change interval.

Figure 11.12 provides a convenient example. It represents the titration of an 0.1000 M aqueous solution of a strong acid with an 0.1000 M aqueous solution of a strong base. It was obtained by magnifying the portion of Fig. 11.6(b) that surrounds the equivalence point, and adding some information about the acid-base indicator methyl orange. A strongly acidic solution of methyl orange is red. On adding a base it remains red until the pH has increased to about 3.1. At that point a tinge of orange can just be detected by the average person under average lighting conditions. As more base is added and the pH increases further, the depth of the orange color increases while that of the red one fades. The color passes through orange-red and red-orange.

At a pH of about 4.4 it is orange, and the average person can see no further change if the pH is increased above that value. The transition interval of methyl orange is said to extend from pH = 3.1 to pH = 4.4, and these values are represented by the horizontal dashed lines in Fig. 11.12.

If a number of replicate titrations are performed, many of them will be stopped when the orange and red colors are equally deep, as they are when the pH is approximately equal to 3.8, halfway between the values at the beginning and end of the transition interval. A few will be stopped when the first change of color is discerned and the pH is equal to 3.1; a few others will be continued until the end of the transition interval is reached and the pH is equal to 4.4. All these points precede the equivalence point, and therefore there will be a negative systematic error. No matter how carefully and reproducibly the titration is made, the volume of base that is used in reaching the end point will always be smaller than it should be. There will also be random errors because the end points of different titrations will occur at different pH-values, and therefore at different values of f.

The *titration error* is defined by the equation

$$\text{titration error} = f_{\text{end point}} - 1. \tag{11.21}$$

The most probable value of the pH at the end point of a titration is equal to the pH at the midpoint of the color-change interval. The titration error at this point is approximately equal to the systematic error of the titration. The values of the pH at the end points of different titrations may range from one end of the color-change interval to the other. The difference between the titration errors at the two ends of this interval yields an estimate of the random error.

The best indicator for a titration is one for which the pH at the midpoint of the transition interval is equal to the pH at the equivalence point. It is the midpoint of the transition interval that is most likely to be taken as the end point. The titration error at that point is equal to zero, and so is the systematic error. If a number of titrations are performed, some will be stopped before that point is reached, and the errors in those titrations will be negative, while others will be continued beyond it and will involve positive errors. On the average the negative and positive errors will counterbalance each other.

A slightly less favorable situation is the one in which the transition interval includes the equivalence point but in which the pH at the midpoint of the transition interval differs from the pH at the equivalence point. If phenol red, whose transition interval extends from pH = 6.8 to pH = 8.4 and has its midpoint at pH = 7.6, is used as an indicator for the titration of Fig. 11.12, a few titrations will be stopped at or even before the equivalence point, but the great majority will be carried beyond it. Negative titration errors will be greatly outnumbered by positive ones, and the average of a number of titrations will be too high: there will be a positive systematic error, though it will be very small because the slope of the titration curve (Fig. 11.12) is very large in this region.

Least favorable of all is the situation that arises with methyl orange, where the entire transition interval lies on one side of the equivalence point. There is a systematic error because the titration error has the same sign in every titration, and it is larger than in the situation described in the preceding paragraph because the midpoint of the transition interval is farther away from the equivalence point.

The random error in all three of these situations, and the systematic error in the second and third of them, is governed by the steepness of the titration curve. The remainder of this section will show how the magnitudes of these errors are estimated in several typical situations.

Let us begin by estimating the random and systematic errors that would result from using methyl orange as an indicator for the titration of Fig. 11.12. The pH is equal to 3.8 at the midpoint of the transition interval, and the systematic error is approximately equal to the titration error at this point. You must know the sign of the titration error before you can calculate its value from the equations given in the preceding sections, for none of those equations is applicable on both sides of the equivalence point. Since the pH is equal to 7.0 at the equivalence point, f must be smaller than 1 at the point where the pH is equal to 3.8, and therefore we choose eq. (11.7a):

$$[S] (= [H_3O^+]) = c_S^0 (1-f) \frac{V_S^0}{V_S^0 + V_R} \qquad (f < 1). \qquad (11.7a)$$

From Fig. 11.12 we can see that the value of f is not much less than 1 at pH $= 3.8$. If f were equal to 1, V_S^0 (the volume of acid being titrated) and V_R (the volume of base added to it) would be equal because the concentrations of the acid and base are equal in this titration. Then $V_S^0/(V_S^0 + V_R)$ would be equal to 1/2. It cannot be very far from the truth to say that $V_S^0/(V_S^0 + V_R)$ is equal to 1/2 at the end point.[†] Solving eq. (11.7a) for the titration error defined by eq. (11.21), and introducing the numerical values $[S] = 10^{-3.8} = 1.6 \times 10^{-4}\ M$, $c_S^0 = 0.1000\ M$, and $V_S^0/(V_S^0 + V_R) = 1/2$, gives

$$\text{titration error} = f_{\text{end point}} - 1 = -\frac{[S]}{c_S^0 \left(\dfrac{V_S^0}{V_S^0 + V_R} \right)} \qquad (f < 1)$$

$$= -\frac{1.6 \times 10^{-4}}{0.1 \times 1/2} = -3 \times 10^{-3}$$

$$= -0.3\%. \qquad (11.22)$$

On the average the result of such a titration would be about 0.3% too low if there were no other systematic error. The indicator is not very satisfactory, because this

---

[†] It would not be equal to 1/2 if the concentrations of the acid and base were different, nor would it be equal to 1/2 in a titration based on the reaction $2\,Ag^+ + CrO_4^{2-} = Ag_2CrO_4(s)$ if the concentrations of the silver- and chromate-ion solutions were the same. However, you can always assume that $V_S^0/(V_S^0 + V_R)$ has the same value that it does at the equivalence point. This does not give an exact result, but you would have no use for an exact result if you could get one.

would almost certainly be the largest single systematic error in a titration performed with good apparatus and reasonable care. Others, such as an error in the concentration of the reagent solution or an error in the calibration of the pipet used in measuring out the acid, would certainly exist. The aim is to choose an indicator for which the systematic error will be so small that the overall accuracy of the titration will not be significantly worse than the accuracy attained in the other manipulations. Chapter 12 will show what this involves: meanwhile a rough guide is that the systematic error introduced by the indicator should not exceed about $\pm 0.05\%$. Methyl orange does not meet this requirement.

The random error introduced by using methyl orange can be estimated by using eq. (11.22) to calculate the titration errors at the two ends of the transition interval. These turn out to be -1.6% at pH $= 3.1$ and -0.08% at pH $= 4.4$. Thus it is possible for the results of two successive titrations to differ by as much as 1.5%, and the relative precision might be as bad as $1.5/2 = \pm 0.8\%$. This is pessimistic because it is very unlikely that half of the titrations would be stopped at one end of the transition while the other half are stopped at the other end. Still, it is no more satisfactory than the accuracy. Other random errors will affect the overall result: one is incurred in reading a buret, and a pipet does not deliver exactly the same volume time after time. As was true of the systematic errors, the aim is to choose an indicator for which the random error is so small that the overall precision of the titration will not be significantly worse than the precision attained in the other manipulations. A rough guide is that the random error introduced by the indicator should not exceed about $\pm 0.05\%$. Methyl orange does not meet this requirement either.

The transition interval of phenol red extends from pH $= 6.8$ to pH $= 8.4$. At its midpoint f exceeds 1, and the titration error may be estimated by using eq. (11.13):

$$\text{titration error} = f_{\text{end point}} - 1 = \cfrac{1}{K_t(r/s)\, c_S^0[S]\left(\cfrac{V_S^0}{V_S^0 + V_R}\right)} \qquad (f > 1)$$

$$= \cfrac{1}{(1 \times 10^{14})(1)(0.1)(10^{-7.6})(1/2)}$$

$$= 8 \times 10^{-6} = 0.0008\%. \qquad (11.23)$$

The systematic error is absurdly small: very special techniques and very hard work would be needed to secure anything like this accuracy in the other steps involved. At pH $= 6.8$ the titration error calculated from eq. (11.22) is -3×10^{-6}, or -0.0003%, and if you look at Fig. 11.5 you will see that even this figure is an overestimate. At pH $= 8.4$ the titration error calculated from eq. (11.23) is 5×10^{-5}, or 0.005%. Two titrations performed with phenol red as the indicator could not disagree by more than 0.0053%, and the relative precision could not be worse than $0.0053/2 = \pm 0.003\%$ if this were the only source of random error. There will be other random errors, and they will swamp this one. On both grounds phenol red is an eminently satisfactory indicator.

The calculations are somewhat different for titrations of other kinds. Suppose that an 0.1000 M aqueous solution of acetic acid, for which $K_a = 1.8 \times 10^{-5}$ mol dm^{-3}, is titrated with 0.1000 M aqueous sodium hydroxide and that phenol red is used as the indicator. The transition interval begins at pH $= 6.8$, where $[H_3O^+] = 1.6 \times 10^{-7}$ M, but you cannot calculate the titration error directly from this figure because it is acetic acid, rather than hydronium ion, that is being titrated. The concentration of hydronium ion is governed by the equations $[H_3O^+] = K_a[HOAc(aq)]/[OAc^-]$ and $[H_3O^+] = K_w/[OH^-]$. Both must be satisfied at every point, but the first is more convenient if the equivalence point has not yet been reached, while the second is more convenient if it has already been passed. The first necessity is to estimate the pH at the equivalence point. The equation for the titration reaction is $HOAc(aq) + OH^- = OAc^- + H_2O(l)$. Comparing this with eq. (11.1) shows that S is $HOAc(aq)$, R is OH^-, and P is OAc^-, and that $s = r = p = 1$. According to Table 11.2 [S] = [R] at the equivalence point and, since V_R at that point will be equal to V_S^0 because the acid and base are equally concentrated and $r = s$ (eq. (11.5)), [P] $= c_S^0/2 = 0.0500\,M$. Combining this information with the expression for K_t gives

$$K_t = \frac{[OAc^-]}{[HOAc(aq)][OH^-]} \left(= \frac{K_a}{K_w} = \frac{[P]}{[S][R]} \right) = \frac{0.0500}{[OH^-]^2} = \frac{0.0500}{(K_w/[H_3O^+])^2}$$

and eventually pcH $= 8.7$. Compare this procedure with Example 6.6 (p. 134).

Hence the beginning of the transition interval precedes the equivalence point. An expression for $[HOAc(aq)]/[OAc^-]$ ($= [S]/]P$) at $f < 1$ may be obtained from Table 11.2:

$$[S]/[P] = (s/p)\frac{1-f}{f} \qquad (f < 1). \tag{11.24}$$

Accordingly $[HOAc(aq)]/[OAc^-] = (1-f)/f = [H_3O^+]/K_a = 1.6 \times 10^{-7}/1.8 \times 10^{-5} = 9 \times 10^{-3}$. Since $(1-f)$ is much smaller than f, f is nearly equal to 1, and the titration error $(f-1)$ is virtually equal to -9×10^{-3} or -0.9%. There is no more reason to calculate a more exact value of f than there was to use an exact description of $V_S^0/(V_S^0 + V_R)$ in dealing with the titration of hydrochloric acid.

In exactly the same way you could find that the titration error is -2×10^{-4}, or -0.02%, at pH $= 8.4$, and you could conclude from this figure and the preceding one that the precision might be as bad as $\pm 0.5\%$. At the midpoint of the transition interval the titration error is -0.001_4, or $-0.1_4\%$. These figures are not entirely satisfactory, and you should try to find another indicator better suited to this titration. If you cannot, the last paragraph of Section 11.9 describes a trick that might make it possible to obtain acceptable results with phenol red.

A titration cannot be highly precise unless the slope of the sigmoidal titration curve is large at its end point, which cannot be true unless the end point lies very close to the equivalence point. Saying the same thing in different words, the random error must be large if the systematic error is large. However, the slope may be small at the end point even if the end point coincides exactly with the equivalence point, and therefore the random error may be either large or small if the systematic error is small. If the systematic error is large, you should look for another indicator. If it is small, you should find whether the random error is also small. If it is, you need look no farther. If the systematic error is small and the random error large, you should look for a way of locating the end point more precisely. You might try to titrate to a particular portion of the color-change interval, or might follow the titration by means of an instrumental technique instead of an indicator.

SUMMARY: Every titration involves both systematic errors and random errors. One systematic error arises from the fact that the average end point may not coincide with the equivalence point. When the titration is made with an indicator, the magnitude of this systematic error can be estimated by calculating the titration error at the midpoint of the color-change interval. Random errors arise from the fact that the color-change interval has a finite width, and their magnitude can be taken as half of the difference between the titration errors at the two ends of the interval.

11.8. The behaviors of indicators

PREVIEW: This section deals with the natures of acid-base and other indicators and with the widths of their color-change intervals.

Methyl orange is a typical acid-base indicator. It is commercially available as sodium salt, which when dissolved in water yields a solution of the ion

$$(CH_3)_2\,N-\!\!\!\left\langle\bigcirc\right\rangle\!\!\!-N\!\!=\!\!N-\!\!\!\left\langle\bigcirc\right\rangle\!\!\!-SO_3^-\qquad(=\textit{In})$$

which is the basic form of the indicator, and which will be represented here by the symbol *In*. The charge is omitted because the basic forms of different indicators have different charges and may be uncharged or even cationic. The ion *In* is colored bea cause it contains an extensive π-electron system, which can absorb energy from a beam of visible radiation. The energy is consumed in raising an electron to an excited state. The frequency or wavelength of the radiation that is absorbed, and therefore the color of the ion, are related to the amount of energy that is needed to excite the electron. Adding a proton to *In* gives the acidic form of the indicator, H*In*:

$$(CH_3)_2\,N-\!\!\!\left\langle\bigcirc\right\rangle\!\!\!\overset{\displaystyle H^+}{-}N\!\!=\!\!N-\!\!\!\left\langle\bigcirc\right\rangle\!\!\!-SO_3^-\qquad(=\textit{H}\textit{In}).$$

The distribution of the π electrons is altered by the addition of the positive charge, and so is the amount of energy that is needed to promote an electron to an excited state. The basic and acidic forms absorb radiation of different wavelengths and have different colors. An ordinary acid-base indicator is a substance for which these colors are visibly different, at least one of them being so intense that it can be discerned

23

even if the concentration of the form having that color is extremely low, and for which the gain and loss of protons occur in a useful range of acidities. The intensity of color is important because the concentration of indicator must be extremely low; if it were not, an appreciable amount of the reagent would be consumed in reacting with the indicator.

The equilibrium between the acidic and basic forms of an indicator is described by the equation

$$\frac{[HIn]}{[In]} = \frac{[H_3O^+]}{K_{a, HIn}} \tag{11.25}$$

where $K_{a, HIn}$ is of course the overall acidic dissociation constant of HIn. If the concentration of hydronium ion is much larger than $K_{a, HIn}$, the concentration of HIn will be much larger than that of In. The color of In will be impossible to detect with the eye, and the solution will appear to have the color of HIn alone. Conversely, if the concentration of hydronium ion is much smaller than $K_{a, HIn}$, the solution will contain much more In than HIn. The color due to the small concentration of HIn will be indetectable, and the solution will appear to have the color of In alone.

There is a range of hydronium-ion concentrations over which the color changes from that of the acidic form of an indicator to that of its basic form. In Section 11.7 this range was called the color-change interval or transition interval of the indicator. The pcH-value at the midpoint of a transition interval is important, and so is the width of the interval.

Equation (11.25) shows that the concentrations of the acidic and basic forms of an indicator are equal if $[H_3O^+] = K_{a, HIn}$ or $pcH = pK_{a, HIn}$. If the two forms are equally intensely colored, an observer whose eye is equally sensitive to their different colors will think that these colors are equally deep. If the acidic form were yellow and the basic one blue, such an observer would see a pure green color if the two concentrations were equal. Of course the color would not be pure green if the acidic form had a very pale yellow color and the basic form a very intense blue one (in which case the blue color would predominate when the concentrations are the same); if the observer is color-blind in either the yellow or the blue part of the spectrum; or if the solution were observed under the yellow light of a sodium vapor lamp, which makes a yellow solution indistinguishable from a colorless one. On the average, however— and the average includes many different indicators, many different observers, and many different lighting conditions—the midpoint of the transition interval of an acid-base indicator can be taken as the point where $pcH = pK_{a, HIn}$.

On neutralizing a solution in which the ratio $[HIn]/[In]$ is so large that the color of In is indetectable, an observer will see no change of color until a certain point is reached. At this point, which is the beginning of the transition interval, the depth of the color due to In is equal to some fraction, say $1/x$th, of the depth of the color due to HIn. If the colors of In and HIn are equally intense, the concentration of In will also be equal to $1/x$th of the concentration of HIn. At the beginning of the transition interval, where

$$pcH = pK_{a, HIn} + \log \frac{[In]}{[HIn]} = pK_{a, HIn} + \log \frac{[HIn]/x}{[HIn]} = pK_{a, HIn} - \log x \quad (11.26a)$$

the color is just barely distinguishable from that of a much more strongly acidic solution.

As the neutralization continues, there comes a point at which the color is indistinguishable from that of pure *In*, or from that of a much more strongly basic solution. This is the end of the color-change interval. Repeating the argument given in the preceding paragraph shows that it will be reached when the concentration of H*In* is equal to $1/x$th of the concentration of *In*, or

$$pcH = pK_{a, HIn} + \log \frac{[In]}{[HIn]} = pK_{a, HIn} + \log \frac{[In]}{[In]/x} = pK_{a, HIn} + \log x. \quad (11.26b)$$

At one end of the transition interval the pcH is equal to $pK_{a, HIn} - \log x$; at the other end it is equal to $pK_{a, HIn} + \log x$. The width of the transition interval is equal to the difference between these values, $2 \log x$.

Table 11.3 lists the acid-base indicators that are most widely used for titrations in aqueous solutions. It gives the colors of their acidic and basic forms and the pH-values at the ends of their transition intervals. Different observers under different lighting conditions would obtain somewhat different values, and those given in Table 11.3 are averages. Here we are primarily concerned with the widths of the transition intervals. Three of the thirteen indicators—2,4-dinitrophenol, phenolphthalein, and thymolphthalein—will be ignored because the acidic form of each is

TABLE 11.3. *Indicators for aqueous acid-base titrations*

Common name	Color in acidic solutions	pH-Values at the ends of the transition interval	Color in alkaline solutions
Thymol blue	red	1.2– 2.8	yellow
2,4-Dinitrophenol	colorless	2.4– 4.0	yellow
Bromophenol blue	yellow	3.0– 4.6	blue
Methyl orange	red	3.1– 4.4	orange
Bromocresol green	yellow	3.8– 5.4	blue
Methyl red	red	4.2– 6.3	yellow
Bromocresol purple	yellow	5.2– 6.8	purple
Bromothymol blue	yellow	6.2– 7.6	blue
Phenol red	yellow	6.8– 8.4	red
Cresol red	yellow	7.2– 8.8	red
Thymol blue	yellow	8.0– 9.6	blue
Phenolphthalein	colorless	8.3–10.0	red
Thymolphthalein	colorless	9.3–10.5	blue

colorless and therefore the above argument does not apply to them without modification. For seven of the other ten indicators the width of the transition interval is 1.6 pH units, and the average width for all ten is also 1.6 pH units. It is safe to conclude that the width of the transition interval, given above as $2 \log x$, is 1.6 pH units for a typical acid-base indicator. This immediately gives $\log x = 0.8$ and $x = 6$, which means that the average person under average conditions can only just detect one color in the presence of another that is six times as deep. Most people think that their eyes are much more sensitive than this.

The results can be summarized very briefly: the pcH-values at the two ends of the transition interval of a typical acid-base indicator are given by

$$\text{pcH} = pK_{a, HIn} \pm 0.8. \tag{11.27}$$

A redox indicator obeys similar principles, but whereas an acid-base indicator is an acid or a base, a redox indicator is the reduced or oxidized form of a redox couple. The two forms must have colors that are visibly different, and at least one of the colors must be so intense that it can be discerned even though the total amount of indicator in a titration mixture is negligible in comparison with the amount of the substance being titrated. The half-reaction of a redox indicator may be represented by the equation $In^{n+} + ne = In$. The color of a solution containing the indicator will be barely distinguishable from that of the pure reduced form if $[In^{n+}] = [In]/6$. It will be barely distinguishable from that of the pure oxidized form if $[In] = [In^{n+}]/6$. These statements are exactly analogous to the corresponding ones about acid-base indicators. Combining them with the Nernst equation yields the following descriptions of the potentials at the two ends of the transition interval of a typical redox indicator:

$$E = E^{0\prime}(In^{n+}, In; \text{ supporting electrolyte}) \pm \frac{0.05}{n}. \tag{11.28}$$

SUMMARY: An indicator is a substance that reacts with the reagent to give a visible color change even when its concentration is extremely low. Its color-change interval is the region in which the ratio of concentrations of two colored forms changes from approximately 6 to approximately 1/6. The color-change interval of an acid-base indicator is approximately given by the equation $\text{pcH} = pK_{a, HIn} \pm 0.8$; for a redox indicator it is approximately given by the equation $E = E^{0\prime}$ (indicator; supporting electrolyte) $\pm 0.05/n$.

11.9. Indicators and titration curves in chelometric titrations

PREVIEW: In a chelometric titration the indicator, ligand, and metal ion may all participate in acid-base equilibria that affect the value of K_t and the systematic and random errors.

Titrations with ethylenediaminetetraacetate and other chelons are usually made with metallochromic indicators. A *metallochromic indicator* is an indicator whose equilibrium and color change are governed by the concentration of a metal ion. The compound

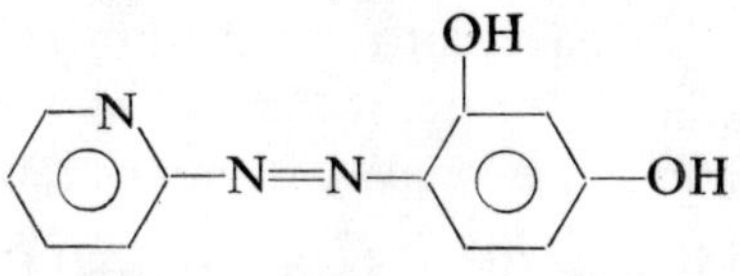

OH

N

—N=N—

—OH

familiarly called PAR, is a typical metallochromic indicator. Like methyl orange, it has an azo (—N=N—) group joining two aromatic rings; unlike methyl orange, its fully deprotonated form

O^-

N

—N=N—

—O^-

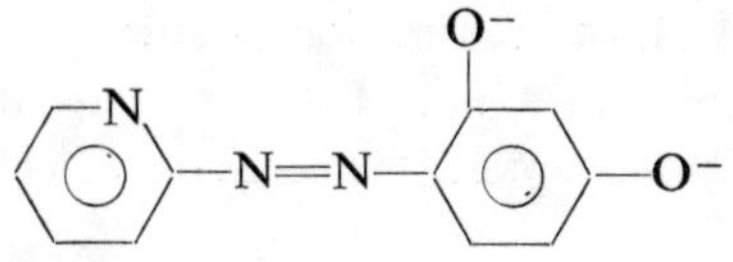

has a structure that enables it to act as a polydentate ligand. Both its electronic structure and its color are affected when it reacts with metal ions or hydronium ions, just as those of the basic form of methyl orange are affected when it reacts with a hydronium ion. The action of a metallochromic indicator is governed by the formation constant K_{MIn} of the metal–indicator complex

$$K_{MIn} = [MIn]/[M][In] \qquad (11.29)$$

where M represents a metal ion and In the fully deprotonated form of the indicator. Ionic charges are omitted to simplify this equation and the ones that follow. If this were the only equilibrium that had to be considered, the values of log [M] at the two ends of the transition interval would be given by

$$\log[M] = \log K_{MIn} \pm 0.8 \qquad (11.30)$$

which resembles eqs. (11.27) and (11.28), and in addition the equilibrium in the titration reaction could be described by the formation constant of the chelonate MY

$$K_{MY} = [MY]/[M][Y] \qquad (11.31)$$

in which Y represents the fully deprotonated form of the chelon being used as the reagent.

Acid-base equilibria are always involved as well, and species such as MOH, HY, H_2Y, HIn, MHY, and $MHIn$ are all common. To take them into account it is convenient to use the notation introduced in Section 7.7:

$$[M]' = [M] + [MOH] + [M(OH)_2] + [MX] + [MX_2] + \ldots$$
$$[Y]' = [Y] + [HY] + [H_2Y] + \ldots$$
$$[MY]' = [MY] + [MHY] + [M(OH)Y] + \ldots$$
$$[In]' = [In] + [HIn] + [H_2In] + \ldots$$

[M]' represents the total concentration of M that has not reacted with the reagent Y, and X represents any other ligand (such as the conjugate base of a buffer system) present in the titration mixture. $[Y]'$ and $[In]'$ are the total concentrations of the reagent and the indicator that have not reacted with the metal ion, and $[MY]'$ is the total concentration of the chelonate. If the pcH and the concentration of X do not change during the titration, these equations can be rewritten as

$$[M] = \alpha_M[M]' \qquad [MY] = \alpha_{MY}[MY]',$$
$$[Y] = \alpha_Y[Y]' \qquad [In] = \alpha_{In}[In]'.$$

Of course the values of the formation functions will depend on the composition, and in particular on the pcH, of the titration mixture.

The titration curve for the titration of M with Y can now be described by equations given in earlier sections of this chapter. One needs only to write $[S] = [M]'$, $[R] = [Y]'$, $[P] = [MY]'$, and

$$K_t = \frac{[P]}{[S][R]} = \frac{[MY]'}{[M]'[Y]'} = \frac{\alpha_M \alpha_Y}{\alpha_{MY}} K_{MY}. \qquad (11.32)$$

In many titrations it is possible to ignore the traces of MHY, M(OH)Y, MXY, and similar complexes, that are formed, so that α_{MY} is approximately equal to 1. Few titrations can be performed under such conditions that the product $\alpha_M\alpha_Y$ has a value anywhere near 1. Chelons are generally fairly strong Lowry–Brønsted bases, and therefore α_Y approaches 1 only at rather high pcH-values. To prevent the precipitation of the hydroxide, hydrous oxide, carbonate, or some other compound of the metal ion being titrated it is usually necessary to add a complexing agent that will react extensively with M, and when this is done α_M becomes much smaller than 1. If the titration is performed at a sufficiently low pcH-value, α_M can be made to approach 1 for all but the most strongly acidic metal ions, but then protonation of the chelon is extensive and α_Y is far below 1. Values of K_t are almost always much smaller than the corresponding ones of K_{MY}.

As a concrete example we may consider the titration of an 0.1000 M solution of copper(II) with 0.1000 M ethylenediaminetetraacetate in a buffered solution having a pcH equal to 7.00. The formation constant of the chelonate CuY is equal to 6×10^{18} mol^{-1} dm^3. Values of α_{Cu} and α_Y are needed to evaluate K_t. The formation constant of CuOH$^+$ is equal to 3×10^6 mol^{-1} dm^3, and if the solution contains no other ligand

$$\alpha_{Cu} = \frac{[Cu^{2+}]}{[Cu^{2+}]+[CuOH^+]} = \frac{1}{1+K_{CuOH^+}[OH^-]} = \frac{1}{1+(3\times10^6)(1\times10^{-7})} = 0.8.$$

The successive values of pK_a for the fully protonated acid H_4Y are $pK_1 = 2.0$, $pK_2 = 2.8$, $pK_3 = 6.2$, and $pK_4 = 10.3$. At pcH = 7.0 only a small fraction of the chelon will be unprotonated:

$$\alpha_Y = \frac{[Y]}{[H_4Y]+[H_3Y]+\ldots+[Y]} = \frac{\beta_4}{[H_3O^+]^4+\beta_1[H_3O^+]^3+\ldots+\beta_4} = 4.4\times10^{-4}.$$

Combining the values of α_{Cu} and α_Y with eq. (11.32), and ignoring the formation of complexes like CuHY and Cu(OH)Y (so that $\alpha_{MY} = 1$), the result is

$$K_t = \frac{(0.8)(4.4\times10^{-4})}{1}(6\times10^{18}) = 2\times10^{15}\ \text{mol}^{-1}\ \text{dm}^3.$$

Section 11.6 showed that the initial concentration of the substance being titrated must be at least 10^3 or 10^4 times as large as its concentration at the equivalence point in order for a titration to be successful. At the equivalence point of this titration, the total concentration [CuY]' of the chelonate will be equal to half of the initial value of [Cu]' because the solution of copper(II) will have been diluted to twice its original volume by the addition of the reagent, and

$$[Cu]'_{f=1} = [Y]'_{f=1} = \left(\frac{[CuY]'_{f=1}}{K_t}\right)^{1/2} = \left(\frac{[Cu]'_{f=0}}{2K_t}\right)^{1/2}. \tag{11.33}$$

Hence

$$[Cu]'_{f=0}/[Cu]'_{f=1} = \sqrt{2K_t[Cu]'_{f=0}}. \tag{11.34}$$

The value of K_t for this titration is so large that the ratio on the left-hand side of eq. (11.34) will be larger than 10^4 unless the initial concentration of copper(II) ion is smaller than about 3×10^{-8} M. Equation (11.34) has a different form from either eq. (11.19) or eq. (11.20), which applied to titrations involving precipitation and other processes in which the activities of the products are constant. This is because the chelonate remains dissolved in the titration mixture. If the concentrations of the solutions and the values of K_t are the same for a chelometric titration and a titration with a reagent that yields a 1 : 1 precipitate with the metal ion, the chelometric one will be more nearly complete at the equivalence point, will yield a more favorable titration curve, and will remain possible down to much lower initial concentrations of the metal ion. Often it is the increasing difficulty of locating the end point, rather than the deterioration of the titration curve, that sets the lower limit to the concentration of a metal ion that can be determined by chelometric titration.

The behavior of the indicator PAR is described by the following formation constants:

$$K_{1,H} = [HIn]/[H_3O^+][In] = 3\times10^{12}\ \text{mol}^{-1}\ \text{dm}^3,$$
$$K_{2,H} = [H_2In]/[H_3O^+][HIn] = 7\times10^5\ \text{mol}^{-1}\ \text{dm}^3;$$
$$\beta_{2,H} = [H_2In]/[H_3O^+]^2[In] = K_{1,H}K_{2,H} = 2\times10^{18}\ \text{mol}^{-2}\ \text{dm}^6,$$
$$K_{3,H} = [H_3In]/[H_3O^+][H_2In] = 2\times10^2\ \text{mol}^{-1}\ \text{dm}^3;$$
$$\beta_{3,H} = [H_3In]/[H_3O^+]^3[In] = \beta_{2,H}K_{3,H} = 4\times10^{20}\ \text{mol}^{-3}\ \text{dm}^9,$$
$$K_{CuIn} = [CuIn]/[Cu^{2+}][In] = 2\times10^{10}\ \text{mol}^{-1}\ \text{dm}^3.$$

These can be combined into a conditional equilibrium constant resembling K_t:

$$K'_{CuIn} = \frac{[CuIn]}{[Cu]'[In]'} = \alpha_{Cu}\alpha_{In}K_{CuIn}. \tag{11.35}$$

At pcH $= 7.0$ the value of α_{In} is given by

$$\alpha_{In} = \frac{[In]}{[H_3In] + [H_2In] + \ldots + [In]} = 3 \times 10^{-6}.$$

Consequently, from eq. (11.35), $K'_{CuIn} = (0.8)(3 \times 10^{-6})(2 \times 10^{10}) = 5 \times 10^4$ mol^{-1} dm^3. The concentration of $CuIn$ will be equal to the concentration of indicator that has not reacted with copper(II) if $[Cu]' = 1/K'_{CuIn} = 2 \times 10^{-5}$ M, or pcCu$' = 4.7$. This is the midpoint of the transition interval, and the two ends of the transition interval will lie at pcCu$' = 3.9$ and $5.5 \, (= 4.7 \pm 0.8)$.

The value of $[Cu]'$ at the equivalence point may be obtained from eq. (11.33); for $[Cu]'_{f=0} = 0.1$ M and $K_t = 2 \times 10^{15}$ mol dm^{-3} it is 5×10^{-9} M, so that pcCu$' = 8.30$. The entire transition interval of the indicator precedes the equivalence point. The titration error is described by eq. (11.22), which can be written

$$\text{titration error} = f_{\text{end point}} - 1 = \frac{[Cu]'_{\text{end point}}}{[Cu]'_{f=0} \dfrac{V^0_{Cu}}{V^0_{Cu} + V_Y}}.$$

With $[Cu]'_{f=0} = 0.1000$ M and $V^0_{Cu}/(V^0_{Cu} + V_Y) = 1/2$, this gives -2.5×10^{-3} or -0.25% for the titration error at the start of the transition interval, -4×10^{-4} or -0.04% for the titration error at its midpoint, and -6×10^{-5} or -0.006% for the titration error at its end.

The transition interval would lie closer to the equivalence point, and all these titration errors would be smaller, if K'_{CuIn} had a larger value. At pcH $= 7.0$ it is much smaller than K'_{CuIn} because α_{In} is small. At a higher pcH-value both α_{In} and K'_{CuIn} would be larger, provided that the pcH is not so high that α_{Cu} becomes very small. Increasing the pcH to 8.0 cause α_{In} to increase from 3×10^{-6} to 3.3×10^{-5}, and it also causes α_{Cu} to decrease from 0.8 to 0.25. The increase of α_{In} overbalances the decrease of α_{Cu}, K'_{CuIn} increases from 5×10^4 mol^{-1} dm^3 to 1.6×10^5 mol^{-1} dm^3, and the titration errors decrease to -0.08% at the start of the transition interval, -0.01% at its midpoint, and -0.002% at its end. These figures are definitely more satisfactory than the corresponding ones at pcH $= 7.0$.

Some chemical reason, such as the presence of another metal ion that would begin to interfere at a higher pcH-value, might forbid raising the pcH above 7 in a practical titration. One might then resort to a trick, which consists of performing the titration in such a way that the end point must lie in the second half of the transition interval. The anion of PAR is red; on coordination with copper(II) it becomes yellow. It will be orange at the midpoint of the transition interval, orange-red or red-orange before it, and yellow-orange or orange-yellow after it. Both the precision and the accuracy would be much improved by titrating to a predominantly yellow color. A refinement of this trick consists of using a comparison solution as an aid in judging when the desired color has been reached. The comparison solution might contain copper(II) at a total concentration of 2×10^{-6} M (pcCu$' = 5.3$), which corresponds to a point that just precedes the end of the transition interval (where pcCu' = 5.5) at pcH = 7.0. It should have the same volume, concentration of indicator, and pcH as the titration mixture does at the end point, and should be contained in a similar vessel. The titration is continued until the color of the titration mixture becomes indistinguishable from that of the comparison solution. This not only confines the end point to the desired part of the transition interval but also decreases the uncertainty in the value of pcCu' at the end point to perhaps ±0.2 or ±0.3 unit.

SUMMARY: The value of K_t in a chelometric titration depends not only on the formation constant of the chelon MY but also on the formation functions α_M and α_Y. Complexation of the metal ion by hydroxide ion and other ligands, and protonation of the chelon, usually cause either α_M or α_Y, or both, to be fairly small. Metallochromic indicators also participate in protonation equilibria, which must be taken into account in calculating the systematic and random errors that their use involves.

11.10. Stepwise titrations

PREVIEW: Many titrations, such as those involving di- and polyfunctional acids (or bases) or stepwise complex formation, occur in two or more successive steps. Their titration curves can be obtained by dividing them into different portions, each corresponding to one of the successive steps.

Only a single step has been involved in each of the titrations considered so far, but there are many titrations in which two or more successive reactions occur. If a difunctional acid H_2B is titrated with a base, such as sodium hydroxide, the reaction $H_2B(aq) + OH^- = HB^- + H_2O(l)$ will occur first, and will be followed by the one $HB^- + OH^- = B^{2-} + H_2O(l)$. If a mixture of two substances is titrated with a reagent that can react with both, the one for which K_t has the larger value will react first; then, as it is consumed by the addition of reagent, the other will begin to react. There will be two equivalence points, and the titration curve will consist of two steps. Under some conditions the steps may merge so that the first equivalence point becomes indetectable, or the second equivalence point may be indetectable because the corresponding value of K_t is too small. For a trifunctional acid or a mixture of three substances there will be three steps on the titration curve and three equivalence points, and again one or more may be indetectable.

Let us consider the titration of 0.1000 M ethylenediamine, which may be represented by the symbol en, with 0.1000 M hydrochloric acid in an aqueous solution. The two successive reactions are

$$en + H_3O^+ = Hen^+ + H_2O(l);$$

$$K_{t,1} = \frac{[Hen^+]}{[en][H_3O^+]} = \frac{1}{K_{a,2}} = 8.5 \times 10^9 \text{ mol}^{-1} \text{ dm}^3, \qquad (11.36a)$$

$$Hen^+ + H_3O^+ = H_2en^{2+} + H_2O(l);$$

$$K_{t,2} = \frac{[H_2en^{2+}]}{[Hen^+][H_3O^+]} = \frac{1}{K_{a,1}} = 7.1 \times 10^6 \text{ mol}^{-1} \text{ dm}^3, \qquad (11.36b)$$

where $K_{a,1}$ and $K_{a,2}$ are the overall acidic dissociation constants of the diprotonated cation H_2en^{2+}. We may suppose that the desired titration curve is a plot of pcH against f.

The value of the ratio $K_{a,1}/K_{a,2}(= K_{t,2}/K_{t,1})$ is 1.2×10^3. This is large enough (see Section 6.8) to permit neglecting the trace of H_2en^{2+} that will be present during the first step of the titration, where en is being neutralized to give Hen^+. At every point in this region there must be some free en that has not yet been consumed, and as long as the concentration of en is appreciable that of H_2en^{2+} cannot be. On similar grounds we shall be able to ignore the trace of en that remains unneutralized during the second step, where H_2en^{2+} begins to appear in substantial concentrations.

To describe the first step we need only combine eq. (11.36a) with an appropriate equation from Table 11.2, where S = en and P = Hen^+. The result is

$$\text{pcH} = pK_{a,2} - \log \frac{[P]}{[S]} = pK_{a,2} - \log \frac{f}{1-f} \qquad (0 < f < 1). \qquad (11.37)$$

This gives curve a in Fig. 11.13, and in particular $\text{pcH} = pK_{a,2}(= \log K_{t,1}) = 9.93$ at $f = 1/2$. As usual, the curve cannot be drawn too near the ends of the range, for eq. (11.37) does not apply either at $f = 0$, where the concentration of Hen^+ is not really zero, or at $f = 1$, where the concentration of en is not really zero. The pcH-values at these two points may be obtained in ways described in Chapter 6. The pcH of an $0.1000\ M$ solution of en is equal to 11.46 (page 147), and the pcH of a solution of the amphoteric substance Hen^+ is equal to 8.39 (eq. (6.38)).

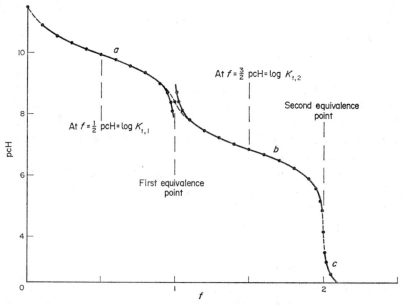

FIG. 11.13. Sigmoidal titration curve for the titration of an $0.1000\ M$ aqueous solution of ethylenediamine with an $0.1000\ M$ aqueous solution of a strong acid.

In the region $1 < f < 2$, the amount of acid that has been added is large enough to convert some, but not all, of the Hen^+ into H_2en^{2+}. In this region the reaction is described by eq. (11.36b), and $S = Hen^+$ while $P = H_2en^{2+}$. When a titration involves only a single step, the amount of P that has been formed at any point is proportional to the value of f at that point, while the amount of S remaining unreacted is proportional to $1-f$. These things are not true here because the region we are considering begins at $f = 1$ instead of at $f = 0$. Hence the amount of P is proportional to $f-1$, and that of S is proportional to $2-f$, so that

$$pcH = pK_{a,2} - \log \frac{f-1}{2-f} \qquad (1 < f < 2). \qquad (11.38)$$

This yields curve b in Fig. 11.13, and in particular $pcH = pK_{a,1} (= \log K_{t,2}) = 6.85$ at $f = 3/2$.

If $f > 2$ the solution contains an excess of hydronium ion ($= R$). Equation (11.9b) assumes that excess reagent begins to appear when f exceeds 1, but here it does not do so until f exceeds 2, and therefore we must write

$$[H_3O^+](= [R]) = c_S^0(f-2) \frac{V_S^0}{V_S^0 + V_R} \qquad (f > 2). \qquad (11.39)$$

which gives curve c in Fig. 11.13. The gap between curves a and b can be bridged with the aid of the point at $f = 1$; to bridge the one between curves b and c we must calculate the pcH at $f = 2$. Here the neutralization of $V_S^0 c_S^0$ millimoles of en has given very nearly $V_S^0 c_S^0$ millimoles of H_2en^{2+}. In titrating V_S^0 cm³ of 0.1000 M en, 2 V_S^0 cm³ of 0.1000 M hydrochloric acid is needed to reach this point, so that the total volume of the titration mixture is 3 V_S^0 cm³ and the concentration of H_2en^{2+} is $V_S^0 c_S^0/3 V_S^0 = c_S^0/3 = 3.3 \times 10^{-2}$ M. It can now be calculated that the pcH is equal to 4.16 at $f = 2$, and the titration curve can be completed by drawing the dashed lines indicated.

There are two ways in which this titration could be made with an indicator. One would be to use an indicator for which the pH at the midpoint of the transition interval was approximately equal to 8.4, thus attempting to locate the first equivalence point. The other would be to use an indicator for which it was approximately equal to 4.2, thus attempting to locate the second equivalence point. The second would give better accuracy and precision because the titration curve is much steeper around the second equivalence point than around the first.

Of course this conclusion reflects the values of $K_{t,1}$ and $K_{t,2}$ in this particular example. The slope at the first equivalence point is governed by the ratio $K_{t,1}/K_{t,2}$; that at the second is governed by $K_{t,2}$. Increasing $K_{t,1}$ would make the first equivalence point better defined but would not affect the second. Increasing $K_{t,2}$ would improve the second, but would make the first one worse. Decreasing $K_{t,1}/K_{t,2}$ decreases the slope at the first equivalence point, and if $K_{t,1}/K_{t,2}$ is sufficiently small the inflection near $f = 1$ disappears altogether so that the curve appears to consist of a single step. All these things are illustrated by Fig. 11.14.

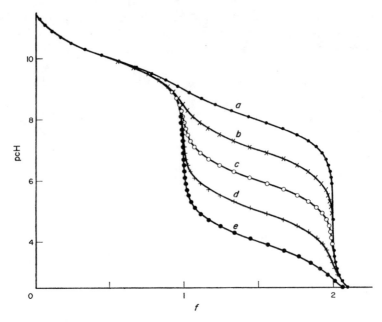

FIG. 11.14. Sigmoidal titration curves for the titrations of 0.1000 M aqueous solutions of difunctional bases with an 0.1000 M aqueous solution of a strong acid. The values of $K_{a,1}$ and $K_{a,2}$ for the diprotonated forms of the bases, and the corresponding values of $K_{t,1}$, $K_{t,2}$, and $K_{t,1}/K_{t,2}$ are

Curve	$K_{a,1}$, mol dm^{-3}	$K_{a,2}$, mol^1 dm^{-3}	$K_{t,1}$, mol^{-1} dm^3	$K_{t,2}$, mol^{-1} dm^3	$K_{t,1}/K_{t,2}$
a	1×10^{-8}	1×10^{-10}	1×10^{10}	1×10^{8}	1×10^{2}
b	1×10^{-7}	1×10^{-10}	1×10^{10}	1×10^{7}	1×10^{3}
c	1×10^{-6}	1×10^{-10}	1×10^{10}	1×10^{6}	1×10^{4}
d	1×10^{-5}	1×10^{-10}	1×10^{10}	1×10^{5}	1×10^{5}
e	1×10^{-4}	1×10^{-10}	1×10^{10}	1×10^{4}	1×10^{6}

Section 7.2 contrasted the one-step overall reaction that occurs when a metal ion reacts with a chelon and the series of stepwise reactions that occur with a ligand capable of occupying only one or two of the coordination positions. The difference between these is largely responsible for the utility of chelons as reagents in titrimetry. To show why this is so, we may construct the titration curves that would be obtained in two different titrations: one of a metal ion M with a chelon Y, and the other of the same metal ion, at the same initial concentration, with a monodentate ligand L. Several assumptions may be made to simplify the construction and comparison of the curves. One is that M has only two coordination positions, so that it reacts with L in only two steps:

$$\text{M} + \text{L} = \text{ML} ; \quad K_1 = K_{t,1} = 1\times10^6 \ \text{mol}^{-1}\,\text{dm}^3,$$

$$\text{ML} + \text{L} = \text{ML}_2; \quad K_2 = K_{t,2} = 1\times10^4 \ \text{mol}^{-1}\,\text{dm}^3;$$

$$\beta_2 = K_1 K_2 = 1\times10^{10} \ \text{mol}^{-2}\,\text{dm}^6.$$

Another is that the formation constant of the chelonate is numerically equal to the overall formation constant of ML_2:

$$M + Y = MY; \quad K_{MY} = 1 \times 10^{10} \text{ mol}^{-1} \text{ dm}^3.$$

A third is that the solutions of M, L, and Y used in the two titrations are all 0.1000 M, and a fourth is that the protonation of L and Y, the acidic dissociation of M, and all other chemical complications can be ignored. Finally, pcM will be plotted against f in each titration.

The chelometric titration curve is easy to construct. You would have S = M, R = Y, and P = MY in the titration reaction. Equation (11.7a) describes the curve almost up to the equivalence point. At the equivalence point the concentrations of unreacted M and Y are equal and the concentration of MY is virtually 0.0500 M. Writing $K_{MY} = 0.05/[M]^2 = 1 \times 10^{10} \text{ mol}^{-1} \text{ dm}^3$ gives $[M] = 2.2 \times 10^{-6} M$ or pcM = 5.65. Beyond the equivalence point you could combine the expression for K_{MY} with eqs. (11.9b) and (11.11b) to obtain

$$[M] = \frac{[MY]}{K_{MY}[Y]} \left(= \frac{[P]}{K_{MY}[R]} \right) = \frac{(p/s)\, c_S^0\, \dfrac{V_S^0}{V_S^0 + V_R}}{K_{MY}(r/s)\, c_S^0(f-1)\, \dfrac{V_S^0}{V_S^0 + V_R}}$$

$$= \frac{1}{K_{MY}(f-1)} \qquad (f > 1), \qquad (11.40)$$

since p, r, and s are equal to 1. The resulting titration curve is shown in Fig. 11.15(a). It is satisfactorily steep at the equivalence point. You would seek a metallochromic indicator for which [M*In*] and [*In*] were equal when $[M] = 2.2 \times 10^{-6}$ M, so that, neglecting protonation of the indicator,

$$K_{MIn} = \frac{[MIn]}{[M][In]} = \frac{1}{2.2 \times 10^{-6}} = 4.5 \times 10^5.$$

Its transition interval would extend from pcM = 4.9 to pcM = 6.5, and the titration errors at these points are only $\pm 0.03\%$.

The curve for the titration of M with L is a little less easy to calculate, not only because two steps are involved but also because [M] does not appear in the expression for $K_{t,2} (= K_2)$. This is in contrast to the titration of ethylenediamine with hydrochloric acid, where the desired concentration, $[H_3O^+]$, does appear in the expression for $K_{t,2}$. You would begin by dividing the curve into three regions. In the first, where $0 < f < 1$, the chief reaction will be M + L = ML. In the second, where $1 < f < 2$, it will be ML + L = ML_2. In the third, where $f > 2$, there will be an excess of L.

The first and third regions are the easiest to handle. In the first, S = M, R = L, and P = ML. Equation (11.7a) can be used, and it gives curve b1 in Fig. 11.15(b). In the third, the usual description provided by eq. (11.9b) has to be modified by writing $f-2$ in place of $f-1$, as was done in obtaining eq. (11.39), because the beginning of

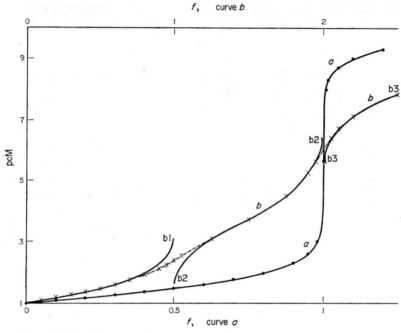

FIG. 11.15. Sigmoidal titration curves for the titration of an 0.1000 M solution of a metal ion M with (a) an 0.1000 M solution of a chelon Y if the formation constant of the chelonate MY is equal to 1×10^{10} mol^{-1} dm^3, and (b) an 0.1000 M solution of a monodentate ligand L that reacts with M to form two complexes: ML, with a formation constant K_1 equal to 10^6 mol^{-1} dm^3, and ML$_2$, with a formation constant K_2 equal to 10^4 mol^{-1} dm^3. The abscissa scales are different for the two curves.

this region lies at $f = 2$ rather than at $f = 1$. To obtain a relation between [L] ($= $ [R]) and [M] in this region we may write [M] = [ML$_2$]/β_2[L]^2. Almost all of the original M will have been converted into ML$_2$ at the beginning of this region, and therefore [ML$_2$] = $c_M^0 V_M^0/(V_M^0 + V_L)$. Combining these equations gives

$$[M] = \frac{[ML_2]}{\beta_2 [L]^2} = \frac{c_M^0 \dfrac{V_M^0}{V_M^0 + V_L}}{\beta_2 \left[c_M^0 (f-2) \dfrac{V_M^0}{V_M^0 + V_L} \right]^2}$$

$$= \frac{1}{\beta_2 c_M^0 (f-2)^2 \dfrac{V_M^0}{V_M^0 + V_L}} \qquad (f > 2). \qquad (11.41)$$

This gives curve b3 in Fig. 11.15(b).

In the second step of the titration S = ML, R = L, and P = ML$_2$. Since [L] = [ML$_2$]/K_2[ML] = [P]/K_2[S], an expression for [L] may be obtained with the aid of the one for [P]/[S] given in Table 11.2. Here f and $f-1$ must be replaced by $f-1$ and $f-2$, respectively, because this region does not begin until $f = 1$ whereas the expression in Table 11.2 assumes that it begins at $f = 0$. The result is

$$[L] = \frac{f-1}{K_2(2-f)} \qquad (1 < f < 2).$$ (11.42a)

However, it is [M] rather than [L] that is wanted. These two concentrations are related by the expression $[M] = [ML]/K_1[L] (= [S]/K_1[L])$. The concentration of ML may be described by eq. (11.7a), again replacing f by $f-1$. Combining all these equations yields

$$[M] = \frac{c_M^0(2-f)\dfrac{V_M^0}{V_M^0+V_L}}{K_1\left[\dfrac{f-1}{K_2(2-f)}\right]} = \frac{K_2}{K_1}c_M^0\frac{(2-f)^2}{f-1}\frac{V_M^0}{V_M^0+V_L} \qquad (1 < f < 2)$$ (11.42b)

which gives curve b2 in Fig. 11.15(b).

As usual, there is a gap between curves b1 and b2 and another between curves b2 and b3. Both may be bridged by plotting the values of pcM at the two equivalence points. At the first equivalence point there is a little M that has not reacted with L to give ML, but the concentration of free L must be very small because the equilibrium constant of the further reaction $ML + L = ML_2$ is fairly large and because the solution contains a fairly high concentration of ML. Any L that did not react with M must have been consumed in forming ML_2, and therefore $[M] = [ML_2]$. Accordingly, since $\beta_2 = [ML_2]/[M][L]^2$, the concentration of L at the first equivalence point is given by

$$[L] = 1/\beta_2^{1/2} \qquad (f = 1).$$ (11.43)

The total concentration of M $(= [M]+[ML]+[ML_2])$ at this point is equal to 0.0500 M; if the concentrations of free M and ML_2 were negligible, that of ML would be 0.0500 M and

$$[M] = \frac{[ML]}{K_1[L]} = \frac{0.0500}{K_1/\beta_2^{1/2}} = 0.0500\left(\frac{K_2}{K_1}\right)^{1/2} = 5\times10^{-3}\ M.$$

This does not quite justify the approximation, and a better result is

$$[M] = \frac{0.0500-2[M]}{K_1/\beta_2^{1/2}} = (0.0500-2[M])\left(\frac{K_2}{K_1}\right)^{1/2} = 4.2\times10^{-3}\ M$$

whence $pcM = 2.38$.

For the second equivalence point similar reasoning gives $[ML] = [L]$, and therefore

$$[M] = [ML]/K_1[L] = 1/K_1 \qquad (f = 2)$$ (11.44)

so that $pcM = 6.00$.

Curve b in Fig. 11.15 is the result. It is much less satisfactory than the one (curve a) for the chelometric titration even though the overall formation constants are numerically equal. There is no pronounced change of slope around the first equivalence point, despite the fact that $K_{t,1}$ $(= K_1)$ is fairly large, because $K_{t,1}/K_{t,2}$ is equal to only 1×10^2. The slope at the second equivalence point is governed by the value of $K_{t,2}$ $(= K_2)$, which is larger and therefore more favorable than $K_{t,1}/K_{t,2}$, but even this slope is not very large. The random errors that might be incurred in attempting to locate each of the equivalence points by using an indicator are easy to estimate from the titration curve. At the first equivalence point pcM $= 2.38$. An indicator for which the midpoint of the transition interval coincided with this value would begin to change color at pcM $= 1.6$, where f is about 0.6, and would continue to change color until pcM $= 3.2$, where f is about 1.3. The results of replicate titrations might differ by as much as $\pm 30{-}40\%$, which is abominably bad. Even at the second equivalence point the precision would be no better than $\pm 5\%$.

This comparison actually underestimates the advantage of the chelometric procedure because it assumes that there are only two steps in the reaction of M with a monodentate ligand. If there were four, with $K_1 = 10^4$ mol^{-1} dm^3, $K_2 = 10^3$ mol^{-1} dm^3, $K_3 = 10^2$ mol^{-1} dm^3, and $K_4 = 10$ mol^{-1} dm^3, the overall formation constant of ML$_4$ would still have a numerical value of 10^{10} (mol^{-4} dm^{12}), but there would be scarcely any inflection point at all on the titration curve. The first three equivalence points would be invisible because the value of $K_{t,i}/K_{t,i+1}$ governing the slope at each would be equal to only 10. The fourth equivalence point would also be invisible because the value of $K_{t,4}$ $(= K_4)$ governing the slope at this point is also equal to 10 (mol^{-1} dm^3). The curve would be almost perfectly smooth and featureless, and even a precision of $\pm 30\%$ would hardly be possible to secure.

SUMMARY: Many titrations involve n successive steps characterized by the values $K_{t,1}$, $K_{t,2}$, $\ldots$, $K_{t,n}$. Titration curves for such titrations may be obtained by joining n segments, with $K_t = K_{t,1}$ for the first segment, and so on to $K_t = K_{t,n}$ for the last. The slope of a sigmoidal titration curve at the ith equivalence point is governed by the ratio $K_{t,i}/K_{t,i+1}$ except for the last equivalence point, where it is governed by $K_{t,n}$. Chelometric titrations give sharper end points than titrations based on successive complex formation even if the overall formation constant of the last complex is numerically equal to the formation constant of the chelonate.

11.11. Exact titration-curve equations

PREVIEW: The equations given in the preceding sections involve approximations that lead to erroneous values near any equivalence point. This section shows how exact equations can be derived for use when they are needed.

You can almost always find out everything you need to know about a titration from a titration curve that has been constructed by the general procedure described above. This involves sketching connections between segments that do not really meet, and was illustrated by Figs. 11.5 and 11.15(b). The resulting curve is naturally somewhat uncertain for a little distance on each side of every equivalence point. If the portion that has to be sketched is relatively short, as it was in Fig. 11.5, its uncertainty causes little discomfort. Even if it is as long as it was in Fig. 11.15(b), you can still tell whether a titration will be satisfactory if it is made with an indicator, and you can still estimate its precision closely enough for any practical purpose. From Fig. 11.15(b) you can tell that the random error caused by trying to find the first equivalence point with the aid of an indicator would be roughly $\pm 35\%$. You would conclude that the attempt is not worth making. You would have reached the same conclusion if you had guessed that the random error would be $\pm 70\%$ or $\pm 15\%$.

Occasionally, however, chemists do need more accurate and more detailed information about the immediate vicinity of an equivalence point than can be provided by curves thus constructed or by the equations that describe them. Titration to the second equivalence point of the curve in Fig. 11.15(b) is not satisfactory if it is made with an indicator, but in a potentiometric titration it might be possible to achieve a precision of ± 0.1 pcM-unit in locating the end point. Would the random error be acceptably small if this were done? The end point of a potentiometric titration is often taken to be the point of maximum slope. Does this coincide with the equivalence point? If not, is the difference acceptably small?

To answer questions like these, one needs equations that describe titration curves more fully and exactly than do the approximate ones we have used. This section will briefly describe how exact equations may be obtained, but it will not discuss their applications because those are matters for more advanced texts.

The simplest possible exact equation is the one that describes a titration based on a reaction such as $H_3O^+ + OH^- = 2\,H_2O(l)$ or $Ag^+ + Cl^- = AgCl(s)$, in which the activity of the product is constant throughout. Let us consider the first of these, and suppose that V_S^0 cm³ of a c_S^0 M solution of hydrochloric acid is titrated with a c_R M solution of sodium hydroxide in water as the solvent. Applying the electroneutrality principle to any possible mixture gives

$$[Na^+] + [H_3O^+] = [Cl^-] + [OH^-] \tag{11.45}$$

of which eq. (6.17) is the special case that corresponds to $f = 0$. Since the concentration of hydronium ion is the one most likely to be of interest, we express the other concentrations in terms of $[H_3O^+]$ and f:

$$\left. \begin{aligned} [Na^+] &= \frac{V_R c_R}{V_S^0 + V_R} = c_S^0 f \frac{V_S^0}{V_S^0 + V_R}\,; \\ [Cl^-] &= c_S^0 \frac{V_S^0}{V_S^0 + V_R}\,; \\ [OH^-] &= \frac{1}{K_t[H_3O^+]}\,. \end{aligned} \right\} \tag{11.46}$$

Substituting these into eq. (11.45) and rearranging gives

$$[S]\,(=[H_3O^+]) = c_S^0(1-f)\frac{V_S^0}{V_S^0 + V_R} + \frac{1}{K_t[S]} \tag{11.47a}$$

24

which you should compare with eq. (11.7a). Making the substitution and rearrangement in slightly different ways would have yielded

$$[R](= [OH^-]) = c_S^0(f-1)\frac{V_S^0}{V_S^0+V_R} + \frac{1}{K_t[R]} \tag{11.47b}$$

and you should compare this with eq. (11.9b). Since all of the statements made in
deriving them are valid regardless of the value of f, eqs. (11.47) are true at every
point on the titration curve. There is no need to divide the curve into separate regions,
as there was with the approximate equations.

   In a titration like that of a weak acid HB with a base, the activity of the product
B^- varies as the titration proceeds, and the expression for K_t has a different form.
If a c_S^0 M aqueous solution of HB is titrated with sodium hydroxide, the electroneutrality equation is

$$[Na^+]+[H_3O^+] = [B^-]+[OH^-]. \tag{11.48}$$

Expressions for the concentrations of sodium and hydroxide ions can be taken from
eqs. (11.46), but the concentration of B^- is related to that of the unneutralized acid
HB by the conservation equation

$$[B^-]+[HB(aq)] = c_S^0 \frac{V_S^0}{V_S^0+V_R}. \tag{11.49}$$

   These equations can be combined in various ways. One approach is to combine
eqs. (11.48) and (11.46) to obtain an expression for $[B^-]$:

$$[B^-] = c_S^0 f\frac{V_S^0}{V_S^0+V_R}+[H_3O^+]-[OH^-] \tag{11.50a}$$

which can be compared with eq. (11.11a). Combining eqs. (11.49) and (11.50a) yields
an expressions for $[HB(aq)]$:

$$[HB(aq)] = c_S^0(1-f)\frac{V_S^0}{V_S^0+V_R}-[H_3O^+]+[OH^-] \tag{11.50b}$$

which can be compared with eq. (11.7a). Finally, eqs. (11.50) can be combined with
the expressions for K_a and K_w to give

$$[H_3O^+] = \frac{[HB(aq)]}{[B^-]}K_a = \frac{c_S^0(1-f)\dfrac{V_S^0}{V_S^0+V_R}-[H_3O^+]+K_w/[H_3O^+]}{c_S^0 f\dfrac{V_S^0}{V_S^0+V_R}+[H_3O^+]-K_w/[H_3O^+]}K_a \tag{11.51}$$

which, like eqs. (11.47), is true regardless of the value of f, and which makes it possible
to discern the conditions under which the much simpler equation obtained from
Table 11.2

$$[H_3O^+] = \frac{1-f}{f}K_a \qquad (0 < f < 1)$$

will be unsatisfactory even within the range for which it is intended.

Exact equations like eqs. (11.47) and (11.51) are most needed around the equivalence points of the titrations they represent. They are more useful in titrations made by instrumental techniques than in titrations made with indicators. The color-change interval of an indicator is so wide that it usually extends into the ranges of validity of the approximate equations, but in instrumental titrations the values of the measured quantity (pH, pM, potential, etc.) may be much more reproducible, and the approximate equations may be misleading because they tend to overestimate the random errors that might be attained.

SUMMARY: An exact equation for any titration curve can be obtained by combining electroneutrality and conservation equations with expressions for the equilibrium constants of all the reactions involved.

Problems

Answers to some of these problems are given on pages 520 – 522.

11.1. Plot the sigmoidal titration curve (pcH against f) for each of the following titrations:

(a) 1.00 M hydrochloric acid with 1.00 M sodium hydroxide.
(b) 0.0100 M hydrochloric acid with 0.0100 M sodium hydroxide.
(c) 1.00×10^{-4} M hydrochloric acid with 1.00×10^{-4} M sodium hydroxide.
Compare the curves.

11.2. Plot the sigmoidal titration curve (pcH against f) for the titration of 0.0100M sodium hydroxide with 0.0100 M hydrochloric acid and compare it with the curve obtained in Problem 11.1(b).

11.3. Plot the sigmoidal titration curve (pcH against f) for each of the following titrations:

(a) 0.100 M chloroacetic acid (ClCH$_2$COOH, pK_a = 2.87) with 0.100 M sodium hydroxide.

(b) 0.100 M p-nitrophenol $\left(O_2N-\!\!\left\langle\bigcirc\right\rangle\!\!-OH, pK_a = 7.15\right)$ with 0.100 M sodium hydroxide.

(c) 0.100 M phenol $\left(\left\langle\bigcirc\right\rangle\!\!-OH, pK_a = 10.00\right)$ with 0.100 M sodium hydroxide.
Compare the curves with each other and with Fig. 11.7 (which represents a similar titration of a monobasic weak acid for which pK_a = 5.00).

11.4. Plot both the sigmoidal titration curve (pcH against f) and the segmented titration curve ($[H_3O^+](V_S^0 + V_R)/V_S^0$ against f) for the titration of 0.100 M hydrochloric acid with an 0.100 M solution of each of the following bases (= R):

(a) ammonia (pK_a = 9.25 for ammonium ion).
(b) pyridine (pK_a = 5.22 for pyridinium ion).
Compare the sigmoidal curves with each other and with Fig 11.6(b), and compare the segmented curves with each other and with Fig. 11.1(b). Which curve do you think is better suited to finding the equivalence point of a titration of an unknown solution of hydrochloric acid?

11.5. Plot the sigmoidal titration curve (pcH against f) for the titration of 0.100 M acetic acid (pK_a = 4.75) with an 0.100 M solution of each of the following bases (= R):

(a) Triethylamine [(C$_2$H$_5$)$_3$N, pK_a = 10.72 for triethylammonium ion].
(b) Ammonia (pK_a = 9.25 for ammonium ion).
(c) Triethanolamine [N(CH$_2$CH$_2$OH)$_3$, pK_a = 7.76 for triethanolammonium ion].
Compare the curves with each other and with Fig. 11.7 (which represents a similar titration with sodium hydroxide).

11.6. Adsorption was ignored in plotting Fig. 11.8(a), but silver iodide does strongly adsorb iodide ions from solutions containing excess iodide, and it strongly adsorbs silver ions from solutions containing excess silver. What effects will these processes have on the titration curve?

11.7. Why is methyl red a better indicator than methyl orange for the titration of Fig 11.12?

11.8. For each of the titrations described in Problems 11.1 through 11.5., select the best of the indicators listed in Table 11.3 (p. 341) and estimate the systematic and random errors that will arise from its use.

11.9. In Section 11.9 it is shown that, in the titration of copper(II) with ethylenediaminetetraacetate using PAR as the indicator, the systematic and random errors are decreased by raising the pcH from 7 to 8. What danger does this change involve if copper(II) forms a complex with the basic constituent of the buffer system employed?

11.10. (a) A solution containing 0.100 M bromide ion and 0.100 M chloride ion is titrated with a solution containing 0.100 M silver ion. Plot the sigmoidal titration curve (pcAg against f, letting $f = 1$ at the bromide equivalence point and 2 at the chloride equivalence point), neglecting coprecipitation of silver chloride with the silver bromide formed in the first step.

(b) How would the curve be altered by the fact that silver chloride actually forms solid solutions with silver bromide?

11.11. (a) Derive an exact equation for the titration of a c_B^0 M solution of the difunctional acid H_2B with sodium hydroxide.

(b) Use the equation to obtain a plot of pcH against f for the titration of 50.0 cm^3 of 0.100 M succinic acid (HOOCCH$_2$CH$_2$COOH, p$K_{a, 1} = 4.20$, p$K_{a, 2} = 5.64$) with 0.200 M sodium hydroxide. Compare the plot with the one obtained by following the procedure described in Section 11.10.

CHAPTER 12

Errors in Scientific Measurements

12.1. Introduction

Three kinds of questions arise in chemistry or any other science. Some examples appeared in Chapter 1. There are some questions to which the answers are verbal and are obtained by observation. "Do these substances react?", "Is their reaction exothermic or endothermic?", and "What is the color of its product?" are questions of this sort. Once the answers to such questions have been discovered, the chemist often proceeds to questions of the second kind, to which the answers are numerical and are obtained by measurement. "What is the percentage of carbon in this substance?", "What is the value of the equilibrium constant of this reaction?", and "At what wavelength in the visible portion of the spectrum does this solution absorb radiation most strongly?" are questions of this sort. Sometimes their answers can be obtained directly; sometimes graphical or numerical analysis of the data is needed. Answering questions of the second kind often leads to asking questions of the third kind, to which the answers may again be verbal but are obtained by induction and inference. "What is the mechanism of this reaction?", "Why is its rate increased by the presence of a phenyl group in one of the reacting molecules?", and "Why does this substance absorb radiation of longer wavelengths than some other substance?" are questions of this sort. Most of them cannot be answered, or even asked, until numerical data are available, but they cannot be answered by any mechanical combination or manipulation of the data.

Psychologists, lawyers, and other students of human observation know that in everyday life it is unusual for different people to observe the same event in quite the same way. Nevertheless it is a fundamental tenet of science that different properly trained observers, working under properly controlled conditions, will obtain exactly the same answer to any properly framed question of the first kind. Questions of the third kind stand at the opposite extreme because it might always be possible to find different answers to them, and because there is no way to assess the probability that any particular answer is the correct one. Different mechanisms can always be found to account for any set of rate data; one mechanism may seem preferable to another, but you can never be completely sure that one mechanism is the right one, and you cannot ever say that it has a 90 or 99% chance of being the right one. Different theories might always be invented to explain any set of observations and numerical values.

At any particular stage there is a general consensus of belief that certain theories are likely to be true, as there once was in the phlogiston theory and as there now is in the atomic theory and the quantum theory. A theory enjoying that reputation is used to interpret and correlate the results of observation and measurement until something comes along that fundamentally conflicts with it.

Questions of the second kind are intermediate between these extremes. They differ from questions of the first kind in that different observers are not expected to obtain —and rarely do obtain—identical numerical answers. An example, and by no means an extreme one, appears in Table 12.1. These numbers have not shaken anyone's

TABLE 12.1 *The atomic weight of tantalum*

The first column gives the year in which the value given in the second column replaced the preceding one in the second column as the internationally accepted value of the atomic weight of tantalum. The values adopted between 1894 and 1953 were based on 16.000 0 for a mixture of all of the isotopes of oxygen in their naturally occurring abundance ratios; those adopted in 1961 and 1970 were based on 12.000 0. . . for the isotope C^{12}. The numbers in parentheses are recalculated to the the older basis to make comparison easier.

Year	Accepted value
1894	182.6
1897	182.84
1900	182.8
1903	183
1907	181
1909	181.0
1912	181.5
1931	181.4
1936	180.88
1953	180.95
1961	180.948 (= 180.955)
1970	180.947_9 (= 180.954_7)

belief in the meaningfulness and constancy of atomic weights, nor have the results in Fig. 12.1 convinced anyone that the speed of light is decreasing as time goes on; the variations are simply ascribed to errors of measurement. Duplicate measurements rarely give the same results even when made by the same observer using the same method and the same apparatus under the most carefully controlled conditions that can be achieved. Questions of the second kind differ from those of the third kind in that the reliability of a numerical result can be expressed quantitatively while that of an interpretation or theory cannot. We can say that there are 99 chances in 100 that the percentage of carbon in a compound lies between 19.75 and 19.80%, or that there are 9 chances in 10 that the value of a rate constant lies between 0.10 and 0.11 s^{-1}. There is no equivalent way to express our confidence in the correctness of a reaction mechanism or the general theory of relativity.

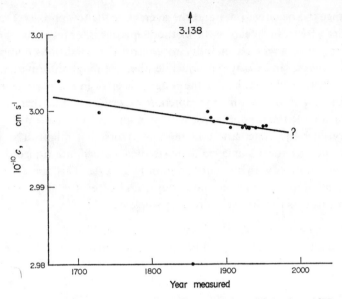

FIG. 12.1. Results of measurements of the velocity of light since 1675.

The measurement of numerical quantities plays a central role in science, and the degree of confidence that can be placed in a numerical result is as important as the result itself. A value that cannot be trusted is of little use in testing an existing theory, supporting a new one, or making a practical decision such as whether to construct a factory or prescribe a medication. This chapter deals with the sources of uncertainty in numerical results and with the confidence that can be reposed in such results.

12.2. Systematic and random errors

PREVIEW: A systematic error is the same in each of a series of identical measurements, while a random error differs from one measurement to another. The errors that may arise in a typical experiment are discussed and classified.

There are two kinds of errors in scientific measurements: *systematic errors* and *random errors*. These were briefly introduced in Section 11.7 and are discussed in more detail here.

Each kind of error can be defined in several different ways. In Section 11.7 it was said that a systematic error would be the same in each of a series of identical measurements, while both the sign and the magnitude of a random error would differ from one such measurement to another. Systematic errors affect the average of a series of measurements by causing each result to be either too high or too low; random errors cause some results to be too high and others to be too low, and can affect the average of a few results, but as the number of results increases the positive errors tend to counterbalance the negative ones and the average tends to approach the truth. Systematic errors arise from defects in the method, apparatus, or reagents; random errors arise from the experimenter's inability to control the conditions completely or to make the readings in an exactly reproducible way. If enough information were available it would be possible to calculate the systematic error in any particular experiment or to predict how it would differ in different experiments, but there is no way of predicting how large the random error will be in any one experiment.

Let us consider the systematic and random errors that might arise in a typical experiment, such as the volumetric analysis of a pure dilute aqueous solution of acetic acid by titration with standard sodium hydroxide. This is a relatively simple experiment: many others involve more steps, and often the steps are much more complex. In this experiment the individual steps are:

1. *Measurement of the volume of acetic acid taken.* If each sample is taken with the same volumetric pipet, there will be a systematic error unless the average volume delivered by the pipet is exactly the same as it is thought to be. It is tempting to trust the figure engraved on the pipet by the manufacturer. For a 25-cm³ pipet of the best quality that figure may be in error by as much as ± 0.025 cm³ at the temperature (usually 25°) at which the manufacturer expected the pipet to be used. The corresponding relative error is ± 0.025 cm³/25 cm³ $= \pm 1 \times 10^{-3}$ or $\pm 0.1\%$. The result might be in error by as much as 0.1% of the true value in either direction for this reason alone. The systematic error would be considerably larger if the pipet were used incorrectly (by adjusting the meniscus to the wrong point or by allowing insufficient time for drainage), if it had been maltreated (by exposing it to acidic fluoride solutions or very concentrated solutions of strong bases, which etch its interior, or by heating it, after which weeks or months may be required for it to return to its original volume), or if the viscosity of the solution were different from that of water.

There will also be random errors because the pipet cannot be filled to exactly the same point or drained in exactly the same way each time it is used. Even the most careful user working under the best conditions could not obtain exactly identical volumes of the sample for successive titrations. The differences among these volumes depend chiefly on the experimenter's skill and care: they might be as large as 0.1 or even 0.2 cm³ or as small as about 0.003 cm³. If the temperature of the solution being pipetted is allowed to vary, the random error will be increased by failure to control it, for an increase of temperature increases the volume of a pipet and also causes the solution to expand. For a soft glass pipet and a dilute aqueous solution, an increase of 1° in the temperature leads to an increase of 0.003% in the volume of the pipet and a decrease of 0.026% in the concentration of the solution, and thus to a decrease of 0.023% in the amount of solute that is taken.

2. *Addition of water.* If there is some carbon dioxide dissolved in the water there will be a systematic error because the carbon dioxide will be neutralized (to hydrogen carbonate ion) along with the acetic acid during the titration. Any variation in the

volume of water used or in the concentration of carbon dioxide will give rise to random errors, and so will any differences of procedure that allow different samples to take up different amounts of carbon dioxide while they are being titrated.

3. *Addition of indicator.* If the indicator is added in its acidic form there will be a positive systematic error because some sodium hydroxide will be consumed in converting part of it to the basic form. If the indicator is added in its basic form there will be a negative systematic error because it will react with the acetic acid to yield the acidic form and some of this will remain in the solution at the end point. Random error can arise from the use of different amounts of indicator in different titrations.

4. *Titration.* The systematic and random errors that are associated with the location of the end point were discussed in Section 11.7. In addition, if the buret is used in exactly the same way in successive titrations (say, by filling it to the 0.00-cm³ graduation before starting each titration), a systematic error will arise from any imperfection in the calibration of the buret. For a good 50-cm³ buret this error might be ± 0.05 cm³, which corresponds to a relative error of $\pm 5\%$ in measuring 1 cm³ of reagent, or to one of $\pm 0.1\%$ in measuring 50 cm³ of reagent. If the volumes of reagent used in different titrations are taken from different portions of the graduated part of the buret, the errors will be different in different titrations unless the buret has a uniform diameter and is uniformly graduated throughout its length. Then the errors will be random even though their magnitudes could be determined by calibrating the buret. There is a random error in each buret reading; it might be as small as ± 0.01 cm³ in careful work, or might be much larger if the readings are made carelessly, if the walls of the buret are greasy in spots, or if the stopcock occasionally leaks.

5. *Calculation.* Any error in the standardization of the reagent is a systematic error in this step because it affects all the results to the same extent. So is any algebraic error in the equation with which the data are combined in calculating the result. Mistakes in arithmetic are usually random errors, but can be made systematic by repeating them.

The list is incomplete but is long enough for our purposes. It indicates that you must usually do something additional to decrease a systematic error, but that doing something more carefully usually serves to decrease a random error. Pipets, burets, and other pieces of volumetric glassware can be calibrated, by weighing the water that they deliver or contain, to decrease the errors caused by relying on the calibrations engraved on them. The systematic error caused by the presence of carbon dioxide in the water used for dilution could be decreased by purifying the water, or by determining the concentration of the carbon dioxide dissolved in it and applying an appropriate correction. The systematic error caused by adding the indicator in either its acidic form or its basic form could be eliminated by neutralizing the solution of the indicator to the pH at the midpoint of the color-change interval. On the other hand, the random error involved in using a pipet can be decreased by taking greater care to ensure that it is clean, that the meniscus is brought to exactly the level of the mark, that there are no droplets of solution on the outside of the tip, that it is allowed to drain for exactly the same length of time whenever it is used, and so on. Whether or not an experimental result is worth the time and effort expended in obtaining it depends very largely on the care that has been taken to identify such sources of error and to minimize their effects.

Altering some detail of a procedure may affect both the systematic and the random errors. One might weigh the reagent that is used in a titration instead of measuring its volume with a buret. This would eliminate the systematic error associated with the calibration of the buret, and replace it with the systematic error resulting from imperfections in the balance and weights. It would also eliminate the random errors of buret reading, and replace them with the random errors of weighing. You would need to know something about both the buret and the weighing device in order to decide whether these changes would be for the better. Replacing a 50-cm^3 buret with an analytical balance should decrease both the systematic and the random errors, but replacing an 0.1-cm^3 microburet with a butcher's scale would make both much worse.

Random errors are much easier to detect than systematic ones. You can detect a random error by simply repeating a measurement, but mere repetition can never reveal the existence of a systematic error. Because this is true, systematic errors often escape notice, and many measurements are thought to be far more accurate than they really are. The sections that follow deal with the analysis of random errors, and assume that systematic errors have already been eliminated. There are three ways of eliminating systematic errors. In each it is necessary to begin by ferreting out every important source of systematic error, which is often the most difficult step. One procedure is to change what is done so that each error is eliminated or made negligible, which is often fairly easy (though it may be troublesome and time-consuming). In the analysis of an acetic acid solution one might calibrate the pipet, buret, and any other volumetric glassware that was used; exercise extraordinary care in standardizing the sodium hydroxide, and perhaps even purify the primary standard that is used; take special precautions to exclude acidic and basic impurities from the water and also from the atmosphere above the solution before and during the titration; and so on. This is the procedure that most chemists usually adopt. Another procedure is to convert the systematic errors into random ones by deliberately making them different in different experiments. Different samples of the acetic acid solution might be taken with different pipets, diluted with different volumes of water obtained from different sources, and titrated with different solutions of sodium hydroxide standardized in different ways and contained in different burets. In this way the systematic errors can be made amenable to the statistical analyses described below. Another is to subject a suitable standard reference material to the same procedure that will be used in analyzing the unknown, and compare the result with the certified value for the standard reference material. Standard reference materials similar in composition to many technically important samples are obtainable from the National Bureau of Standards in the U.S. and from the national standards laboratories of several other countries. Their certified values have been obtained from the results of analyses made by different skilled and careful analysts using different procedures believed to be reliable, and represent the closest approach to the truth that we know how to make. Either of the last two procedures yields an estimate of the certainty of the result that is far more reliable than a chemist's guess at the success of the measures that have been taken to minimize or eliminate systematic errors.

SUMMARY: In a series of identical experiments the systematic errors are the same and can be decreased by changing some detail of the procedure, while the random errors differ and can be decreased by exercising greater care.

12.3. Randomness and the normal law of error

PREVIEW: Random errors are usually assumed to be distributed according to the normal law of error. The properties of the normal error curve are discussed, and the sample standard deviation s is singled out as a measure of the magnitudes of the random errors in a sample of identical measurements. The relative sample standard deviation $s/\bar{x}$ is another measure of dispersion, and is useful in evaluating the precision that has been attained.

The remainder of this chapter is an introduction to statistics, which is the branch of mathematics that deals with random occurrences. Statistics is based on the theory of probability and on the idea that each group of experimental values of any particular kind is a *sample* drawn at random from an infinite number of similar values called the *population*. Note that a statistical sample, which is a group of numbers, is not the same as a chemical sample, which is a portion of the substance being studied or analyzed. In the remainder of this chapter the word "sample" will denote a statistical sample. Three analyses of a solution that really contains acetic acid at a total concentration of 0.100 00 M might give values of 0.1001, 0.0998, and 0.1003 M. The sample consists of these three numbers; the population includes all the numbers that would be obtained by performing the analysis an infinite number of times in exactly the same way. *In the absence of systematic errors*, numbers very close to 0.100 00 would be very common because some of the errors that tend to raise the result would often be nearly counterbalanced by others that tend to lower it. The frequency with which a number appears in the population would decrease as the deviation of the number from 0.100 00 increased. Numbers around 0.0995 or 0.1005 would be more common than those around 0.0980 or 0.1020, and numbers around 0.0900 or 0.1100 would be very rare unless the analyst was very unskilled or careless, or unless the procedure was very defective.

The frequency with which numbers appear in the population is usually assumed to be given by the *normal law of error*

$$\frac{\mathrm{d}N}{N} = \frac{1}{\sqrt{2\pi}\sigma} \exp\left[-\frac{(x-\mu)^2}{2\sigma^2} \right] \mathrm{d}x \tag{12.1}$$

where dN/N is the fraction of the population that consists of numbers in the range between x and $x+dx$, μ is the mean or average of all the numbers in the population, σ is the *standard deviation* of the population, and exp (a) means e^a. Performing a single measurement or analysis is equivalent to drawing a single number from this population, and the probability that the result will lie in the particular range between x' and $x'+dx$ is equal to the value of dN/N for $x = x'$.

Equation (12.1) can be rewritten in a more compact form by making the substitution

$$y = (x-\mu)/\sigma \tag{12.2}$$

which gives

$$\frac{dN}{N} = \frac{1}{\sqrt{2\pi}} \exp\left[-\frac{y^2}{2}\right] dy. \tag{12.3}$$

A plot of dN/Ndy against y is shown in Fig. 12.2 and is called the *normal error curve*. It is symmetrical around $y = 0$ (where $x = \mu$), which means that a positive deviation of any given magnitude is just as probable as a negative one of exactly the same magnitude.

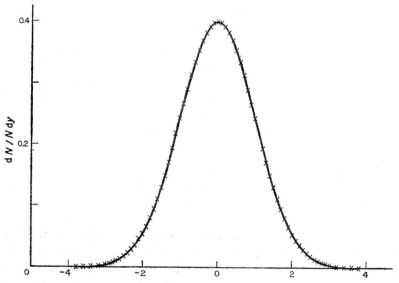

FIG. 12.2. The normal error curve.

The total area under the normal error curve is equal to 1, which means that any single value of y (or x) is certain to lie between $-\infty$ and $+\infty$. Less trivial conclusions are obtained by setting finite limits. The probability that any single value of y will lie between -1 and $+1$ is given by

$$\int_{y=-1}^{y=+1} \frac{dN}{N} = \frac{1}{\sqrt{2\pi}} \int_{-1}^{+1} \exp\left[-\frac{y^2}{2}\right] dy = 0.683$$

which means that 68.3% of the values of x in the population lie between $\mu-\sigma$ and $\mu+\sigma$. Of the remaining 31.7% that lie outside this range, half (15.8$_5$% of the total) are smaller than $\mu-\sigma$, and the rest are larger than $\mu+\sigma$. Table 12.2 lists the corresponding probabilities for ranges having other widths.

TABLE 12.2 *Probability that a single error will lie within or outside a specified range*

Range		Probability that a single value of y or x will lie	
of y	of x	within the range	outside the range
$-0.674 \leqslant y \leqslant +0.674$	$\mu - 0.674\,\sigma \leqslant x \leqslant \mu + 0.674\,\sigma$	0.500	0.500
$-1 \quad\ \leqslant y \leqslant +1$	$\mu - \quad\ \sigma \leqslant x \leqslant \mu + \quad\ \sigma$	0.683	0.317
$-1.645 \leqslant y \leqslant +1.645$	$\mu - 1.645\,\sigma \leqslant x \leqslant \mu + 1.645\,\sigma$	0.900	0.100
$-1.960 \leqslant y \leqslant +1.690$	$\mu - 1.960\,\sigma \leqslant x \leqslant \mu + 1.960\,\sigma$	0.950	0.050
$-2 \quad\ \leqslant y \leqslant +2$	$\mu - \quad 2\sigma \leqslant x \leqslant \mu + \quad 2\sigma$	0.955	0.045
$-2.576 \leqslant y \leqslant +2.576$	$\mu - 2.576\,\sigma \leqslant x \leqslant \mu + 2.576\,\sigma$	0.990	0.010
$-3 \quad\ \leqslant y \leqslant +3$	$\mu - \quad 3\sigma \leqslant x \leqslant \mu + \quad 3\sigma$	0.9974	0.0026
$-3.291 \leqslant y \leqslant +3.291$	$\mu - 3.291\,\sigma \leqslant x \leqslant \mu + 3.291\sigma$	0.9990	0.0010

The quantities that are most interesting are μ and σ: μ because it is the true result, and σ because it describes the precision of the measurements and the certainty with which the truth is known. Unfortunately, neither of these quantities can be calculated from the values in a sample of finite size. The average of those values provides an estimate, but only an estimate, of μ. That average is called the *sample mean*. It is represented the symbol $\bar{x}$ and is defined by the equation

$$\bar{x} = \frac{\sum_i x_i}{n} \tag{12.4}$$

in which x_i represents one of the individual values (x_1, x_2, x_3, etc.) of x, $\sum_i x_i$ denotes the sum of x_i for all the values of i (and is thus equal to $x_1+x_2+x_3+\ldots$), and n is the number of individual values in the sample. If three individual portions of a solution of acetic acid consume 25.00, 25.00, and 25.05 cm³ of sodium hydroxide, the sample mean is equal to $(25.00+25.00+25.05)/3 = 25.01_7$ cm³. As n increases, $\bar{x}$ tends to approach μ, and becomes identical with μ when n is infinitely large. A similar estimate of σ_x is given by the *sample standard deviation* s_x, which is defined by the equation

$$s_x = \sqrt{\frac{\sum_i (x_i - \bar{x})^2}{n-1}}. \tag{12.5}$$

Symbols like σ_x will be used henceforth to represent the standard deviation of the quantity identified by the subscript. For the same sample of values of the volume of sodium hydroxide consumed in titrating portions of a solution of acetic acid, the sample standard deviation is given by

$$s_x = \sqrt{[(25.00-25.017)^2+(25.00-25.017)^2+(25.05-25.017)^2]/2}$$
$$= \sqrt{[2.89\times10^{-4}+2.89\times10^{-4}+1.09\times10^{-3}]/2} = 2.9\times10^{-2} \text{ cm}^3.$$

As n increases, s_x tends to approach σ_x, and becomes identical with it when n is infinitely large.

The quantity $n-1$ in eq. (12.5) is called the number of *degrees of freedom* and is equal to the number of independent values in the sample. There are n values altogether, but they are not all independent because their mean $\bar{x}$ appears elsewhere in the equation. The fact that $\bar{x}$ is already known removes one degree of freedom: $n-1$ values could now be chosen independently, but then they and the value of $\bar{x}$ together would fix the last value.

The accuracy of a set of results is the difference between $\bar{x}$ and the true value, which would be equal to μ if there were no systematic error. The standard deviation σ_x and the sample standard deviation s_x describe the dispersion, scatter, or precision of the values, and are called *measures of dispersion*. There are other measures of dispersion. The mean deviation, which is defined by

$$\text{mean deviation} = \frac{\sum\limits_i (x_i-\bar{x})}{n}$$

is easy to calculate and very popular. Large values of $x_i-\bar{x}$ have much less effect on the mean deviation than they do on s. Their effect on s is large because a large deviation from the mean is very improbable unless the dispersion is large; because their effect on the mean deviation is smaller, the mean deviation tends to underestimate the dispersion and thus to give an excessively favorable impression of the precision that has been achieved. The sample standard deviation gives a much more accurate and honest picture.

Other useful measures of dispersion are the *relative standard deviation* and the *relative sample standard deviation*, of which the latter is often called the *coefficient of variation*. These are defined by the equations

$$\text{relative standard deviation} \qquad = \sigma_x/\mu, \qquad (12.6a)$$
$$\text{relative sample standard deviation} = s_x/\bar{x}. \qquad (12.6b)$$

Their importance arises from the difficulty of interpreting an isolated value of σ_x or s_x. Should a chemist be satisfied with the value $s_x = 2.9\times10^{-2}$ cm^3 for the volume of sodium hydroxide consumed in titrating a portion of a solution of acetic acid? The answer depends on the volume of sodium hydroxide that is used. If this is 100 cm^3, so that the relative sample standard deviation is equal to 2.9×10^{-4} or 0.029%, it is not easy to imagine purposes for which better precision would be needed. If only 0.03 cm³ of sodium hydroxide is used, the same value of s_x corresponds to a relative sample standard deviation of almost 100%, and it is not even possible to say with reasonable assurance that there is any acetic acid in the solution at all.

The *variance* σ_x^2 and the *sample variance* s_x^2 are important in statistical theory but will be avoided in these pages.

SUMMARY: The sample standard deviation $s_x\left(=\sqrt{\sum_i (x_i - \bar{x})^2/(n-1)}\right)$ of a set of results of identical measurements, and the relative sample standard deviation $s_x/\bar{x}$, are the most useful measures of dispersion. If the random errors are distributed in accordance with the normal law of error, the sample standard deviation makes it possible to predict how often a random error of any given magnitude will occur, while the relative sample standard deviation permits making qualitative judgments of the precision attained.

12.4. The standard deviation of the mean

PREVIEW: This section shows how the uncertainty in the average of a set of results is related to the sample standard deviation, and defines the confidence interval.

A single sample of n experimental values contains n numbers drawn at random from the population. Each particular sample of n values has its own mean, $\bar{x}$, but different samples will have different values of $\bar{x}$. For an infinite number of samples, each containing n individual values, the standard deviation of $\bar{x}$ is given by

$$\sigma_{\bar{x}} = \sigma_x/\sqrt{n}. \tag{12.7}$$

Among other things, this means that the values of $\bar{x}$ agree more closely for large samples than for small ones.

In Section 12.3 it was said that the mean of an infinite number of individual measurements of some quantity x will be equal to the true value μ—*in the absence of systematic errors*—and that (Table 12.2) there is a 95% chance that any single value of x will lie between $\mu - 1.96\sigma_x$ and $\mu + 1.96\sigma_x$. Exactly identical statements can be made about any quantity, such as $\bar{x}$. The mean of an infinite number of values of $\bar{x}$, obtained from an infinite number of samples of which each contains n individual values of $\bar{x}$, will be equal to the true value μ. There is a 95% chance that any single value of $\bar{x}$ will lie between $\mu - 1.96\sigma_{\bar{x}}$ and $\mu + 1.96\sigma_{\bar{x}}$. For any one sample we could be 95% sure that $\bar{x}$ lies between these limits—or, in view of eq. (12.7) and in slightly different words,

$$\mu = \bar{x} \pm 1.96\sigma_{\bar{x}}/\sqrt{n} \quad \text{at the 95\% level of confidence.} \tag{12.8a}$$

Other similar statements are easy to construct with the aid of Table 12.2. Since 99% of the values of $\bar{x}$ will lie between $\mu - 2.58\sigma_{\bar{x}}$ and $\mu + 2.58\sigma_{\bar{x}}$,

$$\mu = \bar{x} \pm 2.58\sigma_{\bar{x}}/\sqrt{n} \quad \text{at the 99\% level of confidence.} \tag{12.8b}$$

A decision that *(in the absence of systematic errors)* the true value lay within $2.58\sigma_{\bar{x}}/\sqrt{n}$ of the average for a sample of n individual values would have a 99% chance of being correct.

The value of σ_x is never known exactly, and the value of s_x gives only an approximation to it. Consequently eq. (12.7) cannot be used as it stands, but the probable difference between s_x and σ_x may be taken into account by writing

$$\mu = \bar{x} \pm ts_{\bar{x}} = \bar{x} \pm ts_x/\sqrt{n}. \tag{12.9}$$

The range of values of μ given by this equation is called its *confidence interval*. Values of the quantity t, often called the "Student t", are given in Table 12.3. As n increases and s_x approaches σ_x, the values of t approach the numerical coefficients appearing in eqs. (12.8).

TABLE 12.3 *Values of t for various levels of confidence*

Number of degrees of freedom used in calculating s	Level of confidence			
	90%	95%	99%	99.5%
1	6.31	12.71	63.7	127.3
2	2.92	4.30	9.92	14.1
3	2.35	3.18	5.84	7.45
4	2.13	2.78	4.60	5.60
5	2.02	2.57	4.03	4.77
6	1.94	2.45	3.71	4.32
7	1.89	2.37	3.50	4.03
8	1.86	2.31	3.36	3.83
10	1.81	2.23	3.17	3.58
∞	1.645	1.960	2.576	2.807

Examples 12.1 and 12.2 illustrate the use of this important equation and point out some of the consequences of all these ideas.

Example 12.1. Three 25.00-cm³ portions of a solution of acetic acid were titrated with a solution of sodium hydroxide. The volumes of sodium hydroxide consumed were 25.00, 25.00, and 25.05 cm³. At the 95% level of confidence, within what range would the average of a very large number of similar results lie?

Answer. These numbers appeared in Section 12.3, where it was found that $\bar{x} = 25.017$ cm³ and $s_x = 2.9 \times 10^{-2}$ cm³. With two degrees of freedom (as in the calculation of s_x) and at the 95% level of confidence, Table 12.3 gives $t = 4.30$. Hence the 95% confidence interval is

$$\mu = 25.017 \pm (4.30)(2.9 \times 10^{-2})/\sqrt{3} = 25.017 \pm 0.072 \text{ cm}^3.$$

It is 95% certain that a large number of similar titrations would yield a mean lying between 24.945 cm³ and 25.089 cm³. The width of this range is almost three times as large as the difference between the highest and lowest of the experimental values. You do not obtain much certainty from a small number of measurements.

Example 12.2. In three more titrations like those in Example 12.1, the volumes of sodium hydroxide consumed were 25.05, 25.02, and 24.98 cm^3. On the basis of all six results, what can be said about the mean value that would be obtained from a very large number of titrations?

Answer. For all six results $\bar{x} = 25.017$ cm^3 and $s_x = 2.9 \times 10^{-2}$ cm^3. It is a coincidence that both are the same as in Example 12.1. Five degrees of freedom were used in calculating this value of s_x, and Table 12.3 gives $t = 2.57$ at the 95% level of confidence, so that the 95% confidence interval is

$$\mu = 25.017 \pm (2.57)(2.9 \times 10^{-2})/\sqrt{6} = 25.017 \pm 0.030 \text{ cm}^3.$$

If there are no systematic errors, it is now 95% certain that the true value lies between 24.987 cm^3 and 25.047 cm^3. Doubling the number of data has substantially decreased the width of this range even though the value of s_x is actually slightly larger for the second sample of three results than for the original one.

In the same way it can be concluded that the 99.5% confidence interval is now 25.017 ± 0.056 cm^3. It is therefore well over 99.5% certain that the true value lies inside the range, 25.017 ± 0.072 cm^3, obtained at the 95% level of confidence in Example 12.1. Doubling the number of data has decreased the probability that the truth lies outside that range from 5% to something under 0.5%.

SUMMARY: The standard deviation of the mean, $s_{\bar{x}}$, depends on the sample standard deviation s_x and on the number n of measurements that have been made. Knowing the value of $s_{\bar{x}}$ for a set of measurements enables us to find the confidence interval, which is the range within which the average of a large number of similar measurements (which would coincide with the truth if there were no systematic errors) can be said with any desired degree of confidence to lie.

12.5. The propagation of random errors

PREVIEW: This section shows how the standard deviation of a calculated result R depends on the standard deviations of the quantities $A, B, \ldots$ from which it is calculated.

The preceding section dealt with four interrelated quantities—the number of measurements that have been made, the sample standard deviation, the width of the range within which the true value can be considered to lie, and the certainty that it does actually lie within that range—and with the way in which the first two of these govern the others. One naturally wishes to be as nearly sure as possible that the truth lies within the narrowest possible range. As was shown by Examples 12.1 and 12.2, increasing the number of measurements is one way to satisfy this wish. However, there comes a point at which little further improvement is produced by any reasonable amount of additional work. If n is already fairly large, making it still larger has relatively little effect on either t or $\sqrt{n}$ in eq. (12.9).

The other possible approach is of course to decrease the sample standard deviation s_x. In order to do this efficiently, the chemist must know how s_x depends on the magnitudes of the individual random errors that are combined in it. In titrations of acetic acid with sodium hydroxide, the sample standard deviation $s_{V_{\text{NaOH}}}$ arises from variations among the volumes of acetic acid taken for different titrations, from the uncertainty associated with locating an end point by means of an indicator, and from random errors in the buret readings. Other sources of random error will be ignored in this discussion.

When a result R is some function of two or more independent quantities A, B, ..., the standard deviation of R is related to the standard deviations of A, B, ... by the equation

$$\sigma_R^2 = \left(\frac{\partial R}{\partial A}\right)^2 \sigma_A^2 + \left(\frac{\partial R}{\partial B}\right)^2 \sigma_B^2 + \cdots \tag{12.10a}$$

or, for a finite number of measurements,

$$s_R^2 = \left(\frac{\partial R}{\partial A}\right)^2 s_A^2 + \left(\frac{\partial R}{\partial B}\right)^2 s_B^2 + \cdots \tag{12.10b}$$

Expressions for the partial derivatives in parentheses are obtained by differentiating R with respect to the quantity in question while holding all of the other quantities constant.

For example, if $R = A \pm B$, differentiation with respect to A while holding B constant gives $\partial R/\partial A = 1$, while differentiation with respect to B while holding A constant gives $\partial R/\partial B = \pm 1$. Combining these with eq. (12.10b) yields

$$s_R^2 = s_A^2 + s_B^2 \quad \text{if} \quad R = A \pm B. \tag{12.11}$$

If $R = AB$, then $\partial R/\partial A = B \,(= R/A)$ while $\partial R/\partial B = A \,(= R/B)$, and the result is

$$s_R^2 = \left(\frac{R}{A}\right)^2 s_A^2 + \left(\frac{R}{B}\right)^2 s_B^2.$$

Exactly the same result is obtained if $R = A/B$, for then $\partial R/\partial A = 1/B \,(= R/A)$ while $\partial R/\partial B = -A/B^2 \,(=-R/B)$. In either of these cases a more convenient expression is

$$\left(\frac{s_R}{R}\right)^2 = \left(\frac{s_A}{A}\right)^2 + \left(\frac{s_B}{B}\right)^2 \quad \text{if} \quad R = AB \quad \text{or} \quad A/B. \tag{12.12}$$

Some other consequences of eqs. (12.10) are discussed briefly in Appendix I.

> SUMMARY: Equations (12.10) describe the relationship between the standard deviation (or sample standard deviation) of one quantity R and the standard deviations of the quantities from which R is calculated.

12.6. The design of an experiment

> PREVIEW: This section shows how the design and improvement of an experiment are influenced by the precision that is desired.

This section consists of a single lengthy example that illustrates the use of a number of the ideas presented in the preceding sections. It deals with the experiments that gave the results used in Examples 12.1 and 12.2. Portions of an 0.100 M solution of acetic acid are taken with a 25-cm³ pipet and titrated with a standard 0.100 M solution of acetic acid, and the concentration of acetic acid is calculated from the equation $c_{HOAc} = (V_{NaOH}/V_{HOAc}) c_{NaOH}$. It is convenient to rewrite this as $c_{HOAc} = R c_{NaOH}$, where $R = V_{NaOH}/V_{HOAc}$. The values of V_{NaOH} and V_{HOAc} will be different in different titrations, and their variations will affect the precisions of c_{HOAc} and R. An error in c_{NaOH} will not affect the precision of R, but will affect the overall uncertainty of c_{HOAc}.

The standard deviation of V_{HOAc} will be that associated with the use of a pipet, and we shall estimate it to be $\sigma_{V_{HOAc}} = 0.01$ cm³. According to Table 12.2, this means that 99% of the volumes of an infinite number of portions of a solution taken with the same pipet would be within $2.576 \times 0.01 = 0.026$ cm³ of their mean. A beginner might have some difficulty in achieving this standard deviation, but an experienced and careful chemist could easily attain a smaller one.

The standard deviation of V_{NaOH} is slightly less easy to estimate because there are two things that affect it. Any error in reading the buret, either at the beginning or at the end of the titration, will cause the volume recorded in the chemist's notebook to differ from the volume actually used. If this were the only source of error, we would have $V_{NaOH} = V_f - V_i$, where V_f and V_i are the final and initial readings of the buret, and

$$\sigma^2_{V_{NaOH}} = \sigma^2_{V_f} + \sigma^2_{V_i} = 2\sigma^2_r.$$

where σ_r represents the standard deviation of a single buret reading. In addition, the volume V_{titrn} of sodium hydroxide actually used will vary from one titration to another because different titrations are stopped at different pH-values. If this were the only source of error we would have

$$\sigma^2_{V_{NaOH}} = \sigma^2_{V_{titrn}}.$$

The combination of these two completely independent sources of error gives

$$\sigma^2_{V_{NaOH}} = 2\sigma^2_r + \sigma^2_{V_{titrn}}. \tag{12.13}$$

These two components of $\sigma_{V_{NaOH}}$ can now be estimated separately. Experience indicates that σ_r is roughly 0.015 cm³ in typical student work, meaning that half of the readings are within about $0.674 \times 0.015 = 0.01$ cm³ of the truth, and that 90% of them are within about $1.645 \times 0.015 = 0.025$ cm³ of the truth. An estimate of $\sigma_{V_{titrn}}$ may be obtained from calculations like those in Section 11.7. If 0.100 M acetic acid is titrated with 0.100 M acetic acid, the equivalence point lies at pcH = 8.7. An indicator having the midpoint of its transition interval at pcH = 8.7 will begin to change color at pcH = 7.9 and its color change will end at pcH = 9.5. The titration errors at these two points are -0.070% and $+0.063\%$, respectively. In principle any relative error larger than $\pm 0.07\%$ should be impossible, but one will arise occasionally because of inattention, overanxiety, or the addition of a droplet that is slightly too large just before the end point is reached. To be conservative let us suppose that 90% of the values of V_{titrn} will be within 0.07% of the truth. Then the relative standard deviation of V_{titrn}, which is equal to $\sigma_{V_{titrn}}/V_{titrn}$, will be equal to $7 \times 10^{-4}/1.645 = 4.3 \times 10^{-4}$. The factor 1.645 comes from the third line of Table 12.2. Since $V_{titrn} = 25$ cm^3 under the conditions being assumed here, $\sigma_{V_{titrn}}$ will be equal to $25 \times 4.3 \times 10^{-4} = 1.1 \times 10^{-2}$ cm^3. Combining these estimates of σ_r and $\sigma_{V_{titrn}}$ with eq. (12.13), we obtain

$$\sigma^2_{V_{NaOH}} = 2(0.015)^2 + (0.011)^2 = 4.5 \times 10^{-4} + 1.2 \times 10^{-4} = 5.7 \times 10^{-4} \quad (12.14a)$$

so that $\sigma_{V_{NaOH}} = 2.4 \times 10^{-2}$ cm^3.

We have estimated that $\sigma_{V_{HOAc}} = 0.01$ cm^3 and that $\sigma_{V_{NaOH}} = 0.024$ cm^3. Since both V_{HOAc} and V_{NaOH} are assumed to be equal to 25 cm^3, eq. (12.12) gives

$$\frac{\sigma_R}{R} = \sqrt{\left(\frac{\sigma_{V_{HOAc}}}{V_{HOAc}}\right)^2 + \left(\frac{\sigma_{NaOH}}{V_{NaOH}}\right)^2}$$

$$= \sqrt{\left(\frac{0.01}{25}\right)^2 + \left(\frac{0.024}{25}\right)^2} = \sqrt{1.6 \times 10^{-7} + 9.2 \times 10^{-7}} = 1.04 \times 10^{-3}. \quad (12.14b)$$

Since $R = V_{NaOH}/V_{HOAc} = 1$, σ_R will be equal to 1.04×10^{-3}.

There are now two questions that can be asked:

1. Have all the sources of random error been taken into account?
2. How can the precision be improved?

You can answer the first of these by comparing the estimated value of σ with the values of s obtained when the experiment is actually performed. Here the estimate of σ_R is 1.0×10^{-3} while the data quoted in Examples 12.1 and 12.2 give $s_R = 1.2 \times 10^{-3}$. Such agreement is unusual. You must guess at the values of many of the quantities included in σ, you will sometimes overlook significant sources of random error, and the value of s for a small sample is likely to be quite different from the value of σ for the population from which the sample is drawn. Sometimes s is found to exceed σ by an order of magnitude, or even more. Such a discrepancy usually means that the most important random errors in the experiment have been overlooked. Until you have found them, you cannot hope to improve the results significantly, and therefore you must analyze each step very carefully to make sure that you discover all of the important sources of error.

Once you have done this you can go on to the second question. Make sure that the precision really needs to be improved before you begin to improve it! Do not, in other words, attempt to reach the point where you can be 99.5 per cent sure that the concentration of acetic acid in a solution is between 0.100 06 M and 0.100 08 M if the eventual purpose is merely to decide whether there is any acetic acid in the solution or not.

The general approach is simple. Find the largest random error in the experiment and change the procedure so as to decrease it. If you can make it smaller than some other one, switch your attention to that one and do something to decrease it. Stop when you have decreased the expected value of σ to a satisfactory level.

Evidently you must begin by deciding what a "satisfactory level" is. Let us arbitrarily assume that you need a value for the ratio $R = V_{NaOH}/V_{HOAc}$ that is 99.5 per cent certain to be within 0.10 per cent of the truth, and that you are willing to perform no more than eight titrations to obtain it. Your requirement is that $ts_R/\sqrt{n} \le 1 \times 10^{-3} R$. With $n = 8$, and therefore 7 degrees of freedom, t will be equal to 4.03 at the 99.5% level of confidence, and so you must strive to obtain

$$s_R \le 1 \times 10^{-3} R\sqrt{n}/t = (1 \times 10^{-3})(1)\sqrt{8}/4.03 = 7.0 \times 10^{-4}$$

or (since $R = 1$) $s_R/R = 7.0 \times 10^{-4}$.

Equation (12.14b) shows that the uncertainty in V_{HOAc} has only a negligible effect on σ_R/R. Even if you could eliminate that uncertainty altogether you would still have $\sigma_R/R = \sqrt{9.2 \times 10^{-7}} = 9.6 \times 10^{-4}$, which represents only an insignificant decrease. Improving the precision with which the portions of acetic acid are measured would therefore be a waste of time. The same thing is true of the uncertainty in V_{titrn}: eliminating this in eq. (12.14a) would decrease $\sigma_{V_{NaOH}}$ only from 0.024 cm^3 to 0.021 cm^3. The largest random errors in the procedure are those involved in reading the buret, and these are the ones that must be decreased in order to improve the precision significantly.

How much must they be decreased? If s_R/R cannot exceed 7.0×10^{-4} and if $\sigma_{V_{HOAc}}/V_{HOAc}$ remains unchanged at $4 \times 10^{-4} (= 0.01/25)$, eq. (12.14b) suggesrs that you should aim for

$$\frac{\sigma_{V_{NaOH}}}{V_{NaOH}} = \sqrt{(7.0 \times 10^{-4})^2 - (4 \times 10^{-4})^2} = 5.7 \times 10^{-4}$$

or $\sigma_{V_{NaOH}} = 25 \times 5.7 \times 10^{-4} = 0.014$ cm^3. If $\sigma_{V_{titrn}}$ remains unchanged at 0.011 cm^3 eqs. (12.13) and (12.14a) in turn suggest that you need

$$\sigma_r = \sqrt{\frac{(0.014)^2 - (0.011)^2}{2}} = 0.006 \text{ cm}^3.$$

It would be very difficult, if not impossible, to achieve such precision with a 50-cm^3 buret: it would be much easier to add the sodium hydroxide from a container that you could weigh at the start of the titration and then again when the end point had been reached. A standard deviation of 6 mg or less would have to be attained in the weighings, but no great difficulty would be involved in attaining it.

In this example you could achieve the desired end by making a single change in the original procedure. Often two or more changes are necessary. If the required value of s_R were 5×10^{-4}, you could not quite achieve it even by decreasing σ_r to zero, and then the next thing to tackle would be the precision of locating the end point because this is the second most important source of error. The analysis is useful because it helps you to avoid wasting time on unimportant details while you overlook major problems.

So far there has been only passing mention of the further uncertainty that is introduced by combining the value of R with that of c_{NaOH} to find the concentration of the acetic acid. Since $c_{HOAc} = R c_{NaOH}$, the relative sample standard deviation of c_{HOAc} is given by

$$\left(\frac{s_{c_{HOAc}}}{c_{HOAc}}\right)^2 = \left(\frac{s_R}{R}\right)^2 + \left(\frac{s_{c_{NaOH}}}{c_{NaOH}}\right)^2.$$

Suppose that six standardizations of the sodium hydroxide gave $c_{NaOH} = 0.1001_3$, 0.1000_2, 0.1002_1, 0.1000_8, 0.1002_2, and 0.1001_0 M. Then $\bar{c}_{NaOH} = 0.100\,13\,M$, $s_{\bar{c}_{NaOH}} = 7.7 \times 10^{-5}$, and the overall relative sample standard deviation of c_{HOAc}, on the basis of the six titrations in Examples 12.1 and 12.2, is

$$\frac{s_{c_{HOAc}}}{c_{HOAc}} = \sqrt{(1.2 \times 10^{-3})^2 + \left(\frac{7.7 \times 10^{-5}}{0.100\,13}\right)^2} = \sqrt{1.4 \times 10^{-6} + 6 \times 10^{-7}} = 1.4 \times 10^{-3}.$$

For these data $\bar{R} = 1.000\,68$, and therefore $\bar{c}_{HOAc} = 1.000\,68 \times 0.100\,13 = 0.100\,20\,M$ and $s_{c_{HOAc}} = 1.4 \times 10^{-3} \times 0.100\,20 = 1.4 \times 10^{-4}\,M$. With 5 degrees of freedom and at the 95% level of confidence, so that $t = 2.57$, the concentration of the acetic acid can be said to lie in the range $0.100\,20 \pm (2.57)(1.4 \times 10^{-4})/\sqrt{6} = 0.100\,20 \pm 0.000\,15\,M$.

On an earlier page we asked what would have to be done to obtain, in eight titrations, a value of R that was 99.5% certain to be within 0.1% of the truth. We can now ask a more important question: what would have to be done to obtain a value of c_{HOAc} that was 99.5% certain to be within 0.1% of the truth? You would need $t s_{c_{HOAc}}/\sqrt{n} \leqslant 1 \times 10^{-3} c_{HOAc}$, or $(s_{c_{HOAc}}/c_{HOAc}) \leqslant 1 \times 10^{-3} \sqrt{n}/t$. If you are willing to do only eight titrations, so that $t = 4.03$, the value of $s_{c_{HOAc}}/c_{HOAc}$ cannot exceed $1 \times 10^{-3}\sqrt{8}/4.03 = 7.0 \times 10^{-4}$. The largest error in the experiment is still the one in reading the buret, but even by eliminating all of the random errors in R you could not decrease $s_{c_{HOAc}}/c_{HOAc}$ below $7.7 \times 10^{-4} (= s_{c_{NaOH}}/c_{NaOH})$. Both the precision of the titration and the precision of the standardization would have to be improved.

> SUMMARY: The standard deviation of the result of an experiment depends on the standard deviations of all of the quantities that were combined in calculating it, and does so in such a way that the quantity having the largest standard deviation is the one to which most attention should be paid in trying to improve the overall precision.

12.7. Comparing results

> PREVIEW: This section explains how you can decide, at any desired level of confidence, whether there is a real difference between the sample standard deviations or means of two independent sets of results.

When a measurement is repeated under different conditions, with different apparatus, with a different procedure, or by a different experimenter, there is very little chance that the same result will be obtained. Two sets of measurements may actually have different precisions or yield different results, or may merely appear to do so because individual random errors have combined in different ways. It is often important to know whether the differences between the values of s_x and $\bar{x}$ are real or not—or, in statistical language, whether the samples are drawn from the same population or not.

The F test is a comparison of two sample standard deviations. A value of the quantity F is calculated from the equation

$$F = s_N^2/s_D^2 \tag{12.15}$$

where s_N, the sample standard deviation that appears in the numerator, exceeds s_D, the one that appears in the denominator. It is assumed that the user has no prior knowledge of which of the two sample standard deviations will become s_N and which will become s_D. The value of F is compared with that taken from a table such as Table 12.4. At the stated level of confidence, the calculated value of F will not exceed

TABLE 12.4. *Values of F at the 90 per cent level of confidence*

Number of degrees of freedom used in calculating s_D	Number of degrees of freedom used in calculating s_N				
	2	3	4	5	6
2	19.00	19.16	19.25	19.30	19.33
3	9.55	9.28	9.12	9.01	8.94
4	6.94	6.59	6.39	6.26	6.16
5	5.79	5.41	5.19	5.05	4.95
6	5.14	4.76	4.53	4.39	4.28

the tabulated one if the two samples are drawn from the same population. Table 12.4 pertains to the 90% level of confidence; more extensive tables, including other levels of confidence, are available in the statistical literature and in some handbooks. If a calculated value of F is smaller than the one found in Table 12.4, it can be said, at the 90% level of confidence, that there is no real difference between the two values of s being compared.

Example 12.3. Two different chemists analyze a solution of acetic acid and obtain the following results. Are their precisions significantly different?

Chemist A	0.1009	0.1006	0.1004	0.1010 M	
Chemist B	0.1003	0.1003	0.1005	0.1004	0.1006 M
Chemist A	$\bar{x}_A = 0.1007_3\ M$;		$s_A = 2.75 \times 10^{-4}\ M$		
Chemist B	$\bar{x}_B = 0.1004_2\ M$;		$s_B = 1.30 \times 10^{-4}\ M$		

Answer. Since $s_A > s_B$ we must let $s_A = s_N$ and $s_B = s_D$ in eq. (12.15). Then $F = (2.75 \times 10^{-4})^2/(1.30 \times 10^{-4})^2 = 4.48$. Three degrees of freedom were used in calculating s_N and four were used in calculating s_D; the value of F obtained from the Table is 6.59. The calculated value of F is smaller than the tabulated one. At (and actually well above) the 90% level of confidence, there is no difference between the precisions of the two sets of results.

You can use an F test to answer either of two questions. One is the question asked in Example 12.3: "Is there a real difference between the precisions?" When you ask this question you have no preconceived idea about whether s_A is larger or smaller than s_B, and do not know in advance which of the two you will call s_N and which you will call s_D. When the test is applied in this way it is said to be a *two-sided* or *two-tailed* test. The other question might take the form "Did chemist B achieve a better precision than chemist A?" Here you are testing the expectation that $s_B < s_A$, and plan to let $s_N = s_A$ and $s_D = s_B$ even before you see the numerical values of s_A and s_B. When the test is applied in this way it is called a *one-sided* or *one-tailed* test. If a two-sided test indicates that s_A and s_B are different, either one of them might be larger; in a one-sided test one of the two possibilities is dismissed in advance, and this has the effect of changing the level of confidence from 90% to 95% for the values in Table 12.4.

Another important question is whether the means of two sets of measurements are or are not significantly different. It can be answered by employing the t test. The standard deviation of $\bar{x}_A$ is equal to $\sqrt{s_A^2/n_A}$, while that of $\bar{x}_B$ is equal to $\sqrt{s_B^2/n_B}$ (compare eq. (12.7)). The standard deviation of the difference $\bar{x}_A - \bar{x}_B$ is given by eq. (12.11):

$$s_{\bar{x}_A - \bar{x}_B} = \sqrt{s_{\bar{x}_A}^2 + s_{\bar{x}_B}^2}. \tag{12.16}$$

Equation (12.9) might be rewritten as

$$t = \pm \frac{\bar{x} - \mu}{s_x/\sqrt{n}}. \tag{12.17a}$$

The numerator of the right-hand side is the difference between two means. One, $\bar{x}$, is the mean for a finite sample; the other, μ, is the mean for the population. The denominator is really the standard deviation of the difference between these, which might be described by writing $s_{\bar{x}-\mu} = \sqrt{s_{\bar{x}}^2 + s_\mu^2}$: since $s_{\bar{x}} = s_x/\sqrt{n}$ and since there is no uncertainty in μ by definition, we obtain $s_{\bar{x}-\mu} = s_x/\sqrt{n}$. The $\pm$ sign is needed to make t positive regardless of whether $\bar{x}$ is larger or smaller than μ. In short, eq. (12.17a) is equivalent to

$$t = \left| \frac{\bar{x} - \mu}{s_{\bar{x}-\mu}} \right|. \tag{12.17b}$$

In exactly the same way, the value of t for the difference between the two means $\bar{x}_A$ and $\bar{x}_B$ is given by

$$t = \left| \frac{\bar{x}_A - \bar{x}_B}{s_{\bar{x}_A - \bar{x}_B}} \right| = \left| \frac{\bar{x}_A - \bar{x}_B}{\sqrt{s_{\bar{x}_A}^2 + s_{\bar{x}_B}^2}} \right|. \tag{12.18}$$

The value of t obtained from eq. (12.18) should be compared with one taken from a table for n_A+n_B-2 degrees of freedom and the desired level of confidence. For the data in Example 12.3, where $\bar{x}_A = 0.100\ 73$, $n_A = 4$, $s_{\bar{x}_A} = 2.75 \times 10^{-4}/\sqrt{4} = 1.37 \times 10^{-4}$, $\bar{x}_B = 0.100\ 42$, $n_B = 5$, and $s_{\bar{x}_B} = 1.30 \times 10^{-4}/\sqrt{5} = 5.81 \times 10^{-5}$, eq. (12.18) becomes

$$t = \left| \frac{0.100\ 73 - 0.100\ 42}{\sqrt{(1.37 \times 10^{-4})^2 + (5.81 \times 10^{-5})^2}} \right| = 2.08.$$

There are $4+5-2 = 7$ degrees of freedom, and Table 12.3 gives $t = 1.89$ at the 90% level of confidence and $t = 2.37$ at the 95% level of confidence. The probability is higher than 90%, but not as high as 95%, that there is a real difference between the two means.

> SUMMARY: The F test is a way of deciding whether the precisions of two sets of results are significantly different. It is based on the ratio of the squares of the two sample standard deviations. The t test provides a similar way of deciding whether the means are significantly different.

12.8. Regression analysis

> PREVIEW: Many experiments are performed to permit the evaluation of quantities that appear in an equation relating the dependent variable to the independent one. Regression analysis deals with calculating the best values of such quantities from experimental data.

All the preceding sections of this chapter have dealt with quantities whose actual values do not change from one measurement to the next. Although such quantities are certainly very common, there are many others that depend on the time, temperature, pH, or some other independent variable. Once a dependence has been observed, several stages of increasing complexity may follow. Suppose that the dependent variable y is the solubility of silver chloride and that the independent one x is the concentration of chloride ion present at equilibrium. Having a few data on the relationship between them, one might seek an equation that describes that relationship accurately enough to permit calculating the solubilities in solutions different from those already investigated. Success in this step can save a great deal of experimental labor and enable the experimenter to present the results in a very compact form. Equations of the form $y = a+bx+cx^2+\ldots$ are popular initial choices, but there are many other possibilities, and even the choice of x is not always obvious. If it were taken to be the concentration of chloride ion before the addition of the silver chloride, a reasonably good empirical description of the results might eventually be obtained, but it would not be very helpful in developing a model to account for those results. The currently accepted model is that the solution contains Ag^+, $AgCl(aq)$, $AgCl_2^-$, and so on. The chemist who knew that a similar model had succeeded in accounting for the solubility of silver bromide in solutions of potassium bromide might arrive at this model by simple analogy. Sometimes a model is deduced from a theory; sometimes it comes by inspiration. Occasionally it is suggested by the form of an equation that is found to fit the data: the insight or good fortune that led to $y = a + b/x + cx + \ldots$ for the solubility of silver chloride might stimulate understanding with very little delay. In one way or another an equation that seemed to be both satisfactory and meaningful would eventually be developed. With its aid, chemists could then proceed to use the experimental data to evaluate the numerical constants—equilibrium or rate constants, enthalpy changes, etc.—that appear in the equation. Then it would be possible to study the ways in which those in turn depended on other variables, such as the temperature, and to develop new equations and new models for these dependences.

The statistician's contribution to this process is called *regression analysis*. The problem in regression analysis is to evaluate the parameters $a, b, \ldots$ that appear in some known or assumed functional relationship between y and x, which may be written as

$$y = f(a, b, \ldots, x). \tag{12.19}$$

It is generally assumed that the random errors in x are negligible, which is to say that an independent variable can usually be measured or controlled much more precisely than a dependent one can be measured. It is also common to assume that the random errors in y are normally distributed and are independent of y (or x). Neither of these is necessarily true: there are situations in which close control of x is so difficult that the uncertainty in each value of x approaches or exceeds the uncertainty in the corresponding value of y, and there are situations in which the random error in y is proportional to the value of y.

The basic theorem of regression analysis is that the best values of the parameters $a, b, \ldots$ in eq. (12.19) are the ones that minimize the sum of the squares of the deviations of the calculated values of y from the measured ones.

The simplest form of eq. (12.19) is $y = a$; this is the one to which the preceding sections of this chapter were devoted. The theorem just stated means that the best value of a is the one for which

$$\frac{d\sum_i (y_i - a)^2}{da} = 0.$$

If n individual measurements $(y_1, y_2, \ldots, y_n)$ have been made, this becomes

$$\frac{d\sum_i (y_i - a)^2}{da} = \frac{d\sum_i (y_i^2 - 2ay_i + a^2)}{da}$$

$$= \frac{d[(y_1^2 - 2ay_1 + a^2) + (y_2^2 - 2ay_2 + a^2) + \ldots + (y_n^2 - 2ay_n + a^2)]}{da}$$

$$= (-2y_1 - 2y_2 - \ldots - 2y_n) + (2a + 2a + \ldots + 2a)$$

$$= -2\sum_i y_i + 2n\,a = 0$$

or $a = \sum_i y_i/n$. The best estimate of a is simply the mean of the individual values of y.

The next more complicated case is that in which $y = a + bx$. Here there are two requirements:

$$\frac{d\sum_i [y_i - (a + bx_i)]^2}{da} = 0 \quad \text{and} \quad \frac{d\sum_i [y_i - (a + bx_i)]^2}{db} = 0. \qquad (12.20)$$

Applying an exactly similar procedure yields the expressions

$$\sum_i y_i - na - b\sum_i x_i = 0 \quad \text{and} \quad \sum_i x_i y_i - b\sum_i x_i^2 - a\sum_i x_i^2 = 0$$

which can be solved to obtain the following expressions for b and a:

$$b = \frac{n\sum xy - \sum x\sum y}{n\sum x^2 - (\sum x)^2} \; ; \quad a = \frac{\sum y - b\sum x}{n}. \qquad (12.21)$$

The subscripts have been dropped for neatness but each summation is still understood to include all the individual values.

Example 12.4. A portion of mineral water was analyzed for calcium by precipitating calcium oxalate and igniting this to give calcium oxide. As the calcium oxide was being weighed it absorbed water vapor from the atmosphere, and the following data were obtained:

Time after removal from the desiccator, min ($= x$)	2	3	4	5	6
Weight, g ($= y$)	0.1827	0.1829	0.1833	0.1836	0.1838

What was the weight of the calcium oxide at the instant when it was removed from the desiccator?

Answer. The answer to every such question depends on the form of the function that is assumed. The simplest assumption is that $y = a$ ($= \bar{y} = 0.183\,26$ g). Then $y_{x=0} = 0.183\,26$ g. The sample standard deviation is 4.6×10^{-4} g; at the 95% level of confidence, and with 4 degrees of freedom, $t = 2.78$; the 95% confidence interval is $0.183\,26 \pm (2.78)(4.6 \times 10^{-4})/\sqrt{5} = 0.183\,26 \pm 0.000\,57$ g. However, you should be very skeptical of the whole calculation because each successive value of y is larger than the preceding one. If the values were randomly arranged there would be only one chance in five that the smallest one would be the first one, only one chance in four that the next smallest would be the second, and so on: the probability that the actual distribution arose from pure chance is only $(1/5) \times (1/4) \times (1/3) \times (1/2) = 1/5! = 1/120$. Another way of deciding whether the trend is real is to apply an F test. If the values of y really are increasing, the sample standard deviation will be significantly larger than it would be if they were constant. Suppose that six weighings of an object whose weight was believed to be constant gave $s_y = 1.0 \times 10^{-4}$ g. Then the value of F would be $(4.6 \times 10^{-4})^2/(1.0 \times 10^{-4})^2 = 21.2$. The value of s in the numerator is computed with four degrees of freedom (from five measurements); the one in the denominator is computed with five degrees of freedom (from six measurements). Table 12.4 gives $F = 5.19$, and the two values of F are different at a level of confidence well above 90%.

If the weight y is plotted against the time x, the points seem to fall close to a straight line, suggesting that the relation $y = a + bx$ should give a satisfactory reproduction of the data. The desired value of $y_{x=0}$ is equal to the intercept a. To evaluate a from eqs. 12.21, we calculate $\sum x\,(= 2 + 3 + \ldots) = 20$, $\sum y\,(= 0.1827 + 0.1829 + \ldots) = 0.9163$, $\sum xy\,[= (2 \times 0.1827) + (3 \times 0.1829) + \ldots] = 3.6681$, and $\sum x^2 (= 2^2 + 3^2 + \ldots) = 90$. Then

$$b = \frac{(5)(3.668\,1) - (20)(0.916\,3)}{(5)(90) - (20)^2} = 2.90 \times 10^{-4} \text{ g min}^{-1}$$

and

$$a = \frac{0.916\,3 - (2.90 \times 10^{-4})\,(20)}{5} = 0.182\,10 \text{ g.}$$

This result does not lie within the 95% confidence interval obtained by assuming that $y = a$.

After the best values of a and b have been calculated, it is possible to evaluate the sample standard deviation s_y, which is often called the *sample standard deviation from regression*. It is given by

$$s_y = \sqrt{\frac{\sum_i [y_i - (a + bx_i)]^2}{n-2}} \tag{12.22}$$

which resembles eq. (12.5) but differs from it in two ways. One is that the expected value of y_i is now equal to $a + bx_i$ rather than to the mean of the individual values. The other is that two parameters (a and b) have had to be calculated here in order to find the expected values of y_i whereas only the mean had to be calculated in order to use eq. (12.5), and therefore there are only $n-2$ degrees of freedom instead of $n-1$. Both eq. (12.5) and eq. (12.22) are special cases of the general relationship

$$s_y = \sqrt{\frac{\sum_i [y_i - f(a, b, \ldots, x)]^2}{n-p}} = \sqrt{\frac{\sum (y_{\text{meas}} - y_{\text{calc}})^2}{n-p}} \tag{12.23}$$

where p is the number of parameters appearing in the function that has been chosen; y_{meas} denotes a measured value and y_{calc} the corresponding calculated one. In Example 12.4 the sample standard deviation from regression would be obtained by using the equation $y = 0.182\,10 + 2.90 \times 10^{-4}\, x$ to find the successive values of y_{calc}, and then writing

$$s_y = \sqrt{\frac{(0.1827 - 0.182\,68)^2 + (0.1829 - 0.182\,97)^2 + \ldots}{3}} = 6.0 \times 10^{-5} \text{ g.}$$

In Example 12.4 an F test showed that the sample standard deviation obtained by assuming $y = a$ was significantly different from the expected one. On assuming $y = a + bx$, F becomes equal to $(1.0 \times 10^{-4})^2/(6.0 \times 10^{-5})^2 = 2.73$. With five degrees of freedom in s_N and three in s_D, Table 12.4 gives $F = 9.01$ at the 90% level of confidence. If the two values of F were the same, you could be 90% sure that there was a real difference between the sample standard deviation obtained by assuming $y = a + bx$ and the sample standard deviation you expect. Since the calculated value of F is smaller than the tabulated one, the probability is below 90% that the two values of s are really different. Indeed, the difference between the two values of F is so large that this probability must be far below 90%. You can therefore be reasonably sure that the equation does not misrepresent the data. It is essential to have this assurance because you often do not know that $y = a + bx$ is the correct equation. Indeed, you can sometimes be quite sure that it is not; here it makes absurd predictions about the weight that will be attained after a year or two. Nevertheless it may be a quite adequate description of the limited portion of a complicated curve that is covered by the data you have. You must make sure that this is true before you can trust the values of the parameters that you obtain. Even when you do, there is some danger in using those values to calculate or predict values of y outside the range in which you have made the measurements. We shall return to this point at the end of the next paragraph.

The true values of the intercept α and the slope β are only approximated by the values of a and b obtained from a finite sample. Confidence intervals can be attached to α and β in much the same way that a confidence interval can be attached to the mean of a series of measurements of a constant quantity (Section 12.5). The necessary equations are

$$\alpha = a \pm ts_y \left(1 + \frac{1}{n} + \frac{\bar{x}^2}{\sum_i (x_i - \bar{x})^2} \right), \tag{12.24a}$$

$$\beta = b \pm ts_y \bigg/ \left[(n-2) \sum_i (x_i - \bar{x})^2 \right] \tag{12.24b}$$

where n is the number of measurements and $\bar{x}$ is the mean of the n values of x_1. As usual, the required values of t may be obtained from Table 12.3 for the desired level of confidence. For the data in Example 12.4, $n = 5$, $\bar{x} = 4$, $\sum_i (x_i - \bar{x})^2 = 10$, and the 95% confidence interval for the intercept is given by

$$\alpha = 0.182\ 10 \pm (3.18)(6.0 \times 10^{-5}) \left(1 + \tfrac{1}{5} + \left(\tfrac{4}{10} \right)^2 \right) = 0.182\ 10 \pm 0.000\ 53\ \text{g}.$$

This is really only the intercept of a straight line that provides a satisfactory description of the data that were obtained from $x = 2$ min to $x = 6$ min. Does it correspond to the weight of the calcium oxide at $x = 0$ min? If you assume that it does, you are assuming that the calcium oxide behaved in the same way between $x = 0$ min and $x = 2$ min as it did between $x = 2$ min and $x = 6$ min. That might be true within the precision of your measurements, or it might not. Perhaps some water is adsorbed very quickly at first, and then the rate at which more water is taken up depends on the rate at which molecules of water diffuse from the surface of a particle toward its interior, or from the air outside the weighing bottle to the air inside it. If the initial adsorption is complete within 2 min you will never detect it from data like these, and the true initial weight of calcium oxide may be far smaller than the lower limit of the confidence interval you have calculated. You would use exactly the same procedure to calculate the weight of the calcium oxide after it had been outside the desiccator for 3.5 min, but you would be on very much safer ground here. Interpolation is always much safer than extrapolation.

Other functions, such as $y = a + bx + cx^2$ or $y = a + b/x + cx \log x$, can be handled in much the same way, although the algebra becomes a little more complicated as more parameters are involved. Fitting data to any function, no matter how many terms are involved, is called *linear regression* whenever the function is linear with respect to each of its parameters. Non-linear regression might involve fitting concentration-time data to equations like $c = c^0 \exp(-kt)$, which is the pseudo-first-order rate equation, or evaluating the parameters that appear in eq. (11.47a) or eq. (11.51) by using data on the variation of pH with volume of reagent during a titration.

Some non-linear equations can be converted to linear forms by simple transformations. The equation $c = c^0 \exp(-kt)$ can be gotten into the linear form $y = a + bx$ by making the substitutions $y = \ln c$, $a = \ln c^0$, $b = -k$, and $x = t$. However, it would then be incorrect to use eqs. (12.21) to calculate the values of a and b. Those equations are based on the assumption that s_y is independent of y. Using them would amount to assuming that $s_{\ln c}$ is independent of $\ln c$. If $s_{\ln c} = 0.01$, then the relative sample standard deviation of c is 1%. To obtain a relative precision of 1% when $c = 10^{-2}$ M is not at all the same thing as obtaining a relative precision of 1% when $c = 10^{-4}$ M. It is much more likely that s_c is independent of c. If it is, the values of a and b obtained from eqs. (12.21) will be wrong, and the errors in them will become larger as the data become less precise.

There are two ways of dealing with non-linear equations. One, which is applicable to equations that can be converted to linear forms, is to use *weighted linear regression*. In the situation described in the preceding paragraph, the correct values of a and b are those for which

$$\frac{d \sum_i (c_{\text{meas}} - c_{\text{calc}})^2}{da} = 0 \quad \text{and} \quad \frac{d \sum_i (c_{\text{meas}} - c_{\text{calc}})^2}{db} = 0. \qquad (12.25)$$

These equations are exactly analogous to eqs. (12.20). According to eq. (12.23)

$$\sum_i (c_{\text{meas}} - c_{\text{calc}})^2 = (n-p) s_c^2$$

and similarly

$$\sum_i (y_{\text{meas}} - y_{\text{calc}})^2 = (n-p) s_y^2$$

where y_{meas} denotes the value of y calculated from a measured value of c while $y_{\text{calc}} = a + bx$. The values of s_y and s_c are related by eq. (12.10b):

$$s_y^2 = \left(\frac{\partial y}{\partial c}\right)^2 s_c^2.$$

Since $y = \ln c$, $\partial y / \partial c = 1/c$ and $s_c^2 = c^2 s_y^2$. Combining these equations yields

$$\frac{d \sum_i (c_{\text{meas}} - c_{\text{calc}})^2}{da} = \frac{[d(n-p) s_c^2]}{da} = \frac{d[(n-p)(cs_y)^2]}{da} = \frac{d \sum_i [c(y_{\text{meas}} - y_{\text{calc}})]^2}{da} = 0$$

and similarly

$$\frac{d \sum_i [c(y_{\text{meas}} - y_{\text{calc}})]^2}{db} = 0$$

as the conditions that the values of a and b must satisfy. Letting $y_{calc} = a + bx$, the results are

$$b = \frac{\sum_i c_i^2 \sum_i x_i y_i c_i^2 - \sum_i x_i c_i^2 \sum_i y_i c_i^2}{\sum_i c_i^2 \sum_i x_i^2 c_i^2 - \left(\sum_i x_i c_i^2\right)^2} \; ; \quad a = \frac{\sum_i y_i c_i^2 - b \sum_i x_i c_i^2}{\sum_i c_i^2} . \quad (12.26)$$

These are messier than eqs. (12.21) but still not difficult to handle with a computer or even a pocket calculator. The quantity c_i^2 is called a *weighting factor*. Its effect is to decrease the contributions that are made to the respective sums by the points for which c_i is small and the relative error in c_i is large. The algebraic form of the weighting factor differs from one situation to another but is always governed by the relation between y and the quantity actually measured.

Whether or not this approach is feasible depends on how difficult it is to obtain an expression for the weighting factor. Another approach, which is applicable even to equations that cannot be converted to linear forms, is to employ a computer program that will calculate the value of the measured quantity itself at each point, using the original non-linear equation together with estimated values of the parameters that appear in it, and will adjust those estimates in ways that minimize the sum of the squares of the deviations of the calculated values from the measured ones.

Non-linear regression has been widely used in evaluating the equilibrium constants of complexation and other reactions, but its other chemical applications are just beginning to be explored. It is an extremely powerful aid to the analysis and interpretation of chemical data.

SUMMARY: Regression analysis is based on the theorem that the best values of the parameters appearing in an equation are those that minimize the sum of the squares of the differences between the measured values of the dependent variable and the ones calculated from the equation. Linear regression is applicable to equations that are linear with respect to all their parameters. Some equations that are not can be rewritten in equivalent forms that are, and weighted linear regression can be applied to many of these.

Problems

Answers to some of these problems are given on page 522.

12.1. List the systematic and random errors in a gravimetric determination of the concentration of a solution of hydrochloric acid by precipitating silver chloride from it.

12.2. A chemist made five measurements of the pH of a solution and obtained the following values: 3.87, 3.89, 3.86, 3.88, and 3.86. If no systematic error was made, what can be said about the true pH?

12.3. A clinical laboratory makes four determinations of the concentration of calcium in a patient's blood serum and obtains the following results: 160, 110, 150, and 130 mg dm^{-3}. The normal range of calcium concentrations is considered to extend from 90 to 115 mg dm^{-3}. Estimate the probability that these results lie outside the normal range.

12.4. One procedure for evaluating the rate constant of a certain reaction gave values of 0.100, 0.105, 0.101, 0.109, and 0.100^{-1} in five trials. The procedure was modified and then gave 0.097, 0.099, 0.098, 0.095, 0.096, and 0.097 s^{-1} in six additional trials under the same conditions. Do the means and standard deviations of the two sets of values differ significantly?

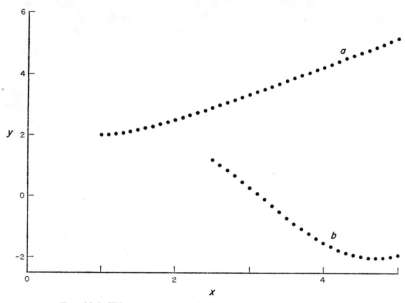

FIG. 12.3. What are the zero intercepts of these two curves?

12.5. Estimate the intercepts (at $x = 0$) of the curves in Fig. 12.3.

12.6. Estimate the values of y at $x = 17$ and $x = 21$ from the following data:

x	18	19	20
y	39.95	39.10	40.08

12.7. Derive eq. (12.7) from eq. (12.10a).

12.8. A solution of sodium thiosulfate was prepared and standardized at frequent intervals. The following results were secured:

Age of the solution, days	0	2	5	7	10
Concentration of thiosulfate, M	0.051 27	0.051 18	0.051 03	0.050 97	0.050 73

(a) What can be said about the rate of its decomposition?

(b) Within what limits can you be 95% certain that the concentration must have fallen on the eighth day after the solution was prepared?

12.9. The following data were obtained in studying the rate of the reaction $A + B \rightarrow C$ in a reaction mixture that contained a large excess of B:

Time, s	10	20	30	40	50
c_A, M	0.002 33	0.001 87	0.001 36	0.000 79	0.000 29

It was concluded that the reaction was zeroth order with respect to A.

(a) What can be said about the initial concentration of A under the conditions of this experiment?

(b) The reaction was later found to be first order with respect to B and it was suggested that the concentration of B in an unknown solution might be found from the value of the pseudo-zeroth-order rate constant obtained in a similar experiment. What certainty could be attained in such a determination?

CHAPTER 13

Activities and Activity Coefficients

13.1. Introduction

Activities were mentioned in Chapter 2 and several of the subsequent chapters, but most of the rate and equilibrium equations in this book have been written with concentrations of solutes in place of their activities. In this chapter we shall examine the relationship between the activity and the concentration of a dissolved substance.

According to Sections 2.9 and 4.2, the solubility of oxygen in pure water is given by

$$O_2(g) = O_2(aq); \quad K = [O_2(aq)]/p_{O_2}. \tag{13.1}$$

At 298 K the value of K is 1.26×10^{-3} mol dm^{-3} atm^{-1}. It decreases with increasing temperature because the process is exothermic (Section 4.2), but at any given temperature eq. (13.1) would lead us to expect the concentration of oxygen in a saturated aqueous solution to depend only on the partial pressure of oxygen in the gas phase. An experimental test of this expectation yields the results shown in Fig. 13.1. We find that the solubility of oxygen is also affected by the presence of a dissolved salt, and

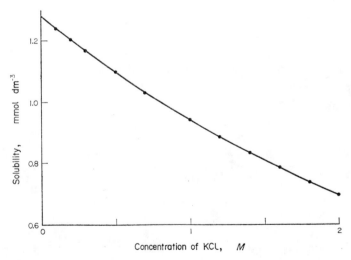

FIG. 13.1. Effect of the concentration of potassium chloride on the solubility of oxygen in aqueous solutions at 298 K. The partial pressure of oxygen is 1 atm.

that it depends on both the nature and the concentration of the salt. Equation (13.1) is incomplete because it provides no explanation of these facts.

Ionic solutes behave in similar ways. Writing

$$AgCl(s) = Ag^+ + Cl^-; \quad K_{AgCl} = [Ag^+][Cl^-] \tag{13.2}$$

to describe the dissolution of silver chloride in an aqueous solution does help us to understand the common-ion effect (Section 5.3). However, it does not enable us to explain why silver chloride is more soluble in dilute solutions of other electrolytes (such as sodium perchlorate and potassium nitrate, with which it does not have an ion in common) than it is in pure water. Nor can we account for this fact by supposing that it is due to the formation of ion pairs like NaCl and $AgClO_4$ in solutions containing silver chloride and sodium perchlorate: not only are these much too unstable to cause the observed increase of solubility, but in addition that increase does not depend on the concentration of sodium perchlorate in the way that it would if this were the correct explanation. Like eq. (13.1), eq. (13.2) is an incomplete description of the behavior we observe.

Equations (13.1) and (13.2) are incomplete because they are too simple. Many other kinds of chemical phenomena exhibit deviations from simple or idealized behavior. The temperature at which water freezes out of an aqueous solution depends on the concentration of the solute. In very dilute solutions, the freezing-point depression ΔT_f (which is defined as the difference between the temperature at which pure water freezes and the temperature at which a solution freezes) is given by

$$\Delta T_f = k \ln N_{solute} \qquad \Delta T_f = k \ln a_{solute} \tag{13.3}$$

where k is a constant of proportionality and N_{solute} is the mole fraction of the solute. In more concentrated solutions eq. (13.3) is only approximately correct: ΔT_f is smaller than we expect it to be in 0.5 M perchloric acid, and is much larger than we expect it to be in 7 M perchloric acid. These results were mentioned in Section 6.4. Boiling-point elevations, vapor pressures, osmotic pressures, and the potentials of electrochemical cells also fail to behave in simple ways. All of this evidence is most conveniently interpreted by envisioning a distinction between the true concentration of a solute and its effective concentration or *activity*.

13.2. Mean ionic activity coefficients

> PREVIEW: This section deals with the relationships between concentration and thermodynamic equilibrium constants and between concentrations and activities. It defines the activity coefficient of an individual substance and the mean ionic activity coefficient of an electrolyte.

The equilibrium constant defined by eq. (13.2) is not a constant: its value is not the same in 0.1 M potassium nitrate as in water. It has still other values in 0.2 M potassium nitrate, 0.1 M calcium nitrate, or 0.1 M aluminum nitrate. If equilibrium constants are written with concentrations, their values differ from one ionic medium to another, and it is therefore necessary to specify the conditions to which any particular value pertains. Such equilibrium constants are called *concentration* constants, *formal* constants, or *conditional* constants.

Another kind of equilibrium constant is the *thermodynamic* constant, which involves activities rather than concentrations. The thermodynamic solubility product of silver chloride is given by the equation

$$K^0_{AgCl} = a_{Ag^+} a_{Cl^-}. \tag{13.4}$$

In this book the superscript zero is used to distinguish a thermodynamic constant from a conditional one. The activities are defined in such a way that a thermodynamic constant is constant no matter what the ionic environment may be.

The activity of any substance is said to be the product of its concentration by a numerical factor called the *activity coefficient*:

$$a_{Ag^+} = y_{Ag^+}[Ag^+]; \quad a_{Cl^-} = y_{Cl^-}[Cl^-] \tag{13.5}$$

which is equivalent to defining an activity coefficient as the ratio of the activity to the concentration. The activity coefficient denoted by the symbol y is called a *molarity activity coefficient*. In some circumstances—of which one is suggested by eq. (13.3) while another will appear in Section 14.3—other descriptions of the composition are more convenient than the concentration, and we may choose to write

$$a_i = f_i N_i \quad \text{or} \quad a_i = \gamma_i m_i.$$

The *mole fractional activity coefficient* or *rational activity coefficient f* is the ratio of the effective mole fraction to the actual one; the *molality activity coefficient* γ (which will be needed, and defined, in Section 14.3) is the ratio of the effective molality to the actual one. Although the values of y, f, and γ are appreciably different in concentrated solutions, they are nearly identical in dilute ones and the differences among them will be ignored here.

Equations (13.2), (13.4), and (13.5) can be combined to give

$$K^0_{AgCl} = (y_{Ag_+}[Ag^+])(y_{Cl^-}[Cl^-]) = y_{Ag^+} y_{Cl^-} K_{AgCl}. \tag{13.6}$$

So that the value of K^0_{AgCl} will remain constant if an electrolyte such as potassium nitrate is added, even though the value of K_{AgCl} varies, we shall say that the added electrolyte affects the value of the product $y_{Ag^+} y_{Cl^-}$. For the solubility product of barium iodate we would write

$$K^0_{Ba(IO_3)_2} = a_{Ba^{2+}} a^2_{IO_3^-} = (y_{Ba^{2+}}[Ba^{2+}])(y_{IO_3^-}[IO_3^-])^2 = y_{Ba^{2+}} y^2_{IO_3^-} K_{Ba(IO_3)_2}. \tag{13.7}$$

We find that the value of the concentration constant $K_{Ba(IO_3)_2}$ can be altered by adding another electrolyte, and attribute this to a variation of the product $y_{Ba^{2+}} y^2_{IO_3^-}$.

Equations (13.7) and (13.8) involve products of activity coefficients. There are many ways in which the values of such products can be calculated from the results of ex-

26•

perimental measurements. Two of them are described in Sections 13.4 and 14.3. We would often like to dissect those values into the individual values of the activity coefficients of the separate ions that they contain, but there is no way of doing this without making some assumption that cannot be proved to be exactly correct except in solutions too dilute to be very interesting. To emphasize these things it is customary to speak of the *mean ionic activity coefficient* $y_\pm$. For silver chloride this is given by

$$y_\pm = \sqrt{y_{Ag^+} y_{Cl^-}}$$

so that eq. (13.6) can be rewritten as

$$K^0_{AgCl} = y^2_\pm [Ag^+][Cl^-] = y^2_\pm K_{AgCl}. \tag{13.8a}$$

For barium iodate the mean ionic activity coefficient is given by

$$y_\pm = \sqrt[3]{y_{Ba^{2+}} + y^2_{IO_3^-}}$$

so that eq. (13.7) can be rewritten as

$$K^0_{Ba(IO_3)_2} = y^3_\pm [Ba^{2+}][IO_3^-]^2 = y^3_\pm K_{Ba(IO_3)_2}. \tag{13.8b}$$

In general, for the electrolyte $C_c A_a$, we define the mean ionic activity coefficient by the equation

$$y_\pm = \sqrt[(c+a)]{y^c_C y^a_A}. \tag{13.9}$$

SUMMARY: The activity coefficient of a substance is the ratio of its activity to its concentration; the mean ionic activity coefficient of an electrolyte is the geometric mean of the activity coefficients of its individual ions. The fact that concentration equilibrium constants depend on the ionic environment is blamed on variations of the activity coefficients.

13.3. The ionic strength

PREVIEW: The ionic strength of a solution depends on the concentrations and charges of all the ions it contains, and is the variable that influences the activity coefficients of the dissolved species.

If a dilute solution of potassium nitrate is saturated with silver chloride, the product $[Ag^+][Cl^-] (= K_{AgCl})$ is found to have a higher value in the saturated solution than it would have at the same temperature in the absence of the potassium nitrate. We interpret this to mean that the potassium nitrate has caused the value of K_{AgCl} to increase by decreasing the value of $y_\pm$. In another solution containing the same concentration of sodium perchlorate, lithium acetate, or any other completely dissociated uni-univalent ("1–1") electrolyte, K_{AgCl} would have the same value as in a solution of potassium nitrate. In still other solutions containing electrolytes such as calcium nitrate and sodium sulfate, K_{AgCl} is larger, so that $y_\pm$ is smaller, than in a solution of a 1–1 electrolyte at the same concentration. Only $c/3$ mol dm^{-3} of an electrolyte containing one divalent ion and one univalent ion is needed to decrease $y_\pm$ just as much as it is decreased by c mol dm^{-3} of a 1–1 electrolyte.

The magnitude of the effect depends on the concentration of the electrolyte, and also on the charges carried by its ions. Both are taken into account in a quantity called the *ionic strength* and defined by the equation

$$\mu = \tfrac{1}{2} \sum_i c_i z_i^2 \tag{13.10}$$

where c_i is the concentration (M) of the ith ion in the solution and z_i is its charge. The factor $\tfrac{1}{2}$ serves to make the ionic strength equal to the concentration for a completely dissociated 1–1 electrolyte. Since the ionic charge is a dimensionless number, the ionic strength has the dimensions of concentration and is always expressed in mol dm^{-3}.

Example 13.1. What is the ionic strength of a c M solution of (a) sodium nitrate or (b) potassium sulfate if c is so small that ion-pair formation is negligible and dissociation can be assumed to be complete?

Answers. (a) In a c M solution of sodium nitrate, complete dissociation gives $c_{Na^+} = c_{NO_3^-} = c$, and the ionic strength is given by

$$\mu = \tfrac{1}{2}[c_{Na^+}z_{Na^+}^2 + c_{NO_3^-}z_{NO_3^-}^2] = \tfrac{1}{2}[(c\times 1^2)+(c\times -1^2)] = \tfrac{1}{2}[c+c] = c\ M.$$

(b) In a c M solution of potassium sulfate, complete dissociation gives $c_{K^+} = 2c$ and $c_{SO_4^{2-}} = c$, and the ionic strength is given by

$$\mu = \tfrac{1}{2}[c_K z_K^2 + c_{SO_4^{2-}}z_{SO_4^{2-}}^2] = \tfrac{1}{2}[(2c\times 1^2)+(c\times -2^2)] = \tfrac{1}{2}[2c+4c] = 3c\ M.$$

Example 13.2. What is the ionic strength of (a) a 1.00×10^{-4} M solution of magnesium sulfate, neglecting ion-pair formation, or (b) an 0.0100 M solution of magnesium sulfate, taking ion-pair formation into account, if the concentration dissociation constant of $MgSO_4(aq)$ in such a solution is equal to 4×10^{-3} mol dm^{-3}?

Answers. (a) Complete dissociation would give $c_{Mg^{2+}} = c_{SO_4^{2-}} = 1.00\times 10^{-4}$ M, and the ionic strength would be

$$\mu = \tfrac{1}{2}[c_{Mg^2}z_{Mg^2}^2 + c_{SO_4^{2-}}z_{SO_4^{2-}}^2] = \tfrac{1}{2}[1.00\times 10^{-4}\times 2^2 + 1.00\times 10^{-4}\times -2^2]$$

$$= \tfrac{1}{2}[4.00\times 10^{-4} + 4.00\times 10^{-4}] = 4.00\times 10^{-4}\ M.$$

(b) Dissociation is far from complete in an 0.0100 M solution. If $c_{Mg^{2+}} = c_{SO_4^-} = c$, then the concentration of the $MgSO_4(aq)$ ion pair is equal to $0.0100 - c$, and

$$K = c_{Mg^2} + c_{SO_4^{2-}}/c_{MgSO_4(aq)} = c^2/(0.01-c) = 4\times 10^{-3}\ \text{mol dm}^{-3}$$

whose solution is $c = 4.63\times 10^{-3}$ M. The ionic strength is given by

$$\mu = \tfrac{1}{2}[4.63\times 10^{-3}z_{Mg^2}^2 + 4.63\times 10^{-3}z_{SO_4^{2-}}^2] = \tfrac{1}{2}[4.63\times 10^{-3}\times 8] = 1.85\times 10^{-2}\ M.$$

As Examples 13.1 and 13.2(a) show, it is easy to calculate the ionic strength of a solution containing a completely dissociated electrolyte. A mixture of completely dissociated electrolytes is equally easy to handle, except that there will be more than two ions whose concentrations have to be included. However, non-ionic substances do not affect the ionic strength, and allowance must therefore be made for incomplete dissociation or ion-pair formation in the fashion illustrated by Example 13.2(b).

> SUMMARY: The ionic strength μ is defined by the equation $\mu = \frac{1}{2}\sum_i c_i z_i^2$, where c_i is the concentra-
> tion and z_i the charge of the ith kind of ion in the solution, and is the variable that
> governs activity coefficients and concentration equilibrium constants.

13.4. Evaluating thermodynamic equilibrium constants and mean ionic activity coefficients

> PREVIEW: This section outlines two ways in which values of thermodynamic equilibrium con
> stants can be obtained from values of concentration constants.

We can now give an operational definition of the thermodynamic equilibrium constant K^0:

$$K^0 = \lim_{\mu \to 0} K \tag{13.11}$$

where K is the concentration constant for the same equilibrium. Suppose that we wished to evaluate the thermodynamic solubility product for calcium sulfate. We could begin by determining the concentration c of either calcium ion or sulfate ion in a saturated solution of calcium sulfate in pure water, and use it to calculate the value of the conditional constant K_{CaSO_4} ($= c^2$) in the saturated solution. Because this solution contains calcium and sulfate ions, its ionic strength is not equal to zero, and therefore the conditional constant is not equal to the thermodynamic one. There are two things we could do:

1. We could determine the values of c in a number of other solutions saturated with calcium sulfate and containing different small known concentrations c' of another electrolyte, such as sodium perchlorate. According to eq. (13.11) the ionic strength of each of these solutions would be equal to $4c+c'$. We could then plot K_{CaSO_4} against the ionic strength (or some function of the ionic strength) and extrapolate to zero ionic strength. The extrapolated value would be equal to $K^0_{CaSO_4}$. After we had done this we could write an equation like eq. (13.8a):

$$K^0_{CaSO_4} = y_{\pm}^2 [Ca^{2+}][SO_4^{2-}] = y_{\pm}^2 K_{CaSO_4} \tag{13.12a}$$

and rearrange it to give

$$y_{\pm}^2 = K^0_{CaSO_4}/K_{CaSO_4}. \tag{13.12b}$$

This would enable us to calculate the value of $y_{\pm}$ in each of the solutions we had used. Another way of evaluating $y_{\pm}$ will be described in Section 14.3.

2. If we had a way of predicting the value of $y_{\pm}$ at the ionic strength of a saturated solution of calcium sulfate in pure water, it would be easy to calculate the value of $K^0_{CaSO_4}$ from eq. (13.12a). Much of the remainder of this chapter will be devoted to showing how values of $y_{\pm}$ can be estimated.

The second of these approaches is much simpler than the first. It is also less accurate, and therefore it should not be used unless an approximate result is all that is needed. Of course the first approach had to be adopted many times in many different situations to obtain experimental values of $y_\pm$ that could be compared with the predicted ones. We could not trust the predictions if we did not know that others like them had always proven to be sufficiently accurate for many purposes.

Equations (13.11) and (13.12a) could be restated in a different way:

$$\lim_{\mu \to 0} y_\pm = \lim_{\mu \to 0} y_i = 1.$$

All mean ionic activity coefficients, and all activity coefficients of individual dissolved ions and non-electrolytes, approach 1 as the ionic strength approaches zero. This means that the activity of any solute is equal to its concentration in an infinitely dilute solution.

SUMMARY: Activity coefficients approach 1 as the ionic strength approaches zero. A thermodynamic equilibrium constant can therefore be evaluated by extrapolating values of the corresponding concentration constant to $\mu = 0$, or by combining them with independently estimated values of the activity coefficients.

13.5. The Debye–Hückel limiting law

PREVIEW: The Debye–Hückel limiting law is an equation that describes the relationship between a mean ionic activity coefficient and the ionic strength of a very dilute solution (up to about $\mu = 0.01\,M$).

At any finite ionic strength the mean ionic activity coefficient may differ from 1. Figure 13.2 shows how the mean ionic activity coefficients of hydrochloric acid and sodium chloride vary with the ionic strength in aqueous solutions at 25°. The reason for plotting $\log y_\pm$ against $\mu^{1/2}$ will soon become apparent. As the ionic strength approaches zero the curves merge and approach the dashed line, which represents the Debye–Hückel limiting law expressed by eqs. (13.14). In extremely dilute solutions—having ionic strengths below about 0.01 M—the value of $y_\pm$ at any particular ionic strength depends only on the ionic charges, and is the same for all 1–1 electrolytes. In more concentrated solutions—having ionic strengths above about 0.01 M—the curves for different electrolytes diverge because other properties of the individual ions become increasingly important.

The Debye–Hückel theory is based on a model in which each dissolved ion is surrounded by an *ionic atmosphere* that has the opposite charge and is spherically symmetrical. Because a cation attracts anions and repels other cations, there will be more anions than cations in its immediate vicinity, and its ionic atmosphere will therefore have a negative charge. The positions of the ions are not fixed, but the probability of finding an anion at any particular distance from a cation at any specific instant depends only on the distance. In a crystal it would depend on the direction as

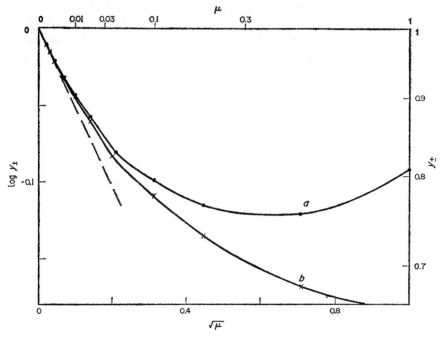

FIG. 13.2. Effects of ionic strength on the mean ionic activity coefficients of (a) hydrochloric acid and (b) sodium chloride in aqueous solutions at 298 K. The dashed line represents eqs. (13.14)

well as on the distance. A cross-section through the ionic atmosphere surrounding a C^+ ion in a solution of the electrolyte CA would look like Fig. 13.3. The radius of the ionic atmosphere, which is denoted by the symbol $1/\varkappa$, decreases as the solution becomes more concentrated and the ions become more tightly packed. Its value in ångstrøm units[†] (1 Å = 0.1 nm = 10^{-10} m) is given by

$$1/\varkappa = 3.04/\mu^{1/2}. \tag{13.13}$$

There is an electrostatic attraction between an ion and the ionic atmosphere that surrounds it, and this attraction tends to stabilize the ion and decrease its free energy. According to eq. (2.7), a decrease of its free energy corresponds to a decrease of its activity, and hence to a decrease of its activity coefficient. If the solution is so dilute that the radius of the ionic atmosphere is very much larger than the radii of the individual ions, so that each of the ions can be represented by a point charge, the mean ionic activity coefficient is given by the *Debye–Hückel limiting law*:

$$\log_{10} y_\pm = -A\,|z_1\,z_2|\,\mu^{1/2}. \tag{13.14a}$$

† Although the ångstrøm] unit will eventually be replaced by the nanometer, it is so firmly embedded in the literature of the Debye–Hückel theory that I have been unable to bring myself to discard it.

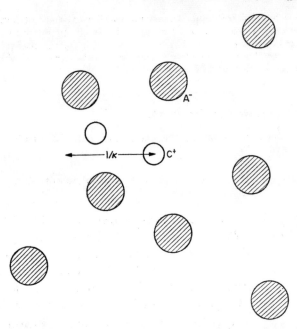

FIG. 13.3. The ionic atmosphere around a C^+ ion in a solution of the electrolyte CA. The radius $1/\varkappa$ of the ionic atmosphere is represented by the arrow. A^- ions (shaded) outnumber C^+ ions (open) at distances smaller than $1/\varkappa$ from the central ion, so that the positive charge of the central C^+ ion is just counterbalanced by the net negative charge of the ionic atmosphere.

The quantities z_1 and z_2 are the charges carried by the ions of the electrolyte for which $y_\pm$ is being calculated, and the factor A is given by

$$A = 1.825 \times 10^6/(\varepsilon T)^{3/2} \qquad (13.14b)$$

where ε is the dielectric constant of the solvent and T is the temperature (K). For aqueous solutions at 298 K, A is equal to 0.511. The term "limiting law" means that although eqs. (13.14) are not *exactly* obeyed at finite ionic strengths, they are more and more closely obeyed as the ionic strength approaches zero.

Figure 13.2 shows that eqs. (13.14) actually provide a very good approximation to the truth in very dilute solutions. For example, at 298 K and $\mu = 5 \times 10^{-3}$ M they predict $y_\pm = 0.920$ for a 1–1 electrolyte, while the experimental values for different 1–1 electrolytes range from about 0.925 to 0.929. For almost any practical purpose the limiting law is entirely satisfactory even up to a somewhat higher ionic strength, say 0.01 M. Above this ionic strength it becomes less and less accurate, and more complicated equations are needed to provide acceptable approximations. These will be discussed in Section 13.7, but first it is convenient to show how the limiting law might be used in evaluating and working with thermodynamic equilibrium constants.

SUMMARY: The Debye–Hückel limiting law assumes each ion to be a point charge surrounded by a spherically symmetrical ionic atmosphere in which oppositely charged ions predominate. It provides a description of $y_\pm$ to which experimental values conform more and more closely as the ionic strength approaches zero, and which can be used in aqueous solutions up to about $\mu = 0.01\ M$.

13.6. Calculations employing the limiting law

PREVIEW: This section gives examples of several typical kinds of calculations using the limiting law.

In this section we shall consider three typical problems involving mean ionic activity coefficients.

1. The solubility of silver bromate ($AgBrO_3$, formula weight 235.8) in pure water at 298 K is 1.92 g dm^{-3}. What is the value of its thermodynamic solubility product?

The concentration of the saturated solution is $1.92/235.8 = 8.14 \times 10^{-3}\ M$. Ion-pair formation can be neglected for a 1–1 electrolyte at this low concentration, and bromate ion should not form a complex of appreciable strength with silver ion. We can therefore take $[Ag^+] = [BrO_3^-] = \mu = 8.14 \times 10^{-3}\ M$. The limiting law gives

$$\log y_\pm = -0.511\,|1\times-1|\,(8.14\times10^{-3})^{1/2} = -0.0461$$

so that $y_\pm = 0.899$. Consequently

$$K^0_{AgBrO_3} = y_\pm^2[Ag^+][BrO_3^-] = (0.899)^2\times(8.14\times10^{-3})^2 = 5.4\times10^{-5}\ mol^2\,dm^{-6},$$

whereas the concentration constant $K_{AgBrO_3}(= [Ag^+][BrO_3^-])$ is equal to 6.6×10^{-5} mol^2 dm^{-6} at this ionic strength.

This was a simple problem because the ionic strength, and hence the mean ionic activity coefficient, could be calculated directly from the information that was given. Often this is not possible, and then we must proceed in the fashion illustrated by the next problem.

2. At 298 K in aqueous solutions the thermodynamic solubility product of thallium(I) iodate, $TlIO_3$, is equal to 3.1×10^{-6} mol^2 dm^{-6}. What is the solubility of thallium(I) iodate in pure water at 298 K?

If the solubility is equal to S mol dm^{-3} we can say that $[Tl^+] = [IO_3^-] = \mu = S$ by the same reasoning as in the preceding problem. We can obtain a preliminary estimate of S by supposing that $y_\pm = 1$, which gives

$$K^0_{TlIO_3} = y_\pm^2[Tl^+][IO_3^-] = y_\pm^2 S^2 \cong S^2 = 3.1\times10^{-6}\ mol^2\,dm^{-6}$$

so that $S = 1.76\times10^{-3}\ M$. If this were correct, the ionic strength would be 1.76×10^{-3} M, and the mean ionic activity coefficient would be given by

$$\log y_\pm = -0.511\,|1\times-1|\,(1.76\times10^{-3})^{1/2} = -0.0214$$

whence $y_\pm = 0.952$. From this result we can obtain a better estimate of S:

$$S = (K^0_{TlIO_3})^{1/2}/y_\pm = (3.1\times10^{-6})^{1/2}/0.952 = 1.85\times10^{-3}\ M$$

which is 5% higher than the first estimate. Repeating the process once more would give $\log y_{\pm} = -0.0220$, $y_{\pm} = 0.951$, and $S = 1.85 \times 10^{-3}$ M. The fact that there are two significant figures in the value of $K^0_{\mathrm{TlIO_3}}$ implies that this value is known to about 1 part in 30. Since S is very nearly proportional to the square root of $K^0_{\mathrm{TlIO_3}}$, its relative uncertainty is about half as large, or 1 part in 60. We are entitled to more than two, but not quite entitled to three, significant figures in S. The first estimate of S is in error by more than 1 part in 60, but the second is as accurate as the information given and the third merely shows that the second is nearly exact.

Section 5.5 showed that problems like these become more complicated when the formation of ion pairs or complexes has to be considered. They are further complicated by the necessity of including the activity coefficients. The following problem is typical.

3. At 298 K in aqueous solutions the thermodynamic solubility product of barium iodate is equal to 1.51×10^{-9} mol^3 dm^{-9} and the thermodynamic dissociation constant of the ion pair $BaIO_3^+$ is equal to 0.10 mol dm^{-3}. What is the solubility of barium iodate in pure water at 298 K?

As in the preceding problem, we can obtain preliminary estimates of the concentrations of the ions by supposing that all of the activity coefficients are equal to 1. If we begin by writing $[IO_3^-] = 2[Ba^{2+}]$ and

$$[Ba^{2+}][IO_3^-]^2 = 4[Ba^{2+}]^3 \cong K^0_{\mathrm{Ba(IO_3)_2}} = 1.51 \times 10^{-9} \ mol^3 \ dm^{-9}$$

we obtain $[Ba^{2+}] = 7.23 \times 10^{-4}$ M and $[IO_3^-] = 1.45 \times 10^{-3}$ M. By combining these values with the equation

$$\frac{[Ba^{2+}][IO_3^-]}{[BaIO_3^+]} \cong K^0_{\mathrm{BaIO_3^+}} = 0.10 \ mol \ dm^{-3}$$

we obtain $[BaIO_3^+] = 1.05 \times 10^{-5}$ M. Now it is possible to estimate the ionic strength:

$$\mu = \tfrac{1}{2}[(7.23 \times 10^{-4} \times 2^2) + (1.45 \times 10^{-3} \times -1^2) + (1.05 \times 10^{-5} \times 1^2)] = 2.18 \times 10^{-3} \ M$$

and the mean ionic activity coefficient defined by eq. (13.9b), which will be represented here by the symbol $y_{\pm, \, \mathrm{Ba(IO_3)_2}}$:

$$\log y_{\pm, \, \mathrm{Ba(IO_3)_2}} = -0.511 \, |2 \times -1| \, (2.18 \times 10^{-3})^{1/2} = -0.0477$$

whence $y_{\pm, \, \mathrm{Ba(IO_3)_2}} = 0.896$.

We must now abandon the approximation $[IO_3^-] = 2[Ba^{2+}]$. Although it would be correct if the reaction $Ba(IO_3)_2(s) = Ba^{2+} + 2IO_3^-$ were the only one that occurred, the reaction $Ba(IO_3)_2(s) = BaIO_3^+ + IO_3^-$ produces iodate ions but not barium ions, and therefore the concentration of iodate ion must be more than twice as large as the concentration of barium ion. Accordingly we begin a second approximation by writing the conservation equations

$$S = [Ba^{2+}] + [BaIO_3^+] \quad \text{and} \quad 2S = [IO_3^-] + [BaIO_3^+] \tag{13.15}$$

where S is the solubility of barium iodate in mol dm^{-3}. Combining these with the equation for the thermodynamic solubility product and then with the results of the first approximation,

$$K^0_{Ba(IO_3)_2} = y^3_{\pm, \, Ba(IO_3)_2}(S-[BaIO_3^+])(2S-[BaIO_3^+])^2$$
$$= (0.896)^3(S-1.05\times10^{-5})(2S-1.05\times10^{-5})^2 = 1.51\times10^{-9}. \quad (13.16)$$

This equation has the solution $S = 8.13\times10^{-4}$ *M*. From eqs. (13.15)

$$[Ba^{2+}] = 8.13\times10^{-4}-[BaIO_3^+] = 8.13\times10^{-4}-1.05\times10^{-5} = 8.02\times10^{-4} \; M$$

and

$$[IO_3^-] = 2\times8.13\times10^{-4}-[BaIO_3^+] = 1.62\times10^{-3} \; M.$$

The thermodynamic dissociation constant of the ion pair is given by

$$K^0_{BaIO_3^+} = \frac{a_{Ba^{2+}}a_{IO_3^-}}{a_{BaIO_3^+}} = 0.10 \text{ mol dm}^{-3}.$$

Multiplying this expression by 1 makes it much easier to manipulate:

$$K^0_{BaIO_3^+} = \frac{a_{Ba^{2+}}a_{IO_3^-}}{a_{BaIO_3^+}}\times\frac{a_{IO_3^-}}{a_{IO_3^-}} = \frac{a_{Ba^{2+}}a^2_{IO_3^-}}{a_{BaIO_3^+}a_{IO_3^-}} = \frac{K^0_{Ba(IO_3)_2}}{y_{\pm, \, Ba(IO_3) \, IO_3}[BaIO_3^+][IO_3^-]}$$
$$= 0.10 \text{ mol dm}^{-3}. \quad (13.17)$$

We can replace the product $a_{Ba^{2+}}a^2_{IO_3^-}$ in the numerator by the thermodynamic solubility product of barium iodate because we are considering a solution saturated with the salt. The mean ionic activity coefficient that appears in the denominator pertains to the 1–1 electrolyte whose cation is the ion pair $BaIO_3^+$ and whose anion is iodate ion. At an ionic strength of 2.18×10^{-3} *M* the value of this mean ionic activity coefficient is given by

$$\log y_{\pm, \, Ba(IO_3) \, IO_3} = -0.511 \, | \, 1\times-1 \, | \, (2.18\times10^{-3})^{1/2} = -0.0239$$

or $y_{\pm, \, Ba(IO_3) \, IO_3} = 0.946$. Equation (13.17) can now be written

$$[BaIO_3^+] = \frac{K^0_{Ba(IO_3)_2}}{y^2_{\pm, \, Ba(IO_3) \, IO_3}[IO_3^-]K^0_{BaIO_3^+}} = \frac{1.51\times10^{-9}}{(0.946)^2(1.62\times10^{-3})(0.10)}$$
$$= 1.04\times10^{-5} \; M.$$

These new values of the concentrations yield

$$\mu = \tfrac{1}{2}[(8.02\times10^{-4}\times2^2)+(1.62\times10^{-3}\times-1^2)+(1.04\times10^{-5}\times1^2)] = 2.42\times10^{-3} \; M$$

and

$$\log y_{\pm, \, Ba(IO_3)_2} = -0.511 \, | \, 2\times-1 \, | \, (2.42\times10^{-3})^{1/2} = -0.0503$$

which corresponds to $y_{\pm, \, Ba(IO_3)_2} = 0.891$. This is so little different from the preceding estimate of it, 0.896, that it is hardly worth while to make another approximation. If one were made, by introducing the values $[BaIO_3^+] = 1.04 \times 10^{-5}$ M and $y_{\pm,\,Ba(IO_3)_2} = 0.891$ into eq. (13.16), it would give $S = 8.18 \times 10^{-4}$ M.

The result is 13% larger than the one obtained in Section 5.2. Only a small fraction of the difference is caused by our having neglected the ion pair $BaIO_3^+$ in Section 5.2 and taken it into account here, for the ion pair is so weak that it is responsible for only a little over 1% of the solubility. Almost all of the difference results from the fact that the mean ionic activity coefficient is smaller than 1, even though barium iodate is not very soluble and the ionic strength of the saturated solution is not very high. The fact that barium ion is divalent has two important consequences: it causes the ionic strength to be much higher than it would be for a 1–1 salt having the same solubility, and it also causes the value of the mean ionic activity coefficient $y_{\pm,\,Ba(IO_3)_2}$ to be much smaller than it would be for a 1–1 electrolyte at the same ionic strength. Each of these effects would be even more important with a salt such as barium sulfate or calcium phosphate, in which both ions are polyvalent.

> SUMMARY: It is easy to take activity coefficients into account if the ionic strength can be calculated from the data available. Otherwise successive approximations to the concentrations, ionic strength, and activity coefficients are usually necessary.

13.7. Activity coefficients at higher ionic strengths

> PREVIEW: Equations more complicated than the Debye–Hückel limiting law are needed to describe the behaviors of mean ionic activity coefficients at ionic strengths above about 0.01 M.

Figure 13.2 showed that the limiting law [eqs. (13.14)] gives values of $y_\pm$ that are very close to the truth if the ionic strength is less than about 0.01 M. In more concentrated solutions the limiting law provides estimates that are too crude to be useful.

According to eq. (13.13) the radius of the ionic atmosphere is about 30 Å (= 3 nm) in a solution having an ionic strength of 0.01 M. The radii of most simple ions—such as the alkali and alkaline earth metal ions, the halide ions, hydronium ion, sulfate and phosphate ions, and many others—are between about 1 and 5 Å in aqueous solutions. These figures are so much smaller than 30 Å that little error is made by considering these ions to be point charges at this ionic strength, which is what is done in deriving the limiting law. However, at ionic strengths above about 0.01 M, the radii of the ions are no longer much smaller than the radius of the ionic atmosphere, and it is impossible to neglect the fact that the ions have finite sizes. Real ions cannot be packed as tightly as point charges: the radius of the ionic atmosphere becomes larger than the value given by eq. (13.13), and this weakens the force that the ionic atmosphere exerts on the ion it surrounds. That force tends to stabilize the ions and decrease the mean ionic activity coefficient, and its weakening causes the decrease of $y_\pm$ with increasing ionic strength to be less rapid than it would be if the ions were point charges. Taking these considerations into account yields the *Debye–Hückel* equation:

$$\log_{10} y_{\pm} = -\frac{A\,|z_1\ z_2|\,\mu^{1/2}}{1 + B\mathring{a}\mu^{1/2}}$$

(13.18a)

where A was described by eq. (13.14b) while

$$B = 50.3/(\varepsilon T)^{1/2}.$$

(13.18b)

For aqueous solutions at 298 K and with the *distance of closest approach* $\mathring{a}$ in ångström units, $B = 0.329$. The significance of $\mathring{a}$ was discussed briefly in Section 5.6. If the ions were rigid spheres, $\mathring{a}$ would be the smallest possible distance between the center of a cation and the center of an anion. However, there are many ions that are neither rigid nor spherical, and it is difficult to give even a qualitative description of $\mathring{a}$ in a solution that contains different kinds of cations and anions. The value of $\mathring{a}$ for any particular electrolyte cannot be predicted, but must be chosen to give the best fit to experimental data.

As the ionic strength approaches zero, the product $B\mathring{a}\mu^{1/2}$ in the denominator of eq. (13.18a) decreases, and at low ionic strengths the denominator is virtually equal to 1. Equation (13.14a) is really only a special case of eq. (13.18a).

Figure 13.4 shows that the values predicted by the Debye–Hückel equation with $\mathring{a} = 5.3$ Å (curve *b*) are clearly superior to the ones predicted by the limiting law (curve *a*), and this equation is widely used up to an ionic strength of about 0.1 *M*. At still higher ionic strengths, up to about 1 *M*, better results are obtained by using the *Hückel equation*, which many chemists call the *extended Debye–Hückel equation*

$$\log_{10} y_{\pm} = -\frac{A\,|z_1\ z_2|\,\mu^{1/2}}{1 + B\mathring{a}\mu^{1/2}} + B\mu$$

(13.19)

in which the added parameter B is called the *salting coefficient*. Curve *c* in Fig. 13.4 shows the values predicted by the Hückel equation with $\mathring{a} = 4.3$ Å and $B = 0.133$ mol^{-1} dm^3, which yield the best fit to the data for hydrochloric acid. The value of B for a particular electrolyte, like that of $\mathring{a}$, has to be obtained empirically, and may lie anywhere between 0 and about 0.2 mol^{-1} dm^3. There is a very rough correlation between the numerical values of B and $\mathring{a}$: as a crude approximation $B = 10^{-4}\,\mathring{a}^3$. If $B = 0$, eq. (13.19) becomes identical with eq. (13.18a), and the value of $y_{\pm}$ decreases continuously as the ionic strength increases. Such behavior is observed with many 1–1 electrolytes, including silver nitrate and ammonium nitrate. For others, including hydrochloric and perchloric acids and lithium chloride, the value of B may become very large in concentrated solutions. In an 8 *M* solution the mean ionic activity coefficient of perchloric acid is equal to 500!

No matter what values of B and $\mathring{a}$ are chosen, the Hückel equation becomes unsatisfactory at ionic strengths above about 1 *M*. Several different but interrelated phenomena become important in concentrated solutions. Ion-pair formation decreases the ionic activities and may be responsible for the low values of $y_{\pm}$ obtained for

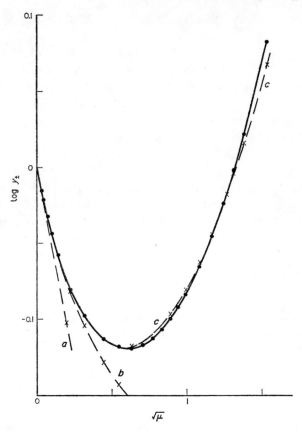

FIG. 13.4. The continuous curve through the solid circles represents experimental data on the variation of the mean ionic activity coefficient of hydrochloric acid in aqueous solutions at 298 K. Curve (a) represents the Debye–Hückel limiting law (eqs. (13.14)); curve (b) represents the Debye–Hückel equation (eqs. (13.18)) with $\mathring{a} = 5.3$ Å; curve (c) represents the Hückel equation (eq. ((13.19)) with $\mathring{a} = 4.3$ Å and $B = 0.133$ mol^{-1} dm^3.

electrolytes like silver nitrate and ammonium nitrate. Ionic hydration raises the activities of the dissolved ions by lowering the activity of free water. Different ions have different effects on the entropies of their solutions: a small and heavily hydrated ion like hydrogen ion tends to decrease the entropy by arranging molecules of water in its hydration shell, whereas a large and nearly unhydrated one like cesium ion merely disrupts the hydrogen-bonded structure of water and thereby increases the entropy. These things affect the activity coefficients of the dissolved ions because the entropy and free energy are related by eq. (2.3). At an ionic strength of 10 M eq. (13.13) gives $1/\varkappa = 1$ Å, which cannot be correct because it is actually smaller than the radii of most anhydrous ions. Even the fundamental assumption of the Debye–Hückel theory, which is that the ionic atmosphere is spherically symmetrical, becomes uncertain in such concentrated solutions. In the face of these complexities it has not yet been possible to develop a satisfactory theory to describe the data obtained at ionic strengths higher than 1 M.

There are many kinds of experiments in which wide ranges of concentration must be covered. To study the silver–chloride complexes and their formation constants, one might want to measure the solubility of silver chloride in solutions containing concentrations of chloride ion ranging from 0.001 M to 3 M. The activity coefficients would have to be taken into account in calculating the formation constants, and difficulties would arise if one simply used solutions of sodium chloride having these different concentrations, because the activity coefficients would change considerably as the concentration of sodium chloride (and therefore the ionic strength) changed. A better expedient is to add another chemically inert electrolyte, such as sodium perchlorate, to make the ionic strengths of all the solutions identical. One solution might contain 0.001 M sodium chloride and 2.999 M sodium perchlorate, while another contained 3 M sodium chloride and no sodium perchlorate. The values of the activity coefficients would be unknown, but they would remain constant as the concentration of chloride ion varied.

SUMMARY: The Debye–Hückel equation involves a parameter $å$ whose value varies from one electrolyte to another, and can be made to yield values of the mean ionic activity coefficient that agree well with experimental data up to about $\mu = 0.1\ M$. The Hückel equation adds a second parameter B and can be used up to about $\mu = 1\ M$.

13.8. Activity coefficients of non-electrolytes

PREVIEW: The activity coefficient of a non-electrolyte depends on ionic strength in a way that can be deduced from the Hückel equation.

Figure 13.1 shows that the solubility of oxygen in water decreases as the ionic strength increases. Because this is true, oxygen will separate from a saturated aqueous solution if a salt is added. This phenomenon is called "salting out" and is exhibited by the great majority of non-electrolytes.

The thermodynamic equilibrium constant for the process is given by

$$O_2(g) = O_2(aq); \quad K^0 = a_{O_2(aq)}/p_{O_2}.$$

For electrolytes we say that the activity and concentration become identical as the ionic strength approaches zero, and it is convenient to say the same thing for non-electrolytes. Let us assign the symbol s^0 to the solubility of oxygen, at a pressure of 1 atm, in pure water. The concentration of oxygen that is dissolved in such a solution is equal to s^0. The ionic strength of the saturated solution is equal to $10^{-7}\ M$ ($= c_{H^+}$ $= c_{OH^-}$), which is so small that we can neglect it and consider the activity of dissolved oxygen to be equal to its concentration. As long as the pressure of oxygen remains unchanged, the activity of dissolved oxygen in any saturated solution must be the same as it is in pure water at the same temperature, so that

$$a_{O_2(aq)} = K^0 p_{O_2} = s^0. \tag{13.20}$$

The activity of dissolved oxygen can also be expressed as the product of an activity coefficient by the concentration of oxygen:

$$a_{O_2(aq)} = y_0 [O_2(aq)]. \tag{13.21}$$

The symbol y_0 is used to denote the molarity activity coefficient (see Section 13.2) of a non-electrolyte. In any solution that is saturated with oxygen, the concentration of oxygen is equal to the solubility s, and eqs. (13.20) and (13.21) can therefore be combined to give

$$s^0 = y_0 s \quad \text{or} \quad \log_{10} y_0 = \log_{10} (s^0/s).$$

Another description of y_0 can be obtained from eq. (13.19). For a non-electrolyte the product of the ionic charges, $z_1 z_2$, is equal to zero, the entire first term on the right-hand side vanishes, and we are left with

$$\log_{10} y_0 = B\mu \; [= \log_{10} (s^0/s)]. \tag{13.22}$$

With the aid of eq. (13.22) it is easy to evaluate B for a sparingly soluble non-electrolyte by determining the effect of ionic strength on its solubility. Other procedures are available for use with non-electrolytes, such as acetic acid and phenol, that are too soluble to be handled in this way, but these will not be discussed here. The value of B turns out to depend on the identity of the non-ionic solute. In aqueous solutions of potassium chloride at 298 K, B is equal to 0.055 for helium, 0.133 for oxygen, and 0.245 for phenol. The value of B also depends on the identity of the electrolyte that is present. For aqueous solutions of ethyl acetate at 298 K it is 0.143 with potassium chloride, 0.088 with lithium chloride, and 0.027 with ammonium nitrate. Occasionally B is negative—it is -0.048 for helium in solutions of perchloric acid—and then the solubility of the non-electrolyte increases on adding electrolyte, which is called "salting in". All these values of B have the units $\text{mol}^{-1}\,\text{dm}^3$.

Unless detailed information about a particular system is available, we usually assume that $B = 0.1$ and expect to get a reasonable approximation. We often have to deal with expressions that contain both the activity coefficient of a non-electrolyte and the mean ionic activity coefficient of an electrolyte. For the equilibrium $AgCl(aq) = Ag^+ + Cl^-$ we would have

$$K^0 = \frac{a_{Ag^+} a_{Cl^-}}{a_{AgCl(aq)}} = \frac{y_\pm^2 [Ag^+][Cl^-]}{y_0 [AgCl(aq)]}.$$

If the ionic strength of the solution is below 0.1 M, $\log y_0$ will be smaller than 0.01, so that y_0 cannot differ from 1 by more than 2.3%. In such solutions the error that we incur by ignoring y_0 is unlikely to exceed the uncertainty in the value of the thermodynamic equilibrium constant that we employ. In more concentrated solutions y_0

must be taken into account, but the error that results from taking B to be equal to 0.1 is unlikely to exceed the uncertainty in the estimate we make of $y_\pm$.

SUMMARY: With rare exceptions, the activity coefficient of a non-electrolyte increases as the ionic strength increases, and therefore the solubilities of most sparingly soluble non-electrolytes decrease as the ionic strength increases. The approximation $\log_{10} y_0 = 0.1\,\mu$ is common.

13.9. Single-ion activity coefficients

PREVIEW: Values can be assigned to the activity coefficients of individual ions and are useful in many chemical situations. This section describes the assumption on which they are based and the reliability they can be expected to have.

The importance of the mean ionic activity coefficient arises from the fact that there is no experimental measurement whose result depends on the activity of a single ionic species. The positions of chemical equilibria, the rates of many chemical reactions, the freezing and boiling points of solutions of electrolytes, and the potentials of electrochemical cells (Section 14.2) can all be described by writing expressions that contain mean ionic activity coefficients, or combinations of activity coefficients that can be converted into mean ionic activity coefficients in the manner illustrated by eq. (13.17).

Sometimes, however, we may wish to consider a restricted part of a chemical situation, such as the potential of one of the two electrodes of an electrochemical cell. An especially important example will be discussed in the following section. Then we often find ourselves compelled to write combinations of ionic activity coefficients that are not equivalent to, and cannot be converted into, mean ionic activity coefficients. This section is devoted to the expedients that have to be adopted in such circumstances.

In the limiting-law region the matter is simple and straightforward. The activity coefficient of the ith ion in a solution of very low ionic strength is given by

$$\log_{10} y_i = -Az_i^2 \mu^{1/2}$$

which can be combined with eq. (13.9) to obtain the limiting law (eq. (13.14)) for the mean ionic activity coefficient.

At ionic strengths above about 0.01 M the matter is anything but straightforward. Up to $\mu = 0.1\,M$ we could certainly write

$$\log_{10} y_i = -\frac{Az_i^2 \mu^{1/2}}{1 + B\mathring{a}\mu^{1/2}} \tag{13.23}$$

on the basis of eq. (13.18). However, the value of $\mathring{a}$ is a property of an electrolyte rather than of the individual ions it contains. Equation (13.23) defines an activity coefficient that depends not only on the properties of the ith ion, in which we are interested, but also on the properties of other ions in the solution. It would be easier to think and talk about an activity coefficient that depended only on the properties of the ith ion. We achieve this by writing

$$\log_{10} y_i = -\frac{Az_i^2 \mu^{1/2}}{1 + B\mathring{a}\mu^{1/2}} \qquad (13.24)$$

in which $\mathring{a}_i$ is a distance that is assigned a fixed value for each particular ion.

Potassium ion is univalent, and so is chloride ion. Each has the same electronic structure as an atom of argon. Their diffusion coefficients in aqueous solutions are almost identical, and they must therefore have nearly identical sizes. It seems reasonable to adopt the *Guggenheim assumption*, which is that the activities of potassium and chloride ions are equal in any aqueous solution of potassium chloride. In view of eq. (13.9) this is equivalent to $y_{K+} = y_{Cl-} = y_{\pm, KCl}$. In view of eqs. (13.18) and (13.24), it is further equivalent to $\mathring{a}_{K+} = \mathring{a}_{Cl-} = \mathring{a}_{KCl}$. The simple picture of the distance of closest approach for potassium chloride is that it corresponds to the sum of the radii of the potassium and chloride ions. Since the Guggenheim assumption amounts to saying that these ions have equal radii, the value of $\mathring{a}_i$ for each of them corresponds, on the basis of the same simple picture, to the diameter of the ion.

Estimates of $\mathring{a}_i$ for different ions can be obtained in the following way. We begin by obtaining values of y_i for the ions at different ionic strengths. At an ionic strength of 0.1 M the mean ionic activity coefficient of potassium chloride is 0.771, and we therefore assume $y_{K+} = y_{Cl-} = 0.771$. Since the activity coefficient of chloride ion at this ionic strength is being imagined to be independent of the composition of the solution, we must take it to be the same in 0.1 M sodium chloride as in 0.1 M potassium chloride. Since $y_{\pm, NaCl}^2 = 0.779^2 = y_{Na+} y_{Cl-}$, we can obtain a value of y_{Na+} by writing $y_{Na+} = y_{\pm, NaCl}^2 / y_{Cl-} = 0.779^2/0.771 = 0.787$. Now we can evaluate y_{Br-} by writing either $y_{Br-} = y_{\pm, KBr}^2 / y_{K+}$ or $y_{Br-} = y_{\pm, NaBr}^2 / y_{Na+}$, of which the first gives $y_{Br-} = 0.775$ while the second gives $y_{Br-} = 0.779$. We can go on to calculate one value of y_{Li+} from the mean ionic activity coefficient of lithium chloride and our value of y_{Cl-}, and another one from the mean ionic activity coefficient of lithium bromide and the average of these two values of y_{Br-}. Eventually we will have a set of single-ion activity coefficients at an ionic strength of 0.1 M. We repeat the whole process at other ionic strengths to find how these vary with ionic strength. Finally we calculate a value of $\mathring{a}_i$ for each ion at each ionic strength, using eq. (13.24), and extrapolate these values to zero ionic strength to eliminate the effects of small correction terms (like the one, $B\mu$, that appears in eq. (13.19)).

In this way Kielland obtained the values of $\mathring{a}_i$ given in Table 13.1. To estimate the activity coefficient of magnesium ion at an ionic strength of 0.1 M, we would find $\mathring{a}_{Mg^{2+}} = 8$ Å in Table 13.1, substitute this value into eq. (13.24), and calculate $y_{Mg^{2+}} = 0.444$. A similar estimate of the activity coefficient of nitrate ion at the same ionic strength would give $y_{NO_3^-} = 0.753$. The two values could be combined to give $y_{\pm, Mg(NO_3)_2}^3 = y_{Mg^{2+}} y_{NO_3^-}^2 = 0.252$, or $y_{\pm, Mg(NO_3)_2} = 0.631$. The experimental value in 0.033 M magnesium nitrate is 0.598. There is no way of checking the values

27*

for the individual ions, because these cannot be measured separately, but comparing calculated and experimental values of $y_\pm$ in the fashion just illustrated shows that the average error in the single-ion values is roughly 5% at this ionic strength. The error increases as the ionic strength increases, and estimates made in this way are not very trustworthy at ionic strengths above about 0.2 M.

TABLE 13.1. *Effective ionic diameters in aqueous solutions at 25° C*

This table gives the values of $\mathring{a}_i$ for a number of common ions as obtained by J. Kielland, *J. Amer. Chem. Soc.* **59**, 1675 (1937). They are intended for use with eq. (13.24). Inorganic ions are listed in the alphabetical order of their formulas; organic ones are alphabetized by name. The common abbreviations "Cit" and "Tart" denote citrate and tartrate, respectively.

Ions	$\mathring{a}_i$, Å
Ag^+, Cs^+, NH_4^+, Rb^+, Tl^+	2.5
Br^-, CN^-, Cl^-, I^-, K^+, NO_2^-, NO_3^-	3
BrO_3^-, H_2Cit^-, ClO_3^-, ClO_4^-, F^-, $HCOO^-$, IO_4^-, MnO_4^-, OH^-, SCN^-	3.5
HCO_3^-, CrO_4^{2-}, $Fe(CN)_6^{3-}$, Hg_2^{2+}, IO_3^-, Na^+, $H_2PO_4^-$, HPO_4^{2-}, PO_4^{3-}, HSO_3^-, SO_4^{2-}	4
CH_3COO^-, CO_3^{2-}, $ClCH_2COO^-$, $HCit^{2-}$, $C_2O_4^{2-}$, Pb^{2+}, SO_3^{2-}	4.5
Ba^{2+}, Cd^{2+}, Cit^{3-}, Cl_2CHCOO^-, $Fe(CN)_6^{4-}$, Hg^{2+}, S^{2-}, Sr^{2+}, $Tart^{2-}$, Cl_3CCOO^-	5
$C_6H_5COO^-$, Ca^{2+}, Co^{2+}, Cu^{2+}, Fe^{2+}, Li^+, Mn^{2+}, Ni^{2+}, $o\text{-}C_6H_4(COO)_2^{2-}$, Sn^{2+}, Zn^{2+}	6
Be^{2+}, Mg^{2+}	8
Al^{3+}, Ce^{3+}, Fe^{3+}, H^+	9
Ce^{4+}, Sn^{4+}	11

Table 13.1 can also be used to obtain values of the distance of closest approach $\mathring{a}$ for use in eq. (13.18a): for the compound CA

$$\mathring{a} = \sqrt{\mathring{a}_{C^{m+}} \mathring{a}_{A^{m-}}}.$$ (13.25)

For hydrochloric acid this gives $\mathring{a} = \sqrt{9 \times 3} = 5.2$ Å, while the value that gives the best fit to experimental data over the range $0 \leqslant \mu \leqslant 0.1$ M is 5.3 Å.

SUMMARY: Values of single-ion activity coefficients can be estimated from the Debye–Hückel equation with the aid of the Guggenheim assumption, and can be made to yield values of the apparent diameters $\mathring{a}_i$ of individual ions.

13.10. Activity coefficients and the pH scale

PREVIEW: Single-ion activity coefficients are used in establishing the pH-values of reference buffers, on which practical measurements of pH are based.

So many different chemical phenomena are affected by the pH that measurements of pH must be made in many different kinds of chemical experiments. Most measurements of pH are made with glass-electrode pH meters. A glass electrode and a reference electrode are immersed in a standard or reference buffer, and the difference E_s

between their potentials is measured. The same electrodes are then immersed in the unknown solution, and the difference E_u between their potentials is measured again. If everything except the potential of the glass electrode were exactly the same in the two measurements, the results would obey the equation

$$E_u - E_s = 0.059\ 16 \log_{10} \frac{a_{H^+,\,u}}{a_{H^+,\,s}} = -0.059\ 16(\text{paH}_u - \text{paH}_s)$$

at 298 K. It would be easy to calculate paH_u, the paH of the unknown solution, if paH_s, the paH of the reference buffer, were known.

Unfortunately we cannot find the paH exactly even for a reference buffer. This section describes how a quantity called the pH_s is evaluated. It is the closest approximation to the paH that we can make.

The most important characteristic of a reference buffer is its buffer capacity, which must be so high that its pH will not be detectably affected by traces of acidic or basic impurities in the materials from which it is prepared or in the atmosphere with which it comes in contact while it is being used. It should be easy to prepare from substances that are readily available in a sufficiently pure state, and it should be stable over long periods of time. As an example we shall consider a solution that is saturated with potassium hydrogen tartrate, which will be represented as KHTart, at 298 K.

The electrochemical cell

$$\text{Ag} \,|\, \text{AgCl}(s),\ \text{NaCl}(c),\ \text{KHTart}(s)\,|\, \text{H}_2(g),\ \text{Pt} \tag{13.26}$$

is obtained by adding a known concentration, $c\,M$, of sodium chloride to a saturated solution of potassium hydrogen tartrate, and immersing a silver–silver chloride electrode and a hydrogen electrode in the mixture. The addition of chloride ion is necessary because the cell must contain an electrode that responds in the theoretical fashion to the activity of some anion, and no such electrode is known for tartrate or hydrogen tartrate ion. Section 14.2 will show that the potential of the cell in (13.26) is given by

$$E_{\text{cell}} = \left(E^0(\text{H}^+, \text{H}_2(g)) - \frac{0.059\ 16}{2} \log \frac{p_{\text{H}_2}}{a_{\text{H}^+}^2} \right) - (E^0(\text{AgCl}(s), \text{Ag}) - 0.059\ 16 \log a_{\text{Cl}^-})$$

$$= E^0(\text{H}^+, \text{H}_2(g)) - E^0(\text{AgCl}(s), \text{Ag}) - \frac{0.059\ 16}{2} \log p_{\text{H}_2} + 0.059\ 16 \log a_{\text{H}^+} a_{\text{Cl}^-}$$

at 298 K. The value of E_{cell} can be measured, and so can the partial pressure of hydrogen gas that is used, while the two standard potentials E^0 are known. A value of the product $a_{\text{H}^+} a_{\text{Cl}^-}$ can be obtained. By combining it with the known concentration c of chloride ion we can calculate the value of the quantity defined by the equation

$$\text{pwH} = -\log a_{\text{H}^+} y_{\text{Cl}^-} (= -\log a_{\text{H}^+} a_{\text{Cl}^-}/c). \tag{13.27}$$

The addition of sodium chloride has altered the ionic strength and the activity coefficients of all the species present in the solution. In particular, the activity of hydrogen ion that appears in eq. (13.27) is not the same as the activity of hydrogen ion in a saturated solution of potassium hydrogen tartrate that is free from chloride ion. These effects can be eliminated by performing similar measurements with other solutions containing different concentrations of sodium chloride, plotting the pwH against c, and extrapolating to $c = 0$. The extrapolated value of the pwH is given the symbol pwH^0 and is defined by the equation

$$\text{pwH}^0 = \lim_{c \to 0} (\text{pwH}) = -\log a_{H^+} y_{Cl^-} \qquad (13.28)$$

where a_{H^+} is now the activity of hydrogen ion in the pure (chloride-free) buffer, and y_{Cl^-} is the activity coefficient that an infinitesimal trace of chloride ion would have in the pure buffer.

We now define a conventional single-ion activity coefficient by the equation

$$\log y'_{Cl^-} = -\frac{0.511 \mu^{1/2}}{1 + 1.5 \mu^{1/2}} \qquad (13.29)$$

(at 298 K). This resembles eq. (13.24) but involves the arbitrary choice of a value close to 4.5 Å for $\mathring{a}_{Cl^-}$. Combining eqs. (13.28) and (13.29) gives

$$\text{pH}_s = -\log (a_{H^+} y_{Cl^-}/y'_{Cl^-}) = \text{pwH}^0 + \log y'_{Cl^-}. \qquad (13.30)$$

If the conventional activity coefficient of chloride ion (y'_{Cl^-}) defined by eq. (13.29) were identical with the true activity coefficient of chloride ion (y_{Cl^-}), both would disappear from the central term of eq. (13.30) and the pH$_s$ would be equal to the paH. We could be confident that the difference between the two activity coefficients was negligible if the buffer was so dilute that its ionic strength lay in the Debye–Hückel limiting-law region, but then the buffer capacity would be so small that the solution would not be very useful. We could improve the buffer capacity by making more concentrated reference buffers, but this would entail increasing the ionic strength and decrease our confidence in the utility of eq. (13.29). The compositions of actual reference buffers are selected as compromises between these two difficulties. Their ionic strengths are generally less than 0.1 M. It is hard to believe that y_{Cl^-} and y'_{Cl^-} could differ very much at ionic strengths as low as this, but there is no assurance that they are the same and there is no way in which that assurance could be obtained. The pH$_s$-values of different reference buffers are internally consistent and are as nearly identical with their paH-values as possible, but in principle the pH$_s$ and the paH are not the same and we do not know how large the difference between them is.

SUMMARY: The pH$_s$-value of a reference buffer solution is the closest approach to its paH that we know how to make. It is obtained by experimentally evaluating the pwH ($= -\log a_{H^+} y_{Cl^-}$) for solutions to which different concentrations of chloride ion have been added, extrapolating to zero concentration of added chloride ion, and combining the extrapolated value with an estimate of the activity coefficient of chloride ion.

Problems

Answers to some of these problems are given on page 522.

13.1. What is the ionic strength of an 0.100 M solution of iron(III) sulfate, $Fe_2(SO_4)_3$
(a) neglecting ion-pair formation?
(b) taking ion-pair formation into account, if the concentration dissociation constant of the $FeSO_4^+$ ion pair in such a solution is equal to 2.8×10^{-3} mol dm^{-3}?

13.2. What is the ionic strength of a solution containing magnesium sulfate at a total concentration of (a) 0.100 M or (b) 1.00 M? Take ion-pair formation into account, assuming that the concentration dissociation constant of $MgSO_4(aq)$ is equal to 4×10^{-3} mol dm^{-3} in each solution.

13.3. The solubility of ammonium picrate

, formula weight 246.1

in pure water at 298 K is 1.18 g dm^{-3}. Calculate the value of its thermodynamic solubility product.

13.4. The solubility of silver nitrite ($AgNO_2$, formula weight 153.9) in pure water at 298 K is 0.347 g dm^{-3}.

(a) Calculate the value of its thermodynamic solubility product, assuming that the reaction $AgNO_2(s) = Ag^+ + NO_2^-$ is the only one that occurs when silver nitrite dissolves in water.
(b) Nitrous acid, HNO_2, is a weak acid and its thermodynamic overall dissociation constant is equal to 7×10^{-4} mol dm^{-3} in aqueous solutions at 298 K. Will the occurrence of the further reaction $NO_2^- + H_2O(l) = HNO_2(aq) + OH^-$ have a significant effect on your answer to part (a)?

13.5. The solubility of calcium sulfate ($CaSO_4$, formula weight 136.1) in pure water at 298 K is 2.080 g dm^{-3}. Calculate the value of its thermodynamic solubility product

(a) neglecting ion-pair formation,
(b) taking ion-pair formation into account, if the thermodynamic dissociation constant of $CaSO_4(aq)$ is equal to 5.2×10^{-3} mol dm^{-3}.

Use eq. (13.18a) to estimate the activity coefficients of calcium and sulfate ions.

13.6. The solubility of magnesium oxalate (MgC_2O_4, formula weight 112.3) in pure water at 298 K is 0.075 g dm^{-3}. Calculate the value of its thermodynamic solubility product

(a) neglecting ion-pair formation,
(b) taking ion-pair formation into account, if the thermodynamic dissociation constant of MgC_2O_4 is equal to 3.7×10^{-4} mol dm^{-3}.

13.7. The solubility of lanthanum iodate [$La(IO_3)_3$, formula weight 663.6] in pure water at 298 K is 1.7 g dm^{-3}.

(a) Calculate the value of its thermodynamic solubility product. Take the formation of the $LaIO_3^{2+}$ ion pair into account, using eq. (5.18) to estimate the value of its thermodynamic dissociation constant.
(b) Calculate the solubility of lanthanum iodate in a solution containing lanthanum perchlorate, $La(ClO_4)_3$, at a total concentration of 0.01 M.

13.8. The thermodynamic dissociation constant of acetic acid in aqueous solutions at 298 K is equal to 1.759×10^{-5} mol dm^{-3}. Estimate the concentration of hydrogen ion in a solution containing

(a) 0.0100 M acetic acid and 0.0100 M sodium acetate,
(b) 0.0100 M acetic acid, 0.0100 M sodium acetate, and 0.0900 M sodium chloride.

Use eq. (13.25) and the values in Table 13.1 to obtain a value of $\mathring{a}$ to combine with eq. (13.18a) in calculating the value of $y_\pm$ for acetic acid.

CHAPTER 14

Potentiometry

14.1. Introduction

This chapter and the following one describe two instrumental techniques that are widely used to analyze solutions, locate the end points of titrations, evaluate equilibrium and rate constants, and obtain many other kinds of information about aqueous and non-aqueous solutions.

Potentiometry is the measurement of the potential of an electrochemical cell. The Nernst equation, which is the foundation of its theory, was discussed in Chapter 8. This chapter deals with the practical use of the Nernst equation and with some other important aspects of potentiometric theory.

The Nernst equation describes the potential of a single electrode, which is a quantity that is impossible to measure. All that can be measured is the difference between the potentials of two electrodes that are combined in an electrochemical cell. There are two kinds of electrochemical cells. In a *cell without liquid junction* both electrodes are immersed in the same solution. A lead storage battery is a cell without liquid junction. Cells without liquid junction are used in evaluating standard potentials and mean ionic activity coefficients; cells with liquid junction are more convenient for most other purposes. In a *cell with liquid junction* the two electrodes are immersed in different solutions. One of these is the solution whose composition or behavior is being studied, and the electrode immersed in this solution is called the *indicator electrode*. The potential of the indicator electrode depends on the activity of the substance that is of interest. The other electrode of a cell with liquid junction is a *reference electrode*, and is so designed and constructed that its potential is constant, reproducible, and independent of the composition of the solution being studied. The reference electrode contains a solution that is different from the one surrounding the indicator electrode, and the two solutions meet at a boundary called the *liquid junction*. Changing the composition of the solution being studied affects the potential of the cell because it alters the potential of the indicator electrode, but it does not affect the potential of the reference electrode.

There are two different ways of using the measured potential of a cell with liquid junction. In *direct potentiometry* a concentration (or a ratio of concentrations) is calculated directly from a measured potential. The measurement of pH with a glass indicator electrode responsive to hydronium ion is one example of direct potentiometry, and the determination of potassium ion in the blood stream by means of an ion-selective electrode responsive to potassium ion is another. Some additional information about the behavior of the indicator electrode is always needed in direct potentiometry. In a measurement of pH with a glass indicator electrode it can be obtained by immersing the indicator and reference electrodes in a buffer solution of known pH and adjusting the pH meter so that it indicates the known pH of the buffer. When the buffer is replaced by a solution of unknown pH, the difference between the potentials of the glass and reference electrodes will change unless this solution has the same pH as the buffer. Any change is translated into a difference of pH-values by the meter, and the pH of the unknown solution can be read directly. In other situations the difference between the formal potential of the indicator electrode and the potential of the reference electrode may be known in advance or may be measured separately, or a calibration curve may be constructed by measuring the potentials obtained with a number of solutions that contain different known concentrations of the substance that is being determined but are otherwise identical with the unknown solutions that must be analyzed.

In a *potentiometric titration* the potential is measured after adding each of a number of known volumes of a standard solution of a reagent to the solution that surrounds the indicator electrode. Usually the aim is to locate the end point of the titration. It can be located graphically after plotting the potential against the volume of reagent, but there are numerical techniques that are faster, easier, and more precise. Potentiometric titration also provides a rapid and convenient way of evaluating formal potentials of half-reactions and conditional equilibrium constants.

14.2. Electrochemical cells and their potentials

PREVIEW: This section explains how electrochemical cells are represented and describes the convention used in describing their potentials.

One way of representing an electrochemical cell was shown in Fig. 8.3. Another, which is more compact and easier to write, is used in this chapter. We would write

$$Pt, H_2(g) \,|\, HCl(c), AgCl(s) \,|\, Ag \tag{14.1}$$

to represent the cell without liquid junction shown in Fig. 14.1. Each of the vertical lines represents a difference of electrical potential that exists at the boundary between two different phases and that is included in the measured potential of the cell. The vertical line on the left corresponds to the potential of the hydrogen electrode; the one on the right corresponds to the potential of the silver–silver chloride electrode. There is also a difference of electrical potential across the boundary between the solid silver chloride and the solution, because the silver chloride acquires a negative charge by adsorbing excess chloride ions, but it is not represented by a vertical line because it does not affect the measured potential: as long as the solution is saturated with silver chloride it does not matter whether any of the solid is actually present. To measure the potential of a cell it must be connected to some measuring device, such as a potentiometer or a high-resistance dc voltmeter, by means of wires attached to its electrodes. The *potential of an electrochemical cell* can be defined formally as the difference between the electrical potentials of two identical conductors, one attached to each of the two electrodes of the cell.

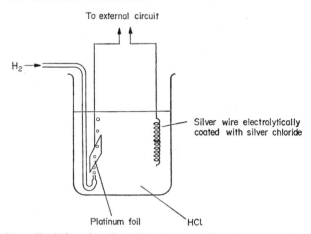

To external circuit

H₂ →

Silver wire electrolytically coated with silver chloride

Platinum foil   HCl

FIG. 14.1. A cell without liquid junction, for which the conventional representation is shown in (14.1). The platinum foil electrode would be "platinized" — electrolytically coated with very finely divided platinum — to increase its surface area and catalyze the half-reaction $H^+ + e = \frac{1}{2}H_2(g)$.

Every electrochemical cell is considered to represent a particular overall reaction, which is called the *net cell reaction*. The reaction may or may not be at equilibrium when its reactants and products have the specified activities, and if it is not it may be either spontaneous or non-spontaneous. So that different chemists will associate the same net cell reaction with any particular cell, and will come to the same decision about whether it is spontaneous under the conditions that prevail in the cell, the so-called Stockholm Convention was adopted by the International Union of Pure and Applied Chemistry in 1953. Some of its features appeared in Section 8.4. Two others of special importance here are:

1. The electrode on the left-hand side of a cell is considered to be the anode of the cell, and the electrode on the right-hand side is considered to be the cathode. The *anode* is defined as the electrode at which oxidation occurs; the *cathode* is the electrode at which reduction occurs.

2. The potential of a cell is equal to the difference between the potential of the electrode on its right-hand side and the potential of the electrode on its left-hand side:

$$E_{cell} = E_{right} - E_{left} = E_{cathode} - E_{anode}.$$

Consequently the process $H_2(g) = 2H^+ + 2e$ is considered to occur at the left-hand electrode on the cell in Fig. 14.1, and the process $AgCl(s) + e = Ag(s) + Cl^-$

is considered to occur at the right-hand electrode. The equation for the net cell reaction is obtained by adding the equations for these two half-reactions, multiplying each by whatever factor may be needed to make the numbers of electrons equal. The result is $H_2(g)+2AgCl(s) = 2Ag(s)+2H^++2Cl^-$.

The potential of the cell is given by

$$E_{cell} = E(AgCl(s), Ag(s))-E(H^+, H_2(g)). \qquad (14.2)$$

It will be positive if the potential of the silver–silver chloride electrode on the right-hand side is more positive than that of the hydrogen electrode on the left-hand side; it will be negative if the silver–silver chloride electrode has the more negative potential. In either case the potential of the cell is related to the change of free energy that accompanies the net cell reaction:

$$\Delta G = -nFE_{cell} \qquad (14.3)$$

where n is the number of electrons that would appear on each side of the balanced equation for the net cell reaction if they had not been dropped in writing it. For the cell in (14.1), and the equation at the end of the preceding paragraph, $n = 2$.

These things mean that the net cell reaction is spontaneous if the electrode on the right-hand side of the cell is the positive electrode, so that E_{cell} is positive. The net cell reaction is non-spontaneous if the electrode on the right-hand side is the negative electrode, so that E_{cell} is negative. If the two electrodes have the same potential, then $E_{cell} = 0$ and the net cell reaction is at equilibrium.

The potentials of the electrodes can be described by writing Nernst equations for them, and combining these with eq. (14.2) gives, at 298 K,

$$E_{cell} = \left[E^0(AgCl(s), Ag(s))-0.059\,16 \log a_{Cl^-}\right]$$
$$- \left[E^0(H^+, H_2(g))-\frac{0.059\,16}{2} \log \frac{p_{H_2}}{a_{H^+}^2}\right]. \qquad (14.4)$$

If each of the substances that is present in the cell has unit activity—which is assumed to be true for solid silver and silver chloride in writing eq. (14.4)—the logarithmic terms on the right-hand side of this equation vanish. The potential of the cell under these conditions is called the *standard potential of the cell*, and is described by the equations

$$E^0_{cell} = E^0(AgCl(s), Ag(s))-E^0(H^+, H_2(g)) \quad [= E^0_{right}-E^0_{left}]$$

and

$$\Delta G^0 = -nFE^0_{cell}(= -RT \ln K^0)$$

where K^0 is the thermodynamic equilibrium constant of the net cell reaction. All of these statements are equivalent to others made in Chapter 8.

SUMMARY: Reduction is considered to occur at the right-hand electrode of a cell, and the potential of a cell is given the same sign as the potential of its right-hand electrode. These arbitrary decisions make it possible to associate each cell with a particular net cell reaction, and to tell whether that reaction is spontaneous under any particular conditions.

14.3. How standard potentials and activity coefficients are evaluated

PREVIEW: This section explains how standard potentials and mean ionic activity coefficients are calculated from the measured potentials of cells without liquid junction.

Equation (14.4) shows how the potential of the cell in (14.1) depends on the activities of hydrogen and chloride ions in the solution with which the cell is filled. This section will show how that equation can be used to evaluate both the standard potential of the silver chloride–silver half-reaction and the mean ionic activity coefficients of hydrochloric acid from experimental data.

The evaluation of a physical constant, such as $E^0(AgCl(s), Ag(s))$ in eq. (14.4), usually involves three steps. The first is to obtain an equation showing how the desired constant is related to other quantities that are known or can be measured. The second is to perform the necessary experimental measurements. The third is to combine the data with the equation to obtain the desired result.

The derivation of eq. (14.4) is part of the first step. It contains the quantity, $E^0(AgCl(s), Ag(s))$, that we wish to evaluate, and it contains the potential of a cell, which can be measured. However, it also involves the activities of hydrogen and chloride ions, which might not be known with sufficient accuracy or might not be known at all. Allowance must often be made for quantities that may be unknown and unpredictable. A way of making it can usually be found if all the terms that are known or measurable are collected onto one side of the equation. We may begin by dropping $E^0(H^+, H_2(g))$ because this is equal to zero by definition. The partial pressure of hydrogen gas at the surface of the hydrogen electrode cannot be measured directly, but it can be calculated from quantities that can be measured. The total pressure exerted on a bubble of gas is equal to the sum of the barometric pressure and the hydrostatic pressure of the column of solution above it, and is also equal to the pressure inside the bubble, which is the sum of the partial pressures of hydrogen and water vapor. Provided that the hydrogen is efficiently mixed with the solution in a separate vessel before being led into the cell, the partial pressure of water will have its equilibrium value at the temperature of the measurement, and will not differ appreciably from the vapor pressure of pure water at the same temperature if the concentration of hydrochloric acid is small.

The activities in eq. (14.4) are unknown, but each can be expressed as the product of a concentration by an activity coefficient. If a solution of hydrochloric acid is prepared by dilution to the mark in a volumetric flask, its concentration c in mol dm^{-3} will be easy to calculate, but it will be very difficult to decrease the relative standard deviation of c below about 0.05%. Equation (14.4) could be rearranged to give the concentration-dependent term $2 \times 0.059\ 16 \log c_{HCl}$ [compare eq. (14.5)]; if the relative standard deviation of c_{HCl} is 0.05%, the standard deviation of $2 \times 0.059\ 16 \log c_{HCl}$ will be 0.026 mV. In careful work all the other terms in the equation could be evaluated more precisely than this, and therefore it is desirable to make up the solutions by weight instead of by volume. Molality (the number of moles of hydrochloric acid per kilogram of solvent) then becomes more convenient than the number of moles per cubic decimeter, and we shall therefore express each ionic activity as the product of a molality m by a molality activity coefficient γ.

Rearranging eq. (14.4) so that all the measurable quantities appear on the left-hand side gives

$$E_{cell} - \frac{0.059\,16}{2} \log p_{H_2} + 0.059\,16 \log m_{H^+} m_{Cl^-}$$

$$= E^0(AgCl(s), Ag(s)) - 0.059\,16 \log \gamma_{H^+} \gamma_{Cl^-}.$$

This can be simplified in several ways. The molalities of the hydrogen and chloride ions will be identical with the molality m_{HCl} of the hydrochloric acid as long as this is not so low that the reaction $AgCl(s) = Ag^+ + Cl^-$ contributes an appreciable additional concentration of chloride ion, and therefore $m_{H^+} m_{Cl^-}$ can be replaced by m_{HCl}^2. The product $\gamma_{H^+} \gamma_{Cl^-}$ can be replaced by the square of the mean ionic activity coefficient $\gamma_\pm$, and finally we can denote the collection of measurable quantities on the left-hand side of the equation by the symbol ε. The result is

$$\varepsilon = E_{cell} - \frac{0.059\,16}{2} \log p_{H_2} + 2 \times 0.059\,16 \log m_{HCl}$$

$$= E^0(AgCl(s), Ag(s)) - 2 \times 0.059\,16 \log \gamma_\pm. \tag{14.5}$$

Since $\gamma_\pm = 1$ at $\mu = 0$, the standard potential would equal the value of ε at $\mu = 0$. Unfortunately that value cannot be measured because m_{HCl} cannot be made smaller than about 0.001 mol kg^{-1} without allowing the solubility of silver chloride to become too large to neglect. What must therefore be done is to find the values of ε for a number of different solutions of hydrochloric acid more concentrated than about 0.001 m, and extrapolate them to $\mu = 0$. Then

$$E^0(AgCl(s), Ag(s)) = \lim_{\mu \to 0} \varepsilon. \tag{14.6}$$

Having thus completed the first of our three steps we could proceed to assemble a cell, prepare the necessary solutions, and obtain the values of ε. In doing these things it would be important to prepare the electrodes in ways that are known to give reproducible and reliable results; and to design the cell in such a way that silver(I), which is produced by the dissolution of silver chloride in the solution surrounding the silver electrode, does not reach the surface of the hydrogen electrode. If it did, it would be spontaneously reduced in the reaction $2\,Ag^+ + H_2(g) = 2\,Ag(s) + 2\,H^+$, the surface of the platinum electrode would become covered with a layer of metallic silver, and its ability to catalyze the hydrogen ion–hydrogen half-reaction would be lost.

As eq. (14.6) indicates, the third step involves an extrapolation. Some of the dangers that beset extrapolations appeared in Section 12.9 and Problems 12.5 and 12.6. You need a model to guide you in constructing the curve that you extrapolate, and it should be one that gives a straight line rather than a curve. In analytic geometry the extrapolation of a straight line is very simple; in science it is much less so because of the random errors of measurement. You can never be completely certain about how to draw a straight line through the points, and your uncertainty leads to an uncertainty in the intercept. According to eq. (12.24a), the uncertainty in the intercept increases as the number of points decreases, as the standard deviation from regression increases, as the absolute value of the slope increases, as the range of values of the independent variable covered by the data becomes narrower, and as the distance covered by the extrapolation becomes longer. The ideal extrapolation would be a very short extrapolation of a very long horizontal straight line that was defined by a great many very precise data. Now you should read p. 383 again.

With these ideas in mind let us return to eqs. (14.5) and (14.6). Our first thought might be to plot ε against m_{HCl}, which gives curve (a) of Fig. 14.2. Only a short

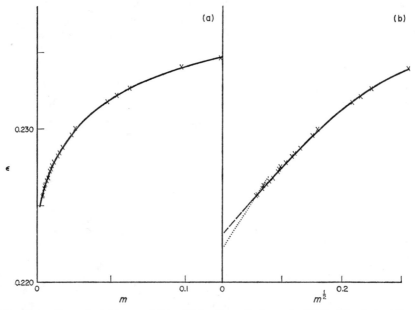

FIG. 14.2. Plots of ε against (a) m and (b) $m^{1/2}$ for the cell shown in (14.1). The dashed line is an attractive extrapolation of curve (b), but it cannot be correct because the slope must approach the slope of the dotted line as m approaches zero. The data shown in this figure and Fig. 14.3 were obtained by H. S. Harned and R. W. Ehlers, *J. Amer. Chem. Soc.* **54**, 1350 (1932).

extrapolation is required, but the curve is so far from linear and so steep that its intercept may not be certain even within 1 mV. This would be a poor reward for the trouble expended in achieving a precision as small as 0.02 mV in the individual measurements. Our second thought might be to employ the Debye–Hückel limiting law as a model: combining the equation $\log \gamma_{\pm} = -0.511 \sqrt{\mu}$ with eq. (14.5) would give

$$\varepsilon = E^0(\mathrm{AgCl}(s), \mathrm{Ag}(s)) + 2 \times 0.059\,16 \times 0.511 \sqrt{\mu}. \tag{14.7}$$

Since μ is almost exactly proportional to m_{HCl} in dilute solutions, this suggests plotting ε against $m_{HCl}^{1/2}$. The result is shown in curve (b) of Fig. 14.2. It is little, if at all, better than curve (a): the points cover a narrower range, and a longer extrapolation is needed. The temptation to draw the dashed line must be resisted, for eq. (14.7) requires the slope shown by the dotted line instead.

Clearly m_{HCl} is a better independent variable than $m_{HCl}^{1/2}$, both because the points cover a wider range of values of m_{HCl} and because only a relatively short extrapolation is needed. The trouble with Fig. 14.2(a) is that it does not employ any model. However, eq. (14.5) can be combined with the Hückel equation to obtain

$$\varepsilon = E^0(\text{AgCl}(s), \text{Ag}(s)) - 2 \times 0.059\,16 \left[\frac{-0.511\,m_{HCl}^{1/2}}{1 + 0.329 \mathring{a} m_{HCl}^{1/2}} + B m_{HCl} \right]$$

and then, after rearrangement

$$\varepsilon - \frac{2 \times 0.059\,16 \times 0.511\,m_{HCl}^{1/2}}{1 + 0.329 \mathring{a} m_{HCl}^{1/2}} = E^0(\text{AgCl}(s), \text{Ag}(s)) - 2 \times 0.059\,16\, B m_{HCl}. \qquad (14.8)$$

The second term on the left-hand side accounts for most of the variation of $\log \gamma_\pm$ as long as m_{HCl} is not too large. Because the value of the salting coefficient B is small, it should be possible to obtain a nearly horizontal straight line by plotting the quantity on the left-hand side against m_{HCl}. It is necessary to choose a value of $\mathring{a}$. Three possible choices are shown in Fig. 14.3. Too large or too small a value of $\mathring{a}$ gives a plot that is slightly curved; the value $\mathring{a} = 4.8$ Å gives a straight line. The intercepts of all three plots are indistinguishable on this scale, and are equal to the desired standard potential. A least-squares fit to eq. (14.8), in which there are three adjustable parameters ($\mathring{a}$ and B in addition to the standard potential), would be chemically equivalent, but statistically superior, to this graphical procedure. Once the standard potential has been evaluated, $\log \gamma_\pm$ can be calculated at each of the experimental points by using eq. (14.5). If the values of $\mathring{a}$ and B have been obtained in the process, either from Fig. 14.3 or from regression analysis, combining them with the Hückel equation makes it possible to calculate the value of $\log \gamma_\pm$ at any desired concentration of hydrochloric acid over the range covered.

There are other experimental ways of evaluating mean ionic activity coefficients, but they do not fall within the scope of this text. Standard potentials can also be evaluated in other ways, which were outlined in Section 8.6. To evaluate a standard potential by the procedure described here, one needs a cell without liquid junction, in which one electrode responds to the activity of the cation, while the other responds to the activity of the anion, of the electrolyte that is used. For example, the standard potential of the couple $Pb^{2+} + 2\,e = Pb(s)$ can be obtained by measuring the potential of the cell $\text{Ag} | \text{AgCl}(s), \text{PbCl}_2(c) | \text{Pb}$ with different concentrations of lead chloride, and combining the data with the value of $E^0(\text{AgCl}(s), \text{Ag}(s))$ taken from Fig. 14.3.

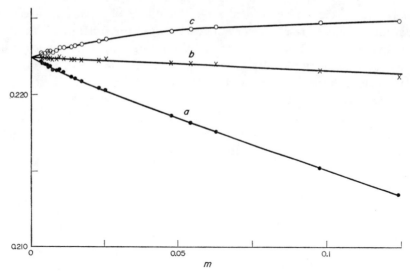

FIG. 14.3. Plots of the quantity on the left-hand side of eq. (14.8) against m_{HCl} with $\mathring{a}$ = (a) 0, (b) 4.8, and (c) 10 Å.

SUMMARY: Standard potentials are evaluated by measuring the potentials of cells without liquid junction over a range of concentrations of electrolyte and extrapolating a function derived from the Hückel equation to zero ionic strength. After the standard potential has been evaluated, it is easy to calculate the mean ionic activity coefficient of the electrolyte at each concentration employed.

14.4. Redox indicator electrodes

PREVIEW: This section describes the two important kinds of indicator electrodes for redox processes: inert electrodes and active electrodes.

Every practical electrochemical cell must contain at least one indicator electrode; cells without liquid junction, like those in Section 14.3, can be regarded as containing two indicator electrodes.

Indicator electrodes are of two different kinds. All those mentioned so far are redox electrodes. The potential of a redox electrode is governed by the reduction and oxidation that occur at the interface between the electrode and the solution surrounding it. Membrane electrodes, often called ion-selective electrodes, operate in a different way and are discussed in later sections.

Redox electrodes can be further divided into two classes: inert and active. An inert indicator electrode is one that is made from a substance that does not appear in the equation for the half-reaction. If the half-reaction is $Fe^{3+} + e = Fe^{2+}$, a platinum electrode would be an inert electrode, serving merely to carry electrons between the electrical circuitry outside the cell and the iron(III) and iron(II) ions dissolved in the solution being studied. An active electrode is made from a substance that does appear in the equation for the half-reaction. If the half-reaction is $Ag^+ + e = Ag(s)$, a silver electrode would be an active electrode, and would be needed to ensure that solid silver was present, at unit activity, at the interface between the electrode and the solution.

Half-reactions that involve only dissolved substances can be studied only with inert electrodes. Some such half-reactions are $Cr(III) + e = Cr(II)$, $Ce(IV) + e = Ce(III)$, and $I_3^- + 2e = 3I^-$. Of course the electrode used in studying such a half-reaction must be a good conductor of electricity and must not react with the solution chemically. However, there are many metals—including stainless steel, tungsten, and tantalum—that fulfill both of these requirements and yet are useless as indicator electrodes because their surfaces are always covered with thin films of oxides that interfere with the transfer of electrons between the metal and the solution. Most inert indicator electrodes are made of platinum. Mercury is used with solutions containing such powerful reducing agents that reactions of the form $Red + H^+ = Ox + \frac{1}{2}H_2(g)$ would occur at appreciable rates at platinum surfaces, and electrodes made from certain forms of carbon are used occasionally.

A platinum indicator electrode may be prepared by coiling several inches of one end of a platinum wire into a helix by winding it around a pencil or glass tube. The coiled portion of the wire is immersed in the solution being studied; the remaining portion is used to make an electrical connection to the external circuit. The coiled portion should be cleaned just before use by immersing it in concentrated nitric acid, rinsing it with water, and igniting it to a bright red heat in an oxidizing flame.

There are two important kinds of active electrodes. One consists of a metal immersed in a solution containing its ions. The electrodes $Ag\,|\,Ag^+$, $Cu\,|\,Cu^{2+}$, and $Hg\,|\,Hg_2^{2+}$ are typical electrodes of this kind.

Such electrodes generally respond to the activities of dissolved cations. To observe the activity of a dissolved anion, one needs an electrode that consists of a metal (such as silver) immersed in a solution saturated with a slightly soluble salt (such as silver chloride) that contains both the cation (silver ion) to which the electrode responds and the anion (chloride ion) that is of interest. In the presence of silver chloride at unit activity, the potential of a silver electrode at 298 K is given by (compare p. 230)

$$\begin{aligned}
E &= E^0(Ag^+, Ag(s)) - 0.059\,16\,\log\,(1/a_{Ag^+}) \\
&= E^0(Ag^+, Ag(s)) - 0.059\,16\,\log\,(a_{Cl^-}/K^0_{AgCl}) \\
&= E^0(Ag^+, Ag(s)) + 0.059\,16\,\log K^0_{AgCl} - 0.059\,16\,\log a_{Cl^-} \\
&= E^0(AgCl(s), Ag(s)) - 0.059\,16\,\log a_{Cl^-}
\end{aligned}$$

so that the electrode responds to the activity of dissolved chloride ion. Similarly, an electrode responsive to the activity of oxalate ion could be made by combining a silver electrode with an excess of solid silver oxalate.

Silver, copper, and other metal electrodes are made in the same general way as platinum electrodes, but they are usually cleaned simply by etching them briefly with

28

4 to 8 M nitric acid and then rinsing them with water. If the solution must be saturated with a sparingly soluble salt, more elaborate preparation is necessary because the salt must be present at the surface of the electrode to ensure that the solubility equilibrium is attained there during the measurement. It is convenient to coat the electrode with the salt by anodic deposition. To make a silver–silver chloride electrode, for example, the clean helical portion of a silver electrode may be immersed in dilute (perhaps 0.1 M) hydrochloric acid and connected to the positive terminal of a 1.5-V dry cell. The negative terminal of the dry cell is connected, through a rheostat and a milliammeter, to a scrap of platinum wire dipped into the acid. The rheostat is adjusted to give a current of 1–10 mA, which is allowed to flow for a few minutes while a thin layer of silver chloride forms on the surface of the silver anode. Too thin a layer is undesirable because it is too easily dissolved or reduced, but too thick a layer would also be undesirable because it would retard the attainment of equilibrium with dissolved chloride ions.

SUMMARY: The metal, or other electronic conductor, from which an indicator electrode is made may or may not participate in the half-reaction that occurs at its surface. If it does, it may respond either to the concentration either of its own ions or to the concentration of the oppositely charged ion of a salt of its own ions in a solution saturated with that salt.

14.5. Reversibility and irreversibility

PREVIEW: This section shows how failure to obey the Nernst equation is identified experimentally, describes its effects in practical work, and explains why it happens.

No redox couple obeys the Nernst equation under all conditions, and many do not obey it under any conditions yet discovered. This section describes several possible reasons why the Nernst equation may not be obeyed.

We must begin by defining obedience to the Nernst equation and saying why it is important. For the half-reaction $aA+bB+\ldots+ne = \ldots yY+zZ$, the Nernst equation is

$$E = E^0 - \frac{RT}{nF}\ln\frac{\ldots a_Y^y a_Z^z}{a_A^a a_B^b \ldots}\left(\text{or}\quad E^{0\prime} - \frac{RT}{nF}\ln\frac{\ldots[Y]^y[Z]^z}{[A]^a[B]^b\ldots}\right)$$

If everything is kept constant except the activity of Y,

$$\frac{\partial E}{\partial \ln a_Y}\left(\text{or}\quad \frac{\partial E}{\partial \ln [Y]}\right) = -y\frac{RT}{nF} \tag{14.9a}$$

while if in another series of measurements everything is kept constant except the activity of A,

$$\frac{\partial E}{\partial \ln a_A}\left(\text{or}\quad \frac{\partial E}{\partial \ln [A]}\right) = +a\frac{RT}{nF}. \tag{14.9b}$$

A half-reaction, couple, or electrode is said to exhibit *Nernstian* or *reversible* behavior whenever equations like eqs. (14.9) are found experimentally to be valid for all of the substances whose activities can be varied. If any one such equation is not obeyed under certain conditions, the behavior is said to be non-Nernstian or irreversible under those conditions.

Either direct potentiometry or potentiometric titration is dangerous or impossible if the couple of interest is irreversible. You could not find the activity or concentration of nickel(II) ion in an aqueous solution by measuring the potential of a nickel electrode, for that potential does not depend on the activity or concentration of nickel(II) ion. Nor could you locate the end point of a titration of nickel(II) ion with ethylenediaminetetraacetate by following the potential of a nickel electrode, because that potential would not change as the concentration of nickel(II) ion changed. Accidental and irrelevant changes of the experimental conditions may affect the potential of an electrode that is behaving irreversibly, and confusion and error are sure to result.

A chemical reaction reaches equilibrium when the rate of the forward process becomes equal to the rate of the backward one, and the expression for the equilibrium constant is always obeyed when these rates are equal. A redox couple reaches equilibrium when the rate at which its oxidized form is being reduced becomes equal to the rate at which its reduced form is being oxidized, and the Nernst equation is always obeyed when these rates are equal. The reduction of the oxidized form gives rise to a cathodic current I_c; the oxidation of the reduced form gives rise to an anodic current I_a. Cathodic currents correspond to a flow of electrons from the external circuit into the indicator electrode and the solution. Anodic currents correspond to a flow of electrons in the opposite direction, and therefore cathodic and anodic currents are considered to have opposite signs: cathodic currents are considered to be negative and anodic ones to be positive. The values of I_c and I_a for any couple depend on the potential of the electrode and on the composition of the layer of solution in contact with its surface. The value of I_c increases exponentially[†] as the potential of the electrode becomes more negative (more strongly reducing), and is proportional to the concentration of the oxidized form Ox in this layer:

$$I_c = -nFAk_h c^0_{Ox}\, e^{-\alpha n_a F(E-E^{0\prime})/RT}. \tag{14.10a}$$

The value of I_a increases exponentially[†] as the potential of the electrode becomes more positive (more strongly oxidizing), and is proportional to the concentration of the reduced form Red at the surface of the electrode:

$$I_a = nFAk_h c^0_{Red}\, e^{(1-\alpha)\, n_a F(E-E^{0\prime})/RT}. \tag{14.10b}$$

---

† The relationship is exponential only if c^0_{Ox} (or c^0_{Red}) is independent of current, which is nearly true only if the current is small. If I_c is appreciable, c^0_{Ox} decreases, and c^0_{Red} increases, because Ox is consumed and Red is produced at the surface of the electrode. Interesting things happen if I_c is so large that c^0_{Ox} is indistinguishable from zero, but they are of no concern in potentiometric measurements.

28*

In these equations n is the number of electrons involved in the overall half-reaction Ox$+ne$ = Red, F is the number of coulombs per faraday, A is the area of the electrode, k_h is a heterogeneous rate constant for the electron-transfer process, c_{Ox}^0 and c_{Red}^0 are the concentrations of Ox and Red at the surface of the electrode, n_a is the number of electrons involved in the rate-determining step and any fast steps that precede it, E is the potential of the electrode, E^0 is the formal potential of the Ox–Red couple, and R and T have their usual meanings. The quantity α, called the transfer coefficient, is a constant whose value lies between 0 and 1 and reflects the shape of the activation-energy barrier for the Ox–Red couple under the conditions being considered.

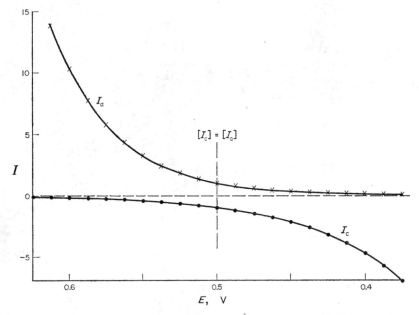

FIG. 14.4. Plots of I_a and I_c against potential for a couple having $\alpha n_a = 0.40$. The dashed vertical line is drawn at the potential where $|I_c| = |I_a|$.

Typical plots of I_c and I_a against potential are shown in Fig. 14.4. There is a particular value of the potential, shown by the dashed vertical line, at which the values of I_c and I_a are numerically equal. If there were no other processes that could occur, the current flowing through the external circuit would be equal to zero at this potential because the current flowing in one direction because of the reduction of Ox would just counterbalance the current flowing in the opposite direction because of the oxidation of Red.

Let us suppose that the half-reaction occurs in a single step, so that $n_a = n$, and that $|I_c| = |I_a|$. Then eqs. (14.10) give

$$nFAk_h c_{Ox}^0 \, e^{-\alpha n_a F(E-E^{0\prime})/RT} = nFAk_h c_{Red}^0 \, e^{(1-\alpha)\, n_a F(E-E^{0\prime})/RT}$$

or

$$\frac{c_{Ox}^0}{c_{Red}^0} = \frac{e^{(1-\alpha) n_a F(E-E^{0'})/RT}}{e^{-\alpha n_a F(E-E^{0'})/RT}} = e^{n_a F(E-E^{0'})/RT}.$$

On taking the logarithm of each side, rearranging, and replacing n_a by n, we can obtain

$$E = E^{0'} - \frac{RT}{nF} \ln \frac{c_{Red}^0}{c_{Ox}^0}. \tag{14.11}$$

But if $I_c = -I_a$ there is no current flowing in the external circuit. Then the concentrations of Ox and Red at the surface of the electrode cannot differ from their concentrations [Ox] and [Red] in the bulk of the solution, and therefore

$$E = E^{0'} - \frac{RT}{nF} \ln \frac{[Red]}{[Ox]}$$

which is of course the Nernst equation. The derivation is a little more complicated if the half-reaction occurs by a mechanism such as

$$Ox + n_a e = Int \quad \text{rate-determining,}$$
$$Int + (n - n_a) e = Red \quad \text{fast equilibrium}$$

(in which Int represents some intermediate), so that n_a is different from n, but the same result is obtained.

Thus the Nernst equation must be obeyed by the potential where $|I_c| = |I_a|$, but it is not obeyed by any other potential. If $|I_c| > |I_a|$, so that Red is being formed more rapidly than it is being consumed, equilibrium has not been attained. In addition the concentration of Red will be larger, and the concentration of Ox will be smaller, at the surface of the electrode than in the bulk of the solution if Red is produced, and Ox is consumed, by the flow of a finite current.

It is not possible to measure either I_c or I_a alone. What is measured is the current I that flows in the external circuit, and every potentiometric measurement is designed to find the potential at which this current is equal to zero. If the reduction of Ox and the oxidation of Red are the only processes that can occur, then the condition

$$I = I_c + I_a = 0$$

could be satisfied only when $|I_c| = |I_a|$. We define the *exchange current* I^0 by the equation

$$I^0 = |I_c| = |I_a|$$

so that it is numerically equal to either the cathodic or the anodic current at the potential where $|I_c| = |I_a|$.

If $|I_c|$ were always equal to $|I_a|$ when there was no current flowing in the external circuit, then the Nernst equation would always be obeyed, and yet the behavior that is generally observed is the one shown by Fig. 14.5. This represents data obtained with the couple $Br_2(aq) + 2e = 2Br^-$, using a platinum indicator electrode in solutions containing identical concentrations of bromide ion but different concentrations of bromine. If the system behaved reversibly, then according to eq. (14.9b) we would find $\partial E/\partial \log [Br_2(aq)] = 0.059\ 16/2 = 0.0296$ V at 298 K. Within the experimental error of measurement this is indeed the slope obtained in region A, which extends up to concentrations of bromine far above those included on this figure. In region B the concentrations of Br_2 are smaller and so is the average slope; in region C, at concentrations below about 3×10^{-7} M, the slope becomes equal to zero and the potential becomes independent of the concentration of bromine.

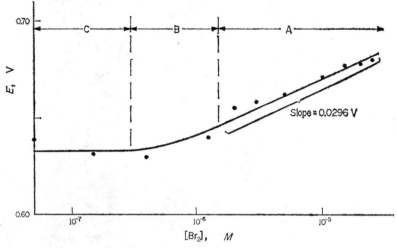

FIG. 14.5. The effect of the concentration of bromine on the potential of a platinum electrode in a solution containing 0.1 M sulfuric acid and 0.2 M sodium bromide. The data were obtained by W. C. Purdy, E. A. Burns, and L. B. Rogers, *Anal. Chem.* **27**, 1988 (1955).

Every indicator electrode behaves in this way, but the boundaries between regions lie at different concentrations for different systems. The potential of this electrode is independent of the concentration of bromine if this is smaller than about 3×10^{-7} M, and it behaves reversibly if the concentration exceeds about 1.5×10^{-6} M. There are only a very few couples for which the corresponding concentrations are smaller than these, but there are a great many for which they are much larger. A few half-reactions, such as

$$Fe^{3+} + e = Fe^{2+} \quad \text{and} \quad Fe(CN)_6^{3-} + e = Fe(CN)_6^{4-},$$

also behave reversibly down to low concentrations, but those that involve atom transfer and structural rearrangement as well, such as

$$SO_4^{2-} + H_2O(l) + 2e = SO_3^{2-} + 2OH^- \quad \text{and} \quad Cr_2O_7^{2-} + 14H^+ + 6e = 2Cr^{3+} + 7\,H_2O,$$

may not behave reversibly under any conditions (which is to say that the concentrations above which they would behave reversibly are so high that they are impossible to attain in practice). Comparing these statements with the ones made on page 241 should make you suspect that reversibility and irreversibility are related to the rates of the processes that occur at the surface of the electrode. These rates are described by the magnitude of the exchange current I^0. If these rates are low—as they will be if the exchange current is small—the couple is likely to behave irreversibly. If they are high, the couple is likely to behave reversibly.

This is because of the form of eqs. (14.10), which say that every oxidizable or reducible substance in a solution gives rise to a finite current at any potential. Every aqueous solution of Ox contains two substances that can be reduced: Ox is one and water is the other [$H_2O(l)+e = \frac{1}{2}H_2(g)+OH^-$]. It may also include others, such as dissolved oxygen and, in acidic solutions, hydrogen ion [$H^+ + e = \frac{1}{2}H_2(g)$]. Similarly, every aqueous solution of Red contains two substances that can be oxidized: Red itself and water [$H_2O(l) = \frac{1}{2}O_2(g)+2H^+ + 2e$].

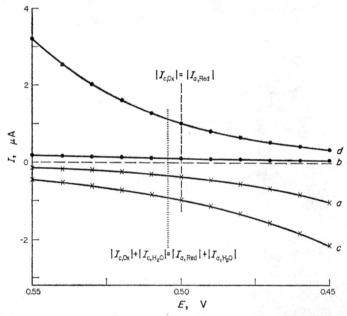

FIG. 14.6. Curves (a) and (b) are plots of current against potential for the reduction of water to hydrogen and the oxidation of water to oxygen, respectively. Curves (c) and (d) are plots of I_c and I_a against potential for the reduction of Ox and the oxidation of Red, respectively, and are copied (and magnified) from Fig. 14.4.

Figure 14.6 shows the result. Curve a represents the cathodic current obtained from the reduction of water to hydrogen, and curve b represents the anodic current obtained from the oxidation of water to oxygen. Curves c and d are copied from Fig. 14.4 and represent the currents obtained from Red and Ox. As in Fig. 14.4, the vertical dashed line shows the potential at which $|I_{c,\,Ox}| = |I_{a,\,Red}|$. However, the net current flowing in the external circuit is given by

$$I = I_{c,\,H_2O} + I_{c,\,Ox} + I_{a,\,Red} + I_{a,\,H_2O} \qquad (14.12)$$

and will not be equal to zero at this potential unless $|I_{c, H_2O}| = |I_{a, H_2O}|$, which could only be true by coincidence. As Fig. 14.6 has been drawn, $|I_{a, H_2O}|$ (curve b) is smaller than $|I_{c, H_2O}|$ (curve a) at the potential where $|I_{c, Ox}| = |I_{a, Red}|$, and therefore I will have a negative value at this potential. Only at a somewhat different potential, which is shown by the dotted line, will the net current I be equal to zero.

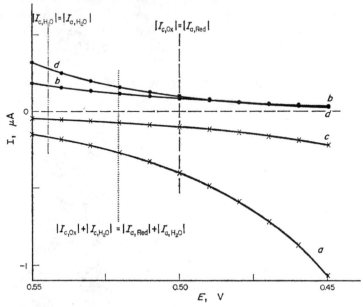

FIG. 14.7. Curves (a) and (b) are identical with curves (a) and (b) in Fig. 14.6, but are magnified by the change of scale on the ordinate axis. Curves (c) and (d) represent currents one-tenth as large at every potential as those represented by curves (c) and (d) in Fig. 14.6.

Figure 14.7 is just like Fig. 14.6 but it is drawn for concentrations of Ox and Red that are just one-tenth as large as for Fig. 14.6. These changes cause each value of $I_{c, Ox}$ and $I_{a, Red}$ in Fig. 14.7 to be just one-tenth as large as at the same potential in Fig. 14.6. Dividing both $|I_{c, Ox}|$ and $|I_{a, Red}|$ by 10 does not alter the potential at which they are equal, and this Nernstian potential is still shown by the dashed line. Now, however, the exchange current for the Ox–Red couple is so small that the couple has much less effect on the measured potential, which approaches the one where $|I_{c, H_2O}| = |I_{a, H_2O}|$. In still more dilute solutions both $I_{c, Ox}$ and $I_{a, Red}$ would be so small that the potential where $I = 0$ would not be changed appreciably by varying them.

Very large values of the exchange current correspond to region A in Fig. 14.5, where reversible behavior is observed. Very small values of the exchange current correspond to region C, where the measured potential is unaffected by the presence of the substance that is of interest. In the transitional region B, where the concentrations of Red and Ox do affect the measured potential but where eqs. (14.9) are not obeyed, the exchange current for the Ox–Red couple is comparable with the cathodic and anodic currents that arise from the solvent or supporting electrolyte and that would flow even if there were no Ox or Red present at all.

Equation (14.11) describes the potential at which $|I_{c, Ox}| = |I_{a, Red}| = I^0$. Substituting this description into eq. (14.10a) (or eq. (14.10b)) gives a description of I^0:

$$I^0 = |I_c| = nFAk_h c_{Ox}^0 \exp \left\{ -\alpha n_a F \left[-\frac{RT}{n_a F} \ln \frac{c_{Red}^0}{c_{Ox}^0} \right] \bigg/ RT \right\}$$

$$= nFAk_h c_{Ox}^0 \left(\frac{c_{Red}^0}{c_{Ox}^0} \right)^\alpha = nFAk_h (c_{Ox}^0)^{1-\alpha} (c_{Red}^0)^\alpha.$$

On the average α is roughly equal to $\frac{1}{2}$. The exchange current decreases as either c_{Ox}^0 or c_{Red}^0 decreases: irreversible behavior is always observed at very low concentrations. Moreover, the exchange current decreases as k_h decreases: irreversible behavior may be observed even at high concentrations if the rate constant for the half-reaction is very small. For the half-reaction $Br_2(aq) + 2e = 2Br^-$ the value of k_h is quite large, and Fig. 14.5 shows that this couple behaves reversibly down to very low concentrations. For the half-reaction $Ni(OH_2)_6^{2+} + 2e = Ni(s) + 6H_2O(l)$ the value of k_h is very small, and irreversible behavior is the result even at the highest concentrations of $Ni(OH_2)_6^{2+}$ that can be attained. The values of I_{c, H_2O} and I_{a, H_2O} must also be considered. If, for example, the potential were so positive that I_{a, H_2O} was very large, the exchange current of the Ox–Red couple would have to be enormous in order to produce Nernstian behavior. Couples that involve powerful oxidizing or reducing agents, such as permanganate and chromium(II) ions, therefore tend to behave reversibly only in very concentrated solutions, if at all. Couples that involve oxidizing and reducing agents of more modest strength, and therefore give potentials such that both I_{c, H_2O} and I_{a, H_2O} are very small, are much more likely to behave reversibly down to very low concentrations.

SUMMARY: The total current flowing through an electrode is the sum of the cathodic and anodic currents for all the processes that occur at its surface. Nernstian behavior is observed only when the cathodic and anodic currents due to the couple of interest, which depend on the concentrations of its oxidized and reduced forms and also on the value of its heterogeneous rate constant for electron transfer, are much larger than those due to other substances.

14.6. Reference electrodes

PREVIEW: This section describes the characteristics of an ideal reference electrode and the construction of several reference electrodes widely used in practical work.

The chief thing we ask of a reference electrode is that it be inconspicuous: we do not want it to have quirks and foibles that distract us from the interesting behavior of the indicator electrode. The following properties would be desirable:

1. It should be easy to prepare from readily available materials.
2. It should reach a known and reproducible potential rapidly, and should retain that potential for a long time in normal use.
3. Its potential should respond rapidly and reversibly to changes of temperature.
4. Its potential should not be detectably altered by the flow of the current drawn by the electrical apparatus with which it is used.
5. Its electrical resistance should be as low as possible.
6. Its use should not alter the compositions of the solutions with which it comes in contact.
7. Its use with common solutions should not involve appreciable liquid-junction potentials (Section 14.11).

The last two of these mean, among other things, that the same solvent should be used in the reference electrode as is present in the solution being studied. If an aqueous reference electrode were used in work with a solution in acetonitrile (CH_3CN), the two solvents would mix while the measurements were being made. Acetonitrile would diffuse into the reference electrode, and would probably change its potential, and at the same time water would diffuse from the reference electrode into the acetonitrile and would probably affect the potential of the indicator electrode. Even apart from the confusion that this would cause, the liquid-junction potential is likely to be so large at the boundary between two different solvents that the measurements may be almost impossible to interpret. Most of this discussion will be devoted to reference electrodes for work with aqueous solutions.

The first, second, and fifth requirements suggest the use of saturated solutions. It is easier to make a saturated solution than an unsaturated one of known composition, and processes that will alter the concentration of an unsaturated solution will not affect that of a saturated one as long as an excess of the solute is present. Combining this with the sixth requirement virtually rules out inert electrodes and electrodes in which a metal is immersed in a solution containing an appreciable concentration of its own ion. Even if an electrode such as $Pt\,|\,Br_2(l)$, $KBr(s)$ or $Ag\,|\,AgNO_3(s)$ met all of the other requirements, its use would cause the solution around the indicator electrode to become contaminated with bromine or silver ion, which are much too likely to react with that solution, or to affect the potential of the indicator electrode, to be tolerated. Reference electrodes are generally active electrodes in solutions saturated with their salts, so that the only solute present at a really high concentration is a relatively innocuous salt such as potassium or sodium chloride. As will be shown in Section 14.11, potassium chloride tends to give lower liquid-junction potentials than most other common salts, and aqueous reference electrodes are therefore usually made with saturated solutions of potassium chloride.

The silver–silver chloride–saturated potassium chloride electrode and the mercury–mercurous chloride ("calomel")–saturated potassium chloride electrode, called the "saturated calomel electrode" or S.C.E., are the two reference electrodes that are most widely used in work with aqueous solutions. The silver–silver chloride electrode has a more reproducible potential and responds more rapidly to changes of temperature, but the solubility of silver chloride in a solution saturated with potassium chloride is so high that care must be taken to avoid contaminating the solutions of interest with dissolved silver(I) in prolonged experiments. Electrodes of both kinds,

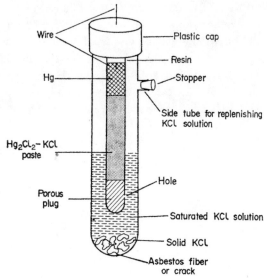

FIG. 14. 8. A common commercially available saturated calomel electrode.

intended for use with glass indicator electrodes and pH meters, can be bought commercially and used in conjunction with any indicator electrode. One common commercial form of the saturated calomel electrode is shown in Fig. 14.8. The mercury and a heavy paste containing mercurous and potassium chlorides are contained in a small inner tube and held in place by a plug of glass wool or some other inert porous material. A small hole in the side of this tube provides an electrical connection to the saturated solution of potassium chloride in the outer tube. A small asbestos fiber may be sealed through the bottom of the outer tube, or the tube may be made with a very tiny crack. The motion of potassium and chloride ions through the fiber or crack conducts electrical current between the electrode and a solution in which it is immersed. In such an electrode the area of the interface between the mercury and the solution is very small, and therefore the potential may change if a current even as large as a few microamperes is allowed to flow. In addition, the resistance R is so large (typically 10^3 to 10^4 ohms) that the IR drop cannot be negligible if the electrical apparatus permits an appreciable current I to flow through the electrode.

Silver–silver chloride and calomel electrodes can be made in many other solvents for use with non-aqueous solutions. Of course their potentials differ from one solvent to another. For example, the potential of the cell

$$Ag\,|\,AgCl(s),\ KCl(s),\ H_2O(l)\,|\,|\,CH_3COOH(l),\ KCl(s),\ AgCl(s)\,|\,Ag$$

is found experimentally to be equal to $+0.23$ V at 298 K, which is to say that the potential of this electrode is 0.23 V more positive in anhydrous acetic acid than it is in water. Most of the difference is due to the fact that the dissociation constant of potassium chloride is only about 10^{-7} mol dm^{-3} in anhydrous acetic acid, so that the activity of chloride ion in a saturated solution of potassium chloride is much smaller in this solvent than in water. There is also a liquid-junction potential of unknown magnitude across the boundary between the two solutions; it is represented by the double vertical line (‖) and will be discussed briefly in Section 14.11. In some solvents the solubility of potassium chloride is so low that lithium chloride is used instead. In others, silver chloride is so soluble, and mercurous chloride is so unstable, that both these electrodes are useless. Reference electrodes like $Cd\,|\,CdCl_2(s)$ and $Hg\,|\,HgCl_2(s)$ are used in liquid ammonia despite what was said above about the disadvantages such of electrodes.

Even in work with aqueous solutions some modification is often desirable. Silver–silver chloride and calomel electrodes made with sodium or lithium chloride are used in work with solutions containing high concentrations of perchlorate ion, for the junction between such a solution and a reference electrode made with potassium chloride would soon become clogged with an impermeable plug of solid potassium perchlorate. In determining halides or in working with solutions of substances that react with halide ions, a reference electrode such as a mercury–mercurous sulfate electrode, $Hg\,|\,Hg_2SO_4(s)$, $K_2SO_4(s)$, might be used to avoid contamination with chloride ion.

SUMMARY: Silver–silver chloride and saturated calomel electrodes are the most widely used reference electrodes in aqueous solutions.

14.7. Electrical apparatus

PREVIEW: The potentials of electrochemical cells can be measured with a potentiometer, which is an instrument so designed that current does not flow through a cell during a measurement. This section describes the construction, operation, and precision of a simple potentiometer.

The essential requirement in a potentiometric measurement is that there should be no current flowing through the cell when the measurement is made. Suppose that a current does flow and that the indicator electrode is the cathode of the cell. Then the half reaction $Ox + ne = Red$ will proceed from left to right at the surface of the electrode: Ox will be consumed and Red will be produced. The concentration of Ox will be a little lower, and the concentration of Red will be a little higher, at the surface of the electrode than in the bulk of the solution some distance away. Each of these differences tends to increase the value of the ratio c_{Red}^0/c_{Ox}^0, and to make the potential of the indicator electrode more negative than it should be. A current flowing in the opposite direction, so that the indicator electrode is the anode, causes the potential of the indicator electrode to be too positive. In either case there is an additional complication due to the *IR* drop through the cell.

A *potentiometer* is a device for applying any desired accurately known e.m.f. to a cell. It is connected to the cell in such a way that its output opposes the potential of the cell, and adjusted so that the two opposing tendencies for current to flow around the circuit just counterbalance each other. Then the e.m.f. obtained from the potentiometer, which we shall call the *applied potential*, is just equal to the potential of the cell at zero current.

Figure 14.9 shows the circuit diagram of a very simple potentiometer. It can be divided into two parts. One is the cell circuit, which consists of the cell itself, the galvanometer G, and the tapping key K. A galvanometer is a sensitive device for indicating whether, and in which direction, a current is flowing. The tapping key protects the galvanometer from the damage that would be done by the prolonged flow of a substantial current. It is depressed for just an instant so that the operator can note the direction in which the galvanometer deflects. The position of the sliding contact S is adjusted accordingly, the key is again depressed momentarily, and the process is continued until no deflection is obtained on depressing the key.

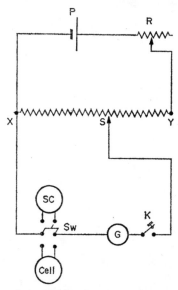

FIG. 14. 9. Schematic diagram of the circuit of a simple potentiometer. The components of the circuit are identified in the text.

The other part of the circuit consists of a d.c. source P (which may be a storage battery or dry cell) whose e.m.f. exceeds the largest value to be measured, a variable resistance R, and a linear voltage divider or *bridge* XSY. Cells having potentials as high as 3 V or so can be prepared with some non-aqueous solvents, but potentials that exceed 1.3 V are very rare with aqueous solutions. The bridge may be as simple as a length of resistance wire of uniform composition and cross-sectional area, having the property that the resistance between either of its ends and the movable contact or slider S is proportional to the corresponding distance. The current flowing through the cell circuit cannot be equal to zero unless the potential of the cell (which reflects the tendency for current to flow in one direction) is just counterbalanced by the applied potential (which governs the tendency for current to flow in the opposite direction). Hence E_{XS} is numerically equal to E_{cell} when the current is equal to zero. When it is, so that the current flowing through XS is the same as the current flowing through SY,

$$\frac{E_{XS}}{E_{XY}} = \frac{I_{XS}R_{XS}}{I_{XY}R_{XY}} = \frac{R_{XS}}{R_{XY}} = \frac{\overline{XS}}{\overline{XY}}.$$

The cell circuit is subjected to the applied potential E_{XS}, and this is equal to the product of the e.m.f. across the entire bridge, E_{XY}, by the ratio of distances $\overline{XS}/\overline{XY}$, which can be read off a scale to which the bridge is affixed. The variable resistance R makes it possible to adjust the value of E_{XY}. If the bridge is provided with a scale having 1500 divisions, it would be convenient for E_{XY} to be 1.500 V, so that each division on the scale would correspond to 1.000 mV.

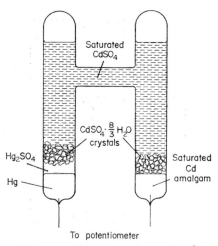

To potentiometer

Fig. 14.10. A "saturated" (with cadmium sulfate) Weston cell.

The standard cell SC is used in making this adjustment. The most widely used standard cell is the Weston cell, whose construction is shown in Fig. 14.10. It consists of a saturated cadmium-amalgam electrode and a mercury–mercurous sulfate electrode in contact with an aqueous solution of hydrated cadmium sulfate, which may or may not be saturated. The equations for the half-reactions at the two electrodes are

$$Cd^{2+} + x\, Hg + 2e = Cd(Hg)_x \quad \text{(saturated amalgam)},$$
$$Hg_2SO_4(s) + 2e = 2\, Hg(l) + SO_4^{2-}.$$

Because the activities of cadmium and sulfate ions are fixed by the solubility (or concentration) of cadmium sulfate, the potential of such a cell depends only on the temperature and pressure. At a pressure of 1 atm it is equal to 1.018 300 V at 293 K and to 1.018 073 V at 298 K if the solution is saturated with cadmium sulfate. The value of E_{XY} is adjusted by connecting the standard cell into the cell circuit by means of the switch Sw, setting the slider S to the point on the bridge that one wants to correspond to the potential of the standard cell, and then adjusting R so that the galvanometer G does not deflect when the key K is tapped. The adjustment should be repeated every 15–30 minutes in careful work because the e.m.f. of a storage battery or dry cell decreases slowly.

The potential of a cell can then be measured in the following way. The cell is connected into the circuit by means of the switch Sw, the slider S is moved to one end of the bridge XY, and the key K is tapped as briefly as possible. Next S is moved to the other end of XY and the key is tapped again. If this causes the galvanometer to deflect in the other direction, S is adjusted so as to decrease the deflection to zero; if it does not, the polarity is wrong and the process should be repeated after interchanging the wires leading from Sw to the cell. The aim is to find a setting of S at which the galvanometer does not deflect in either direction when the circuit is closed by tapping K.

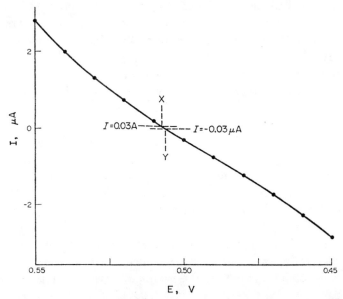

FIG. 14.11. A plot of the total current against potential for the data shown in Fig. 14.6. The dashed lines identify the limits of the range of potentials over which the absolute value of the current is smaller than 0.03 μA.

The precision with which the potential can be measured depends on the characteristics of the cell being studied and on the sensitivity of the galvanometer. Figure 14.11 shows how the total current I flowing through an electrochemical cell varies with the applied potential if only a single redox couple is present. The values of I plotted along the vertical axis were obtained from Fig. 14.6 by using the relationship

$$I = I_c + I_a.$$

As long as the current is very small, the curve is approximately linear and its slope dI/E is roughly constant. The electrical resistance R of the cell can be measured in other ways (which include the use of a Wheatstone-bridge circuit employing alternating current). It is found that the value of dI/dE is never larger than $1/R$, and is often many orders of magnitude smaller.

In every measurement there is a *minimum detectable response*, which is the smallest value of the measured quantity that can just be reliably distinguished from zero. If the pointer of the galvanometer is deflected 1 mm by a current of 0.1 μA, a careful experimenter would probably be able to detect a current of 0.1 μA, but would probably be unable to detect one of 0.01 μA. The minimum detectable response would lie somewhere between these two figures, and we shall arbitrarily imagine that it is equal to 0.03 μA. Then it will just be possible to detect a current of 0.03 μA, but impossible to detect any current smaller than this. In Fig. 14.11 the short dashed horizontal lines represent currents of $+0.03$ μA and -0.03 μA; the positive and negative signs correspond to currents flowing in opposite directions. Over the range of applied potentials from X to Y, the total current is indistinguishable from zero, and the potential of the cell may be found experimentally to lie anywhere within this range, which is 1.2 mV wide. The maximum random error due to this cause might be 0.6 mV.

In a favorable case the slope dI/dE of the current–potential curve may be so high, and the minimum detectable response so low, that the range of uncertainty is only a few microvolts wide. A rather unfavorable case is shown in Fig. 14.12, which bears the same relation to Fig. 14.7 that Fig. 14.11 did to Fig. 14.6. In Figs. 14.7 and 14.12 the exchange current for the couple being studied is smaller than in Figs. 14.6 and 14.11, and the result is that the total current is indistinguishable from zero over a range 5 mV wide. As was explained in Section 14.5, a small value of the exchange current is conducive to irreversible behavior. It also tends to decrease the slope dI/dE of the current–potential curve, and thus to worsen the precision with which the potential can be measured. The result is illustrated by the left-hand portion of Fig. 14.5.

The very simple circuit shown in Fig. 14.9 can be greatly elaborated to improve its accuracy and convenience, but is not well suited to work with glass electrodes and very poorly conducting solutions. The resistance of a typical glass electrode is of the order of 10^8 ohms at 298 K. If a current of 0.03 μA (3×10^{-8} A) were the smallest that could be detected, the IR drop through the electrode might be as large as 3 V even though no current could be observed. The resulting uncertainty of ± 3 V in the measured potential would correspond to an uncertainty of $\pm 3/0.059 = \pm 51$ units in the pH, and a meaningful measurement would be quite impossible to make. Electronic amplification of the current must be employed to permit decreasing the limit of detection to, or below, 10^{-12} A.

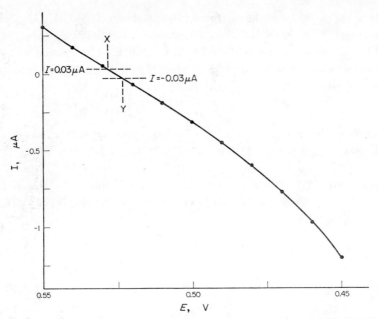

FIG. 14.12. A plot of the total current against potential for the data shown in Fig. 14.7. The dashed lines have the same significance as in Fig. 14.11.

SUMMARY: A potentiometer is an instrument that makes it possible to measure the value of the applied potential that just counterbalances the potential of an electrochemical cell. The precision of measurement depends on the minimum detectable response of the device used to detect the flow of current, and on the slope dI/dE of the current–potential curve for the cell in the vicinity of zero current.

14.8. Analyses by direct potentiometry

PREVIEW: This section emphasizes the errors that arise in analyses by direct potentiometry from uncertainties in measuring the potential, in the value of the liquid-junction potential, and in the formal potential of the redox couple involved.

Suppose that you wish to determine the percentage of potassium hexacyanoferrate (III) that is present as an impurity in a sample of solid potassium hexacyanoferrate(II). If you knew that the hexacyanoferrate(III)–hexacyanoferrate(II) couple behaves reversibly over a wide range of concentrations of the two ions, you might think of dissolving a known weight of the sample in a known volume of some suitable supporting electrolyte (such as 1 M hydrochloric acid, in which the formal potential of the couple is +0.71 V vs. N.H.E.), immersing a platinum indicator electrode and a saturated calomel reference electrode in the solution, and measuring the difference between their potentials. Suppose that this is 0.300 V, with the platinum electrode being the positive electrode.

29

At 298 K the potential of the saturated calomel electrode is 0.2412 V more positive than the potential of the N.H.E.: since the formal potential of the couple is +0.71 V vs. N.H.E., it is +0.47 (= 0.71−0.24) V vs. S.C.E. The measured potential is +0.300 V vs. S.C.E. Substituting these two values into the Nernst equation gives

$$+0.300 = +0.47 - 0.059\,16\,\log\frac{[\mathrm{Fe(CN)_6^{4-}}]}{[\mathrm{Fe(CN)_6^{3-}}]}$$

where $\mathrm{Fe(CN)_6^{4-}}$ and $\mathrm{Fe(CN)_6^{3-}}$ represent the hexacyanoferrate(II) and hexacyanoferrate(III) ions, respectively. Hence $[\mathrm{Fe(CN)_6^{4-}}]/[\mathrm{Fe(CN)_6^{3-}}] = 10^{(-0.17/-0.059\,16)} = 10^{2.87} = 747$; the sample contains 1 mole of hexacyanoferrate(III) for each 747 moles of hexacyanoferrate(II). If the concentration of hexacyanoferrate(II) were known to be $5.0 \times 10^{-3}\,M$, the concentration of hexacyanoferrate(III) would be equal to $5.0 \times 10^{-3}/747 = 6.7 \times 10^{-6}\,M$.

This is simple enough, but it is not quite as straightforward as it looks. Suppose that the sample standard deviation is 0.001 V for a series of measurements of the potential. If we define a quantity R by the equation $R = [\mathrm{Fe(CN)_6^{4-}}]/[\mathrm{Fe(CN)_6^{3-}}]$, the corresponding standard deviation of R is given by

$$s_R^2 = \left(\frac{\partial R}{\partial E}\right)^2 s_E^2. \tag{14.13}$$

The partial derivative $\partial R/\partial E$ is the derivative of R with respect to E if every other quantity is kept constant in the equation relating R and E. Rearranging eq. (14.12) gives

$$R = [\mathrm{Fe(CN_6^{4-}})]/[\mathrm{Fe(CN)_6^{3-}}] = 10^{(E-E^{0\prime})/0.059\,16}.$$

Keeping the formal potential $E^{0\prime}$ and the numerical coefficient constant and differentiating with respect to E gives

$$\frac{\partial R}{\partial E} = 10^{(E-E^{0\prime})/0.059\,16} \times \ln 10 \times \frac{1}{0.059\,16} = \frac{2.303}{0.059\,16}\,R = 38.92\,R.$$

Combining this with eq. (14.13) and the assumed value $s_E = 0.001$ V gives $s_R/R = 0.038\,92$, or roughly 4%. A standard error of 0.001 V in the measured potential produces a relative standard error of 4% in the ratio of the concentrations of hexacyanoferrate(II) and hexacyanoferrate(III) ions, or in the concentration of hexacyanoferrate(III) ion alone if the concentration of hexacyanoferrate(II) ion is known. If n were equal to 2 instead of 1, the value of $\partial R/\partial E$ would be equal to $2.303\,R/(0.05916/2) = 77.84\,R$, and a standard error of 0.001 V in the measured potential would produce a standard error of 8% in the concentration, or ratio of concentrations, being determined.

In calculating the concentration of hexacyanoferrate(III) ion in the solution prepared from our imaginary unknown, we used the literature value, +0.71 V vs N.H.E., for the formal potential of the couple. This value has only two significant figures, and its uncertainty cannot be less than 0.005 V, which corresponds to a relative uncertainty of 20% in the concentration of hexacyanoferrate(III). If n were equal to 2, the relative uncertainty would be 40%.

Such a result might be acceptable or it might not. If it is not, we must evaluate the formal potential more precisely under exactly the same conditions that are used in analyzing the unknown solution. Section 14.10 will describe two ways of doing this.

How precisely can E and $E^{0'}$ be evaluated? That depends on the couple being studied, on the experimental conditions, and on the apparatus and care employed. In Fig. 14.5 the standard deviation of the points from the curve drawn through them is 3.5 mV, but the behavior is irreversible at many of these points and (as was shown by Fig. 14.12) precise values are very difficult to secure in the face of irreversible behavior. Under favorable conditions the standard deviation can be made to be 0.1 mV, or even less. On the average it will lie between these two extremes.

More troubles arise if the sample is more complex than the simple mixture of potassium hexacyanoferrate(II) and potassium hexacyanoferrate(III) that we have been discussing. If it is an aqueous solution it may contain other solutes that affect the pH or the ionic strength, or that form complexes or ion pairs with hexacyanoferrate(II) or hexacyanoferrate(III) ion. Suppose that it is saturated with sodium hydroxide. The data given in Appendix VI for the hexacyanoferrate(III)–hexacyanoferrate(II) couple show why it would be dangerous to combine the potential of an indicator electrode in such a solution with the formal potential of the couple in 0.1 M hydrochloric acid. There would be three things you could do.

One would be to add a very small volume of the sodium hydroxide solution to a much larger volume of 0.1 M hydrochloric acid. In the resulting mixture there would be a little sodium chloride, and the concentration of hydrochloric acid would be a little less than 0.1 M, but if the volumes were sufficiently different you could ignore these facts and use the formal potential of the couple in 0.1 M hydrochloric acid to interpret the potential you measured. The obvious danger is that the dilution might make the concentrations of hexacyanoferrate(III) and hexacyanoferrate(II) ions so small that the couple would no longer behave reversibly.

An alternative would be to use a *standard-addition* procedure. Two portions of the unknown solution are placed in a cell like the one shown in Fig. 8.3. They are connected by a salt bridge, and a platinum indicator electrode (or some other suitable indicator electrode) is immersed in each solution. The difference between the potentials of the two electrodes should be equal to zero. Now a very small and exactly known volume, V_s cm^3, of a standard solution of the substance being determined is added to one of the two solutions, whose volume is known to be V_u cm^3. The difference of potential is measured again and is found to be, say, ΔE V. If the couple is known to behave reversibly over the range of concentrations involved, the value of ΔE obtained on adding a little hexacyanoferrate(III) ion to one solution (the "spiked" one) will be given by

$$\Delta E = 0.059\,16 \log \frac{[\mathrm{Fe(CN)_6^{3-}}]_{\mathrm{spiked}}}{[\mathrm{Fe(CN)_6^{3-}}]_{\mathrm{unspiked}}} = 0.059\,16 \log \frac{(V_u c_u + V_s c_s)/(V_u + V_s)}{c_u}.$$

29*

If V_s is very much smaller than V_u, this becomes very nearly

$$\Delta E = 0.059\ 16 \log \frac{V_u c_u + V_s c_s}{V_u c_u} = 0.059\ 16 \log \left(1 + \frac{V_s c_s}{V_u c_u}\right)$$

from which it is easy to compute c_u if the values of V_s, c_s, and ΔE are known. Keeping V_s small also minimizes the effect of the addition on the ionic strength, pH, and other properties of the solution to which the addition is made. However, an acceptable precision cannot be attained if $V_s c_s$ is too small, and hence the amount that is added should be considerably larger than the amount being determined. The chief advantage of the procedure is that neither the value of $E^{0\prime}$ nor the composition of the supporting electrolyte need be known.

The technique called *differential potentiometry* employs similar ideas, but in a different way. Two identical indicator electrodes are used: one is immersed in the unknown solution and the other is immersed in a "reference" solution identical with the unknown one in every way except that it contains none of the substance being determined. A concentrated standard solution of that substance is added to the reference solution until the difference between the potentials of the two electrodes becomes equal to zero. The concentrations in the reference and unknown solutions are then equal. The value of $E^{0\prime}$ is again immaterial, but the composition of the supporting electrolyte must be known so that a suitable reference solution can be prepared. This disadvantage is offset by the fact that differential potentiometry can be used even if strictly

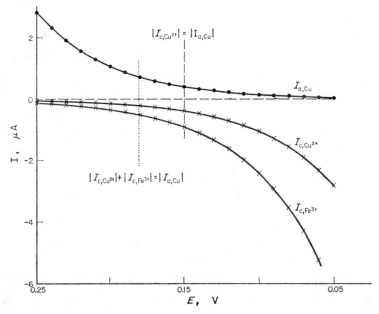

FIG. 14.13. Plots showing the effects of potential on the anodic current for the oxidation of a copper electrode and the cathodic currents for the reductions of Cu^{2+} and Fe^{3+} ions.

Nernstian behavior is not obtained, though its precision naturally becomes poorer as the electrode becomes less responsive to the substance being determined.

The chief disadvantage of direct potentiometry with redox indicator electrodes is that the potentials of such electrodes are affected by all of the reducible and oxidizable substances present in the solution. You might think that the concentration of copper(II) ion in a solution could be determined by measuring the potential of a copper electrode. If the solution also contains some iron(III) ion, Fig. 14.13 shows what may happen. The potential that is sought, where $|I_{c,\,Cu^{2+}}| = |I_{a,\,Cu}|$, is represented by the vertical dashed line. The potential that is measured will be the one where $|I_{c,\,Cu^{2+}}| + |I_{c,\,Fe^{3+}}| = |I_{a,\,Cu}|$, which is represented by the dotted line. If the concentration of iron(III) ion is appreciable, $I_{c,\,Fe^{3+}}$ will also be appreciable. The measured potential will be much more positive than it should be, and the calculated concentration of copper(II) ion will be much larger than the truth. You have to know quite a lot about the composition of a solution before you can trust the result of an analysis by direct potentiometry.

SUMMARY: The form of the Nernst equation is such that small errors of measurement of the potential of a cell and uncertainties in the liquid-junction and formal potentials give rise to large errors in the results of analyses by direct potentiometry. Standard-addition procedures and differential potentiometry mitigate these errors but cannot solve the problems that arise from the presence of other oxidizing and reducing agents.

14.9. Potentiometric titrations

PREVIEW: Potentiometric titrations are more accurate and precise than analyses by direct potentiometry. Their end points may be located by numerical interpolation to find the point of maximum slope on a sigmoidal titration curve, or graphically by means of a Gran plot.

If you can find an indicator electrode that will respond to the concentration of a reactant or product, you can locate the end point of a titration by observing how the potential of the electrode depends on the volume of reagent used. Acid-base titrations involve hydrogen ions and can be followed with glass electrodes (Section 14.12). Redox titrations can be followed with inert indicator electrodes, and many titrations involving precipitation or complex formation can be followed with active indicator electrodes. The indicator electrode and a suitable reference electrode are immersed in the solution being titrated. A measured volume of the reagent is added, some time is allowed for the reaction to reach equilibrium, and the difference between the potentials of the electrodes is measured. The volume and the difference of potential are the coordinates of one point on the titration curve (Chapter 11). More reagent is added and the difference of potential is measured again, yielding the coordinates of another point, and so on. This section explains how the end point can be located from the data obtained.

Potentiometric titrations are more accurate and more precise than analyses by direct potentiometry because many sources of error and uncertainty in direct potentiometry have little or no effect in potentiometric titrations. One of these is the liquid-junction potential, which will be discussed in Section 14.11. You never know the value of the liquid-junction potential exactly, and an uncertainty of 1 mV in its value has exactly the same effect in direct potentiometry as an uncertainty of 1 mV in the measured potential. In a potentiometric titration the only effect of the liquid-junction potential is to shift the whole curve upward or downward along the potential axis: as long as it does not change appreciably in the immediate vicinity of the equivalence point, it cannot affect the calculation of the volume of reagent that is needed to reach that point. In direct potentiometry you must exercise some control over the temperature, both because formal potentials vary with temperature and because you must usually know the numerical value of the coefficient 2.303 RT/nF at the temperature you have used. In a potentiometric titration the values of $E^{0'}$ and 2.303 RT/nF are not needed in locating the end point, and the result could be affected only by a very large change of temperature in the immediate vicinity of the end point. The accuracies and precisions of potentiometric titrations are much more likely to be limited by chemical factors—such as coprecipitation or some other side reaction, or an insufficiently large value of K_t—than by the difficulties that beset direct potentiometry. In a favorable case the limiting factor may be the uncertainty in measuring the amounts of reagent added, or even the uncertainties in the formula weights of the reacting substances.

There are several ways in which the end point of a potentiometric titration can be located. One is to add reagent until the measured potential just attains the value that it is expected to have at the equivalence point. This would be like adding reagent until an indicator changed color. You need to know what the potential will be at the equivalence point, and the effect of an error or uncertainty in its value will depend on the slope of the titration curve around the equivalence point. You may not know enough about the system to enable you to calculate the equivalence-point potential with sufficient accuracy. Even if you can tolerate an uncertainty of 20 mV, you need some assurance that the liquid-junction potential is known within 20 mV. Moreover, the solution is less well poised, and the titration reaction is slower, at the equivalence point than anywhere else. A long time may be needed for the potential to reach a steady value, and titration to the equivalence point may therefore be slow and tedious. It is possible to buy automatic titrators that will perform titrations in this fashion, adding the reagent from a motor-driven syringe until the potential just becomes equal to some preset value: you can often trust a machine with a task you would not be willing to perform yourself.

Usually, however, the end point of a potentiometric titration is taken as the point of maximum slope on the titration curve. In most practical titrations this point almost coincides with the equivalence point. The most convenient and easiest way of locating it is by numerical interpolation.

TABLE 14.1. *Location of the end point of a potentiometric titration by numerical interpolation*

A solution of iron(II), in 1 M sulfuric acid containing a little phosphoric acid, was titrated with a standard solution of cerium(IV) sulfate. The indicator and reference electrodes were a platinum wire and a saturated calomel electrode, respectively. In the first two columns, V_R is the total volume of cerium(IV) solution added and E is the measured potential of the cell.

V_R, cm³	E, V	$\Delta E/\Delta V_R$, mV (0.025 cm³)⁻¹	$\Delta^2 E/\Delta V_R^2$, mV (0.025 cm³)⁻²
9.150	0.644		
		18	
9.175	0.662		2
		20	
9.200	0.682		6
		26	
9.225	0.708		10
		36	
9.250	0.744		25
		61	
9.275	0.805		4
		65	
9.300	0.870		−33
		32	
9.325	0.902		−14
		18	
9.350	0.920		−6
		12	
9.375	0.932		−4
		8	
9.400	0.940		

Volume required to reach inflection point $= 9.275 + \left(\dfrac{4}{4+33}\right)(0.025) = 9.278$ cm³.

Potential at inflection point $= 0.805 + \left(\dfrac{4}{4+33}\right)(0.065) = 0.812$ V.

The first two columns of Table 14.1 show the data obtained near the equivalence point of a titration of iron(II) with cerium(IV). These data are plotted in part (a) of Fig. 14.14. The first 9.150 cm³ of reagent was added all at once; then more reagent was added in equal (0.025-cm³) portions until it was judged that the point of maximum slope had been passed.

Over the interval from $V_R = 9.150$ cm³ to $V_R = 9.175$ cm³, the value of E increases by 18 mV and the value of $\Delta E/\Delta V_R$ is therefore equal to 18 mV (0.025 cm³)⁻¹. Because all of the successive values of ΔV_R are identical, there is no reason to convert these units to mV cm⁻³, kV quart⁻¹, or any others. Approximating this short segment

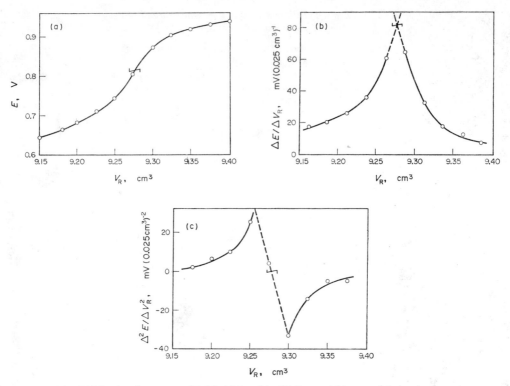

FIG. 14.14. (a) The titration curve of Table 14.1 and its (b) first and (c) second derivatives. Each curve is marked to show the point of maximum slope at $V = 9.278$ cm³.

of the curve by a straight line, we may assign this value of $\Delta E/\Delta V_R$ to the midpoint of the interval, where $V_R = (9.150+9.175)/2 = 9.1625$ cm³. Similarly, the value of $\Delta E/\Delta V_R$ over the next interval is equal to 20 mV (0.025 cm³)⁻¹, which may be assigned to the midpoint of that interval, where $V_R = (9.175+9.200)/2 = 9.1875$ cm³, and so on. The values of $\Delta E/\Delta V_R$ are plotted against V_R in part (b) of Fig. 14.14.

From these values of the first derivative of the titration curve we can calculate values of the second derivative. Over the interval from $V_R = 9.1625$ cm³ to $V_R = 9.1875$ cm³, $\Delta E/\Delta V_R$ changes from 18 to 20 mV (0.025 cm³)⁻¹ while ΔV_R is equal to 0.025 cm³. Hence $\Delta(\Delta E/\Delta V_R)/\Delta V_R$ ($= \Delta^2 E/\Delta V_R^2$) = 2 mV (0.025 cm³)⁻², which is the average value of the second derivative in an interval whose midpoint lies at $V_R = (9.1625+9.1875)/2 = 9.175$ cm³. The arrangement of the numbers in the last two columns of Table 14.1 is intended to help in keeping track of the volumes to which the values of $\Delta E/\Delta V_R$ and $\Delta^2 E/\Delta V_R^2$ pertain. The values of the second derivative are plotted against V_R in part (c) of Fig. 14.14. The last positive value of $\Delta^2 E/\Delta V_R^2$ is at $V_R = 9.275$ cm³; the first positive one is at $V_R = 9.300$ cm³. The value of V_R at the inflection point, where $\Delta^2 E/\Delta V_R^2 = 0$, is found by linear interpolation between these two points.

The curves in Fig. 14.14 are given here only to help you to visualize the process. To plot them for a real titration would waste time and might tempt you to try to find the point where the curve in part (a) was steepest, or the one where the two branches intersected in part (b). The disadvantage of such a graphical procedure is that its result depends on the exact shape of the curve that is drawn through the points. What is sought is simply the value of V_R at which the second derivative is equal to zero, which is most easily found by the linear interpolation shown at the bottom of Table 14.1. The value of the potential at the end point can be found by a similar interpolation: it is chiefly useful as a check on the results of replicate titrations. If it differs very much from one titration to another, something is wrong even if the end-point volumes are exactly the same.

It is not absolutely necessary to add the reagent in exactly equal increments, but the calculations are greatly simplified by doing so. Whether they are equal or not, the increments should be neither too large nor too small. If they are too large, the successive portions of the curves cannot be as accurately represented by straight lines, and the point where $\Delta^2 E/\Delta V_R^2 = 0$ deviates from the true inflection point, where $d^2 E/dV_R^2 = 0$. If only the points at $V_R = 9.150$, 9.225, 9.300, and 9.375 cm^3 had been obtained, the calculated end-point volume would have been 9.262 cm^3 instead of 9.278 cm^3. On the other hand, if the increments are too small, the uncertainty in measuring the individual values of E may become appreciable in comparison with the relatively small values of ΔE, and the second derivative may appear to change sign several times. You must take the slope of the curve into account in choosing the size of the increments you add. A good rule of thumb is that your value of ΔV_R should be between 5 and 10 times as large as the precision with which you expect to be able to find the end-point volume. For the titration of Table 14.1 the slope at the equivalence point is so large that it seemed possible to hope for a precision of about 0.005 cm^3, which corresponds to a relative precision of about 0.05%.

You should not think of trying to add such small increments throughout the whole titration, for so many of them would be needed that the titration would be interminable. If you were to repeat the titration of Table 14.1, the knowledge that the potential was about 0.8 V at the inflection point would enable you to add almost all of the necessary amount of reagent at once, and then to begin adding smaller and smaller portions until the measured potential was not far below 0.8 V. If you had no idea of the potential at the inflection point, you would have to be more cautious to ensure that you did not overrun it, and might find it advantageous to make a rough plot of E against V_R as you went. By comparing the plot with the expected shape of the titration curve you could easily tell how much more reagent you could add at any point without risking the addition of too much.

When you try to titrate to the equivalence-point potential you are concentrating your attention on the region where the titration reaction is slowest and where the indicator electrode is least likely to be stable and to behave reversibly. You can locate the point of maximum slope without making measurements quite so close to the equivalence point, but if you avoid that point by too wide a margin you will sacrifice some of the accuracy and precision to which you are entitled. Either of these techniques is hard to use if the electrode becomes really ill-behaved near the equivalence point.

Figure 14.15 shows an example. There were several difficulties in this titration. Because K_t was only about 10^5 mol^{-1} dm^3, the concentration of calcium ion did not change rapidly around the equivalence point. Because $2.303\,RT/nF$ was only 0.0296 V for the indicator electrode used, the variation of potential was also small. The precision of the measurements was not very high, and the potentials near and beyond the equivalence point drifted slowly as time went on. Such difficulties are not unusual.

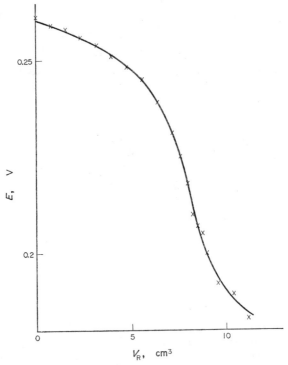

FIG. 14.15. Data obtained in the potentiometric titration of 100 cm^3 of a solution containing calcium ion and having a pcH of 6.3 with a standard 0.1072 M solution of disodium dihydrogen ethylene-diaminetetraacetate using a calcium-ion-selective membrane electrode as the indicator electrode.

It is natural to think of a potentiometric titration curve as being sigmoidal, but "natural" and "wise" are not synonyms. Here it is wiser to deal with a segmented curve obtained from the data. If you solve the Nernst equation

$$E = E^{0\prime} + 0.0296 \log [Ca^{2+}]$$

for the concentration of calcium ion ($=$ S in the equations of Chapter 11), you can obtain

$$[Ca^{2+}] = 10^{(E-E^{0\prime})/0.0296} = 10^{E/0.0296}/10^{E^{0\prime}/0.0296} = k \times 10^{E/0.0296}.$$

From eq. (11.8) you can see that a plot of $[Ca^{2+}](V_S^0+V_R)/V_S^0$ against f (or, more conveniently because we do not know where the equivalence point is, against V_R) should be a straight line if $f < 1$. It would be a waste of time to apply the correction for dilution to these data: since the equivalence point is reached by adding about 8 cm³ of reagent to 100 cm³ of the solution titrated, the value of $(V_S^0+V_R)/V_S^0$ cannot exceed $108/100 = 1.08$ as long as $f < 1$. Ignoring the correction altogether cannot introduce a relative error larger than 8% at any point, and this is no larger than the relative error in $[Ca^{2+}]$ that would result from an error of only 1 mV in the measured potential. If we simply plot $10^{E/0.0296}$ against the volume of reagent we

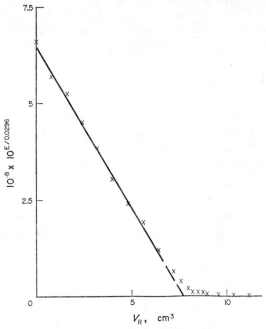

FIG. 14.16. Plot of $10^{E/0.0296}$ against the volume of ethylenediaminetetraacetate solution for the titration of Fig. 14.15.

obtain Fig. 14.16. The first nine points fall on a good straight line. Then there is a region of curvature because K_t is small, but we can ignore it by simply extrapolating the straight line to its intercept on the abscissa axis. This intercept represents the equivalence point. Segmented curves constructed from the data obtained in potentiometric titrations are called *Gran plots*. Other kinds of Gran plots are described in the specialized literature. They have the advantage of emphasizing points that lie some distance away from the equivalence point, and thus of circumventing the problems that may arise at and near the equivalence point.

SUMMARY: Analyses by potentiometric titration circumvent many of the errors that afflict analyses by direct potentiometry, and therefore tend to be more accurate and precise.

There are two recommended techniques for locating the end point of a potentiometric titration: one is to find the point of maximum slope on the sigmoidal titration curve, and the other is to find the point of intersection of the two branches of the segmented curve, called the Gran plot.

14.10. How formal potentials are evaluated

PREVIEW: The formal potential of a couple can be found either by measuring the potentials obtained with a number of solutions containing different concentrations of the oxidized and reduced forms of the couple of interest, or by a potentiometric titration of one of the forms with a reducing or oxidizing agent.

To find the value of a standard potential (Section 14.3), it is necessary to measure the potentials of many cells without liquid junction. Each measurement gives the value of a quantity that includes a mean ionic activity coefficient as well as the standard potential, and these have to be separated by performing an extrapolation to infinite dilution.

It is very much easier to evaluate a formal potential, and there are two ways of doing it. Suppose we wish to find the value of the formal potential of the iron(III)–iron(II) couple in 2 M phosphoric acid. We could prepare a number of solutions, each containing 2 M phosphoric acid but with different concentrations of iron(III) and iron(II). We might vary either the ratio of these concentrations or their sum. It would be best to vary both, keeping it in mind that the sum must always be too small to have any appreciable effect on the ionic strength of the supporting electrolyte or on the concentration of any species present in it. Since 2 M phosphoric acid is only about 10% dissociated into hydronium and dihydrogen phosphate ions, the sum of the concentrations of iron(III) and iron(II) should be kept below 0.01 or even 0.005 M. We would bring each solution to a known temperature, which would be 298 K unless we had some special reason for choosing another temperature, immerse suitable indicator and reference electrodes in it, and measure the difference E between their potentials. For each solution we could calculate a value of the formal potential from the equation

$$E^{0\prime}(\text{Fe(III)}, \text{Fe(II)}; 2\,M\,\text{H}_3\text{PO}_4) = E + 0.059\,16 \log \frac{[\text{Fe(II)}]}{[\text{Fe(III)}]}.$$

Finally we would combine the individual values as suggested by Chapter 12.

An even easier procedure takes advantage of the fact that the ratio of concentrations changes during a potentiometric titration. We might prepare a solution of iron(II) in 2 M phosphoric acid and titrate it potentiometrically with a solution of a suitable oxidizing agent, such as potassium permanganate. In addition to the measurements that were needed for locating the end point, we would need many others made during the first part of the titration, where both iron(II) and iron(III) were present at appreciable concentrations. After we had finished the titration and calculated the volume of reagent that had been consumed in reaching the end point, we would have data resembling those in the first two columns of Table 14.2.

TABLE 14.2. *The formal potential of the iron(III)–iron(II) couple in 2 M phosphoric acid*

1.005 g of $Fe(NH_4)_2(SO_4)_2 \cdot 12 H_2O$ was dissolved in 500 cm³ of $2 M$ phosphoric acid. A large volume of phosphoric acid was used so that its concentration would not change appreciably during the subsequent titration with 0.02 M potassium permanganate. A platinum wire was used as the indicator electrode and a silver–silver chloride–saturated potassium chloride electrode as the reference electrode. The point of maximum slope was found to lie at $V = 25.93$ cm³. The data were obtained by A. Acuna and G. L. Jacobs at the Polytechnic Institute of Brooklyn.

V, cm³	E, V	$0.059\,16 \log$ $[(25.93-V)/V]$	$E+0.059\,16 \log$ $[(25.93-V)/V]\,(= E^{0\prime})$
0	+0.186	—	—
2.0	0.207	$+0.063_6$	$+0.270_6$
4.0	0.220_5	0.043_5	0.264_0
6.0	0.231	0.030_8	0.261_3
8.0	0.240	0.020_8	0.260_8
10.0	0.248	0.012_0	0.260_0
12.0	0.256	0.003_9	0.259_9
14.0	0.264	-0.004_1	0.259_9
16.0	0.272	-0.012_2	0.259_8
18.0	0.281	-0.021_0	0.260_0
20.0	0.292	-0.031_2	0.260_8
22.5	0.312	-0.048_3	0.263_7
24.0	0.330	-0.064_8	0.265_2
25.0	0.353	-0.084_6	0.268_4

Applying the ideas and equations of Chapter 11, we would say that S = Fe(II) and P = Fe(III), and that $p = s$ because equal numbers of iron(III) and iron(II) ions appear on the two sides of the balanced chemical equation. The equation

$$\frac{[P]}{[S]} = \frac{[Fe(III)]}{[Fe(II)]} = (p/s)\frac{f}{1-f} \qquad (f < 1) \qquad (14.15)$$

can be found in Table 11.2. The quantity f is proportional to the volume V of permanganate and is equal to 1 at the equivalence point, where $V = 25.93$ cm³, so that $f = V/25.93$ and $f/(1-f) = V/(25.93-V)$. Combining this with eqs. (14.14) and (14.15) yields

$$E^{0\prime}(Fe(III), Fe(II); 2\,M\,H_3PO_4) = E+0.059\,16 \log\,[(25.93-V)/V] \qquad (14.16)$$

from which we would obtain the values shown in the last column of Table 14.2.

The average of all thirteen of these values is $E^{0\prime} = +0.262_6$ V vs. the silver–silver chloride reference electrode used, and the sample standard deviation is 0.003_6 V. The chemist who is not very observant or thoughtful might go on to calculate that the standard deviation of the mean is $0.003_6/\sqrt{13} = 0.001_0$ V, find that $t = 2.2$ at the 95% level of confidence with 12 degrees of freedom, and conclude that there was a probability of 95% that the true value of $E^{0\prime}$ lies within $0.001_0 \times 2.2 = 0.002_2$ V of the mean $(+0.262_6$ V), or between $+0.260_4$ and $+0.264_8$ V vs. the silver–silver chloride reference electrode used in the measurements.

A more perceptive chemist would see that the values in the last column of Table 14.2 are not randomly distributed, but fall on a smooth curve. In addition, the sample standard deviation is several times too large to be consistent with the care taken in making the measurements. These are signs that some systematic error is involved, and that the situation is more complicated than it seems to be.

One possibility is that the couple does not obey eq. (14.14) in this supporting electrolyte. Perhaps iron(III) forms a binuclear complex, so that the half-reaction should be written $Fe(III)_2 + 2e = 2Fe(II)$. This would lead to the equations

$$[Fe(III)_2] = \frac{1}{2} c^0_{Fe(II)} f \frac{V^0_{Fe(II)}}{V^0_{Fe(II)} + V_{MnO_4^-}},$$

$$[Fe(II)] = c^0_{Fe(II)}(1-f) \frac{V^0_{Fe(II)}}{V^0_{Fe(II)} + V_{MnO_4^-}},$$

and eventually to a relationship that involved a dependence on V that is quite different from the one expressed by eq. (14.16). If this hypothesis produced a much more constant value of $E^{0\prime}$, other kinds of measurements might be undertaken to confirm it and to provide information about the properties and behavior of the dimeric complex.

However, nothing of the sort could account for the fact that the potential at $V = 0$ in Table 14.2 is not very far below the apparent value of $E^{0\prime}$. There must have been an appreciable amount of iron(III) in the initial solution. It might have resulted from an impurity in the iron(II) salt from which the solution was prepared, or from oxidation of some of the iron(II) by atmospheric oxygen or by some oxidizing impurity in the phosphoric acid. Equation (14.15) must have underestimated the concentration of iron(III) at every point because it assumes that the titration reaction is the only source of iron(III). The underestimate would become less serious as V increased and more iron(III) was formed in the reaction with permanganate. A different explanation must be found for the behavior of the last several points. This probably reflects an error in the location of the equivalence point: if the volume of permanganate needed to reach that point were actually a little smaller than 25.93 cm³, eq. (14.16) would overestimate the concentration of iron(II) remaining to be titrated at every point. That overestimate would become less serious as V decreased and more iron(II) remained. On this reasoning, the most accurate and reliable values are those obtained over the range from $V = 10$ to $V = 18$ cm³.

Of course it would be best to repeat the titration, paying special attention to the possible sources of error that we have identified. If we had to rely entirely on the data in Table 14.2, the best value we could select would be $+0.260$ V vs. the reference electrode used—which does not lie within the first chemist's 95% confidence interval. Since the potential of this reference electrode is -0.042 V vs. S.C.E. or $+0.199$ V vs. N.H.E., we might finally say that the formal potential of the iron(III)–iron(II) couple in 2 M phosphoric acid was equal to $+0.218$ V vs. S.C.E. or to $+0.459$ V vs. N.H.E.

SUMMARY: Formal potentials may be evaluated by either direct potentiometry or potentiometric titration. Either is much simpler than the procedure used in evaluating standard potentials. The two kinds of measurements are described, with an emphasis on the interpretation of the data obtained in the titration technique.

14.11. Liquid-junction potentials

PREVIEW: This section deals with the factors that give rise to the liquid-junction potential and govern its magnitude, and with the effects of the liquid-junction potential in practica work.

There is a difference of potential across the boundary, or *liquid junction*, between any two solutions that have different compositions. In the schematic representation of a cell, such a boundary is represented by the symbol ||; the difference of potential across it is called the *liquid-junction potential* and given the symbol E_j. This section describes the cause of the liquid-junction potential, shows how its magnitude can be estimated, and tells why it is important in electrochemical measurements.

The simplest kind of boundary is the one that exists between two solutions that contain the same solvent and solute, but in which the concentrations of the solute are different. An example is the boundary between the two aqueous solutions of hydrochloric acid in the cell

$$Ag \,|\, AgCl(s),\ HCl(0.01\ M) \,||\, HCl(0.1\ M),\ AgCl(s) \,|\, Ag. \qquad (14.17)$$

When these two solutions are brought into contact, hydronium and chloride ions will diffuse from right to left, from the more concentrated solution to the less concentrated one. The rate at which an ion diffuses is described by the value of its *diffusion coefficient D*, which depends on its size and shape, on the viscosity of the solution, and on the temperature, and is also slightly affected by the presence of other ions. It is convenient to ignore the last effect and speak of the value of D in an infinitely dilute solution. Values of D^0, the diffusion coefficient at infinite dilution, are given in Table 14.3 for a number of common ions in aqueous solutions at 298 K.

The abnormally large values for hydronium and hydroxide ions deserve special mention. Hydrogen bonding among molecules of water was mentioned in Section 4.2. It enables a hydronium ion to "move" through an aqueous solution by a process

TABLE 14.3. *Diffusion coefficients of some common ions at infinite dilution in aqueous solutions at 298 K*

Ion	$10^6 D^0$, cm^2 s^{-1}	Ion	$10^6 D^0$, cm^2 s^{-1}
Ag^+	16.48	I^-	20.45
Ba^{2+}	8.47	IO_3^-	10.78
Be^{2+}	6.0	K^+	19.58
Br^-	20.81	Li^+	10.30
BrO_3^-	14.86	Mg^{2+}	7.06
HCO_3^-	11.85	NH_4^+	19.57
CO_3^{2-}	9.23	NO_3^-	19.02
Ca^{2+}	7.92	Na^+	13.34
Cl^-	20.33	OH^-	52.89
ClO_3^-	17.20	Pb^{2+}	9.25
ClO_4^-	17.92	Rb^+	20.72
Cs^+	20.6	SO_4^{2-}	10.65
Cu^{2+}	7.13	Sr^{2+}	7.92
F^-	14.75	Zn^{2+}	7.03
H^+	93.15		

like

$$\text{H—O—H}^+ + \text{O—H—O—H} = \text{H—O—H—O—H—O—H}^+$$
$$\underset{\text{H}}{|} \qquad \underset{\text{H}}{|} \quad \underset{\text{H}}{|} \qquad \underset{\text{H}}{|} \quad \underset{\text{H}}{|} \quad \underset{\text{H}}{|}$$

$$= \text{H—O—H—O} + \text{H—O—H}^+$$
$$\underset{\text{H}}{|} \quad \underset{\text{H}}{|} \qquad \underset{\text{H}}{|}$$

in which bonds are formed and broken, but in which the ion does not have to shoulder molecules of the solvent aside as another kind of ion would have to do in traversing the same distance. Hydroxide ions can be transported by a generally similar mechanism. Equally noteworthy is the sequence $D^0_{Li+} < D^0_{Na+} < D^0_{K+}$. An anhydrous lithium ion is smaller than an anhydrous sodium or potassium ion, but in water it becomes the most highly hydrated of the three. Consequently a hydrated lithium ion is larger than a hydrated sodium or potassium ion, and therefore it has to overcome more viscous resistance in moving through a solution. This was mentioned in Section 7.2, and you can see the same phenomenon in the sequence $D^0_{Be^{2+}} < D^0_{Mg^{2+}} < D^0_{Ca^{2+}} < D^0_{Ba^{2+}}$.

As soon as the boundary HCl(0.01 M) || HCl(0.1 M) is formed, both hydronium and chloride ions will begin to diffuse across it from the right-hand solution to the left-hand one. However, the value of D^0 for the hydronium ion is almost five times as large as that for chloride ion. In the first instant after the boundary is formed, almost five times as many hydronium ions as chloride ions will diffuse across it. The left-hand solution will acquire an excess of positive charge, while an excess of negative charge will remain in the right-hand one. The separation of charge gives rise to a difference of potential, which tends to retard the motion of hydronium ion while it accelerates the motion of chloride ion. Very soon a steady state is reached, in which the difference of potential is just large enough to compensate for the ability of hydronium ion to diffuse more rapidly than chloride ion. Thereafter the two ions will cross the boundary at equal rates. This steady value of the difference of potential is the liquid-junction potential. If the two solutions are left in contact for a very long time, their concentrations will eventually become identical because of the continued diffusion of the ions across the boundary. However, practical liquid junctions (like the asbestos fiber or crack in Fig. 14.8) are designed so that the measurements can be completed long before there is any appreciable change of concentration.

In this situation the electroneutrality principle is not obeyed exactly. By combining our best estimate of the value of E_j with the equations of electrostatics, it can be calculated that the left-hand solution will acquire about 10^{-18} mol dm^{-3} of excess hydronium ion, while an equal excess of chloride ion will remain in the right-hand one. Hence the electroneutrality principle is in error by something like 10^{-14} or $10^{-15}\%$. Because chemical reactions absorb or liberate energy, the law of conservation of mass is also not quite exact: for a typical reaction it may be in error by something like $10^{-7}\%$, Situations in which laws are not obeyed are often more interesting than those in which they are.

The potential of the cell in (14.17) is given by

$$E_{cell} = E_{right} - E_{left} + E_j. \qquad (14.18)$$

The difference $E_{right} - E_{left}$ is negative if the right-hand electrode is the negative electrode of the cell. For the sake of consistency we say that E_j is negative if the solution on the right-hand side is negatively charged, as it is at the boundary HCl(0.01 M) || HCl(0.1 M).

By combining eq. (14.18) with expressions for the potentials of the electrodes, we can obtain, for the cell in (14.17),

$$E_{cell} = \frac{RT}{nF} \ln \frac{a_{Cl-,1}}{a_{Cl-,r}} + E_j$$

where the subscripts "1" and "r" denote the solutions on the left- and right-hand sides. Since E_{cell} could be measured, we could evaluate E_j if we knew the activities of chloride ion in the two solutions. If both are extremely dilute, we could calculate these from the Debye–Hückel limiting law. Unfortunately, the limiting law is not helpful if either solution is concentrated enough to be interesting, in which event we would need to know the value of E_j in order to calculate the activity of chloride ion. We cannot break the circle, and therefore cannot evaluate either E_j or the activity of a single ion without making assumptions and approximations that cannot be rigorously proved and may not be wholly justified. The numerical values of E_j given in the remainder of this section are generally believed to be nearly correct but cannot be proven to be exactly so.

The best estimate we can make gives $E_j = -40$ mV for the boundary

HCl(0.01 M) || HCl(0.1 M). The value is negative because hydronium ion diffuses more rapidly than chloride ion, leaving the right-hand solution with an excess of negative charge, and it is large because these two ions diffuse at very different rates. At the similar boundary KCl(0.01 M) || KCl(0.1 M), both potassium and chloride ions diffuse from right to left. Because the value of D^0 is higher for chloride ion than for potassium ion, chloride ions diffuse more rapidly than potassium ions do, and therefore the right-hand solution acquires an excess of positively charged ions. Hence E_j is positive, but its value is small because the rates of diffusion are not very different: it is only about $+0.4$ mV.

TABLE 14.4. *Liquid-junction potentials for some common boundaries between aqueous solutions at 298 K*

Electrolyte		E_j, mV, across the boundary electrolyte $\|$ KCl(c), when $c =$	
		3.5 M	0.1 M
HCl,	0.01 M	+ 1.4	+ 9.3
	0.1 M	+ 3.1	+26.8
	1 M	+16.6	+56.2
H_2SO_4,	0.05 M	+ 4	+25
	0.5 M	+14	+53
KCl,	0.01 M	+ 1.0	+ 0.4
	0.1 M	+ 0.6	± 0.0
	1 M	+ 0.2	—
KOH,	0.1 M	− 1.7	−15.4
	1 M	− 8.6	−34.2
LiCl,	0.1 M	—	− 8.9
NH_4Cl,	0.1 M	—	+ 2.2
NaCl,	0.1 M	− 0.2	− 6.4
	1 M	− 1.9	−11.2
NaOH,	0.1 M	− 2.1	−18.9
	1 M	−10.5	−45

Table 14.4 gives estimates of E_j for a number of simple boundaries. There are three important conclusions that can be drawn from it:

1. Strongly acidic and strongly alkaline solutions tend to give large values of E_j. This is because hydronium and hydroxide ions diffuse through aqueous solutions much more rapidly than other ions do.

2. With any particular solution on one side of a boundary, the value of E_j almost always decreases on increasing the concentration of potassium chloride in the solution on the other side. Potassium chloride is an important electrolyte because its ions diffuse at nearly, if not quite exactly, equal rates. This is one reason why reference electrodes are usually made with saturated solutions of potassium chloride.

3. With any particular solution of potassium chloride on one side of a boundary, the value of E_j almost always increases on increasing the concentration of the solute in the solution on the other side. The ions of most other electrolytes diffuse at rates that differ more than the rates of diffusion of potassium and chloride ions do, and the difference becomes more prominent as their concentrations increase.

There is also a liquid-junction potential across the boundary between two solutions in different solvents, even if they contain the same solute at the same concentration. Each solvent diffuses into the other, and their mixing is accompanied by a change of free energy; the corresponding difference of potential is described by eq. (14.3). The value of E_j cannot even be estimated in any simple way. Depending on the particular solvents and solutes involved, it may be as high as several hundred millivolts. The resulting difficulty of comparing standard and formal potentials in different solvents hampers our attempts to achieve more thorough and more systematic understandings of ionic solvation and many other processes.

Indeed, the liquid-junction potential contributes a haze of uncertainty to the interpretation of every measurement involving a cell with liquid junction. As a simple example, suppose that a pH meter is set to read 7.00 when it is connected to a glass indicator electrode (Section 14.12) and a saturated calomel reference electrode that are in contact with a buffer known to have that pH-value. Suppose further that the electrodes are withdrawn from that solution and immediately immersed in a second solution about which nothing whatever is known, and that the pH meter then reads 6.50. All that can be said with certainty is that the pH meter reads 6.50 when these things are done. The liquid-junction potential between the unknown solution and the saturated solution of potassium chloride in the reference electrode may have any value within a very wide range. Without having some idea of the sign and magnitude of the liquid-junction potential we cannot tell whether the potential of the indicator electrode in the unknown solution is more positive than, more negative than, or equal to its potential in the standard buffer. The unknown solution may be more acidic than that buffer, but it may also be less acidic, or it may be exactly equally acidic. This is why it is necessary to draw a distinction between the pH, which is the number obtained in such a measurement, and either the pcH or the paH.

The uncertainty decreases if we know something about the composition of the "unknown" solution. If we know that it is an aqueous solution free from non-electrolytes like ethanol and glucose, we can be reasonably sure that the numerical value of E_j does not exceed perhaps 50 mV, though it might be as large as this if the solution is saturated with a very soluble salt such as lithium chloride. If we also know that its ionic strength is below 0.1 M, we can be fairly confident that the value of E_j does not exceed 5 mV or so. Knowing that its composition is almost identical with that of the reference buffer would decrease the uncertainty so much more that we could estimate the activity of hydrogen ion in it with fair confidence. As in innumerable other areas, ignorance and bliss are mutually exclusive.

SUMMARY: The liquid-junction potential arises from differences among the rates at which different substances diffuse across the boundary between two solutions. In practical work it is minimized if the two solutions have the same solvent and if one of them is saturated with potassium chloride, whose ions diffuse at nearly equal rates. However, it cannot be completely eliminated and it is an important source of error in measurements of pH and other analyses by direct potentiometry.

14.12. The glass electrode

PREVIEW: Almost all measurements of pH are made with glass electrodes. This section tells how a glass electrode is constructed and why its potential depends on the pH of a solution in which it is immersed, and describes the errors that can arise in using it.

As was shown in Section 14.5, the potential of a redox indicator electrode depends on the identity and concentration of every reducible or oxidizable substance in the solution surrounding it. According to the Nernst equation, the potential of the redox electrode Pt, $H_2(g) | H^+$ should depend only on the partial pressure of hydrogen gas and the activity of hydrogen ion. However, the Nernst equation will not be obeyed if the solution contains oxidizing agents (such as oxygen or dichromate, iron(III), or copper(II) ions) or reducing agents (such as chromium(II) ion or hydrazine, N_2H_4) at concentrations high enough to yield cathodic or anodic currents that are appreciable in comparison with the exchange current for the hydrogen couple.

We now turn to indicator electrodes of a different type, called *membrane electrodes* or sometimes *ion-selective electrodes*. Oxidation and reduction do not occur at the surfaces of these electrodes, and their potentials are therefore unaffected by foreign oxidizing or reducing agents. The first membrane electrode that was extensively studied and used was the glass electrode, which is now almost the only one used in practical measurements of pH, and which is discussed in this section. An important stimulus to its development was the fact that it could be used to measure the acidities of solutions like the dichromate solutions used in tanning leather, for which the hydrogen electrode gives erroneous and useless results because its potential depends on the activity of dichromate ion as well as on that of hydrogen ion.

The glass electrode is essentially a sealed glass bulb containing a buffered solution of sodium chloride and a silver–silver chloride electrode. By immersing a glass electrode and a reference electrode (such as an S.C.E.) in an unknown solution, one obtains a cell whose potential can be measured:

$$\text{S.C.E.} \, \| \, \text{unknown solution} \, | \, \text{glass} \, | \, Cl^-(a_{Cl-, \, in}), \, H^+(a_{H+, \, in}), \, AgCl(s) \, | \, Ag. \quad (14.19)$$

The measured value of E_{cell} includes five separate differences of potential:

1. The potential of the S.C.E.
2. The liquid-junction potential between the unknown solution and the saturated solution of potassium chloride in the reference electrode.

3. A difference of potential across the boundary between the glass and the unknown solution.
4. A similar difference of potential across the boundary between the glass and the solution with which the bulb is filled.
5. The potential of the silver–silver chloride electrode inside the bulb.

The first and second of these have already been discussed, in Sections 14.6 and 14.11 respectively. The last three are included in what we call the potential of the glass electrode, to which we shall give the symbol E_{glass}.

By comparing the potentials of glass and hydrogen electrodes in buffer solutions having different pH-values, it is found that they depend on the activity of hydrogen ion in exactly the same way over a wide range. That is,

$$E_{\text{glass}} = k + \frac{RT}{F} \ln a_{\text{H}+} = k - \frac{2.303RT}{F} \text{paH}. \tag{14.20}$$

The value of k differs from one glass electrode to another because it depends on the composition of the solution with which the electrode is filled. Allowance is made for this in practical measurements of pH by immersing the glass and reference electrodes in a standard buffer and adjusting the pH meter to read the known pH-value of that buffer. The electrodes are then transferred to an unknown solution. Changing the solution does not affect the potential of the reference electrode or the value of k, and if we optimistically assume that it also does not affect the value of E_j we shall have

$$\Delta E_{\text{cell}} = \Delta E_{\text{glass}} = -\frac{2.303RT}{F}(\Delta \text{pH}).$$

For convenience the scale of the pH meter is calibrated in pH units rather than volts, and a "temperature" control is provided so that the operator can adjust the value of the factor $2.303RT/F$. In effect the meter combines the difference between the two values of E_{cell} with the pH-value of the standard buffer to obtain a value for the pH of the unknown solution.

To explain why the potential of a glass electrode behaves as it does, we imagine the following picture. A simple glass is a network of SiO_4 tetrahedra containing interstitial alkali and alkaline-earth metal ions such as sodium and calcium ions. When it is brought into contact with an aqueous solution, some of the $Si{=}O$ bonds at its surface react with water

$$\sim\!\!\sim\!\! Si{=}O(s) + H_2O(l) = \sim\!\!\sim\!\! Si\!\!\big<^{OH}_{OH}\ (s)$$

giving rise to a "swollen gel layer". The fact that the potential does not vary with pH if the outer surface of the electrode is thoroughly dried shows that the swollen gel layer is essential to the response. The protons contained in the hydroxyl groups can dissociate, and can also be replaced by other ions, such as sodium ions:

$$\sim\!\!\!\sim\!\!\!\sim Si\!\!\big\langle^{OH}_{OH}\,(s)+Na^+ = \sim\!\!\!\sim\!\!\!\sim Si\!\!\big\langle^{ONa}_{OH}\,(s)+H^+.$$

The next paragraph will show how eq. (14.20) can be derived on the basis of this picture. To help you understand that derivation, here is a similar one for a typical redox indicator electrode. Let us suppose that one mole of electrons flows through the cell

$$Hg\,|\,Hg_2Cl_2(s),\ KCl(s)\,||\,Ag^+(a_{Ag^+})\,|\,Ag$$

which is just like the cell in (14.19) except that it contains a different indicator electrode. We ignore the things that happen at the reference electrode and the liquid junction. At the indicator electrode the half-reaction is

$$Ag^+(a_{Ag^+})+e = Ag(s).$$

One mole of silver ion at the activity a_{Ag^+} is reduced to give one mole of metallic silver at unit activity. The accompanying change of free energy is given by eq. (2.8):

$$\Delta G = G_{Ag}-G_{Ag^+}.$$

The metallic silver is in its standard state and its free energy is equal to G^0_{Ag}, but the free energy of silver ion is given by eq. (2.7):

$$G_{Ag^+} = G^0_{Ag^+} + RT\ln a_{Ag^+}.$$

Combining these equations, and using eq. (14.3) to describe the potential that corresponds to ΔG:

$$\Delta G = -nFE = (G^0_{Ag}-G^0_{Ag^+})- RT\ln a_{Ag^+} \tag{14.21}$$

in which $n = 1$ because one mole of electrons is involved. Rearranging eq. (14.21) gives

$$E = \frac{G^0_{Ag^+}-G^0_{Ag}}{F} + \frac{RT}{F}\ln a_{Ag^+} = E^0(Ag^+,\,Ag(s))-\frac{RT}{F}\ln\frac{1}{a_{Ag^+}}$$

which is of course the Nernst equation.

We can account for the behavior of the glass electrode by supposing that the following processes occur if one mole of electrons flows through the cell in (14.19):

1. One mole of hydrogen ion from the "external" solution surrounding the electrode is taken up by the swollen gel layer exposed to that solution. This can be most simply represented by the equation

$$\sim\!\!\!\sim\!\!\!\sim Si\!\!\big\langle^{O^-}_{OH}\,(s)+H^+ = \sim\!\!\!\sim\!\!\!\sim Si\!\!\big\langle^{OH}_{OH}\,(s). \tag{14.22}$$

2. One mole of sodium ion (or perhaps some other ion, depending on the composition of the glass) migrates through the glass from the external swollen gel layer to the internal one.

3. As these positively charged ions arrive at the internal swollen gel layer, they cause a mole of hydrogen ion to be evolved (by the reverse of the reaction in (14.22)) and released into the solution sealed inside the electrode.

4. Finally, one mole of solid silver chloride is reduced at the silver–silver chloride electrode:

$$AgCl(s) + e = Ag(s) + Cl^-$$

The mole of chloride ion counterbalances the positive charge added to the internal solution in the preceding step.

The overall result is

$$\sim\!\!\!\!\sim Si\!\!\begin{array}{c}O^-\\ \\OH\end{array}\!\!(s, out) + H^+(out) + AgCl(s) + e \sim\!\!\!\!\sim = Si\!\!\begin{array}{c}O^-\\ \\OH\end{array}\!\!(s, in) + H^+(in)$$
$$+ Ag(s) + Cl^-(in)$$

where "out" and "in" denote the solutions outside and inside the glass bulb and the swollen gel layers in contact with them. The change of free energy that accompanies this half-reaction is

$$\Delta G = -FE = \left(G_{\sim Si\begin{smallmatrix}O^-\\OH\end{smallmatrix}, \,in} - G_{\sim Si\begin{smallmatrix}O^-\\OH\end{smallmatrix}, \,out}\right) + (G_{Ag(s)} - G_{AgCl(s)})$$
$$+ (G_{H^+, \,in} + G_{Cl^-, \,in}) - G_{H^+, \,out}.$$

The terms inside the first pair of parentheses on the right-hand side can be dropped because the free energy of an active site is not changed by moving it from one place to another. The terms inside the second pair of parentheses are equal (because both silver and silver chloride are present in their standard states) to the corresponding standard free energies, which are constant. The terms inside the third pair of parentheses are also constant because they depend on the composition of the solution with which the bulb is filled, which does not change during the life of the electrode. An expression for $G_{H^+, \,out}$ can be obtained from eq. (2.7):

$$G_{H^+, \,out} = G^0_{H^+} + RT \ln a_{H^+, \,out}.$$

Combining all of the constants in these equations into a single one gives

$$\Delta G = -FE = constant - RT \ln a_{H^+, \,out}$$

or

$$E(= E_{glass}) = -\frac{constant}{F} + \frac{RT}{F} \ln a_{H^+, \,out} = k - \frac{2.303RT}{F} pH_{out} \quad (14.23)$$

which is identical with eq. (14.20).

It has already been said that a glass electrode must be equilibrated with water, so that the swollen gel layer can form on its outer surface, before it can be used. Water is important to the behavior of the glass electrode in other ways. The hydrogen ions in the external solution are hydrated to an extent that depends on the activity of water, and the active $\sim\!\!\sim\!\!\sim Si\!\!\begin{subarray}{l} \diagup O^- \\ \diagdown OH \end{subarray}$ groups in the external swollen gel layer may be hydrated as well. Unless the activity of water in the external solution is equal to 1, the potential of the glass electrode will not obey eq. (14.23), and the pH-value indicated by the meter will misrepresent the acidity of the solution.

The activity of water differs from 1 in any solution that contains a high concentration of solute, regardless of whether the solute is an electrolyte or not. Some typical data are plotted in Fig. 14.17. A pH-value of 0 corresponds to approximately 1 M hydrochloric acid; one of -1 corresponds to approximately 5 M hydrochloric acid.

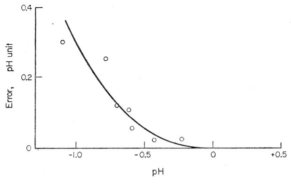

FIG. 14.17. The error of the glass electrode in aqueous solutions of hydrochloric acid at 298 K. The positive sign of the error means that the glass electrode indicates that the measured pH is higher than the true value.

Over this range of concentrations there is a substantial decrease of the activity of water. In 1 M acid the potential of the electrode has almost exactly the value predicted by eq. (14.23), but in 5 M acid it is in error by nearly 20 mV, and therefore the pH of the solution appears to be about 0.3 unit higher than it actually is. The error is called the *water-activity error*. Because it was first discovered in work with fairly concentrated solutions of acids, it is also called the *acid error*, but this name is misleading because it suggests that the error arises only in acidic solutions. Exactly similar errors arise in measuring the pH-values of concentrated solutions of neutral salts, bases, and non-electrolytes. With all such solutions there is also a large liquid-junction potential at the boundary with the saturated aqueous solution of potassium chloride in the reference electrode. The liquid-junction potential introduces an additional uncertainty, but it arises in a very different way from the water-activity error. The liquid-junction potential would afflict any potentiometric measurement, such as one made with a hydrogen electrode instead of a glass electrode, but the water-activity error is peculiar to the glass electrode and could be eliminated by using a hydrogen electrode.

A different kind of error, called the *cation error*, arises in strongly alkaline solutions containing high concentrations of certain cations, among which sodium ion is the worst offender because it is the commonest one for which the error is large. The behavior of the cation error is shown in Fig. 14.18. The error results from the fact

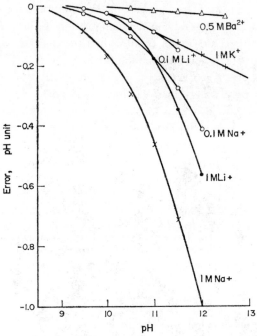

FIG. 14.18. The errors of a typical glass electrode in alkaline aqueous solutions containing various cations at 298 K. The negative sign of the error means that the glass electrode indicates that the measured pH is lower than the true value.

that the ～～Si⟨$^{O^-}_{OH}$ sites in the external swollen gel layer can be occupied by sodium ions, or other cations, which compete with hydrogen ions for the possession of these sites. By combining eq. (14.22) with a similar equation representing the uptake of a sodium ion

$$\sim\!\!\sim Si\!\!\left\langle\begin{array}{c}O^-\\OH\end{array}\right.(s)+Na^+ = \sim\!\!\sim Si\!\!\left\langle\begin{array}{c}ONa\\OH\end{array}\right.(s) \qquad (14.24)$$

we obtain an equation describing an ion-exchange reaction:

$$\sim\!\!\sim Si\!\!\left\langle\begin{array}{c}OH\\OH\end{array}\right.(s)+Na^+ = \sim\!\!\sim Si\!\!\left\langle\begin{array}{c}ONa\\OH\end{array}\right.(s)+H^+.$$

For a typical pH-responsive glass the equilibrium constant of this reaction is very small because sodium ions are much less strongly bound to the active sites than hydrogen ions are. But as the ratio a_{Na+}/a_{H+} increases, more and more of these sites come to be occupied by sodium ions, and the result is that the potential of the electrode becomes less dependent on the activity of hydrogen ion (and more dependent on the activity of sodium ion). The magnitude of the resulting error is discussed in more detail in Section 14.13. It depends on the composition of the glass, and special glass electrodes, which give errors much smaller than the ones shown in Fig. 14.18, are available for work with strongly alkaline solutions.

Finally, glass electrodes give erroneous results in solutions having very small buffer capacities, such as pure water. The reaction described by eq. (14.22) consumes an appreciable fraction of the hydrogen ions that are present at the surface of an electrode exposed to such a solution. The electrode responds to the pH of the layer of solution at its surface, and this is higher than the pH of the solution some distance away. To make matters worse, equilibrium is only very slowly attained under such conditions, and the measured pH drifts upward as time goes on.

> SUMMARY: The response of a glass electrode to pH arises from an acid-base equilibrium in the swollen gel layer at its surface. A glass electrode gives erroneous pH-values in solutions containing high concentrations of solutes, in which the activity of water is much smaller than 1; in alkaline solutions containing other metal ions that can compete with hydrogen ion for the active sites in the swollen gel layer; and in weakly buffered solutions.

14.13. The selectivities of membrane electrodes

> PREVIEW: Glass and other membrane electrodes are selective but not specific: they can respond not only to the ion D being determined but also to an interfering one I that is present as well. An equation relating the potential to the activities of D and I is derived and discussed.

The preceding section described a mechanism that accounts for the Nernstian response of a glass electrode to the activity of hydrogen ion. Four steps are involved: the uptake of hydrogen ion from the solution surrounding the electrode, the motion of some ion through the interior of the glass, the release of hydrogen ion into the solution with which the electrode is filled, and the redox half-reaction that occurs at the silver–silver chloride electrode sealed inside the glass. Each of these steps entails a change of free energy, but only the first of them involves the solution in which the electrode is immersed. For any particular glass electrode the changes of free energy accompanying the last three steps are independent of the composition of that solution, and are lumped together in the value of the constant k in eqs. (14.20) and (14.23).

The preceding section also showed that glass electrodes do not respond to the activity of hydrogen ion alone. Under extreme conditions, where this activity is very much lower than the activity of some other cation M^+, the potential of a glass electrode may also depend on the activity of M^+. Nevertheless, the glass electrode does respond preferentially, or selectively, to hydrogen ion. All membrane electrodes behave in this fashion, and the term "ion-selective electrode" emphasizes that their responses are selective but not exclusive. An electrode that responded to only one ion, regardless of the conditions, would be called *specific* for that ion.

The behavior is so common that it is important to have a description of it. We can obtain one by focusing our attention on the change of free energy that occurs in the external swollen gel layer. We shall use the symbols ΔG_H^0 and ΔG_M^0 to represent the standard changes of free energy accompanying the processes described by eqs. (14.22) and (14.24):

$$\sim\!\!\sim Si\!\!\begin{array}{c} O^- \\ \\ OH \end{array}\!\!(s) + H^+ = \sim\!\!\sim Si\!\!\begin{array}{c} OH \\ \\ OH \end{array}\!\!(s); \quad \Delta G^0 = \Delta G_H^0,$$

$$\sim\!\!\sim Si\!\!\begin{array}{c} O^- \\ \\ OH \end{array}\!\!(s) + M^+ = \sim\!\!\sim Si\!\!\begin{array}{c} OM \\ \\ OH \end{array}\!\!(s); \quad \Delta G^0 = \Delta G_M^0.$$

These equations may be combined to give

$$\sim\!\!\sim Si\!\!\begin{array}{c} OH \\ \\ OH \end{array}\!\!(s) + M^+ = \sim\!\!\sim Si\!\!\begin{array}{c} OM \\ \\ OH \end{array}\!\!(s) + H^+; \; \Delta G^0 = \Delta G_M^0 - \Delta G_H^0$$

$$= -RT \ln K_{M/H}. \quad (14.25)$$

The equilibrium constant $K_{M/H}$ *(sometimes written as* K_H^M*) is called a* selectivity coefficient. *Its value is a measure of the ability of* M^+ *ions to replace hydrogen ions bound to the active sites on the surface.*

If one mole of electrons flows through the electrode, one mole of positive charge must be taken up by the active sites. If hydrogen ions are the only ones that are taken up, one mole of hydrogen ion is consumed, and the potential of the electrode will be given by eq. (14.23). If M^+ ions can also be taken up, the mole of positive charge will consist of x mole of hydrogen ion and $(1-x)$ mole of sodium ion. A solid solution containing both $\sim\!\!\sim Si\!\!\begin{array}{c} OH \\ \\ OH \end{array}$ and $\sim\!\!\sim Si\!\!\begin{array}{c} OM \\ \\ OH \end{array}$ will be formed. If this solution is ideal, the activity of $\sim\!\!\sim Si\!\!\begin{array}{c} OH \\ \\ OH \end{array}$ will be proportional to the fraction of the sites that are occupied by hydrogen ion, or x, while the activity of $\sim\!\!\sim Si\!\!\begin{array}{c} OM \\ \\ OH \end{array}$ will be proportional to $(1-x)$. Hence

$$K_{M/H} = \frac{a_{\sim Si\begin{array}{c} OM \\ OH \end{array}} a_{H^+}}{a_{\sim Si\begin{array}{c} OH \\ OH \end{array}} a_{M^+}} = \frac{(1-x)\,a_{H^+}}{x a_{M^+}}. \quad (14.26)$$

The change of free energy accompanying the uptake of x mole of hydrogen ion is given by

$$G_H = x \left(\Delta G_H^0 + RT \ln \frac{a_{\sim Si\langle^{OH}_{OH}}}{a_{H^+} a_{\sim Si\langle^{O^-}_{OH}}} \right)$$

and that accompanying the uptake of $(1-x)$ mole of M^+ is given by

$$G_M = (1-x) \left(\Delta G_M^0 + RT \ln \frac{a_{\sim Si\langle^{OM}_{OH}}}{a_{M^+} a_{\sim Si\langle^{O^-}_{OH}}} \right).$$

The total change of free energy is the sum of ΔG_H and ΔG_M:

$$\Delta G = x\,\Delta G_H^0 + (1-x)\,\Delta G_M^0 + xRT \ln \frac{a_{\sim Si\langle^{OH}_{OH}}}{a_{H^+} a_{\sim Si\langle^{O^-}_{OH}}} \cdot \frac{a_{M^+} a_{\sim Si\langle^{O^-}_{OH}}}{a_{\sim Si\langle^{OM}_{OH}}}$$

$$+ RT \ln \frac{a_{\sim Si\langle^{OM}_{OH}}}{a_{M^+} a_{\sim Si\langle^{O^-}_{OH}}}.$$

Rearranging the first two terms on the right-hand side, replacing the argument of the third by $K_{M/H}$, and combining the argument of the fourth with eq. (14.26),

$$\Delta G = \Delta G_M^0 - x(\Delta G_M^0 - \Delta G_H^0) + xRT \ln K_{M/H} + RT \ln \frac{K_{M/H} x}{a_{H^+} a_{\sim Si\langle^{O^-}_{OH}}}.$$

The second and third terms on the right-hand side cancel each other by virtue of eq. (14.25), and $K_{M/H}$ and $a_{\sim Si\langle^{O^-}_{OH}}$ can be separated from the last term to give

$$\Delta G = \Delta G_M^0 + RT \ln K_{M/H} + RT \ln \frac{x}{a_{H^+}} - RT \ln a_{\sim Si\langle^{O^-}_{OH}}. \tag{14.27}$$

An expression for x can be obtained by solving eq. (14.26):

$$x = \frac{a_{H^+}}{a_{H^+} + K_{M/H} a_{M^+}}. \tag{14.28}$$

Combining the first two terms on the right-hand side of eq. (14.27) with eq. (14.25), and using eq. (14.28) to eliminate x from the third term, we obtain

$$\Delta G = \Delta G_H^0 - RT \ln (a_{H^+} + K_{M/H} a_{M^+}) - RT \ln a_{\sim Si\langle^{O^-}_{OH}}. \tag{14.29}$$

The last of these terms may be ignored: it represents the disappearance of a mole of unoccupied sites, but these reappear in the internal swollen gel layer as hydrogen ions are released from that layer into the internal solution. Equal but opposite changes of free energy are involved in the disappearance and reappearance of these sites, and there is no effect on the measured potential. Combining eq. (14.29) with eq. (14.3), and using the symbol k to represent all of the constant terms, the final result is

$$E = k + \frac{RT}{F} \ln \left(a_{H+} + K_{M/H} a_{M+}\right)_{out}. \tag{14.30}$$

The subscript "out" has been added as a reminder that we are interested only in the composition of the solution outside the electrode.

Equation (14.30) resembles eq. (14.23). The arguments of the logarithmic terms are different, but the two equations become indistinguishable if a_{H+} is much larger than the product $K_{M/H} a_{M+}$. When this is true the electrode exhibits a Nernstian response to hydrogen ion, and its potential is not affected by changing the activity of M^+. However, even if the value of $K_{M/H}$ is very small—and it is only about 10^{-12} for the electrode used to obtain the data of Fig. 14.17—a_{H+} and $K_{M/H} a_{M+}$ must eventually become comparable if the activity of hydrogen ion is decreased while the activity of M^+ ion remains large. Under conditions such that $K_{M/H} a_{M+}$ is much larger than a_{H+}, the electrode exhibits a Nernstian response to M^+ ion but does not respond to hydrogen ion.

Because the values of $K_{M/H}$ differ for different cations, the effect of an interfering ion M^+ depends on its identity as well as on the ratio of its activity to that of the ion (hydrogen ion) being determined. For example, the curves in Fig. 14.17 show that $K_{Na/H}$ is much larger than $K_{K/H}$ for the electrode used.

The chemist designing, or choosing, a membrane electrode for the determination of one ion, D, in the presence of another one, I, naturally wants the value of $K_{I/D}$ to be as small as possible. Very small values are much more likely if D and I have different charges or sizes than if they have similar ones. It is chiefly because the values of $K_{Na/H}$, $K_{K/H}$, $K_{Li/H}$, etc., are extremely small that the glass electrode is as widely useful for measurements of pH as it is. For most other membrane electrodes the values are less favorable. Ne do not yet know how to make a potassium-selective electrode for which $K_{Na/K}$ is smaller than about 1×10^{-4}. The effect of sodium ion on the potential of such an electrode can be ignored if the ratio $[Na^+]/[K^+]$ is smaller than about 100, and a correction for it can be applied if $[Na^+]/[K^+]$ is even as large as 10^4 provided that the concentration of sodium ion in the solution being analyzed is known with fair certainty, but the electrode becomes useless for the determination of potassium ion if the ratio is very large.

SUMMARY: In a solution containing two ions, D and I, to which it responds, a glass or membrane electrode assumes a potential that depends on their concentrations and on the selectivity coefficient $K_{I/D}$. The value of $K_{I/D}$ depends on the charges and sizes of D and I and varies from one membrane to another.

14.14. Other membrane electrodes

PREVIEW: This section gives qualitative descriptions of several other kinds of membrane electrodes.

During the last decade or two, chemists have developed many other electrodes that behave similarly to the glass electrode but that respond to the activities of many different substances. Some of these, such as copper(II) and sulfate ions, could also be observed with electrodes of the first or second kind, but only at the risk of incurring interferences from oxidizing and reducing agents that might be present. For others, including perchlorate and nitrate ions, membrane electrodes are the only ones available. Electrodes responsive to calcium and potassium ions are useful in clinical chemistry, fluoride-selective electrodes are used to monitor municipal water supplies, and nitrate-selective electrodes serve for analyses of natural waters and fertilizers.

There are a number of different ways in which such electrodes can be constructed. Glass electrodes responsive to a number of univalent cations can be made from glasses having different compositions from those employed in pH-responsive electrodes. The Corning 015 glass used to obtain the data of Figs. 14.17 and 14.18 is a mixture containing 72.2 mole % of SiO_2,[†] 21.4 mole % of Na_2O, and 6.4 mole % of CaO. Replacing most or all of the sodium oxide with lithium and cesium oxides yields even smaller values of $K_{Na/H}$; adding trivalent metal oxides such as aluminum oxide or scandium oxide has the opposite effect, and to so large an extent that a glass containing 71 mole % of SiO_2, 11 mole % of Na_2O, and 18 mole % of Al_2O_3 responds preferentially to sodium ion. Other glasses are known that are selective for each of the other alkali metal ions and for ammonium, silver, and thallium(I) ions. It is possible to make a silver-selective electrode for which the value of $K_{H/Ag}$ is so low that even as little as 10^{-8} M silver ion can be determined in a neutral solution.

Solid-state electrodes are those in which the membrane is made from a crystalline solid, which is usually inorganic. The finely divided solid may be dispersed in just enough silicone rubber or poly(vinyl chloride) to form a rigid matrix without preventing electrical contact between neighboring particles. A single crystal of the solid may be used, or the solid may be fused, or compacted under a very high pressure, to form a disc. Barium sulfate, lanthanum fluoride, and silver iodide are among the solids that have been used. The solid can adsorb its own ions from a solution to which it is exposed: the adsorption of sulfate ion onto a particle of barium sulfate would be analogous to the uptake of hydrogen ion on the $\sim\sim Si\left\langle{}^{O^-}_{OH}\right.$ sites at the surface of a glass electrode. Current is carried through the membrane by the motion of ions, which need not be the same as those adsorbed, and ions are released into the solution surrounding a suitable reference electrode on the other side of the membrane. The

[†] That is, there are 72.2 moles of SiO_2 for every 100 moles of material.

construction and operation of such an electrode are much the same as for a glass electrode. The potential depends on the activity of the adsorbed ion, and is affected by any other ion that can undergo specific adsorption onto the surface. Phosphate ion would interfere in the determination of sulfate at a barium sulfate membrane electrode, and aluminum(III) and iron(III) would interfere—to extents depending on the pH—in the determination of chloride at a silver chloride membrane electrode. The solubility of the solid from which the membrane is made sets a lower limit to the range of concentrations that can be determined even in the absence of any specifically adsorbed ions.

A third kind of ion-selective electrode employs either an organic ion-exchanger or a chelating agent as the active material. This is usually dissolved in a solvent that is immiscible with water, and the solution is immobilized in such a way that it separates two aqueous solutions. One of these is the "external" or unknown solution; the other is the "internal" solution and contains a silver–silver chloride electrode or some other suitable redox electrode. Different manufacturers use different arrangements. The solution may be held in the pores of a fritted glass disc or a cellulose acetate or cellulose nitrate disc with very small pores, or may be trapped between two parallel discs. A typical electrode of this type is a potassium-selective electrode made with a solution of valinomycin in diphenyl ether. Valinomycin is an antibiotic with a complex structure; a molecule of it can wrap itself around to form a cage just the right size to contain a chelated potassium ion. The sizes of a potassium ion and a sodium ion are so different that $K_{\text{Na/K}}$ is only about 10^{-4}. At the boundary between this solution and the external solution, potassium ions enter the organic phase [denoted by "o"] by the process

$$\text{valinomycin}(o) + \text{K}^+(aq) = \text{valinomycin} \cdot \text{K}^+(o).$$

Some anion may accompany the potassium ion to form an ion pair. The general picture is very much like the one given in preceding sections for pH-responsive glass electrodes. So wide a choice of organic reactants is available that electrodes responsive to many different ions can be constructed, and research aimed at the development of better ones is proceeding very actively.

SUMMARY: Membrane electrodes are of three principal kinds: glass electrodes, solid-state electrodes employing solutions of ion-exchangers or chelating agents.

Problems

Answers to some of these problems are given on page 522.

14.1. Identify the net cell reaction that occurs in each of the following cells:

(a) $\text{Cu} \mid \text{Cu}^{2+} \mid\mid \text{Ag}^+ \mid \text{Ag}$.

(b) $\text{Ag} \mid \text{Ag}^+ \mid\mid \text{Cu}^{2+} \mid \text{Cu}$.

(c) $\text{Pt} \mid \text{Fe}^{2+}, \text{Fe}^{3+} \mid\mid \text{Cr}^{3+}, \text{Cr}^{2+} \mid \text{Hg}$

(d) $\text{Ag} \mid \text{Ag}^+ \mid\mid \text{Cl}^-, \text{AgCl}(s) \mid \text{Ag}$.

(e) $\text{Pb} \mid \text{PbBr}_2(s), \text{NaBr}(0.01\ M) \mid\mid$
$\text{NaBr}(0.02\ M), \text{PbBr}_2(s) \mid \text{Pb}$.

14.2. Write a cell that corresponds to each of the following net cell reactions:

(a) $Pb(s) + PbO_2(s) + 2 H^+ + 2 HSO_4^- = 2 PbSO_4(s) + 2 H_2O(l)$.

(b) $AgCl(s) + Br^- = AgBr(s) + Cl^-$.

(c) $Fe(s) + 2 Fe^{3+} = 3 Fe^{2+}$.

14.3. For each of the following cells decide whether the net cell reaction is spontaneous:

(a) $Pt \,|\, Fe^{2+}$ (0.001 M), Fe^{3+} (0.01 M) $||\, Hg_2^{2+}$ (0.05 M) $|\, Hg$.

(b) $Pt \,|\, Hg_2^{2+}$ (0.01 M), Hg^{2+} (0.01 M) $||\, Hg_2^{2+}$ (0.01 M) $|\, Hg$.

14.4. At 298 K the potential of the cell $Pt \,|\, H_2$ (1 atm), $HCl(c)$, $AgCl(s) \,|\, Ag$ is 0.5727 V when $c = 1.129 \times 10^{-3}$ M and is 0.6353 V when $c = 3.288 \times 10^{-4}$ M. Find the standard potential of the half-reaction $AgCl(s) + e = Ag(s) + Cl^-$ and calculate the mean ionic activity coefficient of hydrochloric acid in a 1.129×10^{-3} M solution.

14.5. On page 452 it was said that reference electrodes often contain saturated solutions of potassium chloride because the diffusion coefficients of potassium and chloride ions are nearly equal. The diffusion coefficients of potassium and nitrate ions, rubidium and bromide ions, and lead and carbonate ions are even more nearly equal than those of potassium and chloride ions. Why are reference electrodes not made with potassium nitrate, rubidium bromide, or lead carbonate instead of potassium chloride?

14.6. At 298 K the formal potential of the iron(III)–iron(II) couple in 1 M hydrochloric acid is $+0.459$ V vs. S.C.E. A sample of reagent-grade iron(II) sulfate ($FeSO_4 \cdot 7 H_2O$, formula weight 278.1) is analyzed by dissolving 0.500 g of it in enough 1 M hydrochloric acid to give 100.0 cm^3 of solution. A platinum indicator electrode and a saturated calomel reference electrode are immersed in the solution and the potential of the platinum electrode is found to be $+0.320$ V vs. S.C.E. What percentage of iron(III) sulfate [$Fe_2(SO_4)_3$, formula weight 399.9] did the iron(II) sulfate contain?

14.7. A silver indicator electrode and a saturated calomel electrode are immersed in 100 cm^3 of an unknown solution of silver ion, and the potential of the indicator electrode is found to be $+0.400$ V vs. S.C.E. Exactly 1.00 cm^3 of an 0.100 M solution of silver nitrate is added, and the potential of the indicator electrode is then found to be $+0.416$ V vs. S.C.E.

(a) What was the concentration of silver ion in the unknown solution if no other oxidizing agent was present?

(b) In view of Fig. 14.13, what could be said about the concentration of silver ion if it were suspected that the unknown solution might also contain some iron(III)?

14.8. The following data were obtained in the potentiometric titration of a solution of sodium bromide with a solution of silver nitrate, using a silver indicator electrode and a saturated calomel reference electrode:

V_{AgNO_3}, cm^3	39.60	39.70	39.80	39.90	40.00	40.10
E_{Ag}, V vs. S.C.E.	0.090	0.102	0.148	0.197	0.238	0.260

Find the volume of silver nitrate added at the end point.

14.9. The following data were obtained in a potentiometric titration of a solution of osmium(II) with a solution of potassium hexacyanoferrate(III) ($=R$), using a platinum indicator electrode and a silver–silver chloride reference electrode:

V_R, cm^3	5.00	10.00	15.00	20.00	25.00
$E_{indicator}$, V vs. Ag|AgCl	$+0.126$	$+0.150$	$+0.165$	$+0.181$	$+0.203$

The end point was found to lie at $V_R = 32.38$ cm^3.

(a) To what oxidation state was the osmium oxidized in the reaction that occurred, assuming that the indicator electrode behaved reversibly?

(b) What other data would be useful in answering the preceding question, and how would you use them?

(c) What was the formal potential of the osmium couple under the conditions of the titration?

CHAPTER 15

Spectroscopy

15.1. Introduction

Spectroscopy is the branch of science that is concerned with the absorption and emission of electromagnetic radiation. Its applications in chemistry are of a great many different kinds. Many of our ideas about the structures of atoms and the compositions of stars are obtained from atomic emission spectroscopy, which is based on the radiation emitted by atoms and ions exposed to energy that causes molecules to dissociate and raises electrons to excited states. Much insight into the arrangement of atoms and groups of atoms in organic and inorganic compounds is obtained from infrared absorption spectroscopy, which is based on the effects of radiation on the vibrational and rotational energies of molecules and ions. These and other spectroscopic techniques provide ways of investigating molecular structures, identifying and characterizing chemical substances and determining the proportions in which they are present in complex mixtures, studying the natures of chemical bonds, and obtaining a wealth of information about the behaviors of chemical systems. Much of modern chemical theory was constructed to explain the results of spectroscopic measurements of various kinds.

This chapter is a very condensed introduction to chemical spectroscopy. It begins with brief descriptions of electromagnetic radiation and the electromagnetic spectrum. The absorption and emission of radiation are processes in which a chemical substance exchanges energy with a beam of radiation or an electromagnetic field, and the different kinds of spectroscopy are classified according to the processes that they involve at the atomic or molecular level. Most of the rest of the chapter is devoted to the absorption of visible and near-ultraviolet radiation. This is not only a typical branch of spectroscopy but also one for which the apparatus is relatively inexpensive, readily available, and easy to use, and in addition it is applicable in many ways to studies of the rates and equilibria of chemical reactions of the kinds considered in earlier chapters.

15.2. The electromagnetic spectrum

PREVIEW: This section summarizes the properties of electromagnetic radiation and describes the processes that are responsible for the interaction of matter with radiation in different portions of the spectrum.

Visible light is one of many different kinds of electromagnetic radiation. Others include ultraviolet and infrared radiation, X-radiation, gamma radiation, microwave and radio-frequency radiation, and the radiation emitted by a wire through which an alternating current is flowing.

Radiation of all these types can be propagated even in the absence of a material medium, and travels through a vacuum at the velocity c^0, which is equal to 2.9979×10^{10} cm s^{-1}. It has two components: an alternating or sinusoidal electric field, and an alternating magnetic field that is in phase with the electric field but at right angles to it in space. The electric field is the one that is important in spectroscopy. Like any other sinusoidal phenomenon, a beam of radiation has two important characteristics: its period and its amplitude. The *period* is the time that elapses between the instants at which two successive maxima pass any given point; the *frequency* ν is the reciprocal of the period and is equal to the number of oscillations per second. The *wavelength* λ is the distance between two successive maxima, and is related to the frequency by the equation

$$\nu\lambda = c$$

where c is the velocity of the radiation in the medium (air, glass, solution, etc.) through which the radiation is passing.

The velocity c varies from one medium to another, and its value in any particular medium is given by

$$c = c^0/n$$

where c^0 is the velocity in a vacuum and n is the *index of refraction* or *refractive index* of the material. The index of refraction is defined as being equal to 1 for a vacuum; in any material medium its value exceeds 1 and varies with the frequency. This variation is responsible for the ability of a prism to disperse a beam of light into its constituent frequencies and is called *dispersion*. Some typical values of n, at the frequency (5.087×10^{14} s^{-1}) corresponding to the familiar intense yellow radiation emitted by excited sodium atoms, are 1.000 266 for dry air at a pressure of 760 torr, 1.3329 for water, 1.361 for ethanol, and 1.500 for a saturated aqueous solution of sucrose, all at 298 K. The frequency of an electromagnetic wave depends only on the properties of the source from which the wave is obtained, and is independent of the medium through which the wave travels, but the wavelength also depends on the nature of the medium and on the temperature and pressure. Frequency is therefore more fundamental than wavelength, but wavelengths are widely used nonetheless. You should always assume, unless something is said to the contrary, that any wavelength is a wavelength in air (or in a vacuum, which is the same thing within the accuracies of most instruments).

The unit of frequency is the reciprocal second or hertz (Hz). Wavelengths might be expressed in centimeters; this is convenient in microwave spectroscopy, where the wavelengths of interest range from a few hundredths of a centimeter to a few tens of centimeters, but is unwieldly in other parts of the spectrum. The yellow radiation emitted by excited sodium atoms has a frequency of 5.087×10^{14} Hz and a wavelength of 5.893×10^{-5} cm in air. The latter is more easily expressed as 589.3 nm (1 nm = 10^{-9} m = 10^{-7} cm); wavelengths in the visible and ultraviolet portions of the spectrum are usually given in nanometers. The millimicron (mμ) is an older name for the nanometer, and ångstrøm units (1 Å = 0.1 nm) are still used in emission spectroscopy. Another quantity of some importance is the *wavenumber* σ, which is the number of waves per centimeter in a vacuum and is given by

$$\sigma = 1/\lambda_{vac}$$

where λ_{vac} is the wavelength in a vacuum.

The visible portion of the spectrum extends from about 7.5×10^{14} to 3.75×10^{14} Hz, or 400 to 800 nm. Absorption and emission in this region result from processes that involve outer-shell or valence electrons, and that will be discussed in Sections 15.5 and 15.10. Such processes are also responsible for absorption and emission in the ultraviolet region, and therefore the visible and ultraviolet regions are always considered together. The ultraviolet portion of the spectrum extends from about 10^{16} to 7.5×10^{14} Hz, or 30 to 400 nm, and is generally divided into two parts: the near ultraviolet, from about 1.5×10^{15} to 7.5×10^{14} Hz or 200 to 400 nm, and the far or vacuum ultraviolet from 10^{16} to 1.5×10^{15} Hz or 30 to 200 nm. The vacuum ultraviolet is so called because oxygen and nitrogen absorb strongly in this region, so that it can be studied only with instruments from which air is evacuated.

The fact that any particular kind of matter absorbs or emits radiation having only certain frequencies—which will be discussed in Section 15.5—played a primary role in the development of the quantum theory. This asserts that the internal energies of molecules, atoms, and ions are quantized, which is to say that they can assume only certain definite discrete values corresponding to different states. (The translational or kinetic energy associated with motion from one point to another in space is not quantized and is of no importance in spectroscopy.) The lowest value of the internal energy is said to correspond to the ground state, and the higher values are said to correspond to excited states. Absorption and emission result from transitions among these different states. If sodium chloride is introduced into a flame it decomposes into atoms of sodium and chlorine. Most of the sodium atoms remain in the ground state, but a few are raised to an excited state by absorbing energy from the flame. There is an equilibrium between the atoms in the ground state and those in the excited state. At any instant some atoms in the ground state are absorbing energy and being promoted to the excited state, while an equal number of atoms in the excited state are emitting energy and decaying back to the ground state. As they decay they lose an amount of energy that is equal to the difference between the energies of the excited and ground states. This energy appears as electromagnetic radiation having a frequency of 5.087×10^{14} Hz and a wavelength of 589.3 nm. Emission at this frequency and wavelength is characteristic of sodium atoms, and its intensity under carefully controlled conditions is proportional to the concentration of sodium atoms in the flame. This is the principle of flame emission spectroscopy or flame photometry. The reverse transition can be observed by illuminating the flame with a beam of light having this frequency and wavelength. Energy is absorbed from the beam by the atoms of sodium in the ground state: the atoms are raised to the excited state and the intensity of the beam decreases. The fact that energy is absorbed at this frequency and wavelength shows that sodium atoms are present in the flame and the extent of absorption depends on their concentration. This is the principle of flame absorption spectroscopy. The energy that is needed to produce and excite the sodium atoms may be obtained from sources other than flames, and "atomic emission spectroscopy" and "atomic absorption spectroscopy" are more general names for the families of techniques that embody these general principles. As applied to the detection and determination of sodium, all these techniques depend on the fact that sodium atoms can neither emit nor absorb radiation having a wavelength of, say, 587 or 591 nm.

To account for such facts it is convenient to imagine that electromagnetic radiation is composed of packets of energy called *photons*, and that each photon has an energy E given by

$$E = h\nu \quad \text{J photon}^{-1}$$

where ν is the frequency of the radiation and h is Planck's constant, 6.6262×10^{-34} J s photon^{-1}. The decay of a sodium atom from the excited $3p$ state to the ground $3s$ state is accompanied by the emission of a photon having an energy of 6.6262×10^{-34} J s photon$^{-1} \times 5.087 \times 10^{14}$ s$^{-1} = 3.371 \times 10^{-19}$ J photon^{-1}; the absorption of a photon having the same energy converts an atom of sodium from the ground state to the excited state. The difference between the energies of these two states is thus 3.371×10^{-19} J atom^{-1}. Multiplying by Avogadro's number N ($= 6.0222 \times 10^{23}$ mol^{-1}) gives

$$E = hN\nu = 3.9904 \times 10^{-10} \nu \quad \text{J mol}^{-1}. \tag{15.1}$$

For this particular transition $E = 2.030 \times 10^5$ J mol^{-1} or 203.0 kJ mol^{-1}. The corresponding figures are approximately 600 kJ mol^{-1} at $\nu = 1.5 \times 10^{15}$ Hz ($\lambda = 200$ nm) and 150 kJ mol^{-1} at $\nu = 3.75 \times 10^{14}$ Hz ($\lambda = 800$ nm), which are the extremes of the range which visible and near-ultraviolet spectroscopy is concerned. Many chemical reactions involve comparable changes of energy.

Energy changes of this order of magnitude correspond to processes involving outer-shell or valence electrons, such as the $3s$ electron of a sodium atom in the ground state. These are the electrons that are affected by chemical combination. The difference of energy between the ground and excited states of a copper(II) ion dissolved in an aqueous solution is altered by substituting a chloride ion for one of the molecules of water in its coordination shell, and the difference of energy between the ground and excited states is much smaller for a carbonyl (C=O) group than for a C—O single bond. Consequently $Cu(OH_2)_4^{2+}$ and $CuCl(OH_2)_3^{+}$ absorb at different frequencies and wavelengths, and so do acetaldehyde

$$\left(CH_3C\diagup_{H}^{O}\right) \quad \text{and its hydrate} \quad \left(CH_3CH\diagup_{OH}^{OH}\right).$$

This suggests why visible and near-ultraviolet absorption spectroscopy is so valuable to the chemist studying dissolved substances and their reactions.

Processes that involve inner-shell electrons correspond to larger differences of energy and therefore to radiation having higher frequencies and shorter wavelengths. X-ray absorption, emission, and fluorescence are based on such processes and occur in the range of frequencies from about 10^{16} Hz, which is the upper end of the ultra-violet region, to 10^{20} Hz; the corresponding range of wavelengths is 30–0.003 nm or 300–0.03 Å. Inner-shell electrons are almost unaffected by chemical bonding, and X-ray spectroscopy therefore furnishes information about the identities and propor-tions of the atoms in a sample regardless of the ways in which they are combined.

Nuclear transitions involve still larger differences of energy and correspond to gamma radiation at frequencies above about 10^{20} Hz. Like inner-shell electronic transitions, nuclear transitions are independent of the chemical states of the atoms in which they occur. They do, however, depend on the structure of the nucleus, and therefore the frequency of the gamma radiation differs for different isotopes of the same element. Physicists employ gamma-ray spectroscopy to study the properties of atomic nuclei; chemists usually combine it with neutron activation. A sample is exposed for some time to a beam of neutrons, some of which are absorbed to give unstable isotopes that decay by emitting gamma rays. An excited nucleus of any particular kind emits a gamma ray having a particular energy as it decays to the ground state, and different kinds of nuclei that are present in the sample after activation can be identified by measuring the energies of the gamma rays that are emitted.

Going in the other direction, toward lower energies, from the visible–ultraviolet region at which we began, the infrared region is generally considered to extend from about 3.75×10^{14} to 3×10^{12} Hz (0.8 to 100 μm). The processes that are responsible for absorption in this range involve energy differences ($hN\nu$ in eq. (15.1)) ranging from about 150 to 1.2 kJ mol^{-1}, which are much lower than those for electronic transitions, and which correspond to transitions between different vibrational energy states. In any molecule that contains two or more atoms, these oscillate around their mean positions. A diatomic molecule, such as hydrogen fluoride, has a single vibration-al mode in which the atoms alternately approach and recede from each other. The frequency with which they do so depends on their masses, and also on the prop-erties of the bond between them. As the masses of the atoms increase, the frequency of the vibration tends to decrease: it is lower for hydrogen chloride or deuterium fluoride than for hydrogen fluoride. A molecule of hydrogen fluoride has a dipole moment because the electron density on the fluorine atom is greater than that on the hydrogen atom. Because the dipole moment decreases as the atoms approach each other, and increases as they move farther apart, there is an oscillating electric field around the molecule. The field can absorb radiation of the same frequency, and as it does so the amplitude of the vibration increases. In a polyatomic molecule there are a number of vibrational modes. Transitions from one mode to another are not accompanied by the absorption or emission of radiation unless they involve a change of dipole moment, but almost all polyatomic molecules can absorb at many different frequencies and therefore give complex and highly characteristic infrared absorption spectra. Infrared spectroscopy is a powerful technique for identifying and characterizing molecules, and is indispensable to organic chemists.

The infrared portion of the spectrum is divided into three regions that require somewhat different apparatus and experimental techniques and that furnish different kinds of information. The *near infrared* extends from about 3.75×10^{14} to 1×10^{14} Hz (0.8 to 3 μm), and is chiefly concerned with the vibrations of hydrogen atoms in C—H, N—H, O—H, and similar bonds. These occur at relatively high frequencies because the hydrogen atom is so light. The *sodium chloride region,* so called because it is amenable to study with instruments employing prisms and sample holders made of sodium chloride, extends from about 1.2×10^{14} to 2×10^{13} Hz (2.5 to 15 μm) and is chiefly concerned with vibrations involving carbon, oxygen, and nitrogen atoms, such as those in C—C, C=C, C—N, C—O, C=O, and N—O bonds. Since the individual bonds in a molecule are not completely isolated from each other, similar molecules generally give infrared spectra that are similar but recognizably different. The similarities enable the chemist to identify the kinds of bonds that are present in a molecule of unknown structure; the differences make it possible to distinguish closely related compounds and to analyze complex mixtures. The *far infrared,* from about 2×10^{13} to 7.5×10^{12} Hz (15 to 40 μm), deals with the still slower vibrations that occur in bonds joining groups of atoms or that involve halogen or metal atoms. Of these three regions it is the sodium chloride region that has been the most studied and is still the most used.

The far infrared region is being extended beyond 7.5×10^{12} Hz (40 μm) by the development of new optical materials and instrumental components; the microwave region involves a different technology. It extends from roughly 1×10^{12} to 5×10^{8} Hz (300 μm, or 0.03 cm, to 60 cm). Absorption in the microwave region arises from rotations of asymmetrical molecules around their centers of gravity and from rotations of groups of atoms around the bonds joining them to other groups. Successive rotational energy levels differ by only small increments of energy, ranging from 400 J mol^{-1} at $v = 1 \times 10^{12}$ Hz to 0.2 J mol^{-1} at $v = 5 \times 10^{8}$ Hz. Microwave spectroscopy has not become very popular because of the complexity and cost of the apparatus needed to study it.

Nuclear magnetic resonance spectroscopy and electron-spin resonance spectroscopy are based on the absorption of energy by a nucleus or an unpaired electron that is spinning in a magnetic field. In nuclear magnetic resonance spectroscopy the differences between spin-energy levels lie in the radio-frequency portion of the spectrum, between about 10^8 and 10^5 Hz; in electron-spin resonance spectroscopy they lie in the microwave region, between about 10^{11} and 10^{10} Hz. Not all nuclei have the properties that nuclear magnetic resonance spectroscopy requires: some that do are ^1H, ^{14}N, ^{19}F, ^{31}P, and ^{35}Cl. The proton is the one that has received most attention, and proton magnetic resonance spectroscopy is of enormous value in the study of organic compounds because of the insight that it provides into the environments that surround hydrogen nuclei, which are very common in such compounds. Electron-spin resonance spectroscopy is more limited in scope, but is useful for detecting and studying free

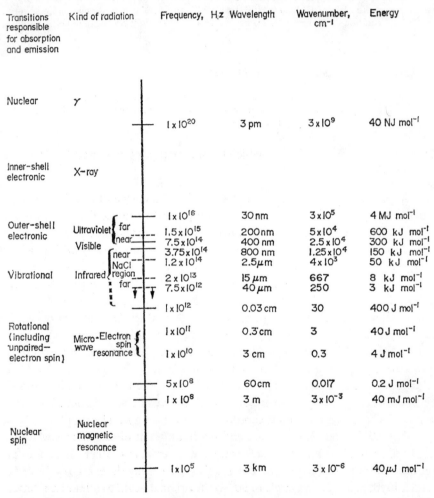

Transitions responsible for absorption and emission	Kind of radiation	Frequency, Hz	Wavelength	Wavenumber, cm^{-1}	Energy
Nuclear	γ				
		1×10^{20}	3 pm	3×10^9	40 NJ mol^{-1}
Inner-shell electronic	X-ray				
		1×10^{16}	30 nm	3×10^5	4 MJ mol^{-1}
Outer-shell electronic	Ultraviolet { far	1.5×10^{15}	200 nm	5×10^4	600 kJ mol^{-1}
	{ near	7.5×10^{14}	400 nm	2.5×10^4	300 kJ mol^{-1}
	Visible	3.75×10^{14}	800 nm	1.25×10^4	150 kJ mol^{-1}
	{ near NaCl	1.2×10^{14}	2.5 μm	4×10^3	50 kJ mol^{-1}
Vibrational	Infrared { region	2×10^{13}	15 μm	667	8 kJ mol^{-1}
	{ far	7.5×10^{12}	40 μm	250	3 kJ mol^{-1}
		1×10^{12}	0.03 cm	30	400 J mol^{-1}
Rotational (including unpaired-electron spin)	Micro- { Electron	1×10^{11}	0.3 cm	3	40 J mol^{-1}
	wave { spin resonance	1×10^{10}	3 cm	0.3	4 J mol^{-1}
		5×10^8	60 cm	0.017	0.2 J mol^{-1}
		1×10^8	3 m	3×10^{-3}	40 mJ mol^{-1}
Nuclear spin	Nuclear magnetic resonance				
		1×10^5	3 km	3×10^{-6}	40 μJ mol^{-1}

FIG. 15.1. The portion of the electromagnetic spectrum that is of interest in chemical work, showing the frequencies, wavelengths, and wavenumbers of different kinds of radiation, the physical processes that are responsible for their adsorption and emission, and the energies corresponding to radiation of various frequencies.

radicals (such as the one,

$$CH_3—\overset{\overset{\displaystyle \ddot{O}}{|}}{\underset{\displaystyle \cdot}{C}}—CH_3,$$

that appeared on page 243) and transition-metal ions having unpaired d and f electrons.

All these ideas are summarized in Fig. 15.1, which represents the electromagnetic spectrum and the techniques that are concerned with its different regions.

SUMMARY: Electromagnetic radiation is characterized by its frequency, wavelength, or wavenumber, or by the energy of a single photon or of a mole of photons. Processes of interest to chemists occur at frequencies ranging from about 10^5 Hz (4×10^{-5} J mol^{-1}) to more than 10^{20} Hz (4×10^{10} J mol^{-1}), and include transitions among spin-energy, rotational, vibrational, electronic (both outer-shell and inner-shell), and nuclear energy levels.

15.3. Absorption, emission, and re-radiation

PREVIEW: Matter can either absorb electromagnetic radiation or, if it is raised to an excited state, emit it. Absorption may be followed by radiative decay, which gives rise to fluorescence or phosphorescence.

There are three kinds of ways in which the interaction of matter and electromagnetic radiation can manifest itself.

1. *Absorption.* A beam of radiation of a certain frequency and wavelength is passed through a sample. If the sample contains some kind of molecule, atom, or ion for which the difference between the energy in an excited state and the energy in the ground state is just equal to the energy of a photon in the beam, energy will be absorbed from the beam. In absorption spectroscopy the energy of a beam that has passed through a sample is compared with the energy of a similar beam that has not.

2. *Emission.* If a sample contains some kind of molecule, atom, or ion that can be raised to an excited state by absorbing heat or some other form of energy that is not electromagnetic, decay back to the ground state (or to a lower excited state) may be accompanied by the emission of electromagnetic radiation. Emission spectroscopy involves measuring the energy of the emitted radiation.

3. *Re-radiation.* A molecule, atom, or ion in a sample is first raised to an excited state by absorbing energy from a beam of electromagnetic radiation. It must then return to the ground state, and it can do so in several ways. The simplest is by emitting a photon. Different photons are emitted in different directions: a few along the path of the original beam, a few back toward its source, and others at various angles to the beam. Fluorescence spectroscopy involves measuring the energy of the radiation that is emitted, usually at an angle of 90° to the original beam to minimize interference from radiation that is transmitted or reflected. Usually the lifetime of the excited state is between about 10^{-9} and 10^{-7} s, but in some molecules the direction of spin of an electron in the excited state may undergo reversal. Then the electron is no longer

paired with the other electron in the orbital from which it was excited, and decay back to the ground state becomes unlikely and therefore slow: the excited state may have a lifetime as long as several seconds. Eventually a photon may be emitted as it decays; this is called *phosphorescence*. Phosphorescence spectroscopy involves measuring the energy of the radiation that is emitted some time after the interruption of the beam impinging on the sample. There are also ways in which the excited state of a molecule can decay by interacting with other nearby molecules, such as those of a solvent. These do not involve the emission of radiation, and they compete so efficiently with the processes that give rise to fluorescence and phosphorescence that there are relatively few molecules that fluoresce, and fewer still that phosphoresce even at the temperature of liquid nitrogen (about 75 K), where collisions with neighboring molecules are both less frequent and less energetic than at higher temperatures.

SUMMARY: A molecule, atom, or ion can absorb energy from a beam of radiation by undergoing excitation to a higher energy level, or can be excited thermally and emit radiation as it decays to a lower energy level. Energy that has been absorbed may be re-radiated as the excited state decays, and the decay may be nearly instantaneous (fluorescence) or delayed (phosphorescence).

15.4. Apparatus for spectroscopy

PREVIEW: This section describes the functions and arrangement of the different components of apparatus designed for the measurement of absorption, emission, and re-radiation.

To observe these different kinds of interactions, it is necessary to use instruments that are differently arranged. Microwave absorption spectroscopy, visible–ultraviolet absorption spectroscopy, atomic absorption spectroscopy, infrared absorption spectroscopy, and X-ray absorption spectroscopy all employ the basic arrangement shown in Fig. 15.2. A *source* is needed to provide radiation of the frequency and

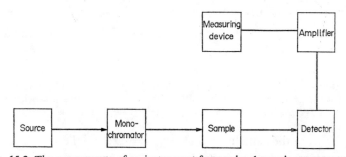

FIG. 15.2. The components of an instrument for use in absorption spectroscopy.

wavelength at which the absorption is to be measured. If the source emits radiation having just a single frequency and wavelength, that radiation is called *monochromatic*. The use of monochromatic radiation has the advantage of simplifying the relation between the concentration of a species that can absorb radiation and the fraction of the energy of the beam that is absorbed; this will be discussed in Section 15.8. On the other hand, a source giving monochromatic radiation is inconvenient if one wishes to study the way in which the absorption by a sample varies with frequency or wavelength. A common compromise is to employ a source that emits continuous radiation, together with a *monochromator* that rejects all of this radiation except the portion that lies in a very narrow band of frequencies and wavelengths. In a visible–ultraviolet spectrophotometer the source might be an incandescent light bulb with a tungsten filament, and the monochromator might include a prism or grating to disperse the radiation from the source into a continuous spectrum, from which a suitably narrow band is selected by means of a narrow slit. Figure 15.3 shows one possible arrangement of a prism monochromator. A beam of radiation obtained from the source is first passed through an entrance slit to exclude stray light, and arrives at one surface of the prism. There it undergoes refraction, which changes its direction. As the frequency increases and the wavelength decreases, the index of refraction of the prism increases, and the change of direction increases. Refraction occurs again as the radiation leaves the prism: the dashed and dotted lines in Fig. 15.3 show the paths that are followed by radiation having two different frequencies or wavelengths, and also show how the exit slit passes a portion of the dispersed radiation while blocking off the rest. The range of frequencies and wavelengths that passes through the exit slit can be changed by rotating the prism.

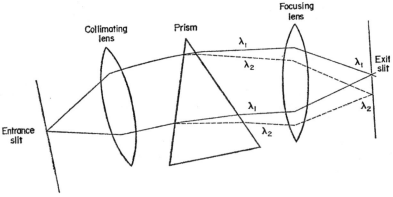

Fig. 15.3. A prism monochromator. Solid lines show the path followed by radiation having the wavelength λ_1; dashed lines show the one followed by radiation having a shorter wavelength λ_2. The wavelength λ_2 can be focused on the exit slit by rotating the prism.

On emerging from the exit slit the beam passes through the sample and then to a *detector*. The nature of the detector, like that of the source, depends on the portion of the electromagnetic spectrum being investigated. A Geiger counter might be used in X-ray absorption spectroscopy, or a thermocouple in infrared absorption spectroscopy; in visible-ultraviolet absorption spectroscopy a phototube or a photomultiplier tube is most common. Each yields an electric current that varies linearly with the power (energy per unit area) of the beam, and the current is amplified and measured or recorded.

In absorption spectroscopy radiation from some separate source is passed through the sample; in emission spectroscopy the source is the sample itself. Emission techniques employ the basic arrangement shown in Fig. 15.4. The substance being studied must be at least partially converted into an excited state that will emit radiation as it decays. If the lifetime of the excited state is sufficiently long, the excitation may be completed before the measurement is begun. In gamma-ray spectroscopy, where excitation is accomplished by irradiation with neutrons or other subatomic particles, the excited nuclei may have half-lives of a few minutes or even days, long enough to permit transferring the sample from the point at which it was activated to a gamma-ray spectrometer. At the other extreme, the lifetime of an atom or ion excited in a flame or an electric arc is only about 0.01 microsecond, and excitation and measurement must therefore be simultaneous. The radiation emitted by the sample is passed through a monochromator and allowed to fall on a detector.

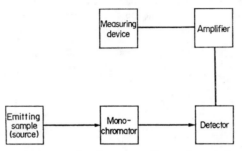

FIG. 15.4. The components of an instrument for use in emission spectroscopy.

Fluorescence techniques employ the arrangement shown in Fig. 15.5, which permits the detection and measurement of radiation that is emitted at right angles to the beam by which excitation is accomplished, but which otherwise resembles the one shown in Fig. 15.2. Phosphorescence is detected and measured in much the same way, but with the addition of a mechanical device—which might be as simple as the one shown in Fig. 15.6—that periodically interrupts the exciting beam and then, after some delay, permits the emitted radiation to reach the monochromator and detector. The delay permits phosphorescence to be distinguished from fluorescence, and can be adjusted to aid in resolving the phosphorescence due to different species that have different lifetimes.

In detecting and determining very small amounts of material, techniques involving emission and re-radiation have an inherent advantage over those involving absorption. A very dilute solution of an absorbing species absorbs very little energy from a beam of radiation passing through it. To discover whether absorption has occurred at all, one must compare the powers of two rather intense beams. To detect emission and re-radiation one must compare the power of one very weak beam with that of

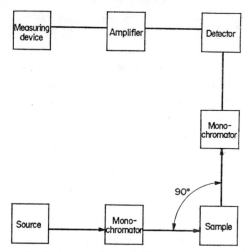

FIG. 15.5. The components of an instrument for use in fluorescence spectroscopy. Two monochromators are shown. The first disperses the radiation obtained from the source and selects a particular narrow band of wavelengths for exciting the fluorescence; the second disperses the fluorescent radiation and is used to study the dependence of its intensity on its wavelength or frequency.

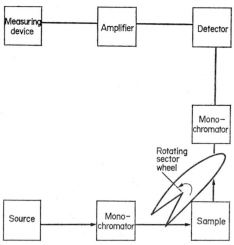

FIG. 15.6. The components of an instrument for use in phosphorescence spectrometry. In the position shown the rotating sector wheel allows the passage of radiation from the first monochromator to the sample, but blocks the path from the sample to the second monochromator. When it has rotated through 180°, radiation from the first monochromator cannot reach the sample but the phosphorescent radiation from the sample is allowed to pass to the second monochromator.

another whose power is extremely low and may be equal to zero. The second comparison is much easier, and therefore measurements of emission and re-radiation can often be made successfully at much lower concentrations than those of absorbance. In visible-ultraviolet absorption spectroscopy there are few molecules that absorb strongly enough, and few instruments that are stable enough, to permit detecting the absorbance of a 10^{-7} M solution, but by fluorescence techniques it is possible to detect quinine and many other fluorescent molecules in solutions as dilute as 10^{-8} or even 10^{-9} M.

Whether or not this inherent advantage can be realized in any particular case depends on the efficiencies of the processes involved. When an atom of some element in a flame, or a molecule of quinine dissolved in a solution, is raised to an excited state by absorbing a photon, the probability that it will return to the ground state by emitting one is high enough to make re-radiation easier to detect than absorption. Dissolved permanganate ions absorb very strongly at a wavelength of 515 nm $(5.82 \times 10^{14}$ Hz), but they then return to the ground state by non-radiative processes; if any photons are emitted, they are too few to be detected. In such a case measurements of absorption are the only ones that can be made. Even when emission does occur, its intensity depends on the efficiency of the excitation process. This may be so low that hardly any of the atoms or molecules are in the excited state, and then emission will be so weak that absorption measurements will be more sensitive.

SUMMARY: Instruments for spectroscopic measurements have five components: a source that emits radiation, a monochromator that selects a particular narrow band of frequencies or wavelengths, a cell that contains the sample, a detector that converts radiant energy into electrical energy, and a device that measures or records the electrical signal obtained from the detector. These components are arranged differently in different instruments and for different purposes.

15.5. Atomic and molecular spectra

PREVIEW: This section explains why atomic and nuclear spectra consist of lines whereas molecular spectra consist of bands.

There are two extreme conditions under which absorption can be observed. One is typified by the atoms in a flame, and the other by molecules or ions dissolved in a solution. The former is simpler for two reasons. One is that atoms are simpler than molecules or solvated ions. The other is that a flame, or a gas at a very low pressure, is so rarefied that interactions among the particles in it are much weaker than the solute-solvent interactions in solutions.

Some of the energy levels of a sodium atom are shown in Fig. 15.7. In the ground state there is one $3s$ electron. The lowest excited state is one in which this electron has been raised to a $3p$ orbital. Actually there are two $3p$ states having slightly different energies. In one, the spin of the $3p$ electron creates a magnetic field that is aligned with the one produced by the orbital motions of the other electrons in the atom; in the other, these two magnetic fields are opposed. One of the $3p$ states results from the absorption of radiation having $\nu = 5.0833 \times 10^{14}$ Hz and $\lambda = 589.59$ nm; the other is obtained from absorption at $\nu = 5.0885 \times 10^{14}$ Hz and $\lambda = 589.00$ nm. By absorbing more energy, an atom in the ground state can be excited to a higher

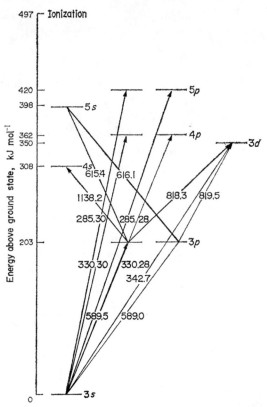

FIG. 15.7. A portion of the energy-level diagram for a sodium atom. The number accompanying
each arrow shows the wavelength (in nm) that corresponds to the transition.

energy level. The absorption of a photon having $\nu = 9.09 \times 10^{14}$ Hz ($\lambda = 330.3$ nm)
raises the original $3s$ electron to a $4p$ state, and a photon having $\nu = 1.05 \times 10^{15}$ Hz
($\lambda = 285.3$ nm) would raise it to a $5p$ state. A sufficiently energetic photon could
remove the electron completely, producing a sodium ion, but as this requires
8.232×10^{-18} J atom^{-1} or 4957 kJ mol^{-1} the photon would have to have $\nu = 1.24 \times 10^{16}$
Hz and $\lambda = 24.1$ nm.

Other transitions can also be observed. A small fraction of the sodium atoms in
a flame are converted to one of the $3p$ states by absorbing energy from the flame.
In that state they can absorb photons having $\nu = 2.63 \times 10^{14}$ Hz ($\lambda = 1138$ nm) and
undergo excitation to a $4s$ state. Figure 15.7 shows some of the other possible transitions as well.

The result is the atomic absorption spectrum shown in Fig. 15.8. To the chemist a
spectrum is a plot showing how the extent of absorption, emission, or re-radiation
varies with frequency, wavelength, or wavenumber. Spectra can be classified according
to the processes (absorption, emission, etc.) that they represent, the portions of the
electromagnetic spectrum (infrared, visible–ultraviolet, etc.) that they cover, the

natures of the transitions (electronic, vibrational, etc.) that are responsible for them, and the kinds of chemical substances (atomic or molecular) that give rise to them. The information that is contained in a spectrum can be used in a number of different ways. Knowing that a sample does absorb or emit or fluoresce at certain frequencies and wavelengths but not at others, a chemist can tell which substances it contains. The frequencies at which a particular kind of atom or molecule absorbs give a picture of its energy levels. Energy-level diagrams, like the one in Fig. 15.7, are constructed on the basis of spectral data, and are responsible for most of our ideas about atomic and molecular orbitals. By measuring the extent of absorption at a certain frequency or wavelength under carefully standardized conditions it is possible to find how much of the element or molecule causing the absorption is present in the sample.

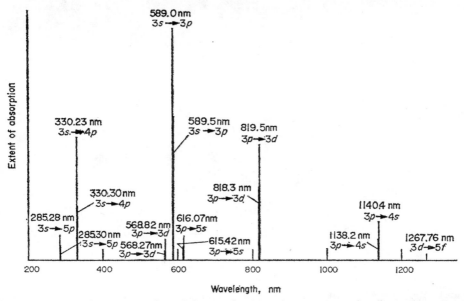

FIG. 15.8. Schematic representation of the atomic absorption spectrum of sodium. The ordinate scale is arbitrary and greatly compressed: if it were drawn to scale several of these transitions would be indetectable.

Atomic (and nuclear) spectra consist of lines like the ones shown in Fig. 15.8. Absorption (or emission) can occur at certain frequencies but not at others, and these frequencies are very sharply defined. Molecular spectra differ from atomic spectra in several ways. Figure 15.9 shows a portion of the energy-level diagram for a molecule. Only two of the electronic energy states are shown here: the ground state and the lowest excited state. The difference between their energies might be roughly 300 kJ mol^{-1}, which corresponds to $\nu = 7.5 \times 10^{14}$ Hz and $\lambda = 400$ nm. For each of the electronic energy states there are a number of different vibrational states, whose energies typically differ by only about 30 kJ mol^{-1}. Transitions among the different vibrational energy levels of the ground electronic state would be caused by photons having $\nu = 7.5 \times 10^{13}$ Hz and $\lambda = 4$ μm, which is in the infrared portion of the spectrum. For each of the vibrational states in turn there are a number of different rotational states, and the difference between the energies of neighboring rotational states might be roughly 3 kJ mol^{-1}.

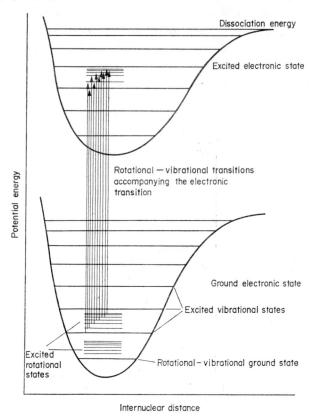

FIG. 15.9. A portion of the energy diagram for a molecule, showing how radiation of many different frequencies (which correspond to the different lengths of the arrows) can be absorbed in effecting vibrational and rotational transitions as well as electronic ones.

The molecule will absorb in the vicinity of $\nu = 7.5 \times 10^{14}$ Hz ($\lambda = 400$ nm) because that corresponds to the difference between the two electronic energy levels. However, the absorption will not be confined to a single wavelength as it is with atoms. The absorption of 300 kJ mol^{-1} would raise the molecule to the excited electronic state if it remained in its lowest vibrational and rotational energy states. It can absorb 330 kJ mol^{-1} by undergoing a transition from the lowest vibrational energy level to the next higher one at the same time that it undergoes the electronic transition, or 270 kJ mol^{-1} if the vibrational energy level decreases during the transition. Changes of rotational state can occur as well. If excitation to the next higher rotational state accompanies the changes of electronic and vibrational energy levels, absorption will occur at 303, 333, and 273 kJ mol^{-1}. Absorption of 297, 327, and 267 kJ mol^{-1} would result from decay to the next lower rotational state during the transition.

This is the general pattern that is shown by the molecular spectra of gases at low pressures. It is illustrated by Fig. 15.10, where the broad peak in curve (a) is successively resolved into a series of narrow peaks representing changes of vibrational state and then into still narrower peaks that represent changes of rotational state. The appearance of curve (a) results from the fact that the rotational and vibrational components occur at frequencies too closely spaced to be resolved on the scale employed for curve (a), and all the different frequencies corresponding to a particular kind of electronic transition are combined into a peak of finite width.

Increasing the pressure changes the spectrum because it brings molecules closer together and enables them to interact. Details are obscured even in curves obtained

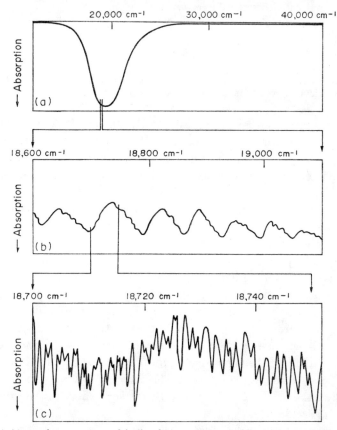

FIG. 15.10. (a) Absorption spectrum of iodine from 12 000 to 40 000 cm^{-1} (833 to 250 nm), showing a single electronic absorption band; (b) a magnified view of the portion of that band extending from 18 600 to 19 000 cm^{-1} (537.6 to 523.6 nm), showing the vibrational band structure; (c) a more highly magnified view of the portion extending from 18 700 to 18 750 cm^{-1} (534.75 to 533.33 nm), showing the rotational fine structure. Redrawn by permission from E. J. Bair, *Introduction to Chemical Instrumentation*, McGraw-Hill Book Co., Inc., New York, 1962.

at high resolution: the peaks become broader and merge into bands. Carrying this to extremes by condensing the gas to a liquid, or dissolving it in a solvent, yields spectra of the type shown in Fig. 15.10(a). This is by far the commonest kind of molecular absorption spectrum in the visible and near-ultraviolet region, because most molecular and ionic substances, both inorganic and organic, are either liquid or solid at ordinary temperatures and are most conveniently studied in solutions. It would certainly not be practical to vaporize a dye, a vitamin, a polymer, or a complex ion in order to obtain its absorption spectrum. All of the spectra in the rest of this chapter will be band spectra resembling the one in Fig. 15.10(a).

> SUMMARY: Atomic and nuclear spectra consist of lines representing sharply defined differences
> of energy, but molecular spectra in the visible and ultraviolet consist of bands because
> the fundamental electronic transitions are accompanied by changes of closely spaced
> vibrational and rotational energy levels.

15.6. Beer's law

> PREVIEW: A relation between the extent of absorption and the concentration of the absorbing
> species is derived and several terms associated with it are defined.

Thus far it has been convenient to speak loosely about the "extent of absorption" but now a more exact term is needed. Beer, Lambert, and Bouguer between 1850 and 1855 studied the ways in which the power P of a beam of radiation depends on the thickness b and concentration c of a solution through which the beam passes. Their observations can be rationalized in the following way.

We imagine that each particle of the absorbing species has a certain cross-sectional area A_m over which it may absorb a photon, and that a photon passing outside this area cannot be absorbed. The area A_m may differ from the true cross-sectional area because a photon may not be absorbed unless it strikes a certain part of the absorbing particle, or may be absorbed if it strikes a nearby molecule of the solvent that is closely associated with the absorbing particle. In a dilute solution the sum of the effective areas will be equal to nA_m, where n is the number of absorbing molecules. If that number is increased by dn, the sum of the effective areas will be increased by $A_m\,dn$. There will be an increase of the fraction $-dP/P$ of the photons that are absorbed by the solution; the minus sign means that the number of photons decreases as the beam passes through the solution. This increase will be proportional to the increase of the fraction of the total area A_t that is covered by the effective areas of the absorbing molecules:

$$-\frac{dP}{P} = k\,\frac{A_m\,dn}{A_t}$$

where k is a constant of proportionality. Integration gives

$$-\ln P = kA_m n/A_t + \gamma.$$

The volume V of the solution is equal to the product of its total area A_t by its thickness b, so that $A_t = V/b$ and

$$-\ln P = kA_m bn/V + \gamma$$

while the ratio n/V, which is equal to the number of absorbing particles present in the volume V, is proportional to the concentration c of the absorbing species, or

$$n/V = k'c$$

where k' is another constant of proportionality. Hence

$$-\ln P = kk'A_m bc + \gamma.$$

The constant of integration γ can be evaluated by noting that absorption cannot occur if $A_m bc = 0$; then the power P of the beam emerging from the solution will be the same as the power P_0 of the beam that entered the solution. Hence $\gamma = -\ln P_0$, and replacing the product $kk'A_m$ with a single constant of proportionality a' gives the final result:

$$-\ln \frac{P}{P_0} = a'bc. \tag{15.2}$$

This is the Beer–Lambert–Bouguer law, usually called Beer's law for short.

Beer's law is the fundamental quantitative law of absorption. It applies to gases, liquids, and solids, and to absorption in any part of the electromagnetic spectrum. The following definitions and statements are based on it:

1. The *transmittance* T is defined as the ratio of the power P of the beam that emerges from the sample to the power P_0 of the beam that entered it:

$$T = \frac{P}{P_0} = e^{-a'bc} = 10^{-abc} \tag{15.3}$$

where $a = 0.4343\, a'$. The transmittance is equal to 1 if the sample does not absorb at all. It decreases as a, b, or c increases, and would be equal to zero if the sample were completely opaque. It is often expressed as a percentage, and the percent transmittance ranges from 100 for a sample that is completely transparent to 0 for one that is completely opaque.

2. The *absorbance* A is defined by the equation

$$A = -\log_{10} T = -\log_{10} \frac{P}{P_0} = abc. \tag{15.4}$$

The absorbance is equal to zero for a completely transparent solution and would be infinitely large for a solution that was completely opaque. Absorbance is more useful than transmittance for quantitative purposes because it is proportional to the concentration c if the values of a and b are fixed. "Optical density" is an older name for absorbance.

32*

3. The *path length b* is the thickness of the solution through which the beam passes. It is usually expressed in centimeters and depends on the geometry of the cell in which the solution is contained. Cells for which $b = 1$ cm are the most common in visible and ultraviolet absorption spectroscopy. However, cells for which b is as high as 100 cm have been used to study the spectra of gases at very low pressures, as for example in detecting low concentrations of NO_2 in polluted atmospheres, and cells for which b is as low as 0.01 cm have been used with concentrated solutions of strongly absorbing species.

4. The *absorptivity a* depends on the identity of the absorbing species and also on the frequency or wavelength of the radiation employed. Every substance absorbs more strongly at some wavelengths than at others because a varies with wavelength, and different substances have different absorption spectra because the variations of a with wavelength are different. The numerical value of a for any particular substance at any particular wavelength depends on the units in which it is expressed, and since the product abc is dimensionless these depend in turn on the units in which b and c are expressed. The natural units of c are mol dm^{-3} or mmol cm^{-3}. When c is expressed in these units and b is expressed in cm, the units of a become $dm^3\ mol^{-1}\ cm^{-1}$ or, much more sensibly and conveniently, $cm^2\ mmol^{-1}$. The value of a expressed in these units is called the *molar absorptivity* and given the symbol ε, and eq. (15.4) is often written as

$$A = \varepsilon bc \tag{15.5}$$

to indicate that these units are being used. Other units are sometimes more convenient: if, for example, the formula weight of the absorbing substance were unknown, it would be impossible to calculate the value of c in mmol cm^{-3}, and then c might be expressed in g cm^{-3} and a in $cm^2\ g^{-1}$.

SUMMARY: Beer's law asserts that the absorbance of a sample is proportional to its concentration and that the constant of proportionality is equal to the product of its path length by its absorptivity. The absorptivity depends on the identity of the absorbing species and on the frequency or wavelength at which the absorbance is measured.

15.7. The measurement of absorbance

PREVIEW: This section describes the problems that can arise in experimental measurements of absorbance and discusses ways of eliminating or mitigating them.

The absorbance of a sample is defined in terms of two quantities, P_0 and P, of which neither can be measured directly. If the sample is a liquid or a gas, it must be confined in a cell. As Fig. 15.11 shows, the incident beam, having a power equal to P_0', must traverse a wall of the cell before it can enter the solution. A little of its power is lost by reflection at the boundary between the air and the wall, a little more is lost by absorption as it passes through the wall, still more may be lost by scattering due to inhomogeneities in the material from which the wall is made, and more yet is lost by reflection at the boundary between the wall and the solution. The power P_0 that is considered in Beer's law is the power that remains after these losses. Next the beam passes through the sample, and P in Beer's law denotes its power as it arrives at the opposite wall. Reflection, absorption, and possibly scattering occur again, and the power P' of the beam that emerges into the air is smaller than P.

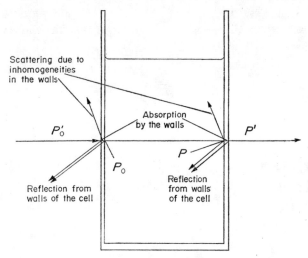

FIG. 15.11. Losses of radiant power from a beam passing through a sample contained in an absorption cell.

To make matters worse, P will be smaller than P_0 even if the sample does not contain any of the absorbing species that is being studied. The solvent itself may absorb power from the beam, especially at short wavelengths. Pyridine absorbs strongly at wavelengths below about 305 nm; benzene absorbs at wavelengths below about 280 nm; even water absorbs at wavelengths below about 210 nm. Often other reagents —acids, bases, buffering and complexing agents, and so on—are present as well, and these may also absorb, as may the impurities that accompany them. Unless very special precautions have been taken, suspended solid matter, such as dust and silica, may also be present, and then there will be some loss by scattering.

All of these unavoidable but uninteresting problems can be ignored if we define P and P_0 in slightly different ways. We shall take P to be the power of a beam after it has passed through a cell containing the sample, and P_0 to be the power of an originally identical beam that has passed through an identical cell containing the same solvent, reagents, and impurities but none of the absorbing molecules or ions that are of interest. Losses by reflection, unwanted adsorption, and scattering affect P and P_0 in the same proportion, and the ratio P/P_0 gives the transmittance of the species being studied.

Two cells must therefore be used to measure the transmittance or absorbance. Some typical absorption cells are shown in Fig. 15.12. For accurate work these must have plane parallel windows, so that the path length b is exactly defined, and must also have reflection, absorption, and scattering characteristics. They are usually bought as matched pairs selected to have indistinguishable values of b. They may be made of glass for use at wavelengths above about 325 nm, but because glass absorbs strongly at lower wavelengths they must be made of quartz or fused silica if they are to be used below 325 nm. Less expensive cells are used for many purposes, and even test tubes can be used if their diameters are much larger than the width of the beam and if they are always carefully aligned in the same way.

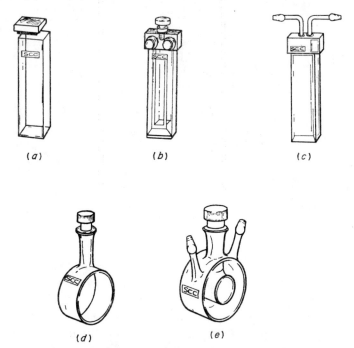

FIG. 15.12. Some typical absorption cells for visible–ultraviolet spectrophotometry: (a) an ordinary rectangular cell available in path lengths as short as 1 mm, (b) a water-jacketed cell allowing for circulation of water at a constant temperature, (c) a flow-through cell for the continuous analysis or monitoring of the liquid (e.g., a reaction mixture) in an external stream or container, (d) a cylindrical cell available in path lengths as long as 10 cm, (e) a water-jacketed cylindrical cell. Reproduced by courtesy of the Scientific Glass Apparatus Co. (Bloomfield, N. J.).

There are two kinds of spectrophotometers (which many chemists call spectrometers): single-beam and double-beam. Figure 15.2 represented a single-beam instrument. To measure the transmittance or absorbance of a solution with such an instrument one must first adjust it so that it will read $T = 0$ when there is no power reaching the detector. This is done by closing a shutter in the path of the beam, so that the detector is in the dark, and adjusting an electrical control that compensates for the current resulting from the thermal emission of electrons by the anode of the detector, as well as for imperfections in the electronic circuit that amplifies the current obtained from the detector. A *reference cell*, containing only the solvent and reagents, is then placed in the path of the beam, the shutter is opened so that the beam strikes the detector, and a second control is adjusted so that the meter or other indicating device reads $T = 1$. This control changes the gain of the amplifier and compensates, not only for the loss of power as the beam passes through the reference cell, but also for the fact that both the intensity of the source and the sensitivity of the detector vary with wavelength. This adjustment must be repeated each time the wavelength is changed, and oftener if the line voltage may vary or if the source and detector are not perfectly stable. Finally the *sample cell*, which contains the substance of interest, is substituted for the reference cell. The indicating device then reads the transmittance directly. Most instruments have provisions for indicating the absorbance as well, or instead.

In a double-beam instrument the beam emerging from the monochromator is split into two beams. One way to do this is to direct it onto a semicircular mirror that is rotated around the center of the semicircle. During half of each rotation the beam is reflected by the mirror and directed through one cell; during the other half the beam is uninterrupted by the mirror and is directed through the other cell. In one of several possible arrangements, the beams emerging from the two cells are reflected onto a single detector. If $P = P_0$, so that $T = 1$, the two emerging beams will have equal powers, and the output of the detector will be the same regardless of which path has been followed by the beam striking its surface at any instant. If P and P_0 are different, a periodically changing electrical signal will be obtained, and its amplitude will depend on the value of T. The output of the detector is fed to an a.c. amplifier, which ignores the d.c. signal presented to it when P and P_0 are equal and amplifies only the alternating signal that results from a difference between them, and thus yields an a.c. output that can be made proportional to T. Double-beam instruments are especially useful for recording spectra automatically because variations of the intensity of the source and the sensitivity of the detector are cancelled out by the double-beam arrangement, thereby eliminating the necessity of readjusting the instrument at each different wavelength.

SUMMARY: Because of losses at the walls of the cell and absorption by the solvent and other constituents of the solution, it is convenient in practical measurements of the absorbance to take P_0 as being the power of a beam after it has passed through a reference cell that is identical with the sample cell but does not contain the substance being studied or determined. Allowance for losses in the reference cell is made differently with single and double-beam spectrometers.

15.8. Deviations from Beer's law

PREVIEW: Despite the prediction of Beer's law that A/c should be constant, it is often found to vary. This section describes the reasons why it may do so.

Not every real chemical system obeys Beer's law: the ratio of absorbance to concentration may vary in either of the two ways represented in Fig. 15.13. *Negative deviations*, for which A/c decreases as c increases, are more frequent than *positive deviations*. There are three different kinds of reasons for such deviations.

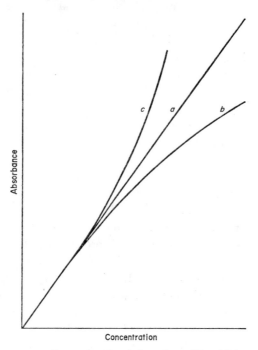

FIG. 15.13. Curve (a) represents adherence to Beer's law. Curve (b) exhibits negative deviations, and curve (c) exhibits positive deviations.

1. *Real deviations* represent actual failures of the law, and occur in solutions containing high concentrations of the absorbing species. Equation (15.5) is a simplified version of a more complete equation

$$A = \frac{n}{(n^2+2)^2}\, \varepsilon' bc \qquad (15.6)$$

where n is the index of refraction of the solution. The constants ε and ε' in eqs. (15.5) and (15.6) have different numerical values, but they have the same significance. If the total concentration of solute is less than about 0.01 M, the value of n is unlikely to differ appreciably from that for the reference solution, but at higher concentrations the variation of n becomes significant and causes A/c to decrease as c increases. Another problem is that, in a dilute solution, each molecule or ion of the absorbing species is separated from the one nearest to it by many molecules of the solvent. In more concentrated solutions the molecules of the solute are closer together. Their interactions are not the same as the solute-solvent interactions. The change of environment affects the value of ε at any particular wavelength, and usually also causes the absorption band to shift along the wavelength axis. These problems are unlikely to be troublesome except with substances for which ε is so small that the absorbance is inconveniently small unless the product bc is large. With such substances it is better to use a cell for which b is large than to try to work with concentrated solutions.

2. *Instrumental deviations* reflect the limitations and defects of the instrument. The radiation reaching the detector may include some *stray light,* which has not passed through the sample cell. A finite value of the transmittance will then be obtained even if the sample is completely opaque, and the absorbance cannot increase above a limiting value no matter how high c may be. Negative deviations from Beer's law result. The detector and amplifier may be non-linear, and therefore the measured

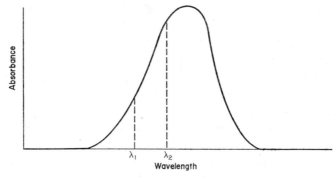

FIG. 15.14. A typical absorption band, showing a range of wavelengths over which A/c varies largely with wavelength.

electrical signal may not be accurately proportional to the desired quantity. A third problem is that the beam in a real spectrophotometer is not monochromatic but always includes a range of wavelengths. Imagine that a solution having the absorption spectrum shown in Fig. 15.14 is exposed to a beam containing radiation of all of the wavelengths from λ_1 to λ_2. Absorption will be much more efficient at λ_2 than at λ_1, and most of the power of the emerging beam will be concentrated at wavelengths near λ_1. If a little more solute is added, it can absorb only the energy that was not absorbed by the solute already present, which is the energy that it absorbs least efficiently; the ratio $\Delta A/\Delta c$ will have a smaller value for the second portion of solute than for the first. The slope of a plot of absorbance against concentration decreases as the concentration increases; negative deviations from Beer's law again result. They are less serious if the absorbance is measured at the peak of an absorption band, where the wavelength is equal to the *wavelength of maximum absorption* λ_{max} (and where the molar absorptivity is represented by the symbol ε_{max}), than if it is measured at a wavelength on either side of the band. At the peak the absorptivity is almost independent of wavelength over the range of wavelengths contained in the beam, which is called its *band width.* On the side of a band this cannot be even approximately true unless the band width is very small. The severities of all these problems decrease as the instrument becomes more sophisticated and more expensive.

3. *Chemical deviations.* 2,6-Dinitrophenol,

is an acid-base indicator. Its acidic form is colorless (and therefore does not absorb at wavelengths above about 400 nm) and has an overall dissociation constant that is equal to 2.0×10^{-4} mol dm^{-3} in water at 298 K. Its basic form is yellow and has $\lambda_{max} = 460$ nm. In a solution in which its total concentration was 0.01 M the indicator would be about 13% dissociated, and the equilibrium concentration of the basic form would be about 1.3×10^{-3} M. In an 0.001 M solution it would be about 36% dissociated, and the equilibrium concentration of the basic form would be about 3.6×10^{-4} M. The ratio of the absorbance to the total concentration of the indicator is m uch smaller for the more concentrated solution, and it decreases as the concentration increases: there is a negative deviation from Beer's law. On the other hand, there would be a positive deviation from Beer's law if the absorbance were measured at a wavelength where the acidic form absorbed but the basic one did not. This sort of deviation is called a chemical deviation from Beer's law. Chemical deviations arise when the absorbing species is involved in a chemical equilibrium whose position depends on the total concentration of some substance, and when the value of A/c is calculated from that total concentration rather than from the equilibrium concentration of the absorbing species. The chemical deviation for 2,6-dinitrophenol would disappear if the solution were buffered, for then the concentration of the basic form would become proportional to the total concentration of the indicator. In other systems a cure may be difficult or even impossible to find.

> SUMMARY: There are three kinds of deviations from Beer's law: real deviations, which reflect approximations made in deriving the law and changes in the environments of the absorbing particles as the concentration increases; instrumental deviations, which reflect the limitations and defects of the spectrometer; and chemical deviations, which reflect shifts in the positions of equilibria involving the absorbing species.

15.9. Plotting absorption spectra

PREVIEW: This section describes the different ways in which absorption spectra can be plotted.

Figure 15.15 shows several different ways in which absorption spectra can be plotted. Part (a) shows plots of absorbance against frequency for two solutions containing different concentrations of the acidic form of the indicator methyl red.

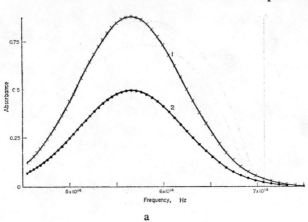

a

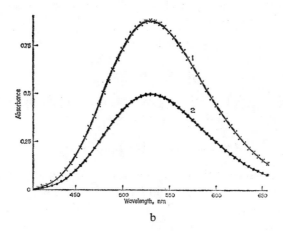

b

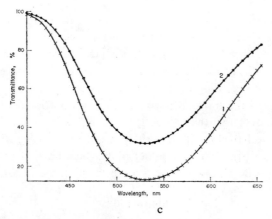

c

FIG. 15.15. Three common ways of plotting absorption spectra: (a) absorbance against frequency, (b) absorbance against wavelength, (c) transmittance against wavelength. The remainder of this figure appears on p. 494.

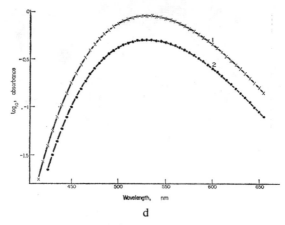

d

FIG. 15.15 *cont.* A fourth common way of plotting absorption spectra. (d) the logarithm of the absorbance against wavelength. The numbers beside each curve identifies the solution for which the curve is drawn. Curves 1 all refer to the same solution; curves 2 refer to another slightly more dilute one.

When only a single electronic transition is involved, such a plot is symmetrical around ist peak and can be represented by the equation

$$A = A_{max} \exp\left[-(v - v_{max})^2/k\right]$$

where A_{max} is the absorbance at the maximum, where the frequency is equal to v_{max}, A is the absorbance at any frequency v, and k is a constant whose value governs the width of the band. This equation describes a curve whose general shape is identical with that of the normal error curve (Fig. 12.2). Most visible–ultraviolet spectrophotometers are calibrated in wavelength rather than frequency, and plots of absorbance against wavelength, like those in part (b) of the figure, are therefore easier to construct and much more common. Plots of transmittance against wavelength for the same two solutions are shown in part (c). Although these are less useful for quantitative purposes because the transmittance is not proportional to concentration, they are obtained from some recording instruments because it is easier to design an instrument that will record the transmittance (which is proportional to the power of the beam reaching the detector) than to design one that will record the absorbance (which varies linearly with the logarithm of the power). Plots of log A against wavelength are shown in part (d). Unlike the other plots, these have identical shapes for the two solutions. Since Beer's law can be written in the form

$$\log A = \log \varepsilon b + \log c$$

changing c merely displaces such a plot in a direction parallel to the ordinate axis, but does not affect the shape of the plot because this reflects the variation of log ε

with wavelength. Consequently it is easier to identify the substance that is responsible for a particular absorption spectrum by inspecting a plot of log A against wavelength (or frequency) than by using a plot of any other kind.

SUMMARY: Most absorption spectra are plots of absorbance against wavelength or frequency but some instruments yield plots of transmittance rather than absorbance, and plotting log A against wavelength or frequency makes it easier to identify the absorbing species.

15.10. Electronic absorption by metal ions and complexes

PREVIEW: This section discusses the processes that are responsible for the absorption of visible and ultraviolet radiation by metal ions and their complexes.

Transitions involving d and f electrons are responsible for absorption by inorganic substances in the visible and near ultraviolet. Few common inorganic ions have unpaired f electrons, which are easier to promote to higher energy levels than the paired ones in f orbitals that are completely filled, but each of the elements (including titanium, vanadium, chromium, manganese, iron, cobalt, nickel, copper, molybdenum, tungsten, and platinum) in the transition series of the periodic table has at least one ionic species in which the d orbitals are incompletely filled. In an isolated ion of this sort there are five d orbitals having equal energies, and radiation is neither absorbed nor emitted as an electron moves from one of these orbitals to another. A dissolved ion is coordinated with molecules of water, or with molecules or ions of other ligands. The electron pair donated by a ligand repels the electrons in nearby d orbitals, rendering these orbitals less stable than others that are farther away. Now the ion must absorb energy if an electron in one d orbital is to be promoted to another in which its energy is higher. The difference between the energies of the d orbitals—and hence the frequency and wavelength of the radiation that will be absorbed—depends on the way in which the ligands are arranged around the metal ion, and also on the strength of the electrostatic field created by each molecule or ion of the ligand. The latter is called the *ligand-field strength*. The order of increasing ligand-field strength for some common ligands is

$$I^- < Br^- < SCN^- < Cl^- < OH^- < H_2O < NH_3 < CN^-.$$

Coordination with ammonia causes the energies of different d orbitals to differ more than coordination with water does, so that the complexes

and

have wavelengths of maximum absorption of about 680 and 810 nm, respectively.

Absorption of this kind is sometimes called *ligand-field absorption*. In ligand-field absorption the electron undergoing the transition is primarily associated with the metal ion, both before and after the transition occurs. It is also possible for an electron that is primarily associated with the metal ion to be excited to a state in which it is primarily associated with a molecule or ion of the ligand, or vice versa: absorption due to such processes is called *charge-transfer absorption* and resembles a redox reaction. In order for charge-transfer absorption to occur, either the metal ion or the ligand must be an oxidizing agent and the other must be a reducing agent. The wavelength of maximum absorption depends on the strength of both. The monochloroiron(III) complex, $[Fe^{3+}Cl^-]^{2+}$, from which molecules of water are omitted for simplicity, can absorb a photon to give an excited state resembling $[Fe^{2+}Cl^0]^{2+}$. Much energy is needed to bring this about because a chloride ion is a very weak reducing agent, and the absorption occurs in the ultraviolet. In the corresponding thiocyanate complex, $[Fe^{3+}(NCS^-)]^{2+}$, the ligand is a much stronger reducing agent, much less energy is needed to form the excited state, and the absorption occurs in the visible portion of the spectrum and is responsible for the familiar red color of the complex.

Charge-transfer absorption is especially valuable to the chemist working with dilute solutions because it leads to very high values of the molar absorptivity. It is convenient to speak of ε_{max}, the molar absorptivity at the wavelength of maximum absorption. For the ligand-field absorption of an ion like $Co(OH_2)_6^{2+}$ or $Ni(OH_2)_6^{2+}$, ε_{max} might be approximately equal to 5 cm^2 mmol^{-1}. A 1 mM solution, containing 1×10^{-3} mmol cm^{-3}, would have a maximum absorbance equal to 0.005 in a cell for which b was 1 cm. This is just large enough to be distinguished from zero with a good spectrophotometer. The *limit of detection* is defined as the lowest concentration of a substance that can just be detected at, say, the 90% level of confidence, and is the lowest concentration at which any useful information about that substance can be obtained with a particular technique, method, or instrument. If $\varepsilon_{max} = 5$ cm^2 mmol^{-1} the limit of detection by spectrophotometry is roughly 1 mM if a 1-cm cell is used. For FeNCS^{2+}, which exhibits charge-transfer absorption, ε_{max} is approximately equal to 5×10^3 cm^2 mmol^{-1}, and the limit of detection would be 1 μM with a 1-cm cell. In charge-transfer absorption the value of ε_{max} may be as high as 5×10^4 cm^2 mmol^{-1}.

A third kind of absorption arises from transitions involving electrons that are primarily associated with the ligand. Most spectrophotometric procedures for determining the concentrations of metal ions in dilute solutions employ strongly absorbing organic ligands. Coordination alters the electronic energy levels of the ligand, and the absorption spectrum of the complex differs from that of the free ligand. Acid-base indicators and the metallochromic indicators used in chelometric titrations (Section 11.9) illustrate the principle. In a typical procedure for the spectrophotometric determination of copper, a solution of copper(II) is treated with an excess of the ligand 2,9-dimethyl-1,10-phenanthroline where each R represents an atom of hydrogen, and then with a weak reducing agent to form the copper(I) chelate

The reagent absorbs strongly in the ultraviolet for reasons that will be described in Section 15.11, but less energy is needed to excite an electron when it is coordinated with copper(I), and the wavelength of maximum absorption for the chelate is about 455 nm. The concentration of the chelate can be increased by a factor of 50 to 100 by extracting its chloride, or some other ion-association complex, from a large volume of aqueous solution into a much smaller volume of a solvent such as *n*-hexanol, $CH_3(CH_2)_5OH$. Since ε_{max} is about 8×10^3 cm^2 mmol^{-1}, even a very low concentration of copper can be detected and determined. The limit of detection can be lowered further by replacing each of the hydrogen atoms at R with a phenyl group, which increases λ_{max} to 480 nm and ε_{max} to 1.4×10^4 cm^2 mmol^{-1}.

SUMMARY: The electron that is excited when a metal ion or complex absorbs visible or ultraviolet radiation may be a *d* or *f* electron that remains associated with the metal ion (ligand-field absorption), may be excited to a state in which it is primarily associated with a ligand (charge-transfer absorption), or may be primarily associated with a ligand both before and after excitation. The second and third are the most useful in work with dilute solutions because molar absorptivities are higher for them than for ligand-field absorption.

15.11. Absorption by organic molecules

PREVIEW: The absorptions of visible and ultraviolet radiation by organic molecules arise from excitations of sigma, pi, and non-bonding electrons. For any particular molecule the wavelength of maximum absorption and the value of ε_{max} depend on the functional groups that are present and on the way in which they are arranged.

Experimental data on the absorption of visible and ultraviolet radiation by organic molecules are interpreted and rationalized by considering that there are three kinds of electrons in such molecules:

1. Sigma (σ) electrons, which occur in single bonds, such as C—C, C—H, or O—H bonds.
2. Pi (π) electrons, which occur in double and triple bonds, such as C=C, C≡C, C=O, and C=N bonds. A double bond contains two σ electrons and two π electrons; a triple bond contains two σ electrons and four π electrons.
3. Non-bonding (n) electrons, which are not involved in the formation of bonds and are associated with atoms of oxygen, nitrogen, the halogens, and other elements in groups V–VII of the periodic table.

A molecule of acetic acid contains electrons of all three kinds:

A single bond contains two σ electrons. A cross-section through a single bond is represented by Fig. 15.16(a), where the depth of the shading is roughly proportional to the probability that there will be an electron at the corresponding point at any particular instant. The two π electrons in a double bond are represented by Fig. 15.16(b). If the axis of the σ orbital containing the two σ electrons lies in the plane of this page, the two lobes of the σ orbital lie above and below the page, as shown in Fig. 15.16(c). In general, π electrons have higher energies than σ electrons, and n electrons have energies that are higher still.

For each of the two kinds of bonding orbitals, σ and π, there is an antibonding orbital into which the electrons can be excited by absorbing energy. The antibonding orbitals are designated by the symbols σ^* and π^* and are represented by Fig. 15.16(d) and (e). There are thus five kinds of orbitals, and they give rise to four common kinds of transitions: $\sigma \to \sigma^*$, $n \to \sigma^*$, $n \to \pi^*$, and $\pi \to \pi^*$.

1. $\sigma \to \sigma^*$ *Transitions.* These are the only transitions that are possible in a molecule like that of ethane (CH_3CH_3), in which all of the bonds are single bonds and in which there are no non-bonding electrons. The difference between the energies of a σ orbital and the corresponding σ^* orbital is so large that compounds of this sort do not absorb at wavelengths above 180 nm. They do absorb at wavelengths below 180 nm, but absorption in this region is difficult to study because molecular oxygen and nitrogen also absorb at these wavelengths, and it is so nearly independent of molecular structure that it has very little practical importance.

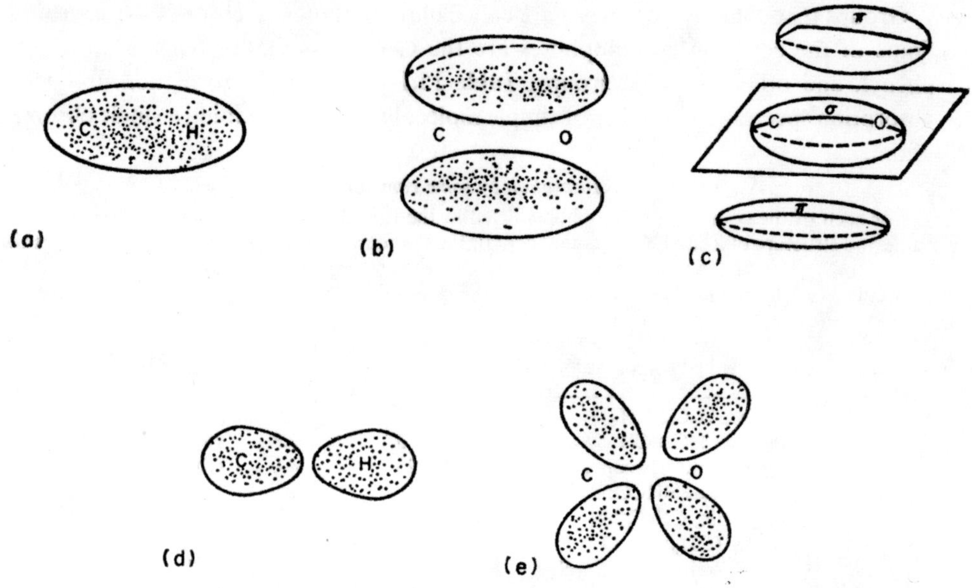

FIG. 15.16. Cross-sections through (a) a sigma bond, (b) a pi bond, (c) a double bond, (d) a σ^* antibonding orbital, and (e) a π^* antibonding orbital. The depth of shading increases with the probability of finding an electron at the point represented.

2. $n \rightarrow \sigma^*$ *Transitions.* Saturated compounds that contain non-bonding electrons absorb at wavelengths between about 150 and 250 nm. Methanol (CH_3OH) and dimethyl ether (CH_3OCH_3) both have $\lambda_{max} = 184$ nm; methylamine (CH_3NH_2) has $\lambda_{max} = 215$ nm; trimethylamine [$(CH_3)_3N$] has $\lambda_{max} = 227$ nm. The values of ε_{max} for the $n \rightarrow \sigma^*$ transitions in these and similar compounds usually lie between about 100 and 2500 cm^2 mmol^{-1}.

Like $\sigma \rightarrow \sigma^*$ transitions, $n \rightarrow \sigma^*$ transitions have little practical importance, but for a different reason. The molecules of many common solvents also undergo $n \rightarrow \sigma^*$ transitions and therefore absorb in the same range of wavelengths. To observe such a transition, and to measure the corresponding absorbance, one must therefore dissolve the substance of interest in a saturated hydrocarbon or another solvent whose molecules contain only σ electrons. Such solvents are non-polar, and may not dissolve a sufficiently high concentration of a polar substance that one wishes to study.

3. $n \rightarrow \pi^*$ *and* $\pi \rightarrow \pi^*$ *Transitions.* Transitions of these two kinds are the most widely useful because they occur in the conveniently accessible portion of the spectrum, between about 220 and 750 nm, where many common solvents are nearly or completely transparent. The $n \rightarrow \pi^*$ transitions are the less interesting of the two because the values of ε_{max} for such transitions rarely exceed 100 cm^2 mmol^{-1}; for $\pi \rightarrow \pi^*$ transitions they are not often below 10^3 cm^2 mmol^{-1}, and may be as high as 10^5 cm^2 mmol^{-1}.

33

Groups that contain π electrons are called chromophores or chromophoric groups. Some of the common chromophores are the C=C, C=O, C=N, N=N, —NO₂, phenyl, and other aromatic groups. Table 15.1 gives some information about the absorption spectra of a number of simple molecules, each containing a single chromophore.

TABLE 15.1. *Absorption by some common chromophores*

For each of the common chromophores listed in the first column, the second column gives the name and structure of a simple compound containing that chromophore. The third column gives the value(s) of λ_{max} for that compound, the fourth gives the corresponding value(s) of $\log_{10} \varepsilon_{max}$, and the fifth tells whether the data were obtained with the gaseous compound or with a solution of it in the solvent named.

Chromophore	Typical compound	λ_{max}, nm	$\log_{10} \varepsilon_{max}$	Conditions
C=C	ethylene, $H_2C=CH_2$	175	4.0	vapor
C=C	acetylene, HC=CH	173	3.8	vapor
C=N	acetone semicarbazone, $(CH_3)_2C=NHCONH_2$	225	4.0	ethanol
C=N	acetonitrile, $CH_3C=N$	< 160	—	vapor
C=O	acetone, $(CH_3)_2C=O$	264	1.2	water
C=S	carbon disulfide, S=C=S	318	2.0	CCl_4
C=C=O	ketene, $CH_2=C=O$	325	1.2	hexane
C=N=N	diazomethane, $CH_2=N=N$	338	0.7	ethanol
⬡	benzene, C_6H_6	203, 254	3.9, 2.4	ethanol+methanol
⬡⬡	naphthalene, $C_{10}H_8$	220, 275	5.0, 3.8	ethanol
N=N	azomethane, $CH_3N=NCH_3$	344	1.0	H_2O
N=O	1-chloro-1-nitrosoethane, $CH_3CH\overset{Cl}{\underset{N=O}{\big<}}$	319, 648	2.4, 2.8	petroleum ether
N=N (azoxy, O)	azoxymethane, $CH_3N=NCH_3$ (O)	221	3.8	vapor
ONO	butyl nitrite, $CH_3CH_2CH_2CH_2ONO$	218, 357	3.2, 1.7	ethanol
NO_2	nitromethane, CH_3NO_2	277	1.3	petroleum ether

In general, a molecule containing one chromophoric group will have an absorption band corresponding to a $\pi \rightarrow \pi^*$ transition somewhere in the visible or ultraviolet portion of the spectrum, and both its wavelength of maximum absorption and its molar absorptivity will be characteristic of the chromophoric group that it contains. This means that the visible–ultraviolet absorption spectra of 1-butene ($CH_3CH_2CH=CH_2$), 1-pentene ($CH_3CH_2CH_2CH=CH_2$), and 1-hexene ($CH_3CH_2CH_2CH_2CH=CH_2$) are virtually identical. Each of these compounds contains the C=C group but no other chromophore; when dissolved in a saturated hydrocarbon such as *n*-hexane, each has a wavelength of maximum absorption close to 180 nm, and the molar absorptivity of each is close to 1.2×10^4 cm² mmol⁻¹.

By comparing the spectrum of any of these compounds with the spectrum of a compound known to contain the C=C group, it would be easy to identify that group as the one responsible for the absorption. In order to make such a comparison the spectrum of the unknown compound should be obtained in the same solvent that was used to obtain the reference spectrum, for the values of λ_{max} and ε_{max} often differ from one solvent to another because of solute-solvent interactions. These can range from electrostatic interactions between the electrons in the absorbing molecule and the electric fields of nearby molecules of a polar solvent to chemical reactions such as

$$\text{CH}_3\text{C}\underset{\text{H}}{\overset{\text{O}}{\big<}}(aq) + \text{H}_2\text{O}(l) = \text{CH}_3\underset{\text{H}}{\overset{\text{OH}}{\text{C}}}\!\!-\!\text{OH}(aq).$$

The visible–ultraviolet spectrum of a molecule makes it possible to identify the chromophoric group that is responsible for the absorption, but the molecule itself cannot be identified in this way. The spectra of 1-butene, 1-pentene, and 1-hexene in any particular solvent are so nearly identical that none of them can be distinguished from the others, or from any closely related compound such as 2-butene ($\text{CH}_3\text{CH}=\text{CHCH}_3$).

If a molecule contains two or more chromophoric groups, they can be arranged in either of two ways. If they are isolated—separated by two or more single bonds— they behave so nearly independently that the absorbance at any wavelength can be obtained by adding the separate absorbances of the two chromophoric groups at that wavelength. Thus the spectrum of a c M solution of the compound $\text{CH}_2=\text{CHCH}_2\text{CH}_2\text{CH}_2\text{NO}_2$ in a non-polar solvent such as n-hexane would be indistinguishable from that of a solution containing c M $\text{CH}_2=\text{CHCH}_3$ and c M $\text{CH}_3\text{CH}_2\text{NO}_2$ in the same solvent. There would be one band, for which λ_{max} was about 180 nm and ε_{max} was about $1.2\times10^4\,\text{cm}^2\,\text{mmol}^{-1}$, corresponding to the $\pi \to \pi^*$ transition of the C=C group; and there would be another, with λ_{max} equal to about 280 nm and ε_{max} equal to about 20 $\text{cm}^2\,\text{mmol}^{-1}$, corresponding to the $n \to \pi^*$- transition of the $-\text{NO}_2$ group. Similarly, the compound $\text{CH}_2=\text{CHCH}_2\text{CH}_2\text{CH}=\text{CH}_2$ would give one band having λ_{max} equal to about 180 nm and ε_{max} equal to about 2×10^4 $\text{cm}^2\,\text{mmol}^{-1}$. The value of λ_{max} would be the same as for propene ($\text{CH}_2=\text{CHCH}_3$), but that of ε_{max} would be approximately twice as large because there are twice as many C=C bonds in each molecule.

Conjugation, which can be represented as an alternation of single and double bonds,[†] has marked effects on the electronic structures of the groups it involves, and

---

† This would not be a good definition of conjugation, because it suggests that there are two different kinds of carbon–carbon bonds in a compound like $\text{CH}_2=\text{CH}-\text{CH}=\text{CH}_2$. It is less misleading to represent this compound as $\text{CH}_2\cdots\text{CH}\cdots\text{CH}\cdots\text{CH}_2$, suggesting that the π electrons are not confined to separate bonds but are spread throughout the whole system. Similarly, benzene is better

represented as ⬡ than as ⬡ .

these effects have many different kinds of consequences. Propene can react with bromine in the following way:

$$CH_2{=}CHCH_3 + Br_2 = \underset{\underset{Br}{|}}{CH_2}{-}\underset{\underset{Br}{|}}{CHCH_3}$$

The compound $CH_2{=}CHCH_2CH_2CH{=}CH_2$ (1,5-hexadiene), which was mentioned in the preceding paragraph, reacts with bromine in an exactly similar way:

$$CH_2{=}CHCH_2CH_2CH{=}CH_2 + Br_2 = \underset{\underset{Br}{|}}{CH_2}{-}\underset{\underset{Br}{|}}{CHCH_2CH_2CH}{=}CH_2$$

A second molecule of bromine can be added across the other double bond, giving $CH_2Br{-}CHBr{-}CH_2CH_2{-}CHBr{-}CH_2Br$. On the other hand, the conjugated compound $CH_2{=}CH{-}CH{=}CH_2$ (1,3-butadiene) behaves differently. Bromine atoms are added at the outer ends of the conjugated system rather than to adjacent carbon atoms:

$$CH_2{=}CH{-}CH{=}CH_2 + Br_2 = \underset{\underset{Br}{|}}{CH_2}{-}CH{=}CH{-}\underset{\underset{Br}{|}}{CH_2}$$

In 1,5-hexadiene the two isolated $C{=}C$ groups behave independently, but in 1,3-butadiene the conjugated system behaves as a single unit. Another effect of conjugation is illustrated by the following values of K_a:

Compound	CH_3CH_2COOH	$CH_2{=}CHCOOH$	$CH_2{=}CHCH_2CH_2CH_2COOH$
K_a, mol dm^{-3}	1.34×10^{-5}	5.6×10^{-5}	1.90×10^{-5}

(in water at 298 K)

A $C{=}C$ group has a much smaller effect on the ability of a $-COOH$ group to donate a proton to a molecule of water if it is some distance away than it does if the $C{=}C$ and $C{=}O$ bonds are conjugated.

Conjugation displaces visible–ultraviolet absorption bands toward longer wavelengths. Whereas $\lambda_{max} = 184$ nm for the non-conjugated compound 1,5-hexadiene, it is 217 nm for the conjugated one $CH_2{=}CHCH{=}CHCH_2CH_3$ (1,3-hexadiene). The magnitude of the shift increases as the length of the conjugated system increases: for $CH_2{=}CHCH{=}CHCH{=}CH_2$ (1,3,5-hexatriene) $\lambda_{max} = 250$ nm. On the other hand, the value of ε_{max} depends chiefly on the number of π electrons in the molecule but is hardly affected by the way in which these are arranged, and is therefore insensitive to conjugation. The compounds 1,5-hexadiene and 1,3-hexadiene, of which the latter is conjugated while the former is not, both have $\varepsilon_{max} = 2.0\times10^4$ cm^2 mmol^{-1}, and the values of ε_{max} for the bands corresponding to the $n \to \pi^*$ transitions in the compounds

$$\underset{\underset{CH_3CCH_2CH_2CH_3}{}}{\overset{\overset{O}{\|}}{}}, \quad \underset{\underset{CH_3CCH_2CH{=}CH_2}{}}{\overset{\overset{O}{\|}}{}}, \quad \text{and} \quad \underset{\underset{CH_3CCH{=}CHCH_2}{}}{\overset{\overset{O}{\|}}{}}$$

are all equal to 30 ± 3 cm^2 mmol^{-1} although the third of these is conjugated while the first two are not.

Even though a substituent atom or group would not produce absorption if it were present alone, it can affect the electronic structure of a chromophoric group. Such atoms or groups are called *auxochromes*. Some common auxochromes are halogen atoms and the methyl, hydroxyl, methoxyl (—OCH$_3$), and amino (—NH$_2$) groups. One effect of substituting a halogen atom for one of the hydrogen atoms in the methyl group of acetic acid was mentioned on page 146; another is to change the values of λ_{max} and ε_{max}. In a non-polar solvent such as *n*-hexane, λ_{max} = 204 nm and ε_{max} = = 40 cm^2 mmol^{-1} for acetic acid, but λ_{max} = 280 nm and ε_{max} = 400 cm^2 mmol^{-1} for iodoacetic acid, ICH$_2$COOH. Usually, as in this example, both λ_{max} and ε_{max} are increased by the presence of an auxochrome, but exceptions are fairly common. 1,4-Benzoquinone has two absorption bands and they are affected differently:

Compound	O=⟨⟩=O	O=⟨⟩=O Cl
λ_{max}, nm	245	258
ε_{max}, cm^2 mmol^{-1}	2.5×10^4	2.0×10^4
λ_{max}, nm	437	323
ε_{max}, cm^2 mmol^{-1}	20	8×10^2

All this can be summarized by saying that many organic compounds have fairly characteristic absorption spectra in the visible and ultraviolet. It is generally easy to distinguish between two compounds if they contain different chromophores: for example, if R is some saturated group, the nitroso compound RNO will have λ_{max} roughly equal to 680 nm whereas the corresponding nitro compound RNO$_2$ will have λ_{max} equal to about 280 nm. Small differences between similar conjugated systems are also easy to detect: λ_{max} is 420 nm for the compound C$_6$H$_5$(CH=CH)$_6$C$_6$H$_5$, but is 435 nm for C$_6$H$_5$(CH=CH)$_7$C$_3$H$_5$. It is even possible to distinguish among many closely related compounds that differ only in the auxochromes they contain: fluorobenzene (C$_6$H$_5$F), chlorobenzene (C$_6$H$_5$Cl), phenol (C$_6$H$_5$OH), and aniline (C$_6$H$_5$NH$_2$) have absorption bands for which the values of λ_{max} are 254, 264, 270, and 280 nm, respectively. However, substituent atoms or groups cannot usually be detected if they are separated from a chromophoric group by two or more single bonds: the compounds C$_6$H$_5$CH$_3$, C$_6$H$_5$CH$_2$Cl, and C$_6$H$_5$CHCl$_2$ all have identical values of λ_{max}, and so do the compounds

SUMMARY: Other kinds of transitions are also possible, but $\pi \to \pi^*$ transitions are the most important in the visible–ultraviolet spectroscopy of organic molecules. Functional groups that contain π electrons (chromophoric groups) confer distinctive spectral features on otherwise non-absorbing molecules, but these features may be modified by the presence of auxochromes, which are functional groups that do not contain π electrons but alter the electronic distribution in a chromophoric group. The spectral features of two chromophoric groups are additive if they are isolated from each other, but are changed markedly by the interactions resulting from conjugation.

15.12. Spectrophotometric analysis

PREVIEW: This section summarizes the techniques and calculations associated with quantitative analysis by visible–ultraviolet absorption spectroscopy.

Measurements of absorbance in the visible–ultraviolet region can be used in many different ways to determine both inorganic and organic substances. Some preliminary treatment of the sample may be necessary, depending on whether the substance of interest absorbs in this region, and also on whether the other substances are transparent at the wavelength where the absorbance is measured. The simplest situation is the one in which the substance being determined has a large value of ε at a wavelength where no other constituent of the sample absorbs at all. Then one need only measure the absorbance and calculate the desired concentration from it.

The limit of detection in such an analysis is given by

$$c^* = A^*/\varepsilon b$$

where A^* is the smallest value of the absorbance that can be reliably distinguished from zero—that is, the minimum detectable response—and b is the path length for the cell with which this absorbance is measured. Typically $b = 1$ cm, but many spectrophotometers can accommodate cells for which b is as large as 5 cm. The smallest detectable absorbance depends on the care that is taken to eliminate suspended solid particles from the solution and to keep dust and fingerprints off the windows of the cell, and it also depends on the ability of the spectrophotometer to discriminate between the large and nearly equal powers of the beams transmitted by the cells containing the sample and reference solutions. If A^* is taken to be 0.005, $c^* = 10^{-3}/\varepsilon$ if $b = 5$ cm. Since the value of ε may be as high as 10^5 cm^2 mmol^{-1}, the limit of detection c^* may be as small as 10^{-8} M, which is well below the limits of detection for many other techniques. Of course the figure would be much less favorable if the value of ε were smaller: near the opposite extreme, the molar absorptivities of the trivalent ions of the rare earth elements, such as dysprosium and ytterbium, are no larger than about 2 cm^2 mmol^{-1} in aqueous solutions, and therefore these ions could not be detected at concentrations below about 5×10^{-4} M.

There are two ways of determining a small concentration of a substance for which the value of ε is small. One is to obtain that substance in a smaller volume of solution, which can sometimes be done by evaporating the solvent or by solvent extraction or some other technique of separation. The other, which is usually easier, is to transform it into a more strongly absorbing species by causing it to react with a suitable reagent. A weakly absorbing metal ion might be made to react wich a chelating agent of the type mentioned in the last paragraph of Section 15.10 or, if it has oxidizing properties, with a weakly reducing ligand such as iodide or thiocyanate ion to give a complex that exhibits charge-transfer absorption. Acetone, for which ε_{max} is only 16 cm^2 mmol^{-1}, can be converted to a derivative having $\varepsilon_{max} = 2 \times 10^4$ cm^2 mmol^{-1} by treating it with excess 2,4-dinitrophenylhydrazine:

$$\begin{array}{c} H_3C \\ \\ H_3C \end{array}\!\!\!\!C{=}O + H_2N{-}NH\!\!\left\langle\bigcirc\right\rangle\!\!NO_2 = \begin{array}{c} H_3C \\ \\ H_3C \end{array}\!\!\!\!C{=}N{-}NH\!\!\left\langle\bigcirc\right\rangle\!\!NO_2 + H_2O$$

$$\qquad\qquad\qquad\qquad NO_2 \qquad\qquad\qquad\qquad\qquad\qquad NO_2$$

Even a substance that does not absorb at all can be determined if it can be made to react stoichiometrically with another that does. Fluoride ion reacts with the yellow titanium(IV)–peroxide complex, giving the colorless complex TiF^{3+}, and the decrease of absorbance that accompanies this reaction can be correlated with the concentration of fluoride.

Most spectrophotometric analyses are performed by measuring the absorbance at the wavelength of maximum absorption. This is partly because a small error in wavelength has much less effect on the absorbance at the peak of a band, where the absorbance is nearly independent of wavelength, than it does on the side of a band. Moreover, as was said in Section 15.8, the finite band width of the beam gives rise to negative deviations from Beer's law unless the absorbance is nearly independent of wavelength over the range of wavelengths contained in the beam. Sometimes, however, the sample contains an impurity that absorbs at the wavelength of maximum absorption of the substance being determined. If the absorption bands of the substance being determined and the impurity do not overlap too much, there may be some other wavelength where the former absorbs but the latter does not, and then it is easiest to measure the absorbance at that wavelength.

It is necessary to know whether Beer's law is obeyed and, if so, over what range of concentrations. Experimentally this is investigated by measuring the absorbances of a number of solutions containing different known concentrations of the substance being determined. In these measurements the cell, the spectrophotometer, and the experimental conditions—the concentrations of reagents, pH, temperature, etc.— must all be the same as in the actual analysis. If Beer's law is obeyed, a plot of absorbance against concentration will be a straight line passing through the origin, while a plot of A/c against concentration will be a horizontal straight line. The second of these is a much more sensitive test of adherence to the law. When the law is obeyed, the concentration c_u of an unknown solution can be calculated from its absorbance A_u by means of the equation

$$c_u = \frac{A_u}{(A/c)} \qquad\qquad (15.7)$$

where (A/c) is the average value of the ratio of absorbance to concentration obtained from the data on known solutions. If Beer's law is not obeyed, the concentration c_u corresponding to the measured absorbance A_u can be read off a plot of absorbance against concentration constructed from the data on known solutions, but because this always entails a loss of precision it is better to find and, if possible, cure the cause of the deviation from Beer's law.

Sometimes you can determine two or more substances in the same solution. This is easy to do if their absorption bands do not overlap. You can then measure the absorbance at the wavelength of maximum absorption for each of the separate substances, and calculate the corresponding concentrations from equations just like eq. (15.7), using one set of standard solutions of each substance to evaluate (A/c) for that substance. It is less easy when the bands do overlap. Then the spectrum of a mixture might look like curve c in Fig. 15.17, while the spectra of the individual components look like curves a and b. You can measure the absorbance of the mixture at each of the two separate wavelengths of maximum absorption λ_1 and λ_2, and known solutions of the individual components are used to find the values of (A/c) for each component at both wavelengths under exactly the same conditions. The absorbance A_{m, λ_1} of the mixture at the wavelength λ_1 is given by

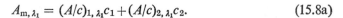

$$A_{m, \lambda_1} = (A/c)_{1, \lambda_1}c_1 + (A/c)_{2, \lambda_1}c_2. \qquad (15.8a)$$

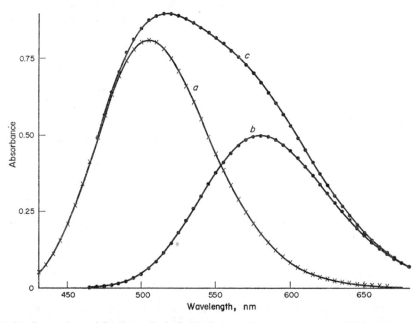

FIG. 15.17. Curves (a) and (b) show the individual absorption spectra of two different substances; curve (c) is the spectrum of a mixture of the two at the same concentrations. Note that the wavelength of maximum absorption on curve (c) is not the same as on curve (a), and that curve (c) does not have a maximum at the wavelength of maximum absorption on curve (b).

At the other wavelength, λ_2, the absorbance $A_{\mathrm{m}, \lambda_2}$ of the mixture is given by

$$A_{\mathrm{m}, \lambda_2} = (A/c)_{1, \lambda_2} c_1 + (A/c)_{2, \lambda_2} c_2 \tag{15.8b}$$

and both c_1 and c_2 can be evaluated by solving these two simultaneous equations. As far as the algebra goes, it is possible to determine three or more absorbing substances at once in the same way, and this is quite feasible if the absorption bands are so well separated that each of the components is chiefly responsible for one of the absorbances measured for the mixture. The precision deteriorates if the bands overlap very much or if the products $(A/c)_{i, \lambda_i} c_i$ for the different substances at their respective wavelengths of maximum absorption are widely disparate.

SUMMARY: In addition to whatever steps may be necessary to bring the concentration of the absorbing species into a conveniently measurable range, quantitative spectrophotometric analysis entails careful evaluation of A/c for known solutions. Two or more substances can be determined simultaneously if their absorption bands do not overlap too much.

15.13. Spectrophotometric titrations

PREVIEW: Spectrophotometric titrations are described and compared with analyses by ordinary spectrophotometry.

A spectrophotometric titration is a titration that is followed by measuring the absorbance of the solution being titrated. Usually it is possible to choose a wavelength at which the absorbance is due to just a single substance. This may be the substance that is titrated (S, in the notation of Chapter 11), the reagent (R), a product (P or Q) of the reaction between them, or an indicator that is added to the solution. A typical spectrophotometric titration is that of an acidic solution of dichromate ion, $Cr_2O_7^{2-}$, with a standard solution of arsenic(III) acid, $HAsO_2$. The absorbance at 350 nm is entirely attributable to dichromate ion, and is proportional to its concentration if Beer's law is obeyed. The titration curve is described by eqs. (11.8): it is a plot of the quantity $A(V_{Cr}^0 + V_{As})/V_{Cr}^0$ against V_{As}, where V_{Cr}^0 is the volume of the dichromate solution at the beginning of the titration and V_{As} is the total volume of the reagent that has been added at the point where the absorbance is equal to A. The titration curve is segmented and _-shaped, and its two linear segments intersect at the equivalence point. Other shapes are of course obtained by measuring the absorbance due to the reagent or a product.

It is very easy to perform a spectrophotometric titration. The solution being titrated is placed in a suitable clear-walled vessel, such as a beaker, in the path of the light beam in a spectrophotometer, and the absorbance is measured at a wavelength chosen in advance from the spectra of the reactants and products. A suitable volume of the reagent is added, the mixture is stirred, and a little time is allowed to ensure that the reaction has reached equilibrium. The absorbance is measured again, more reagent is added, and the process is repeated until the end point has been passed and enough points have been obtained. It is desirable to have at least half a dozen readings on each side of the end point. The titration curve is constructed by plotting $A(V_S^0 + V_R)/V_S^0$ against V_R, where V_S^0 is the volume of the solution titrated and V_R is the volume of reagent added. The two linear branches of the curve are extrapolated to their point of intersection, and the value of V_R at this point is read off the abscissa scale and considered to represent the end point.

Figures 11.10 and 11.11 showed that there will be a region of curvature around the end point unless the value of K_t is extremely large. Points in this region are of no use in locating the end point; this is in contrast to potentiometric titrations, where points close to the equivalence point are essential. Points just outside the region of curvature are very useful, however, because the precision improves as the extrapolation becomes shorter. If the end point of a titration is expected to lie at $V_R = 4 \text{ cm}^3$, one might add the reagent in 0.5-cm^3 increments until a total of 6 to 8 cm^3 has been used. When the titration curve is plotted it is easy to identify, and ignore, points that lie in the region of curvature if there is one, and the extrapolations from the remaining points are as short as they can conveniently be made.

It was said in Section 14.9 that potentiometric titrations are more accurate and more precise than analyses by direct potentiometry. For very similar reasons, spectrophotometric titrations are more accurate and more precise than analyses by the techniques described in the preceding section. In a spectrophotometric analysis there may be an error if the path length, the temperature, or the composition of the supporting electrolyte is different from what the analyst thinks it is, but none of these things can affect the location of the end point in a spectrophotometric titration unless it changes while the titration is being performed. In a spectrophotometric analysis one must know the value of A/c for the substance being determined, and any error in this ratio will lead to an error in the concentration that is calculated from the measured absorbance. In a spectrophotometric titration the value of A/c governs the slope of one of the line segments, but it need not be known and does not affect the location of the end point. In a spectrophotometric analysis the measured absorbance will be too high if there is a little dust or other solid matter suspended in the solution of the sample, and then the calculated concentration is almost certain to be too high. In a spectrophotometric titration the presence of suspended solid matter would cause all of the measured absorbances to be too high, but this would merely displace the curve along the ordinate axis without affecting the location of the end point.

Values of the relative sample standard deviation, s_c/c, in spectrophotometric analysis are usually about 0.02, or 2%, under favorable conditions. [There are special techniques and exceptionally good (and correspondingly expensive) instruments that can provide still better precision. On the other hand, even a relative precision of 2% may be impossible to obtain if Beer's law is not obeyed, if the concentration of the substance being determined is only a little higher than the limit of detection, or if the sample is so complex that a great deal of chemical pretreatment is required.] In spectrophotometric titrations relative precisions of 0.2 to 0.5% are typical.

SUMMARY: Spectrophotometric titrations yield segmented titration curves and are more accurate and precise than ordinary spectrophotometric analyses because they are unaffected by many changes that would affect the measured absorbance of a single solution.

15.14. Spectrophotometric studies of chemical equilibria

PREVIEW: Spectrophotometry is a versatile and powerful technique for the study of chemical equilibria, and a number of different ways of using it have been devised.

Many different techniques have been devised for studying chemical equilibria with the aid of spectrophotometric measurements. They all involve preparing a number of mixtures containing the reactants in different proportions, measuring their absorbances, and correlating these with the known concentrations of the reactants. The details depend on the nature of the equilibrium and on the number of substances that absorb at the wavelength used. Three common possibilities will be discussed here:

1. There is only one equilibrium, and it involves only one absorbing substance.
2. There is only one equilibrium, and it involves both an absorbing reactant and an absorbing product.
3. There are two or more stepwise equilibria, and they involve two or more absorbing substances.

The reaction between bismuth(III) and ethylenediaminetetraacetate ($= Y^{4-}$) in a strongly acidic solution is an example of the first possibility. If we were just beginning to study the reaction and had no prior information about it, we might represent it by the equation

$$m\text{Bi}^{3+} + y\text{Y}^{4-} = \text{Bi}_m\text{Y}_y^{(3m-4y)+} \tag{15.9}$$

Since protonation of the chelon is certain to be extensive, ignoring it implies that all of our measurements will be made with solutions containing some fixed concentration of hydronium ion. In 1 M perchloric acid the chelonate is the only one of these substances that absorbs at 265 nm. The first thing we would want to know about the complex is its formula, and this can be found by the *molar ratio method*. A number of mixtures are prepared, containing equal concentrations of one reactant but different concentrations of the other. The ratio of the two concentrations may vary from about 0.1 to 10 or 20. The absorbance of each mixture is measured at some suitable wavelength, and the absorbances are plotted against the ratio of the numbers of moles of the two reactants. A curve resembling curve (a) in Fig. 15.18 may be obtained. An entirely similar curve could be obtained by performing a spectrophotometric titration, which would be both faster and easier, because it would eliminate the necessity of handling a great many different mixtures, unless the formation of the complex was extremely slow.

If it has been obtained under properly chosen conditions, a *molar ratio plot* like the one in Fig. 15.18 will consist of two linear segments, as you would expect from the fact that it is really nothing more than a spectrophotometric titration curve in disguise. The two segments can be extrapolated to their point of intersection, whose coordinate on the abscissa axis gives the ratio in which the reactants combine to form the product. Here the point of intersection lies at a mole ratio of 1, and this shows that the simplest possible formula of the product is BiY^-, although the actual formula might of course be $Bi_2Y_2^{2-}$ or $Bi_3Y_3^{3-}$. If the point of intersection lay at a mole ratio of 0.5, the product would contain 0.5 mole of ethylenediaminetetraacetate for each mole of bismuth, and its simplest possible formula would be Bi_2Y^{2+}. At this stage you could not tell whether the complex was protonated or not: other experiments would be needed to tell whether it was BiY^- or $HBiY(aq)$ under these conditions.

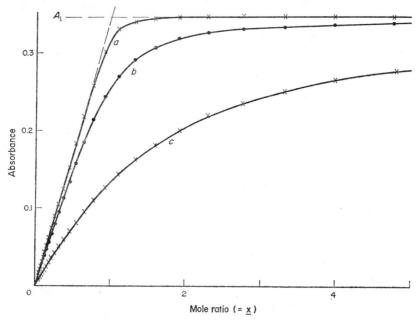

FIG. 15.18. Molar ratio plots for $c(= [M]+[MY]) = 1.00 \times 10^{-4}\ M$, $\varepsilon b = 3.50 \times 10^3\ cm^3\ mmol^{-1}$, and $K = (a)\ 1.00 \times 10^6$, (b) 1.00×10^5, and (c) $1.00 \times 10^4\ mol^{-1}\ dm^3$.

There may be a region of curvature centered around the point of intersection on a molar ratio plot. If there is, the data obtained in that region can be used to evaluate the conditional equilibrium constant of the reaction. Let us consider a point at which x mole of ethylenediaminetetraacetate is present for each mole of bismuth(III), and at which the measured absorbance is equal to A, and let us say that there is c mol dm^{-3} of bismuth(III) in the mixture. Conservation equations give

$$[Bi^{3+}]+[BiY^-] = c, \tag{15.10a}$$
$$[Y]'+[BiY^-] = xc. \tag{15.10b}$$

As in Sections 7.7 and 11.9, the symbol [Y]' is used to denote the total concentration of the ligand that is not bound to the metal ion. As more and more ligand is added, the reaction becomes more and more nearly complete, the concentration of the chelonate approaches c, and the absorbance approaches the limiting value A_l shown by the horizontal dashed line in Fig. 15.22. At the point where the mole ratio is equal to x, the absorbance is smaller than A_l because the reaction is incomplete. If Beer's law is obeyed, the ratio A/A_l is equal to the fraction of the bismuth(III) that has been converted into the chelonate, so that $[BiY^-] = (A/A_l)c$. Combining this with eqs. (15.10) and with the expression for the conditional formation constant of the chelonate, we obtain

$$K = \frac{[BiY^-]}{[Bi^{3+}][Y]'} = \frac{(A/A_l)}{[(1-(A/A_l))][x-(A/A_l)]\,c}$$

It is easy to compute a value of K from the values of A/A_l and x at each of the points in the region of curvature.

Unless you have some prior information about the reaction, you will have to perform some preliminary experiments so that you can select conditions suitable for making the final measurements. Curves (b) and (c) of Fig. 15.18 show why this is necessary. Curve (a) was drawn for $Kc = 100$; curves (b) and (c) show what would be obtained if Kc were equal to 10 or 1. The value of Kc for the reaction between bismuth(III) and ethylenediaminetetraacetate could be increased either by increasing the value of c or by using less strongly acidic solutions, which would increase the values of α_Y and K (see Section 11.9). If Kc is very large, the reaction will be so nearly complete at every point that A/A_l will be almost equal to either x or 1, whichever is smaller. Then the point of intersection on the plot will be very easy to locate, but the values of K computed will be very uncertain. As the value of Kc decreases, the region of curvature becomes wider, and [as on curve (b)] the value of A_l may not be attained even with the largest excess of reagent used. If Kc is very small, it may even be impossible [as it is on curve (c)] to find the point of intersection of the linear segments with reasonable certainty.

The *slope-ratio method* is better suited to the study of reactions for which K is so small that it is inconvenient or impossible to achieve values of Kc sufficiently high for success with the molar ratio method. In solutions that contain identical large concentrations of ethylenediaminetetraacetate and different much smaller total concentrations c_{Bi} of bismuth(III), the formation of the complex will be nearly complete even if K is small. On the basis of eq. (15.9) we can write

$$dc_{Bi_mY_y}/dc_{Bi} = 1/m.$$

If Beer's law is obeyed, a plot of the absorbances of such solutions against c_{Bi} will have a slope S_{Bi} that is given by

$$S_{Bi} = dA/dc_{Bi} = \varepsilon b(dc_{Bi_mY_y}/dc_{Bi}) = \varepsilon b/m.$$

In a second set of solutions containing different small concentrations c_Y of ethylene-diaminetetraacetate and a constant and much larger concentration of bismuth(III), we would have

$$dc_{Bi_mY_y}/dc_Y = 1/y$$

and the slope S_Y of a plot of the absorbances of these solutions against c_Y would be given by

$$S_Y = \varepsilon b/y.$$

The ratio of the two slopes gives the stoichiometry of the reaction:

$$S_{Bi}/S_Y = y/m.$$

For the reaction between bismuth(III) and ethylenediaminetetraacetate we would find $S_{Bi}/S_Y = 1$ and conclude that $y = m$. Accordingly the complex might be BiY^- (or $Bi_2Y_2^{2-}$, etc.).

The slope-ratio method fails if two different reactions take place. If either Bi_2Y^{2+} or BiY_2^{5-} could be formed, the first would predominate in the solutions containing excess bismuth(III) and would govern the value of S_Y, while the second would predominate in the solutions containing excess ligand and would govern the value of S_{Bi}. The values of S_Y and S_{Bi} would pertain to different products, and their ratio would not correspond to the composition of either one.

A third method for investigating the stoichiometry of the reaction is the *method of continuous variations*. In using this method we would prepare a series of solutions in which the sum of the concentrations of the two reactants was kept constant while the ratio of their concentrations differed from one solution to another. Let us represent the constant sum by c and define a quantity f such that $c_Y = fc$. Since $c_{Bi} + c_Y = c$, we would have $c_{Bi} = (1-f)c$. If the formula of the complex is unknown, its formation constant has to be represented by the expression

$$K = \frac{[Bi_mY_y^{(3m-4y)+}]}{[Bi^{3+}]^m([Y]')^y} . \tag{15.11}$$

At equilibrium in any one mixture

$$[Bi^{3+}] = c_{Bi} - m[Bi_mY_y^{(3m-4y)+}] = (1-f)c - mC \tag{15.12a}$$

and

$$[Y]' = c_Y - y[Bi_mY_y^{(3m-4y)+}] = fc - yC \tag{15.12b}$$

where C is the equilibrium concentration of the chelonate. The absorbance will be equal to zero if f is either 0 or 1 because the chelonate is the only species that absorbs at 265 nm and because none of the chelonate can be formed in the absence of either one of the reactants. At some intermediate value of f the concentration of the chelonate will attain a maximum value, and so of course will the absorbance. The maximum will be the point where $dC/df = 0$. Combining eqs. (15.12) with eq. (15.11) yields

$$K[(1-f)c - mC]^m[fc - yC]^y = C.$$

On differentiating this with respect to f and setting dC/df equal to zero, the result is

$$y/m = f/(1-f).$$

For the reaction between bismuth(III) and ethylenediaminetetraacetate the maximum is found to lie at $f = 0.5$, so that $f/(1-f) = 1$ and $y = m$, which corresponds to BiY^- as the simplest formula. If the maximum were found to lie at $f = 0.67$, we would have $f/(1-f) = 2$ and would conclude that the simplest formula of the complex was BiY_2^{5-}. Like a molar ratio plot, a continuous-variations plot will consist of two line segments if the reaction is virtually complete in every one of the mixtures studied, and these line segments will intersect sharply at the maximum. If the pH of the solution or the concentrations of the reactants are changed so that the reaction becomes less complete, the plot will become curved in the vicinity of the maximum. If the curvature is not too extensive we can calculate values of K from the data obtained in this region after we have found the formula of the complex from the location of the maximum. However, if the reaction is too incomplete the entire plot will be a smooth curve and it may even be difficult to decide, for example, whether the maximum lies at $f = 0.67$ (so that $y/m = 2$) or at $f = 0.75$ (so that $y/m = 3$).

So far the discussion has been confined to the first of the three situations listed at the beginning of this section. The second, in which there is only one equilibrium but in which both a reactant and a product absorb, may be illustrated by the behavior of an acid-base indicator. Methyl orange was mentioned in Section 11.8. Its acidic form gives an absorption band for which $\lambda_{max} = 522$ nm, and its basic form gives one for which $\lambda_{max} = 464$ nm. Spectra of solutions containing identical total concentrations of methyl orange, but having different pH-values, are shown in Fig. 15.19. In strongly acidic solutions (curve a) the concentration of the basic form of the indicator is negligible and only the absorption of the acidic form can be detected; at much higher pH-values (curve d) only the absorption of the basic form can be detected. There is an intermediate range, in which both forms are present at appreciable concentrations. This range is analogous to the color-change interval, but is two or three times as wide because absorbance measurements can reveal the presence of a much smaller concentration of one form in a solution containing an excess of the other than could be detected by visual observation.

At 522 nm the molar absorptivity of the acidic form is larger than that of the basic form; at 464 nm the molar absorptivity of the acidic form is smaller than that of the basic form. At some intermediate wavelength the molar absorptivities of the two forms will be equal. The absorbance of any solution at that wavelength will be given by

$$A = \varepsilon_{HIn}bc_{HIn} + \varepsilon_{In}bc_{In} = \varepsilon b(c_{HIn} + c_{In}) = \varepsilon bc$$

where ε is the common value of the two individual molar absorptivities, and c is the total concentration of the indicator. The value of A depends on this total concentration, but it does not depend on the separate values of c_{HIn} and c_{In} and is therefore

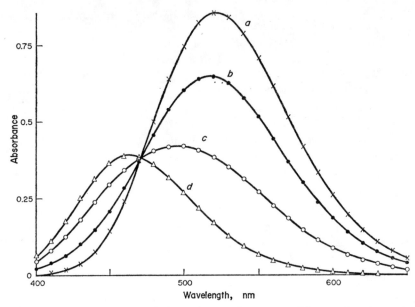

Fig. 15.19. Absorption spectra of solutions of methyl orange at different pH-values. The total concentration of methyl orange was the same in each solution and the pH was (a) 1.3, (b) 3.1, (c) 3.8, and (d) 6.2.

independent of pH. The *isosbestic point* is the point at 472 nm where all of these curves intersect because the two molar absorptivities are equal at 472 nm.

The appearance of an isosbestic point on a family of spectra like the one in Fig. 15.19 is important because it enables us to conclude that the sum of the concentrations of the two species responsible for the two absorption bands was constant throughout the experiment. This could hardly be true if either of those species were involved in a side reaction. If the reaction $H_2In = H^+ + HIn$ were also occurring, the sum of the concentrations of HIn and In would decrease as the pH decreased, and there could not be an isosbestic point unless there was some wavelength at which the molar absorptivities of H_2In, HIn, and In all happened to be equal. That would be such a coincidence that it does not deserve serious consideration. We therefore interpret an isosbestic point as meaning that there is only one equilibrium over the range of conditions investigated.

To calculate the value of the equilibrium constant we need equations like eqs. (15.8). For any particular mixture, the absorbance at 522 nm, which is the wavelength of maximum absorption for the acidic form, is given by

$$A_{m,\,522} = \varepsilon_{HIn,\,522}bc_{HIn} + \varepsilon_{In,\,522}bc_{In}. \qquad (15.13a)$$

The absorbance of the same mixture at 464 nm, the wavelength of maximum absorption for the basic form, is given by

$$A_{m,\,464} = \varepsilon_{HIn,\,464}bc_{HIn} + \varepsilon_{In,\,464}bc_{In}. \qquad (15.13b)$$

The molar absorptivities of the acidic form at these two wavelengths can be obtained from the spectrum (curve a) of a solution so acidic that the concentration of H*In* is indistinguishable from the known formal concentration c, and the molar absorptivities of the basic form can be obtained from the spectrum (curve d) of a solution so basic that the concentration of *In* is indistinguishable from c. Using the subscript "a" (for "acid") to denote the first of these spectra and the subscript "b" (for "base") to denote the second, and assuming that Beer's law is obeyed,

$$\varepsilon_{\text{H}In,\,522}b = A_{\text{a},\,522}/c \quad \text{and} \quad \varepsilon_{\text{H}In,\,464}b = A_{\text{a},\,464}/c$$

$$\varepsilon_{In,\,522}b = A_{\text{b},\,522}/c \quad \text{and} \quad \varepsilon_{In,\,464}b = A_{\text{b},\,464}/c$$

Combining these with eqs. (15.13), and letting $c_{\text{H}In} + c_{In} = c$, we can eventually obtain

$$\frac{c_{In}}{c_{\text{H}In}} = \frac{A_{\text{a},\,522} - A_{\text{m},\,522}}{A_{\text{a},\,522} - A_{\text{b},\,522}} \times \frac{A_{\text{b},\,464} - A_{\text{a},\,464}}{A_{\text{b},\,464} - A_{\text{m},\,464}} \, .$$

An alternative but entirely equivalent approach would be to calculate values of the ligand number $\bar{n}$, which is the average number of protons bound to each *In* group in a particular mixture (see Section 7.6), and which is defined by

$$\bar{n} = c_{\text{H}In}/c.$$

The value of $\bar{n}$ can be calculated from either of the equalities

$$\bar{n} = \frac{A_{\text{b},\,464} - A_{\text{m},\,464}}{A_{\text{b},\,464} - A_{\text{a},\,464}} = \frac{A_{\text{m},\,522} - A_{\text{b},\,522}}{A_{\text{a},\,522} - A_{\text{b},\,522}} \, .$$

Once $\bar{n}$ has been calculated, the corresponding value of $c_{In}/c_{\text{H}In}$ is easily found from the equation

$$\frac{c_{In}}{c_{\text{H}In}} = \frac{1 - \bar{n}}{\bar{n}} \, .$$

If we knew both the concentration of hydrogen ion in a mixture and the value of the ratio $c_{In}/c_{\text{H}In}$, we could find the value of the concentration constant $K_{\text{a, H}In}$ in that mixture:

$$K_{\text{a, H}In} = c_{\text{H}^+}(c_{In}/c_{\text{H}In}).$$

If we knew the values of $K_{\text{a, H}In}$ at a number of different and low ionic strengths, we could extrapolate them to an ionic strength of zero to obtain the value of the thermodynamic constant. Unfortunately, the overall dissociation constant of methyl orange is only about 3×10^{-4} mol dm^{-3} in aqueous solutions at 298 K, which means that a solution containing equal concentrations of the acidic and basic forms would contain only about 3×10^{-4} M hydrogen ion. Such a solution could be prepared by mixing methyl orange, hydrochloric acid, and a neutral electrolyte such as potassium chloride for adjusting the ionic strength, but its buffer capacity would be extremely low. If there were a little acidic or basic impurity in the methyl orange or the potassium chloride, or if the solution were exposed to atmospheric carbon dioxide, the concentration

34

of hydrogen ion might change appreciably. It would therefore be wiser to employ a buffered solution, and to find its acidity by measuring its pH. The concentration of hydrogen ion cannot be obtained exactly from the result of a pH-measurement, but *if* the assumptions described in Section 13.10 are accurate and *if* the liquid-junction potential is the same in measuring the pH of the solution of methyl orange as it is in standardizing the pH meter against the reference buffer, the measured pH-value will be indistinguishable from the paH. This will enable us to find the value of a *mixed constant* K^* (so called because it contains both concentrations and activities):

$$K^*_{a,\,HIn} = \frac{a_{H^+}c_{In}}{c_{HIn}} = \frac{[H^+][In]}{[HIn]}\,y_{H^+} = \frac{a_{H^+}a_{In}}{a_{HIn}} \times \frac{y_{HIn}}{y_{In}}$$

$$= K_{a,\,HIn}y_{H^+} = K^0_{a,\,HIn} \times \frac{y_{HIn}}{y_{In}}.$$

Mixed constants are fairly common because they are so easy to calculate from the results of experimental measurements. Like formal potentials, however, they pertain only to the particular conditions under which they are measured, and vary with experimental conditions that affect the values of the activity coefficients that they involve.

The third of the three situations described at the beginning of this section would be exemplified by the behavior of thymol blue. Thymol blue is an acid-base indicator that has three colored forms. The fully protonated one, which may be represented as H_2In, has $\lambda_{max} = 544$ nm; the intermediate one HIn has $\lambda_{max} = 430$ nm; and the basic one In has $\lambda_{max} = 596$ nm. The values of pK_1 and pK_2 are 1.65 and 9.2, respectively, and these are so far apart that the two equilibria are virtually independent. There is one isosbestic point in acidic solutions, where the concentration of In is far too small to detect, and there is another in alkaline solutions, where the concentration of H_2In is negligible. It would be easy to study these two equilibria separately by the procedure outlined in the preceding few paragraphs. We could begin by preparing three solutions, each containing the same total concentration of thymol blue. One might contain 2 M hydrochloric acid, which would convert nearly all of the indicator to H_2In. Another might contain 0.1 M sodium hydroxide, and would have In as the only form present at an appreciable concentration. The third might be a buffer having a pH-value around 5 or 6, where HIn is the only important form. From the spectra of these solutions it would be easy to obtain values of the molar absorptivities needed in equations like eqs. (15.13), or of the values of A/c needed in equations like eqs. (15.8). The spectra of solutions buffered at pH-values between about 0.5 and 2.5 could then be used to obtain a value of the mixed constant K^*_1, while a value of K^*_2 could be obtained from the spectra of solutions having pH-values between about 8 and 10.5.

Figure 15.20 shows another possibility. It represents the stepwise formation of the iron(III)–thiocyanate complexes $FeNCS^{2+}$, $Fe(NCS)_2^+$, and $Fe(NCS)_3$. All three of these exhibit charge-transfer absorption. Their wavelengths of maximum absorption are not very far apart, and neither are the values of their formation constants. The fact that there is no isosbestic point shows that more than one equilibrium is involved, but data like these are difficult to analyze because there is no way of preparing a solution that contains either $Fe(NCS)_2^+$ or $Fe(NCS)_3$ alone. Ways have been devised for deducing the number of complexes, and the formation constant of each, from spectra like these, but they are too complex to be discussed here.

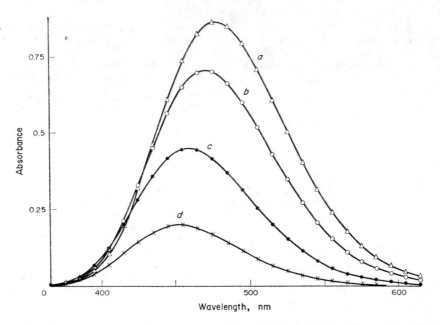

FIG. 15.20. Absorption spectra of solutions containing 9.2×10^{-5} M iron (III) and the following total concentrations of thiocyanate: (a) 0.5, (b), 0.1, (c) 0.02, and (d) 0.005 M. Curves (a) and (b) intersect at 429 nm, curves (a) and (c) intersect at 420 nm, and curves (b) and (c) intersect at 407 nm.

SUMMARY: Spectrophotometric techniques for elucidating the stoichiometry of a reaction and evaluating its equilibrium constant include the molar ratio method, the slope-ratio method, the method of continuous variations, and the analysis of mixtures that contain two or more absorbing species, such as the starting material and one or more of the products.

Problems

Answers to some of these problems are given on page 522.

15.1. A 1-cm layer of a certain solution of potassium permanganate has a transmittance of 0.800 at 525 nm. What would be the transmittance of a 5-cm layer of the same solution at the same wavelength?

15.2. A number of solutions of 1,3-butadiene, $CH_2 = CHCH = CH_2$, in hexane are prepared and found to have the following transmittances at 217 nm in 1-cm cells:

34*

Concentration, M	$T, \%$
4.53×10^{-6}	81.2
8.22×10^{-6}	68.5
1.07×10^{-5}	61.1
1.37×10^{-5}	53.2
2.28×10^{-5}	35.0

(a) Is Beer's law obeyed?

(b) What is the molar absorptivity of 1,3-butadiene in hexane solutions?

15.3. The molar absorptivity of ascorbic acid in aqueous solutions is equal to $8.1 \times 10^{3} \, cm^{2} \, mmol^{-1}$ at 265 nm, and the formula weight of ascorbic acid is 176.1.

(a) What will be the absorbance at 265 nm of a solution containing $1.73 \times 10^{-5} \, M$ ascorbic acid?

(b) What will be the absorbance at 265 nm of a solution containing $5.00 \, mg \, dm^{-3}$ of ascorbic acid?

(c) What is the concentration of ascorbic acid in a solution whose absorbance is 0.200 at 265 nm, assuming that ascorbic acid is the only substance present that absorbs at that wavelength?

15.4. The following table gives the absorbances of a number of standard solutions of cobalt(II) in a concentrated thiocyanate medium under certain experimental conditions:

Total concentration of cobalt(II), M	Absorbance
2.00×10^{-4}	0.033
5.00×10^{-4}	0.081
1.25×10^{-3}	0.195
3.12×10^{-3}	0.440
7.81×10^{-3}	0.855

(a) Is Beer's law obeyed by these data?

(b) Estimate the total concentration of cobalt(II) in a solution whose absorbance, measured under the same experimental conditions, is 0.300.

15.5. The molar absorptivity of the yellow anion of 2,6-dinitrophenol is equal to $5.5 \times 10^{4} \, cm^{2}$ $mmol^{-1}$ at 460 nm in an aqueous solution. Using other data given in part 3 of Section 15.8, plot the absorbance of 2,6-dinitrophenol at 460 nm against the total concentration of 2,6-dinitrophenol in an aqueous solution containing that compound as the only solute.

15.6. Dichromate $(Cr_2O_7^{2-})$ and chromate ions take part in the reaction $Cr_2O_7^{2-} + H_2O(l) = 2 \, CrO_4^{2-} + 2 \, H^{+}$ in aqueous solutions. The wavelength of maximum absorbance is 375 nm for chromate ion and 349 nm for dichromate ion. Predict the shape of a plot of absorbance against the total concentration of dichromate in solutions prepared by dissolving potassium dichromate in pure water if the absorbance is measured at

(a) 375 nm.   (b) 349 nm.

15.7. On the basis of the facts stated in the preceding problem, suggest at least two ways of obtaining a linear plot of absorbance against the total concentration of dichromate ion.

15.8. The concentration of copper(II) in a solution containing sulfuric acid was determined in the following way. Exactly 5 cm^3 of the solution was transferred to an absorption cell, and its absorbance was found to be 0.157 at the wavelength of maximum absorption for copper(II) ion. Exactly 1 cm^3 of an 0.010 00 M solution of copper(II) sulfate was added, and the absorbance of the mixture was found to be 0.156 at the same wavelength. What was the concentration of copper(II) in the original solution?

15.9. A reagent was added to a certain solution and the absorbance was measured at various times thereafter, with the following results:

Elapsed time, min	5	10	30	55	75	95	135	240
A	0.153	0.282	0.395	0.469	0.480	0.469	0.416	0.353

What would the absorbance have been immediately after mixing if the formation of the absorbing species had been instantaneous? State explicitly any assumption you make about the kinetics of the decomposition of the absorbing species.

15.10. The wavelengths of maximum absorption for the acidic and basic forms of methyl red are 528 and 400 nm, respectively. A solution containing methyl red at a total concentration of 1.22×10^{-5} M and 0.1 M hydrochloric acid had an absorbance of 0.077 at 400 nm and an absorbance of 1.738 at 528 nm. Another solution containing methyl red at a total concentration of 1.09×10^{-5} M and 0.1 M sodium bicarbonate had absorbances of 0.753 and 0.000 at the same wavelengths. A third solution of methyl red had a pcH of 4.18, an absorbance of 0.166 at 400 nm, and an absorbance of 1.401 at 528 nm. The same absorption cell was used for all of the measurements. What is the value of the overall acidic dissociation constant of methyl red, and what was the total concentration of methyl red in the third solution?

15.11. Three solutions of an indicator HIn were prepared and their absorbances were measured under identical conditions, with the following results:

Composition	A
1×10^{-5} M HIn, 0.1 M NaOH, 0.1 M NaCl	0.286
2×10^{-5} M HIn, 0.1 M NaOH, 0.1 M NaCl	0.572
1×10^{-5} M HIn, 0.2 M NaOH	0.446

The absorbance was entirely due to the anion In^-. Compute the value of the overall acidic dissociation constant of HIn.

15.12. A newly synthesized organic acid was known to be monobasic and to have a formula weight of 297. Exactly 10 mg of the pure acid was weighed into the absorption cell used in securing the data of Problem 15.10. It was dissolved in 5 cm^3 of 0.1 M sodium perchlorate; then exactly 0.150 cm^3 of 0.1000 M sodium hydroxide and one drop of a dilute methyl red solution were added. After thorough mixing, the absorbances of the resulting solution were measured and found to be 0.464 at 400 nm and 0.927 at 528 nm. What is the value of the overall acidic dissociation constant of the acid? (Use the answer to Problem 15.10 in the calculations.)

15.13. A certain metal ion M is reported to form a 1:1 complex with a ligand L. At the wavelength of maximum absorption by the complex, where neither M nor L exhibits measurable absorption, a solution containing M at a total concentration of 1.00 M and L at a total concentration of 5.00×10^{-4} M has an absorbance exactly equal to that of a solution containing M at a total concentration of 5.00×10^{-3} M and L at the same total concentration. What is the value of the dissociation constant of the complex?

15.14. Some time after the data given in Problem 15.13 were secured, it began to seem doubtful that the formula of the complex really was ML, and the following measurements were made in an attempt to resolve this question. Four solutions containing known total concentrations of M and L were prepared, and their absorbances were measured at the wavelength of maximum absorption by the complex. The compositions and absorbances of the solutions were as given in the following table. Unless these data show that the complex is indeed ML, find its actual dissociation constant under the conditions of Problem 15.13. These new measurements were made with a 5-cm absorption cell. Find the molar absorptivity of the complex.

Total concentration of M, M	1.00	1.00	3.89×10^{-4}	5.21×10^{-4}
Total concentration of L, M	2.33×10^{-4}	5.91×10^{-4}	0.98	0.98
Absorbance	0.127	0.334	0.443	0.593

15.15. Iron(III) and mercury(II) are known to react with a certain chelating agent Y to form chelates having the formulas FeY_3^{3+} and HgY_2^{2+}. The iron(III) chelate is known to obey Beer's law and its formation constant is equal to 3.1×10^{19} mol^{-1} dm^3. A solution containing Y and iron(III), each at a total concentration of 1.00×10^{-4} M, has an absorbance of 0.740 at a wavelength where the iron(III) chelate absorbs but where the mercury(II) chelate, unchelated iron(III) and mercury(II), and the chelating agent do not absorb. Another solution, identical with the first one but also containing mercury(II) at a total concentration of 1.00×10^{-4} M, has an absorbance of 0.465 under the same conditions. Calculate the formation constant of the mercury(II) chelate.

Answers to Problems

2.2. 1.588 mg.

2.5. 38.79 J K^{-1} mol^{-1}.

2.7. (a) No.

(b) 330.1 K = 57.0 °C.

2.10. (b) $K_1 = [CuCl^+]/[Cu^{2+}][Cl^-]$.

(c) $K = 1/[Cu^+][Cl^-]$.

(j) $K = 1$.

2.11. (e) $K = [HCO_3^-]/p_{CO_2}$.

2.12. (b) $MgCO_3(s) + H_2O(l) =$
$= Mg^{2+} + OH^- + HCO_3^-$.

(d) $2Fe^{3+} + Fe(s) = 3 Fe^{2+}$.

3.1. (d) (i) Third order; (ii) zeroth order.

3.2. (b) $1.5 \times 10^{-5} s^{-1}$.

(c) 0.041 M.

3.6. (a) 7×10^{-4} s.

(b) $Co(OH_2)_4Cl_2^+$
$= Co(OH_2)_4Cl^{2+} + Cl^-$ rate-determining

$Co(OH_2)_4Cl^{2+} + OH^-$
$= Co(OH_2)_4ClOH^+$ fast.

3.11. 182.8 kJ mol^{-1}.

4.1. (a) $D = 14$.

(b) 20.8 mg.

(c) 7.

5.2. 2.6×10^{-8} mol^2 dm^{-6}.

5.5. 3.2×10^{-7} mol^2 dm^{-6}.

5.8. (c) $2.0 \times 10^{-4} M$.

5.11. $5.73 \times 10^{-3} M = 0.780$ g dm^{-3}.

6.1. 5.26×10^{-10} mol dm^{-3}.

6.3. 0.18 mol dm^{-3}.

6.6. (c) pcH = 8.35.

6.7. (e) $K_s = 3.4 \times 10^{-15}$

6.10. (c) 6.62.

6.15. $[H_3O^+] = 5.13 \times 10^{-2} M$,
$[HC_2O_4^-] = 5.12_5 \times 10^{-2} M$,
$[C_2O_4^{2-}] = 5.4 \times 10^{-5} M$.

6.18. 4.76.

6.22. 1.47×10^{-6} mol dm^{-3}, virtually all as BeOH$^+$.

7.1. 0.17 (17%).

7.3. (b) $1.02 \times 10^{-4} M$.

7.5. (d) A value of 2 for d log S/d log [Cl$^-$] over a wide range of chloride-ion concentrations must reflect the occurrence of a reaction for which the left-hand side of the balanced equation is TlCl(s) + 2 Cl$^-$ =.

7.8. $[H_3O^+] = 4.79 \times 10^{-4} M$,
$[G^-] = 2.32 \times 10^{-3} M$,
$[HG(aq)] = 7.52 \times 10^{-3} M$,
$[Zn^{2+}] = 8.43 \times 10^{-4} M$,
$[ZnG^+] = 1.57 \times 10^{-4} M$.

8.1. (c) $2 MnO_4^- + C_2O_4^{2-} + 2 Ba^{2+} + 4 OH^-$
$= 2 BaMnO_4(s) + 2 CO_3^{2-} + 2 H_2O(l)$.

8.2. (b) 4×10^{36} mol^{-12} dm^{36}.

8.4. 1×10^7 mol^{-1} dm^3.

8.9. 1.0×10^{-12} mol^2 dm^{-6}.

8.10. 39 mol^{-1} dm^3.

9.2. (a) 1.157 millimoles.

(b) 1.157 millimoles.

(c) 0.038 56 M.

9.7. (c) 2.000 millimoles.

(f) 1.435 millimoles.

9.13. 0.5375 g.

10.3. The first one, because it will contain occluded sodium ions. The second precipitate will be contaminated with adsorbed sodium ions, but their number will be relatively small because lead sulfate is a crystalline precipitate having only a small specific area.

11.1. See the accompanying figure. As the concentrations of the acid and base decrease, the slope in the vicinity of the equivalence point decreases. A relative prevision of about 1.5% is just possible in titrating 10^{-4}M solutions with the aid of an indicator.

11.5. See the accompanying figure. As the base becomes weaker, the slope in the vicinity of the equivalence point decreases. A relative precision of about 1.2% could be achieved in titrations with triethylamine (curve a) if a properly selected indicator were used to locate the end point, but a relative precision of about 20% is the best that could be attained in similar titrations with triethanolamine (curve c).

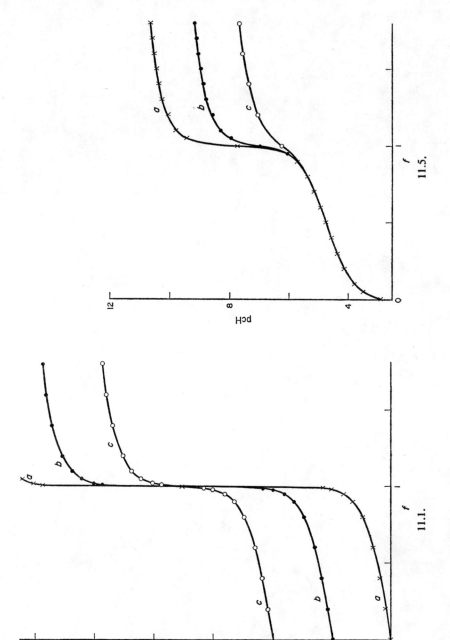

11.11. (a) $f =$

$$\dfrac{\left(\dfrac{K_1[H^+]+2K_1K_2}{[H^+]^2+K_1[H^+]+K_1K_2}\right) c_a^0 - \left([H^+] - \dfrac{K_w}{[H^+]}\right)}{c_a^0 + \left([H^+] - \dfrac{K_w}{[H^+]}\right) \dfrac{c_a^0}{c_b}}$$

There are many alternative forms, but this is the one that is easiest to use in constructing theoretical titration curves.

12.5. (a) $y = x + 1/x$.

(b) $y = 2 \sin x$.

12.6. x = atomic number, y = atomic weight.

13.3. 1.95×10^{-5} M (using the Debye–Hückel limiting law).

13.5. (a) 4.44×10^{-5} mol^2 dm^{-6}.

(b) 2.58×10^{-5} mol^2 dm^{-6}.

13.8. (a) 1.448×10^{-5} M.

14.1. (a) $2 \, Ag(s) + Cu^{2+} = 2 \, Ag^+ + Cu(s)$.

14.2. (b) $Ag \mid AgBr \ (s), \ Br^-(c_1) \mid \mid Cl^-(c_2), \ AgCl(s) \mid Ag$.

14.6. 0.32%.

14.8. 39.83 cm^3.

15.1. 0.328.

15.4. (b) $2.0_0 \times 10^{-3} M$.

15.8. $1.04 \times 10^{-2} M$.

15.11. 3.07×10^{-14} mol^2 dm^{-6}.

Significant Figures

THE result of a scientific measurement or calculation is expressed by a number. How many digits should the number contain? It depends on the precision with which the measurement or calculation has been made and the confidence that may be placed in the result. In writing any number you should include all the digits you are sure of, and the first uncertain one in addition. The total number of digits you include is called the number of significant figures.

If, for example, you believe that the distance between two points is somewhere between 13 and 15 cm, you should express it as 14 cm, using two significant figures. The reader will know that the "1" in the tens place is reliable but that you are uncertain about the "4" in the units place. A more careful measurement might enable you to say that the distance is 14.23 cm, which contains four significant figures instead of two and would convey your certainty that the distance is between 14.2 and 14.3 cm but would also mean that it might be 14.22 or 14.24 cm.

Students sometimes have difficulty in deciding whether a zero is or is not a significant figure. If you have decided that the distance between two points should be expressed as 14 cm, you would convey exactly the same information by stating it as 0.14 m or 0.000 14 km. In both of these the zeros serve only to locate the decimal point. However, you should not convert it to 140 mm, because that would imply that you knew it to be between 139 and 141 mm. You could instead say that it is 1.4×10^2 mm. A zero at the left-hand side of a number is not a significant figure; one at the right-hand side of a number is a significant figure.

You might have to add three distances, say 30.1, 1.04, and 0.1759 cm. If you use a pocket calculator, you will obtain 31.3159 cm for their sum. Since the first of the three distances has an uncertainty of 0.1 cm, the sum must be at least equally uncertain, and it would be misleading to express it as anything but 31.3 cm. You can save time by identifying the first digit that is uncertain in any of the numbers you add, and rounding off all the other numbers to discard everything beyond the place in which that digit lies. Here the first uncertain digit is the "1" in the tenths place of the "30.1", and therefore you could write $30.1 + 1.0 + 0.2 = 31.3$ cm. Subtraction follows exactly the same rule: the difference 30.1−0.2293 could be abbreviated to $30.1 - 0.2 = 29.9$.

Multiplication and division follow a different rule. If a solution contains 1.04 mole of potassium thiocyanate, of which one mole weighs 97.18 g, a calculator will tell you that there is 101.0672 g of potassium thiocyanate in it. Not all of these figures are significant, for the solution might contain 1.03 mole or 1.05 mole of the salt, and these amounts correspond to weights ranging from 100.095 g to 102.039 g. Since the uncertainty is about 1 g, the proper statement is that it contains 101 g. Similarly, if you weigh out an amount of potassium thiocyanate that you describe as 174 g because the "4" is uncertain and the weight might be 173 g, the number of moles is equal to 174/97.18; a pocket calculator will give 1.790 49... for the quotient and you should round it off to 1.79.

To make sure that you use the right number of significant figures in expressing a product or quotient, you should consider the *relative precisions* of the numbers you use in the multiplication or division. The relative precision of a number is obtained by dividing the uncertainty in the number by the number itself. If you are correct in saying that a solution contains 1.04 mole of potassium thiocyanate, your precision of measurement was 0.01 mole, and the relative precision is 0.01/1.04 or approximately 0.01. If the weight of a mole of potassium thiocyanate is correctly stated as 97.18 g, its relative precision is 0.01/97.18 or approximately 1×10^{-4}. The relative precision of the weight of 1.04 moles of the salt ($= 1.04 \times 97.18$) is the sum of these separate relative precisions, or virtually 0.01. The product is 101.0672 g, and its relative precision is 0.01; the uncertainty is equal to $101.0672 \times 0.01 = 1$ g. You would therefore round off the result to 101 g. Similarly, if the weight of potassium thiocyanate used in an experiment is expressed as 174 g, its uncertainty is 1 g and its relative precision is $1/174 = 5.7 \times 10^{-3}$. To obtain the number of moles of potassium thiocyanate you divide the weight by 97.18 g, in which the precision is 0.01 g and the relative precision is 1×10^{-4}. The sum of the two relative precisions is 5.8×10^{-3}, and this is the relative precision of the quotient, 1.790 49... mol. The uncertainty is therefore $5.8 \times 10^{-3} \times 1.790 49... = 0.01$ mole, and the quotient should be expressed as 1.79 mole.

These rules are sometimes abbreviated by saying that the number of significant figures in an answer should be equal to the smallest number of significant figures in any of the numbers combined in calculating it. You can judge the merits of the abbreviation after you have considered the following examples:

$$80.36 + 0.11 = 80.47 \qquad \text{(four significant figures, not two)}$$
$$29.3025 - 0.1914 = 29.1111 \qquad \text{(six significant figures, not four)}$$
$$99/97.18 = 1.02 \qquad \text{(three significant figures, not two)}$$
$$1.08 \times 90.85 = 98 \qquad \text{(two significant figures, not three)}$$

In other kinds of calculations the rules are different. There is a general discussion of the propagation of errors in Section 12.5; here we shall merely note two special cases that arise in many examples:

(a) The number 1.00 has three significant figures and might have an uncertainty of one unit in the second decimal place. The corresponding uncertainty in its square root is 0.005, since $\sqrt{1.01} = 1.005$ while $\sqrt{0.99} = 0.995$. It is the third decimal place that is uncertain in the square root, and you would be entitled to write $\sqrt{1.00} = 1.000$· As a general guide, the relative precision of the square root of a number is half as large as the relative precision of the number itself. In $\sqrt{32}$, the two significant figures imply that the "32" is known to within 1 unit, so that its relative precision is 1/32. The relative precision of the square root is 1/64, and a pocket calculator gives its value as 5.66. Since $5.66 \times 1/64$ is close to 0.1, the result is properly expressed by writing $\sqrt{32} = 5.7$.

(b) The logarithm of 1.00 is equal to 0.000 while that of 1.01 is equal to 0.004. If a number has a relative precision of 0.01, its logarithm is uncertain to 0.004 unit. Saying the same thing differently, if dx is the uncertainty in a number x and $d \log x$ is the corresponding uncertainty in its logarithm, the relative precision of x is equal to dx/x, and

$$\frac{d \log x}{dx/x} = \frac{0.4343 \, d \ln x}{dx/x} = 0.4343$$

since $d \ln x = dx/x$. Conversely, an uncertainty of 0.004 unit in $\log x$ corresponds to a relative precision of 0.01 in x. If you say that pcH $= 5.042$, using three figures to the right of the decimal point, you are implying that you know the concentration of hydrogen ion within a relative precision of only 0.0025 (actually 0.0023), which is unlikely because few equilibrium constants are known accurately enough to permit calculations so reliable. Hence values of the pcH and similar quantities (such as pK) should generally be given to only two decimal places. You are invited to convince yourself that the number of significant figures is irrelevant: the statements "pcH $=$ $= 0.07$" and "pcH $= 13.07$" both mean that the concentration of hydronium ion has a relative precision of 0.023, even though the first gives only one significant figure while the second gives four.

The precision of a result can be expressed more accurately by stating the uncertainty in the last significant figure. If a value is believed to lie between 18.43 and 18.47, writing it as 18.45 ± 0.02 conveys exactly what is known about it. You will occasionally see this carried as far as 20.75 ± 0.15, where the "7" is actually the last significant figure but where it would be slightly misleading to round the number off to either 20.7 to 20.8 or to round the precision off to either 0.1 or 0.2.

Standard Enthalpies of Formation, Entropies, and Free Energies of Formation of Some Common Substances at 298 K

Substance	H_f^0 (kJ mol^{-1})	S^0 (J K^{-1} mol^{-1})	G_f^0 (kJ mol^{-1})
Ag(s)	0.00	42.72	0.00
Ag$^+$(aq)	105.90	73.93	77.11
AgBr(s)	−99.50	107.11	−93.68
AgCl(s)	−127.03	96.11	−109.70
AgI(s)	−62.38	114.22	−66.32
Ag$_2$O(s)	−30.59	121.71	−10.84
Al(s)	0.00	28.33	0.00
Al^{3+}(aq)	−524.7	−313.4	−481.2
α-Al$_2$O$_3$(s)	−1669.79	51.00	−1576.41
Ba^{2+}(aq)	−538.36	13	−561.7
BaCl$_2$(aq)	−873.28	121	−823.0
Ba(OH)$_2$(aq)	−998.22	−8	−875.3
Br$_2$(l)	0.00	152.3	0.00
Br$^-$(aq)	−120.92	80.71	−102.80
C(diamond, s)	1.88	2.43	2.89
C (graphite, s)	0.00	5.69	0.00
CH$_3$Br(g)	−35.6	245.8	−25.9
CH$_2$Br$_2$(g)	−4.2	293.6	−5.9
CHBr$_3$(g)	25	331.3	15.9
CBr$_4$(g)	50	358.2	36.0
CH$_3$Cl(g)	−82.0	234.2	−59.0
CH$_2$Cl$_2$(g)	−87.9	270.6	−59
CHCl$_3$(g)	−100.4	296.5	−67
CCl$_4$(g)	−106.7	309.4	−64
CH$_4$(g)	−74.85	186.19	−50.79
C$_2$H$_2$(g)	226.75	200.82	209.20
C$_2$H$_4$(g)	52.28	219.83	68.12

APPENDIX II *(cont.*

Substance	H_f^0 (kJ mol^{-1})	S_f^0 (J K^{-1} mol^{-1})	G_f^0 (kJ mol^{-1})
$C_2H_6(g)$	−84.67	229.49	−32.89
C_6H_6 (benzene, g)	82.93	269.20	129.66
C_6H_{12} (cyclohexane, g)	−123.14	298.24	31.76
$CH_3OH(g)$	−201.17	237.6	−161.9
(l)	−238.57	126.8	−166.2
$CH_3COOH(l)$	−487.0	159.8	−392.5
$CO(g)$	−110.52	197.91	−137.27
$CO_2(g)$	−393.51	213.64	−394.38
$C_2O_4^{2-}(aq)$	−824.2	51.0	−674.9
$Ca^{2+}(aq)$	−542.96	−55.23	−553.04
$CaC_2O_4 \cdot H_2O(s)$	−1669.8	156.0	−1508.8
$Ca(OH)_2(aq)$	−1002.8	−76.2	−867.64
$Cl_2(g)$	0.00	222.97	0.00
$Cl^-(aq)$	−167.46	55.10	−131.17
$Cu(s)$	0.00	33.30	0.00
$Cu^{2+}(aq)$	64.39	−98.7	64.98
$CuO(s)$	−155.2	43.5	−127.2
$Fe^{2+}(aq)$	−87.9	−113.4	−84.9
$Fe^{3+}(aq)$	−47.7	−293.3	−10.54
$H_2(g)$	0.00	130.58	0.00
$H_2O(g)$	−241.84	188.74	−228.61
(l)	−285.85	69.96	−237.19
(s)	−291.86	47.96	−
$I_2(g)$	62.26	260.58	19.37
(s)	0.00	116.73	0.00
$I^-(aq)$	−55.94	109.37	−51.67
$Mg(s)$	0.00	32.51	0.00
$MgO(s)$	−601.83	26.78	−569.57
$Mg(OH)_2(s)$	−924.66	63.14	−833.75
$Na^+(aq)$	−239.66	60.25	−261.88
$NaBr(aq)$	−360.58	141.00	−364.68
$NaCl(aq)$	−407.11	115.48	−393.04
$O_2(g)$	0.00	205.03	0.00
$OH^-(aq)$	−229.94	−10.54	−157.30
$Pb^{2+}(aq)$	1.63	24.31	21.3
$PbCl_2(aq)$	−333.26	131.4	−286.65
PbO (red, s)	−219.24	67.8	−189.33
(yellow, s)	−217.86	69.5	−188.49
S (monoclinic, s)	0.30	32.55	0.10
(rhombic, s)	0.00	31.88	0.00
$SO_2(g)$	−296.90	248.53	−300.37
$SO_3(g)$	−395.18	256.23	−370.37
$Tl^+(aq)$	5.77	127.2	−32.45
$TlBr(s)$	−172.4	111.3	−166.1
$TlCl(s)$	−204.97	108.4	−184.9
$TlI(s)$	−50.2	236.4	−83.3
$Zn(s)$	0.00	41.63	0.00
$Zn^{2+}(aq)$	−152.42	−106.48	−147.21

APPENDIX III

Solubility Products of Some Common Salts in Aqueous Solutions at 298 K

The symbol M is used to denote mol dm^{-3}; the solubility product of silver bromide is equal to $5.2 \times 10^{-13}\ M^2 = 5.2 \times 10^{-13}\ (\text{mol dm}^{-3})^2 = 5.2 \times 10^{-13}\ \text{mol}^2\ \text{dm}^{-6}$.

Salt	Solubility product	Salt	Solubility product
AgBr	$5.2 \times 10^{-13}\ M^2$	CoC_2O_4	$6 \times 10^{-8}\ M^2$
$AgBrO_3$	$5.2 \times 10^{-5}\ M^2$	CuBr	$5.2 \times 10^{-9}\ M^2$
AgCN	$1.2 \times 10^{-16}\ M^2$	CuCN	$3.2 \times 10^{-20}\ M^2$
Ag_2CO_3	$8.1 \times 10^{-12}\ M^3$	CuC_2O_4	$2.3 \times 10^{-8}\ M^2$
$AgOOCCH_3$	$4.4 \times 10^{-3}\ M^2$	CuCl	$1.2 \times 10^{-6}\ M^2$
$Ag_2C_2O_4$	$3.5 \times 10^{-11}\ M^3$	CuI	$1.1 \times 10^{-12}\ M^2$
AgCl	$1.8 \times 10^{-10}\ M^2$	$Cu(IO_3)_2$	$7.4 \times 10^{-8}\ M^3$
Ag_2CrO_4	$1.1 \times 10^{-12}\ M^3$	CuSCN	$4.8 \times 10^{-15}\ M^2$
AgI	$8.3 \times 10^{-17}\ M^2$	$Fe(OH)_3$	$4 \times 10^{-38}\ M^4$
$AgIO_3$	$3.0 \times 10^{-8}\ M^2$	Hg_2Br_2	$5.8 \times 10^{-23}\ M^3$
Ag_2S	$6 \times 10^{-50}\ M^3$	Hg_2CO_3	$8.9 \times 10^{-17}\ M^2$
AgSCN	$1.0 \times 10^{-12}\ M^2$	$Hg_2C_2O_4$	$2 \times 10^{-13}\ M^2$
Ag_2SO_4	$1.6 \times 10^{-5}\ M^3$	Hg_2Cl_2	$1.3 \times 10^{-18}\ M^3$
$Al(OH)_3$	$2 \times 10^{-32}\ M^4$	Hg_2I_2	$4.5 \times 10^{-29}\ M^3$
$Ba(BrO_3)_2$	$3.2 \times 10^{-6}\ M^3$	$Hg_2(SCN)_2$	$3.0 \times 10^{-20}\ M^3$
$BaCO_3$	$5.1 \times 10^{-9}\ M^2$	Hg_2SO_4	$7.4 \times 10^{-7}\ M^2$
BaC_2O_4	$2.3 \times 10^{-8}\ M^2$	$KBrO_3$	$5.7 \times 10^{-2}\ M^2$
$BaCrO_4$	$1.2 \times 10^{-10}\ M^2$	$KClO_4$	$1.1 \times 10^{-2}\ M^2$
$Ba(IO_3)_2$	$1.5 \times 10^{-9}\ M^3$	KIO_3	$5.0 \times 10^{-2}\ M^2$
$BaSO_4$	$1.3 \times 10^{-10}\ M^2$	$MgCO_3$	$1 \times 10^{-5}\ M^2$
$CaCO_3$	$4.8 \times 10^{-9}\ M^2$	MgC_2O_4	$1 \times 10^{-8}\ M^2$
CaC_2O_4	$4 \times 10^{-9}\ M^2$	$Mg(OH)_2$	$1.8 \times 10^{-11}\ M^3$
$Ca(IO_3)_2$	$7.1 \times 10^{-7}\ M^3$	$NiCO_3$	$6.6 \times 10^{-9}\ M^2$
$Ca(OH)_2$	$1 \times 10^{-7}\ M^3$	$PbBr_2$	$3.9 \times 10^{-5}\ M^3$
$Ca_3(PO_4)_2$	$2.0 \times 10^{-29}\ M^5$	$Pb(BrO_3)_2$	$2 \times 10^{-2}\ M^3$
$CaSO_4$	$3.0 \times 10^{-6}\ M^2$	PbC_2O_4	$4.8 \times 10^{-10}\ M^2$
CdC_2O_4	$9 \times 10^{-8}\ M^2$	$PbCl_2$	$1.6 \times 10^{-5}\ M^3$
CdS	$2 \times 10^{-28}\ M^2$	$PbCrO_4$	$1.8 \times 10^{-14}\ M^2$
$Ce(IO_3)_3$	$3.2 \times 10^{-10}\ M^4$	PbF_2	$2.7 \times 10^{-8}\ M^3$

APPENDIX III *(cont.)*

Salt	Solubility product	Salt	Solubility product
PbI_2	$7.1 \times 10^{-9} \ M^3$	$Sr(IO_3)_2$	$3.3 \times 10^{-7} \ M^3$
$Pb(IO_3)_2$	$3.2 \times 10^{-13} \ M^3$	$SrSO_4$	$3.2 \times 10^{-7} \ M^2$
$Pb_3(PO_4)_2$	$7.9 \times 10^{-43} \ M^5$	$TlBr$	$3.4 \times 10^{-6} \ M^2$
PbS	$1 \times 10^{-28} \ M^2$	$TlBrO_3$	$8.5 \times 10^{-5} \ M^2$
$Pb(SCN)_2$	$2.0 \times 10^{-5} \ M^3$	$Tl_2C_2O_4$	$2 \times 10^{-4} \ M^3$
$PbSO_4$	$1.6 \times 10^{-8} \ M^2$	$TlCl$	$1.7 \times 10^{-4} \ M^2$
$RaSO_4$	$4.3 \times 10^{-11} \ M^2$	Tl_2CrO_4	$9.8 \times 10^{-13} \ M^3$
$RbClO_4$	$2.5 \times 10^{-3} \ M^2$	TlI	$6.5 \times 10^{-8} \ M^2$
$SrCO_3$	$1.1 \times 10^{-10} \ M^2$	$TlIO_3$	$3.1 \times 10^{-6} \ M^2$
SrC_2O_4	$1.6 \times 10^{-7} \ M^2$	$TlSCN$	$1.7 \times 10^{-4} \ M^2$
$SrCrO_4$	$3.6 \times 10^{-5} \ M^2$	$ZnCO_3$	$1.4 \times 10^{-11} \ M^2$
SrF_2	$2.5 \times 10^{-9} \ M^3$	ZnC_2O_4	$2.8 \times 10^{-8} \ M^2$

Overall Dissociation Constants of Some Common Acids in Aqueous Solutions at 298 K

All the values given here have the units mol dm^{-3} (M).

Acid	Formula	Overall dissociation constant (mol dm^{-3})
Acetic acid	CH_3COOH	1.75×10^{-5}
Acrylic acid	$CH_2{=}CHCOOH$	5.53×10^{-5}
Ammonium ion	NH_4^+	5.69×10^{-10}
Anilinium ion	$C_6H_5NH_3^+$	2.54×10^{-5}
Arsenic(III) acid	$HAsO_2$	6×10^{-10}
Arsenic(V) acid	H_3AsO_4	$K_1 = 6 \times 10^{-3}$
		$K_2 = 2 \times 10^{-7}$
		$K_3 = 4 \times 10^{-12}$
Benzoic acid	C_6H_5COOH	6.14×10^{-5}
Boric acid	HBO_2	5.83×10^{-10}
Bromoacetic acid	$BrCH_2COOH$	1.25×10^{-3}
n-Butyric acid	$CH_3CH_2CH_2COOH$	1.52×10^{-5}
Carbon dioxide	CO_2	$K_1 = 4.45 \times 10^{-7}$
		$K_2 = 4.7 \times 10^{-11}$
Chloroacetic acid	$ClCH_2COOH$	1.36×10^{-3}
Citric acid	$\underset{\displaystyle CH_2COOH}{\overset{\displaystyle CH_2COOH}{HO{-}C{-}COOH}}$	$K_1 = 7.45 \times 10^{-4}$
		$K_2 = 1.73 \times 10^{-5}$
		$K_3 = 4.02 \times 10^{-7}$
Dichloroacetic acid	$Cl_2CHCOOH$	3.3×10^{-2}
Ethanolammonium ion	$^+H_3NCH_2CH_2OH$	2.1×10^{-10}
Ethylenediamine-tetraacetic acid	$(HOOCCH_2)_2NCH_2CH_2N(CH_2COOH)_2$	$K_1 = 1.0 \times 10^{-2}$
		$K_2 = 1.6 \times 10^{-3}$
		$K_3 = 5.33 \times 10^{-7}$
		$K_4 = 1.13 \times 10^{-11}$
Ethylenediammonium ion	$^+H_3NCH_2CH_2NH_3^+$	$K_1 = 1.4 \times 10^{-7}$
		$K_2 = 1.2 \times 10^{-10}$
Formic acid	$HCOOH$	1.77×10^{-4}

Acid	Formula	Overall dissociation constant (mol dm^{-3})
Hydrocyanic acid	HCN	4.93×10^{-10}
Hydrofluoric acid	HF	5×10^{-4}
Hydrogen peroxide	H_2O_2	$K_1 = 2 \times 10^{-12}$
Hydrogen sulfide	H_2S	$K_1 = 6 \times 10^{-8}$
		$K_2 = 1 \times 10^{-14}$
Hypochlorous acid	HOCl	3×10^{-8}
Iodic acid	HIO_3	1.66×10^{-1}
Iodoacetic acid	ICH_2COOH	6.68×10^{-4}
Malonic acid	$HOOCCH_2COOH$	$K_1 = 1.40 \times 10^{-3}$
		$K_2 = 2.01 \times 10^{-6}$
Oxalic acid	HOOCCOOH	$K_1 = 5.36 \times 10^{-2}$
		$K_2 = 5.42 \times 10^{-5}$
Periodic acid	HIO_4	2.3×10^{-2}
Phenol	C_6H_5OH	1.00×10^{-10}
Phosphoric acid	H_3PO_4	$K_1 = 7.11 \times 10^{-3}$
		$K_2 = 6.34 \times 10^{-8}$
		$K_3 = 4.2 \times 10^{-13}$
o-Phthalic acid		$K_1 = 1.12 \times 10^{-3}$
		$K_2 = 3.91 \times 10^{-6}$
		1.34×10^{-5}
Propionic acid	CH_3CH_2COOH	
Pyridinium ion		6.0×10^{-6}
Quinolinium ion		1.31×10^{-5}
Succinic acid	$HOOCCH_2CH_2COOH$	$K_1 = 6.2 \times 10^{-5}$
		$K_2 = 2.32 \times 10^{-6}$
Sulfuric acid	H_2SO_4	$K_2 = 1.20 \times 10^{-2}$
Tartaric acid		$K_1 = 9.20 \times 10^{-4}$
		$K_2 = 4.31 \times 10^{-5}$
		1.73×10^{-8}
Triethanolammonium ion	$^+HN(CH_2CH_2OH)_3$	
Triethylenetetrammonium ion	$^+H_3NHC_2CH_2^+NH_2CH_2CH_2^+NH_2CH_2CH_2NH_3^+$	$K_1 = 5 \times 10^{-4}$
		$K_2 = 2 \times 10^{-7}$
		$K_3 = 6.3 \times 10^{-10}$
		$K_4 = 1.3 \times 10^{-10}$

Stepwise Formation Constants of Some Common Complexes in Aqueous Solutions at 298 K

EACH number in the body of this table is the logarithm (to the base 10) of the formation constant of a complex containing the metal ion M that appears above it in the first row and the ligand L that appears in the first column on the same line. The number "2.3", which appears in the column containing silver ion on the line containing hydroxide ion, means that $\log_{10}[AgOH(aq)]/[Ag^+][OH^-] = 2.3$, so that $K_1 = [AgOH(aq)]/[Ag^+][OH^-] = 2 \times 10^2$ (mol^{-1} dm^3).

If more than one complex is formed, the logarithms of their formation constants are given in order: that for ML appears first, then that for ML$_2$, and so on. The numbers "6.2; 1.8", which appear in the column containing copper(II) cupric ion on the line containing oxalate ion, mean that $\log_{10}[CuC_2O_4(aq)]/[Cu^{2+}][C_2O_4^{2-}] = 6.2$ and $\log_{10}[Cu(C_2O_4)_2^{2-}]/[CuC_2O_4(aq)][C_2O_4^{2-}] = 1.8$, so that
$K_1 = [CuC_2O_4(aq)]/[Cu^{2+}][C_2O_4^{2-}] = 1.6 \times 10^6$ (mol^{-1} dm^3) and
$K_2 = [Cu(C_2O_4)_2^{2-}]/CuC_2O_4(aq)][C_2O_4^{2-}] = 6.3 \times 10^1$ (mol^{-1} dm^3).

Occasionally the product of two or more stepwise formation constants is known but their individual values are not. When this is true, a value of the logarithm of an overall formation constant is given and identified. The entry "4.4; $\beta_3 = 15.5$; 0.1", which appears in the column containing zinc ion on the line containing hydroxide ion, means that $\log_{10}[ZnOH^+]/[Zn^{2+}][OH^-] = 4.4$, $\log_{10}[Zn(OH)_3^-]/[Zn^{2+}][OH^-]^3 = 15.5$, and $\log_{10}[Zn(OH)_4^{2-}]/[Zn(OH)_3^-][OH^-] = 0.1$, so that
$K_1 = 2.5 \times 10^4$ (mol^{-1} dm^3), $\beta_3 = K_1K_2K_3 = 3 \times 10^{15}$ (mol^{-3} dm^9), and
$K_4 = 1.3$ (mol^{-1} dm^3).

APPENDIX V *(cont.)*

Ligand	Metal ion							
	Ag^+	Ba^{2+}	Ca^{2+}	Cu^{2+}	Fe^{3+}	Hg^{2+}	Pb^{2+}	Zn^{2+}
Acetate	0.4; −0.2	0.4	0.5	2.2; 1.1	−	$\beta_2 =$ 8.4	2.7; 1.5	1.0
Ammonia	3.2; 3.8	−	−0.2; −0.6; −0.8; −1.1; −1.3; −1.7	4.3; 3.7; 3.0; 2.3; −0.5	−	8.8; 8.7; 1.0; 0.8	−	2.4; 2.4; 2.5; 2.1
Bromide	4.2; 3.0; 1.8; 0.3	−	−	0.3	−0.3; −0.5	9.0; 8.3; 1.4; 1.3	−	1.2; 0.7; $\beta_4 =$ 3.0
Chloride	2.7; 2.1; 0.7; 0.5	−0.1	−	0.1; −0.6	1.5; 0.6; 1.0	5.3; 7.5; 1.1; 1.0	1.6; $\beta_3 =$ 1.8	−0.2; −0.4
Citrate	−	2.8	3.2	14.2	11.8	−	5.7	−
Cyanide	$\beta_2 =$ 21.1; 0.9	−	−	−	$\beta_6 =$ 31	18.0; 16.7; 3.8; 3.0	−	$\beta_3 =$ 17.5; 2.7
EDTA	−	7.9	10.7	18.8	25.1	22.1	17.9	16.5
Ethylene- diamine	4.7; 3.0	−	−	10.7; 9.3	−	$\beta_2 =$ 23.4	−	6.0; 4.8; 2.2
Hydroxide	2.3	0.6	1.3	6.5	1.1; 10.7	10.3	6.2	4.4; $\beta_3 = 15.5$; 0.1
Oxalate	0	2.3	3.0	6.2; 1.8	9.4; 6.8; 4.0	−	$\beta_2 =$ 6.5	4.9; 2.7
Pyridine	2.0; 2.1	−	−	2.4; 1.9; 1.1; 0.6	−	5.1; 4.9; 0.4	−	1.0; 0.5
Tartrate	−	1.6	1.8	3.0; 2.1	−	−	−	2.7
Thiocyanate	$\beta_2 =$ 8.2; 1.1; 0.7	−	−	−	2.1; 1.3; 0.3	$\beta_2 =$ 17.5; $\beta_4 =$ 21.2	1.1; 1.4	1.6
Triethanol- amine	2.3; 1.3	−	−	4.2	−	6.9; 6.2	−	2.0
Triethylene- tetramine	2.3; 1.3	−	−	20.1	−	25.0	10.4	11.9

Standard and Formal Potentials and Equilibrium Constants of Some Common Redox Couples in Aqueous Solutions at 298 K

The second column of this table gives the standard potential of the couple and the third gives its thermodynamic equilibrium constant K; the fifth column gives its formal potential and the sixth gives its conditional equilibrium constant K', in the supporting electrolyte described in the fourth column. In the third and sixth columns the units of K and K' are omitted to save space. The significances of K and K' are explained in Section 8.10.

Half-reaction	Standard potential, V vs. N.H.E.	K	Supporting electrolyte	Formal potential, V vs. N.H.E.	K'
$Ag^+ + e = Ag(s)$	$+0.7994$	3.25×10^{13}	$1\ M\ HClO_4$	$+0.792$	2.44×10^{13}
			$1\ M\ H_2SO_4$	$+0.77$	1.0×10^{13}
$AgBr(s) + e = Ag(s) + Br^-$	$+0.071$	15.9			
$AgCl(s) + e = Ag(s) + Cl^-$	$+0.2224$	5.75×10^3			
$AgI(s) + e = Ag(s) + I^-$	-0.152	2.70×10^{-3}	$1\ M\ KI$	-0.137	4.83×10^{-3}
$H_3AsO_4(aq) + 2\ H^+ + 2\ e = HAsO_2(aq) + 2\ H_2O(l)$	$+0.559$	7.90×10^{18}	$1\ M\ HCl$ or $HClO_4$	$+0.577$	3.20×10^{19}
$Br_2(l) + 2\ e = 2\ Br^-$	$+1.087$	5.60×10^{36}			
$2\ BrO_3^- + 12\ H^+ + 10\ e = Br_2(l) + 6\ H_2O(l)$	$+1.52$	8.5×10^{256}			
$Cd^{2+} + 2\ e = Cd(s)$	-0.402	1.60×10^{-7}			
$Ce(IV) + e = Ce(III)$	M	M	$1\ M\ HCl$	$+1.8$	2.7×10^{30}
			$1\ M\ HClO_4$	$+1.70$	5.4×10^{28}
			$1\ M\ HNO_3$	$+1.60$	1.1×10^{27}
			$1\ M\ H_2SO_4$	$+1.44$	2.2×10^{24}
$Cl_2(g) + 2\ e = 2\ Cl^-$	$+1.359$	8.77×10^{45}			
$Cr(III) + e = Cr(II)$	-0.41	1.2×10^{-7}	$5\ M\ HCl$	-0.40	1.7×10^{-7}

APPENDIX **VI** *(cont.)*

Half-reaction	Standard potential, V vs. N.H.E.	K	Supporting electrolyte	Formal potential, V vs. N.H.E.	K'
$Cr_2O_7^{2-}+14\,H^++6\,e =$					
$\quad 2\,Cr^{3+}+7\,H_2O(l)$	$+1.33$	7.7×10^{134}	$0.1\,M$ HCl	$+0.93$	2.1×10^{94}
			$1\,M$ HCl	$+1.00$	2.6×10^{101}
			$1\,M$ HClO$_4$	$+1.025$	9.0×10^{103}
			$0.1\,M$ H$_2$SO$_4$	$+0.92$	2.0×10^{93}
$Cu^{2+}+2\,e = Cu(s)$	$+0.337$	2.47×10^{11}			
$Cu(II)+e = Cu(I)$	$+0.153$	3.85×10^2	$1\,M$ NH$_3$+	$+0.01$	1.5
			$1\,M$ NH$_4^+$		
			$1\,M$ KBr	$+0.52$	6.2×10^8
$Fe(III)+e = Fe(II)$	$+0.771$	1.08×10^{13}	$0.5\,M$ HCl	$+0.71$	1.0×10^{12}
			$1\,M$ HClO$_4$	$+0.735$	2.65×10^{12}
			$1\,M$ H$_2$SO$_4$	$+0.68$	3.1×10^{11}
$Fe(CN)_6^{3-}+e = Fe(CN)_6^{4-}$	$+0.356$	1.04×10^6	$1\,M$ HCl	$+0.71$	1.0×10^{12}
$2\,H^++2\,e = H_2(g)$	±0.0000	1.000	$1\,M$ HCl or HClO$_4$	$+0.005$	1.48
$Hg^{2+}+2\,e = 2\,Hg(l)$	$+0.792$	5.95×10^{26}	$1\,M$ HClO$_4$	$+0.776$	1.71×10^{26}
$2\,Hg^{2+}+2\,e = Hg_2^{2+}$	$+0.907$	4.60×10^{30}			
$Hg_2Cl_2(s)+2\,e = 2\,Hg(l)+2\,Cl^-$	$+0.2680$	1.15×10^9			
$I_2(s)+2\,e = 2\,I^-$	$+0.536$	1.32×10^{18}			
$2\,IO_3^-+12\,H^++10e =$					
$\quad I_2(s)+6\,H_2O(l)$	$+1.19$	1.4×10^{201}			
$MnO_4^-+8\,H^++5e = Mn^{2+}+$					
$\quad 4\,H_2O(l)$	$+1.51$	4.2×10^{127}			
$H_2O_2(aq)+2\,H^++2\,e = 2\,H_2O(l)$	$+1.77$	6.9×10^{59}			
$O_2(g)+4\,H^++4e = 2\,H_2O(l)$	$+1.229$	1.25×10^{83}			
$PbO_2(s)+3H^++HSO_4^-+2e =$					
$\quad PbSO_4(s)+2\,H_2O(l)$	$+1.68$	6.2×10^{56}			
$Pb^{2+}+2\,e = Pb(s)$	-0.126	5.5×10^{-5}			
$PbSO_4(s)+2\,e = Pb(s)+SO_4^{2-}$	-0.356	9.22×10^{-13}	$1\,M$ H$_2$SO$_4$	-0.29	1.6×10^{-10}
$Sn(II)+2\,e = Sn(s)$	-0.140	1.85×10^{-5}	$1\,M$ HCl	-0.19	3.8×10^{-7}
$Ti(IV)+e = Ti(III)$			$1\,M$ HCl	-0.09	3.0×10^{-2}
			$2\,M$ H$_2$SO$_4$	$+0.12$	1.1×10^2
$Tl(III)+2\,e = Tl(I)$	$+1.28$	1.9×10^{43}	$1\,M$ HCl	$+0.78$	2.4×10^{26}
$Tl^++e = Tl(s)$	-0.336	2.09×10^{-6}			
$Zn^{2+}+2\,e = Zn(s)$	-0.7628	1.63×10^{-26}			

Index